Effects of Emerging Chemical Contaminants on Water Resources and Environmental Health

Victor Shikuku
Kaimosi Friends University College, Kenya

A volume in the Advances in Environmental Engineering and Green Technologies (AEEGT) Book Series

IGI Global
DISSEMINATOR OF KNOWLEDGE

Published in the United States of America by
IGI Global
Engineering Science Reference (an imprint of IGI Global)
701 E. Chocolate Avenue
Hershey PA, USA 17033
Tel: 717-533-8845
Fax: 717-533-8661
E-mail: cust@igi-global.com
Web site: http://www.igi-global.com

Library of Congress Cataloging-in-Publication Data

Names: Shikuku, Victor, 1986- editor.
Title: Effects of emerging chemical contaminants on water resources and environmental health / Victor Shikuku.
Description: Hershey, PA : Engineering Science Reference, [2020] | Includes bibliographical references and index. | Summary: ""This book explores current issues dealing with the occurrence, toxicology, and abatement of emerging contaminants. It also examines new technologies and developments in sample preparation and detection of contaminants in water resources."-- Provided by publisher.
Identifiers: LCCN 2019034149 (print) | LCCN 2019034150 (ebook) | ISBN 9781799818717 (hardcover) | ISBN 9781799818724 (paperback) | ISBN 9781799818731 (ebook)
Subjects: LCSH: Emerging contaminants in water. | Water--Pollution. | Water--Purification. | Environmental health.
Classification: LCC TD420 .E364 2020 (print) | LCC TD420 (ebook) | DDC 628.1/68--dc23
LC record available at https://lccn.loc.gov/2019034149
LC ebook record available at https://lccn.loc.gov/2019034150

This book is published in the IGI Global book series Advances in Environmental Engineering and Green Technologies (AEEGT) (ISSN: 2326-9162; eISSN: 2326-9170)

British Cataloguing in Publication Data
A Cataloguing in Publication record for this book is available from the British Library.

Advances in Environmental Engineering and Green Technologies (AEEGT) Book Series

Sang-Bing Tsai
Zhongshan Institute, University of Electronic Science and Technology of China, China & Wuyi University, China
Ming-Lang Tseng
Lunghwa University of Science and Technology, Taiwan
Yuchi Wang
University of Electronic Science and Technology of China Zhongshan Institute, China

ISSN:2326-9162
EISSN:2326-9170

Mission

Growing awareness and an increased focus on environmental issues such as climate change, energy use, and loss of non-renewable resources have brought about a greater need for research that provides potential solutions to these problems. Research in environmental science and engineering continues to play a vital role in uncovering new opportunities for a "green" future.

The **Advances in Environmental Engineering and Green Technologies (AEEGT)** book series is a mouthpiece for research in all aspects of environmental science, earth science, and green initiatives. This series supports the ongoing research in this field through publishing books that discuss topics within environmental engineering or that deal with the interdisciplinary field of green technologies.

Coverage

- Radioactive Waste Treatment
- Contaminated Site Remediation
- Cleantech
- Alternative Power Sources
- Biofilters and Biofiltration
- Industrial Waste Management and Minimization
- Waste Management
- Renewable Energy
- Green Technology
- Green Transportation

IGI Global is currently accepting manuscripts for publication within this series. To submit a proposal for a volume in this series, please contact our Acquisition Editors at Acquisitions@igi-global.com or visit: http://www.igi-global.com/publish/.

The Advances in Environmental Engineering and Green Technologies (AEEGT) Book Series (ISSN 2326-9162) is published by IGI Global, 701 E. Chocolate Avenue, Hershey, PA 17033-1240, USA, www.igi-global.com. This series is composed of titles available for purchase individually; each title is edited to be contextually exclusive from any other title within the series. For pricing and ordering information please visit http://www.igi-global.com/book-series/advances-environmental-engineering-green-technologies/73679. Postmaster: Send all address changes to above address.

Titles in this Series

For a list of additional titles in this series, please visit:
https://www.igi-global.com/book-series/advances-environmental-engineering-green-technologies/73679

Physiological Aspects of Imprinting and Homing Migration in Salmon Emerging Research and Opportunities
Hiroshi Ueda (Hokkaido University, Japan)
Engineering Science Reference • © 2020 • 200pp • H/C (ISBN: 9781799820543) • US $175.00

Artificial Intelligence and IoT-Based Technologies for Sustainable Farming and Smart Agriculture
Pradeep Tomar (Gautam Buddha University, India) and Gurjit Kaur (Delhi Technological University, India)
Engineering Science Reference • © 2020 • 350pp • H/C (ISBN: 9781799817222) • US $215.00

Predicting, Monitoring, and Assessing Forest Fire Dangers and Risks
Nikolay Viktorovich Baranovskiy (National Research Tomsk Polytechnic University, Russia)
Engineering Science Reference • © 2020 • 417pp • H/C (ISBN: 9781799818670) • US $235.00

Waste Management Techniques for Improved Environmental and Public Health Emerging Research and Opportunities
Sang-Bing Tsai (Zhongshan Institute, University of Electronic Science and Technology of China, China & Wuyi University, China) Zhengxi Yuan (Zhongshan Institute, University of Electronic Science and Technology of China, China) Jian Yu (Research Center for Environment and Sustainable Development of China Civil Aviation, Civil Aviation University of China, China) and Xuexin Liu (Capital University of Economics and Business, China)
Engineering Science Reference • © 2020 • 229pp • H/C (ISBN: 9781799819660) • US $190.00

Handbook of Research on Smart Computing for Renewable Energy and Agro-Engineering
Valeriy Kharchenko (Federal Scientific Agroengineering Center VIM, Russia) and Pandian Vasant (Universiti Teknologi Petronas, Malaysia)
Engineering Science Reference • © 2020 • 537pp • H/C (ISBN: 9781799812166) • US $295.00

Emerging Developments and Environmental Impacts of Ecological Chemistry
Gheorghe Duca (Academy of Sciences of Moldova, Moldova) and Ashok Vaseashta (ICWI/NJCU, USA & Academy of Sciences of Moldova, Moldova)
Engineering Science Reference • © 2020 • 390pp • H/C (ISBN: 9781799812418) • US $225.00

IGI Global
DISSEMINATOR OF KNOWLEDGE

701 East Chocolate Avenue, Hershey, PA 17033, USA
Tel: 717-533-8845 x100 • Fax: 717-533-8661
E-Mail: cust@igi-global.com • www.igi-global.com

Editorial Advisory Board

Table of Contents

Detailed Table of Contents

The ever-increasing production of plastics and concomitant poor plastic waste disposal systems explain the recent rising concerns over the occurrence of microplastics in freshwater resources. Microplastics are presently recognized as emerging contaminants owing to the increasing reports on their occurrence in the environment and the associated toxicological effects. This chapter discusses the recent trends in the monitoring of microplastics in freshwater resources, the toxicological effects of microplastics, and the sampling and analysis techniques available for detection and quantification. The challenges in analysis and comparison of various studies and future prospects have also been highlighted.

Aquatic ecosystems are pivotal resources that nurture diverse life forms apart from providing different ecosystem services. Global pollution, directly and indirectly, depletes the quality and standards of these resources and hampers the animals residing there. The incomplete combustion of all sorts of organic substances found in nature produces and release an emerging group of contaminants known as polycyclic aromatic hydrocarbons (PAHs). There are over a hundred different kinds of PAHs known and 16 amongst them are regarded as priority pollutants including phenanthrene (PHE). PHE is abundantly found in the aquatic environment and poses a higher risk to animals. It causes a vast array of toxicities in aquatic animals including genotoxicity, cardiotoxicity, transgenerational toxicity, neurotoxicity, developmental toxicity, and potentially induces oxidative stress and behavioral alterations. However, many areas of PHE toxicity in aquatic organisms are yet to be properly understood and management measures are yet to be initiated.

This chapter introduces readers to the background of emerging contaminants by defining emerging contaminants and telling their history and their corresponding effects. It describes the dynamic properties of emerging contaminants such as advection and dispersion, chemistry, and their reactivity behavior. Lastly, it tells the analytical methodologies on sample preparation such as solid phase extraction and solid-phase micro-extraction, detection and quantification of organic ECs, and it proposes future perspectives of emerging contaminants.

Emerging contaminants (ECs), which constitute a group of chemicals, such as personal care products, food additives, and endocrine disruptors, are not commonly monitored and have the potential to cause adverse effects on the environment and humans. In water, even at low concentrations, they pose risks to environmental health. Several technologies have been developed to determine these compounds; the most common ones are liquid-liquid extraction (LLE), solid-phase extraction (SPE), solid-phase microextraction (SPME), stir-bar sorptive extraction (SBSE), and dispersive liquid-liquid microextraction (DLLME) with analysis by gas and liquid chromatography (GC and LC) coupled to mass spectrometry (MS). This chapter analyzes these methodologies and highlights their possibilities and limitations and compiles the most recent advances in this field.

Development of sensors for pharmaceuticals has become very essential. This is due to the need to monitor the release and toxicological effects of pharmaceuticals into the environment. In this work, the authors explored bismuth sulphide (Bi2S3) nanorods and graphene as photoactive material for constructing a photoelectrochemical (PEC) aptasensor for sulfadimethoxine (SDM) detection, exhibiting high sensitivity, stability, and reproducibility. In another experiment, Mo-doped BiVO4 (Mo-BiVO4) and graphene nanocomposites were explored as photoactive material to construct a visible light-driven photoelectrochemical biosensor. Graphene in the nanocomposites was very essential in immobilizing streptomycin aptamer through π-π stacking interaction. Finally, graphene doped CdS (GR-CdS) synthesized via one pot hydrothermal technique and gold nanoparticles (Au NPs) were employed to construct a PEC aptasensor for diclofenac (DCF).

This chapter presents an overview of over 529 articles on designs, models, laboratory setups, and applications of electrochemical processes from 1973 to 2017 with particular attention paid to the removal of emerging environmental pollutants. The chapter demonstrates that electrochemical and advanced oxidation processes are efficient despite the economic implications of the technologies. The electrodes

below the 10 ppb and fluoride less than 1.5 ppm as prescribed by WHO. The removal efficiency was after 60-70 minutes with recyclability of 11 cycles. The nanocomposite worked well in all pH ranges 6.5-8.5. A filter cartridge biopolymer metal oxide nanocomposite constituting of template aluminium homogenized in the aggregated network of chitosan was developed as an adsorbent for fluoride from the water with better adsorption limit.

Preface

The delicate balance between water quality and quality of life, both human and aquatic, is summarized in the popular phrase 'water is life'. As such, the existence and sustenance of life on earth, our only home, is dependent on the use and misuse of water. Of particular importance is water pollution. The subject of water pollution, though has been discussed and debated within and across national boundaries for centuries, is not anywhere near to die away. Instead, it is growing in momentum with each upcoming generation. With each subsequent generation, new water quality compromising processes and sources emerge, creating new challenges. In the recent past, chemical compounds ubiquitous in the environment have received considerable attention. These include, personal care products, pharmaceutical compounds, microplastics, perfluoroalkyl substances, nanoparticles among others. Collectively, they have come to be known as 'chemicals of emerging concern' (CEC) or emergent contaminants (EC). The revelation of their occurrence in the environment has been made possible by the continuing development of robust analytical tools for detection and quantification. Though their transport and fate may not be completely understood, toxicity studies have shown these emergent contaminants to be toxic even at low concentrations and may induce development of drug-resistant bacterial strains. These reports have sparked debate over the occurrence of these chemicals in the aquatic environment and the implications on environmental health. The answer to the 'common' man's question "is my drinking water safe?" is increasingly becoming elusive. The recalcitrance of these emergent contaminants to the conventional wastewater and drinking water treatment processes has further opened fields for research in water treatment technologies. Several technologies have been proposed in the recent years to sequester emerging contaminants from water. This book explores these issues and the chapters have been organized in a logical sequence. The first three chapters describe the occurrence of emergent contaminants in various water resources, the untold underlying implications and way forward is proposed. Chapter 4 describes the toxicity studies relating to chemicals of concern in aquatic environment. From Chapter 5 to Chapter 7, the reader is furnished with the principles and applications of various analytical tools for sample preparation, detection and quantification of CECs. Chapters 8-9 deal with water treatment technologies for the removal of CECs from water. Their strengths and inherent limitations are concisely presented. Chapter 10 is an evaluation and recommendation of pollution control and abatement strategies. Chapters 11-13 deal with modeling in removal of water contaminants. The models described are not limited to the water-pollutants described but extend to other contaminants as well.

Suggestions that can improve the future editions of this book are highly welcome. I hope you enjoy the reading.

Victor O. Shikuku
Kaimosi Friends University College, Kenya

Acknowledgment

First, I would like to thank all the people who were part of this project and saw it develop from concept to reality. Your support through reading, writing, reviewing and proofreading is acknowledged. I appreciate my family's unwavering support of my pursuits as much as it takes away valuable time that would have been spent together. It is from the interactions with Prof. Zanella and Prof. Osmar at LARP-Brazil that inspired my interest in this subject. Their inspiration is recognized. Dr. Orata, Dr. Defo, Dr. Abosede and Dr. Kemmerich, I appreciate your persistent interest and readiness to step in at any point of this project. Last but not least, I thank the Almighty God for His providence throughout the course of this project.

Thank you all.

Victor O. Shikuku
Kaimosi Friends University College, Kenya

Chapter 1

Chemicals of Emerging Concern in Surface and Wastewater: A Perspective of Their Fate Within the Lake Victoria Catchment Area of Kenya

Francis Orata
Masinde Muliro University of Science and Technology, Kenya

ABSTRACT

Rapid technological advancement in the pharmaceuticals and chemical industry has led to synthesis of compounds used for health/personal care and industrial products in large amounts. These chemicals of emerging concern (CECs) are consequently released into the environment through industrial emissions, disposal processes, and during use and application. Rapid population growth and urbanization within the Lake Victoria catchment region has inserted tremendous pressure on the environment and its resources, thus resulting to potential point and diverse sources of CECs introduction to the environment. Improper waste disposal and conventional wastewater treatment technology that are practiced in the catchment have not helped in prevention and removal of CECs and other pollutants from the environment. This chapter evaluates the occurrence of CECs mainly in surface and wastewater within the Lake Victoria catchment of Kenya and informs on the fate and diverse health effects that come with their presence in the environment.

INTRODUCTION

New artificial chemicals are produced in large amounts to meet the increasing industrial demands. The rapid discovery and production of these chemicals may hinder their monitoring in the environment through development of appropriate analytical techniques and development of analytical methods. Consequently, these chemicals of emerging concern (CECs) may enter the environment and cause known or suspected adverse ecological and/or human health effects. Current practices and technologies for water and wastewater treatment in Africa are mainly conventional and are insufficient to ensure safe water for

DOI: 10.4018/978-1-7998-1871-7.ch001

use and basic sanitation (Orata, 2018). The situation therefore is a "recipe" for various contaminants and pollutants present in a wide range of water bodies, some of which are sources of drinking water.

Over the past decade, the rapid development in analytical instrument sensitivity has led to wide detection of up to pg/L levels of pollutants in the environment. Some of these detected pollutants may have not been monitored before or may just have been discovered. Therefore, the CEC terminology acknowledges the existence of harmful environmental agents whose identities, occurrences, hazards, and effects are not sufficiently understood (Halden, 2015). There is a need to understand how CECs interact in and with the environment. CECs are insufficiently regulated, and even some chemicals which were reported much earlier such as dichlorodiphenyltrichloroethane (DDT) can still be considered as CECs because of the continuing interest that they attract among researchers.

Types of Chemicals of Emerging Concern

Chemicals of Emerging Concern (CECs) can be classified into many known micro-pollutant classes that include, pesticides, nanoparticles, pharmaceuticals (Gonzalez-Naranjo et al., 2013), personal care products, poly and per- fluorinated substances, trace and radioelements, industrial chemicals, combustion products, chlorinated solvents, biocides, flame retardants and even rare metal complexes among others. The presence of CECs in the environment can be from diverse sources which can be categorized as synthetic or naturally occurring.

Within Lake Victoria catchment area, the CECs that have been determined in various environmental matrices include endocrine disrupting chemicals (EDCs), poly and per-fluorinated substances (PFASs), personal care products, pharmaceutically active compounds (PhACs), persistent organic pollutants such as pesticides, poly-aromatic hydrocarbons (PAHs) among others. However, a large number of CECs such as, trichloroacetic acid (TCAA), nitrosodimethylamine, methyltert-butyl ether, trichloroethylene, perchlorate, 1,4-dioxane, prions, triclocarban, triclosan and nanomaterials have not been determined within the Lake Victoria catchment.

Endocrine disrupting chemicals are among the most common CECs that are common in wastewater (Liu *et al.*, 2011; Manickum *et al.*, 2011; Xu *et al.*, 2012) especially from industrial effluents. These EDCs are bioactive at very low concentrations and alter function(s) of the endocrine system of an organism and consequently harming it (Pothitou & Voutsa, 2008). Some of the most common EDCs are Bisphenol A, Nonylphenol, Nonylphenol Ethoxylates, Octyphenol, and 17α- ethynylestradiol (EE2) which is a bioactive PhACs that is present in birth control pills (Von Saal *et al.*, 2012).

The sources of various classes of CECs are varied and largely depend on the type of anthropogenic activities that take place in respective areas, as well as exposure routes for the CECs in to different environmental matrices. Example, in the case of PhACs, Kumar et al., (2010) reported that the emission of pharmaceuticals per person measured in wastewater in residential areas is about 10% of the emission measured in hospital wastewater. In modern urban centres, wastewater from various sources such as hospitals, households, agricultural and industrial premises is a CECs exposure media that contain micro-pollutants. The wastewater is often channeled to WWTPs, which acts as an exposure route (Orata, 2018). The CECs are subsequently released to other aquatic ecosystems as effluent and also often as sludge (Orata, 2018).

Assessment of the occurrence and fate of CECs within the Lake Victoria aquatic resources is critical, considering the economic importance of Lake Victoria resources. This chapter mainly focuses on surface and waste water because of its relevance as fate of most pollutants and their role as intermediate in abiota and biota interactions. This chapter evaluates the occurrence, concentrations and fate of CECs in various surface water and wastewater within the Lake Victoria catchment area of Kenya. Reports on the occurrence of CECs in other environmental matrices are also discussed in this chapter.

Lake Victoria Catchment Area, Kenya

The Lake Victoria catchment of Kenya has major towns namely Kisii, Homabay, Kisumu, Kakamega, Mumias, Webuye, Bungoma, and Eldoret among others. The catchment also has four major river basins namely: Nzoia, Nyando, Yala and Sondu Miriu. The main rivers are indicated in Figure 1. The rivers discharge (m^3s^{-1}), and their percentage (%) basin in the Kenyan side of the Lake Victoria catchment area are 11.4 m^3s^{-1}, 3.5% for Sio, 115.3 m^3s^{-1}, 35.0% for Nzoia, 37.6 m^3s^{-1} 11.4% for Yala, 18.0 m^3s^{-1} 5.5% for Nyando, 3.7 m^3s^{-1} 1.1% for North Awach, 5.9 m^3s^{-1} 1.8% for Sondu-Miriu, 42.2 m^3s^{-1} 12.8% for South Awach, 58.0 m^3s^{-1} 17.6% for Kuja-Migori, 37.5 m^3s^{-1} 11.4% for Mara, respectively. The total discharge is 778.3 m^3s^{-1} from the catchment. The Kenyan side comprises 42.4% of the whole of Lake Victoria catchment. The other rivers within the Lake Victoria catchment area in Kenya are Awach Kano (clustered as North Awach) and Awach Tende and Awach Kibuon (clustered as South Awach). There are also several seasonal rivers and streams originating from areas with high rainfall. Figure 1 shows the river basins and the major urban centers within the catchment.

Source of Chemicals of Emerging Concerns And Their Introduction Into Lake Victoria Catchment Area Environment

Reports on quantified concentrations of CECs within the Lake Victoria catchment area of Kenya indicate that the potential sources of CECs are diverse and varied. There could also be point-sources of the CECs that may arise from discrete locations whose inputs into aquatic systems can often be defined. Examples include industrial effluent streams, municipal sewage treatment plants and combined sewage storm water overflows, mining, waste disposal sites and buried septic tanks, as reported elsewhere by (Kalantzi and Alcock, 2012; Thomaidis et al., 2012; Siddique et al., 2012). According to Thomaidis et al., (2012), diffuse pollutants on the other hand originate from poorly defined sources; often over a broad geographical scales and which include agricultural runoffs, storm-water and urban runoffs, leakage from urban sewerage systems and diffuse aerial deposition. Generally, CECs have mainly been determined in sediments and river water within the catchment. Various studies have also been done to access pollutants in wastewater treatment plants (WWTPs) of which the results gave a wide range of determined CECs levels in the influent as well as effluent and sludge. The WWTPs within towns in the catchment do not use advanced treatment technologies, especially for the effluent which are suspected to contain CECs (Orata 2018). The CECs can therefore easily be transferred to receiving waters such as rivers. Table 1 shows the major river basins in the Lake Victoria catchment, and the probable contributor source of CECs within the respective river basins.

For example, PhACs may enter the aquatic environment through excretion in human and animal urine and feces, and also through flushing of unused medications, household uses, or bathing, and other nanoparticles in the environment (Bloutsos and Giannopoulos, 2011; Smeti et al., 2011). Hospitals are a major source of PhACs within the Lake Victoria catchment area (Kimosop et al., 2016) and contribute a significant amount of PhACs to the environment through their effluent. In the study by Kimosop *et al*, (2016), antibiotics were found to be largely spread in WWTPs situated within the vicinities of hospitals

Figure 1.

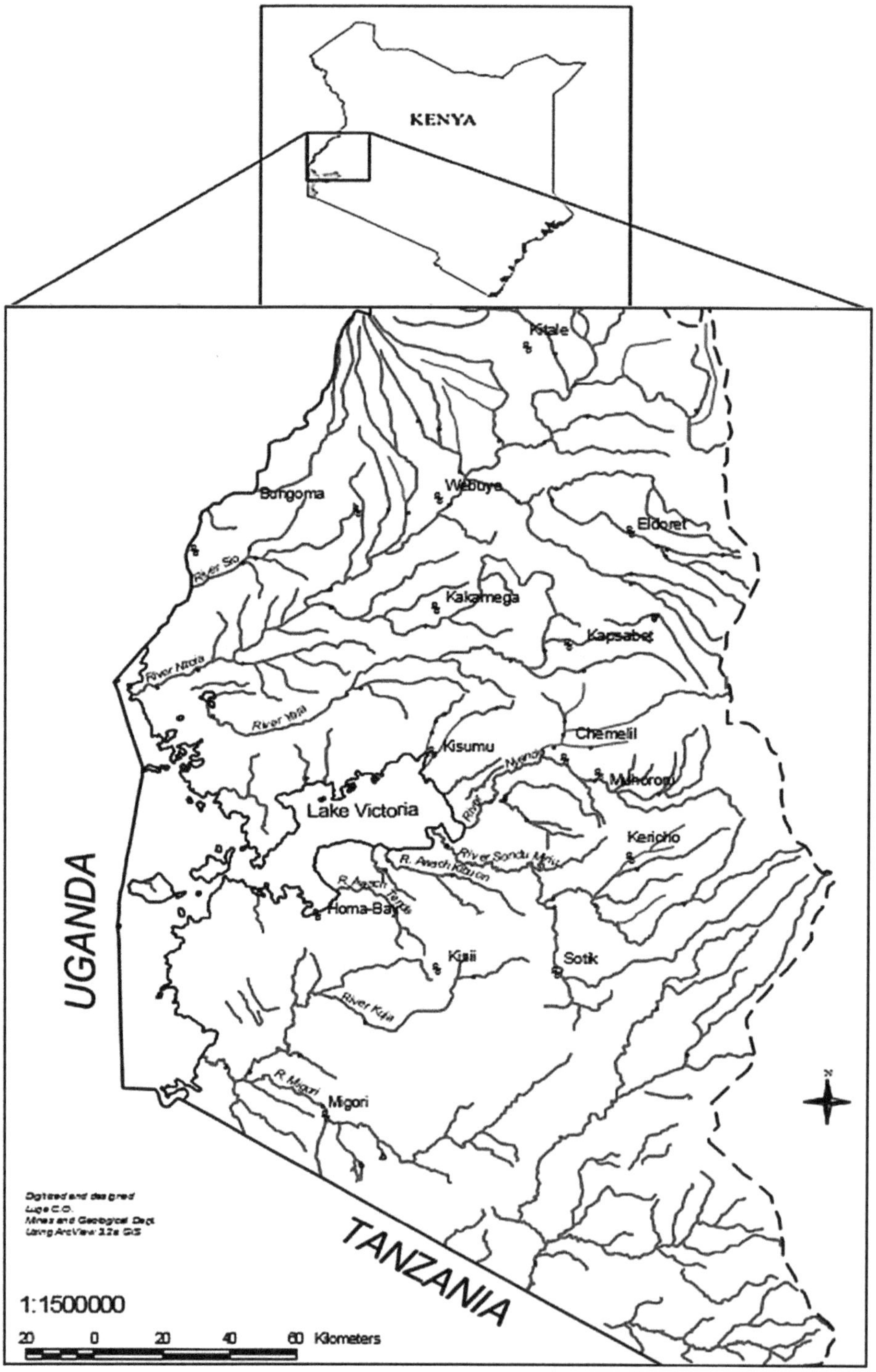
KENYA
Kitale
Bungoma
Webuye
Eldoret
River Sio
Kakamega
Kapsabet
River Nzoia
River Yala
Kisumu
Chemelil
Muhoroni
Lake Victoria
Kericho
Homa-Bay
Kisii
Sotik
UGANDA
River Kuja
R. Migori
Migori
TANZANIA
1:1500000
20 0 20 40 60 Kilometers

Table 1. Probable sources of CECs reported in selected river basin

Basins	CECs	Probable source/s
Nzoia river basin	Pesticide residues	Agricultural activities, municipal waste,
	PhACs,	Municipal effluent, WWTPs
	Dioxin and PCBs	Pulp and paper mill' Industrial/sewage discharges
Nyando River Basin	Pesticide residues	Agricultural activities, municipal waste
	PhACs,	Municipal effluent, WWTPs
Sosiani River	Pesticides	Agricultural activities
	Dioxin and PCBs	municipal waste, municipal effluent
Yala River Basin	Pesticide residues	Agricultural activities
Mara River basin	Pesticide residues and antibiotic resistant microbial populations	Agricultural activities Sewage discharges
Winam Gulf of Lake Victoria, and its tributaries	PAHs, PFASs, Endocrine disrupting compounds, PhACs and Personal care products	petroleum fuel spillages, cottage industries, motor vehicle exhaust, asphalt roads and road tar, fires of all types, WWTPs
	Dioxin and PCBs	municipal waste, municipal effluent

and that receives influent from municipal households. Other CECs that may be contained in hospital effluents are radionuclides, solvents, endocrine disrupting microbial populations and disinfectants that are used for medical purposes often in a wide range of concentrations. Shala and Foster (2010) reported that by-products from pharmaceutical and personal care products in urban settings' wastewater treatment facilities can contribute of between 1 to 2 micrograms/litre (μg/L) of a single compound. Poor solid waste disposal also contributes significantly to CECs release in the atmosphere. Combustion of solid waste is a major source of CECs account for nearly 80% of air emissions (Tangri, 2003). CECs that are pesticides in nature are purposefully introduced into the environment to control pests and diseases.

Distribution and Fate of Chemicals Of Emerging Concerns Within Lake Victoria Catchment Area

The fate of organochlrorinated CECs are influenced by the nature environmental matrices and organism within the site of CEC exposure. In addition, the physical and chemical properties of the individual CEC greatly affect their mobility and partitioning. Table 2 indicates the CECspreviously studies, the matrices and the study site within the Lake Victoria catchment area of Kenya.

These properties can be used to assess the physical hazards and to understand or predict the CECs environmental fate, its toxicity to human and its eco-toxicity. Physical properties of CECs such as solvation (phase partitioning, solubility, and colligative properties) and molecular attributes (molecular weight, molecular structure, and type of bonds) influence their mobility.

The physical and chemical properties can be used to predict the persistence, bioaccumulation, and biotransformation of CECs among other fates. For example, some PhACs can persist in the environment and, either via the food chain or drinking water; make their way back to humans. The CECs can therefore be transferred to many organisms depending on their trophic level. Consequently, the process of CECs transfer between organisms can cause bioaccumulation and biomagnification of the levels of

Table 2. Summary of CECs Studied in the Lake Victoria catchment of Kenya

Chemicals of Emerging Concern	Matrices	Sampling site	References
Dioxin – like PCBs and PCDD/Fs	Sediment core samples	Winam Gulf of Lake Victoria at Kisumu, Homabay and Mbita	Omwoma et al (2015)
Carbofuran (2,3-dihydro-2,2-dimethylbenzofuran-7yl methylcarbamate)	soil	Bunyala Irrigation Scheme within Nzoia River basin	Onunga et al (2015)
Halogenated and non-halogenated hydrocarbons, residual hydrocarbons (bendiocarb, benzene hexachloride (BHC), carbaryl, cypermethrin, decis, deltamethrin, diazinon, dieldrin, DDT, DDD, DDE, malathion, propoxur, sumithion, 5-phenylrhodanine, 1,3,5-trichlorobenzene, 1-(2phenoxybenzyl) hydrazine)	Water, sediments and papyrus reed plant materials	Kigwal/Kimondi wetland along River Yala, Nyando and Nzoia wetland ecosystems	Mule et al (2015)
Perfluoroalkyl acids (PFAAs) and their homologues (perfluorooctanoic acid (PFOA) and perfluorooctane sulfonate (PFOS).	Wastewater and sludge	Lake Victoria Basin (WWTP in Bungoma, Busia, Kakamega, Kisii and Mumias	Chirikona et al (2015)
Per-fluoroalkyl substances (PFOS, PFOA)	Sediments	Lake Victoria Gulf Basin	Orata et al (2011)
Per-fluoroalkyl substances (PFOS, PFOA)	River water and Lake Water	Winam gulf and its tributaries	Orata et al (2008)
Per-fluoroalkyl substances (PFOS, PFOA)	Fish, (*Lates niloticus, Oreochromis niloticus*)	Winam gulf of Lake Victoria, Kenya	Orata et al (2008)
Chlorsulfuron	soil in wheat Growing Regions Within the Mara River Basin	Mara River Basin, Kenya	Ogunah et al (2016)
PAHs	Air, gas	Kisumu city area	Lisouza et al. (2011) Bowa et al (2009).
PhACs, antibiotics	Wastewater, Sediments, Sewage Sludge	WWTPs and wastewater streams	Kimosop et al (2016)
Aldrin	Lake Victoria	Aquatic plants	Abong'o, et al (2018)
Diuron	Soil	Nzoia basin	Kimosop et al, (2012)

CECs in organisms. Residues of CECs have been detected in food, drinking water supplies as well as in export products such as fish, fruits and horticultural produce (Madadi et al., 2005, Twesigye et al., 2011). Example, the two commercially important fish species in Lake Victoria, Nile perch (*Lates niloticus*) and Nile tilapia (*Oreochromis niloticus*) belonging to different trophic levels were collected from the Napoleon Gulf and Thurston Bay in in the Ugandan side of Lake Victoria and analyzed to determine the concentrations of polychlorinated dibenzo-p-dioxins (PCDDs), polychlorinated dibenzofurans (PCDFs) and dioxin-like polychlorinated biphenyls (dl-PCBs (Ssebugere et al, 2013). The concentrations of analytes ranged from 0.07 to 0.59 $ngKg^{-1}$ fresh weight (fw) and 0.3 - 19.0 $ngKg^{-1}$ in *Lates niloticus* and 0.06 - 0.18 and 0.2 - 15.7 $ngKg^{-1}$ in *Oreochromis niloticus*, for $\sum$PCDD/Fs and $\sum$dl-PCBs, respectively (Ssebugere et al, 2013). Improper disposal of hazardous chemical wastes in urban municipalities within the Lake Victoria catchment area pose a major threat that can result in potential human exposures to polychlorinated dibenzodioxins (PCDDs) and/or polychlorinated dibenzofurans (PCDFs), polychlorinated biphenyls (PCBs) and related compounds. In the environment, CECs can bio-transform and partitioning between different environmental matrices. For example, Abong'o et al (2018) reported the presence organochlorine pesticides in Lake Victoria water as well as in the weeds (average concentration 15.519 ± 3.756 μg kg^{-1}), obtained from the lake. The results showed that aldrin tended to accumulate in the weeds. Atmospheric deposition is also an important route through which pollutants are transported through and enter the environment. Study by Arinaitwe at al., (2016), suggested that backward air trajectories suggested transboundary and local emission sources of selected pesticides analyzed. In the study, an average concentrations of total endosulfan, total DDT related compounds and hexachlorocyclohexanes ranged from 12.3 - 282, 22.8 -130 and 3.72 - 81.8 pg m^{-3}, respectively, for all the sample sets.

Atmospheric prevalence of residues of persistent organic pollutants (POPs) increased with fresh emissions of endosulfan, DDT and lindane. Hexachlorobenzene (HCB), pentachlorobenzene (PeCB) and dieldrin were also detected in air samples. It was also observed that transformation products such as pentachloroanisole, 3, 4, 5-trichloroveratrole and 3, 4, 5, 6-tetrachloroveratrole, prevalent within the study area. Figure 2 indicates the fate of CECs within the various segments of the environment.

The comparable levels of PFAAs obtained in sludge from hospital WWTPs indicates that the discharge from hospital effluent contributes significantly to the amounts of PFAAs released to the municipal water systems and probably to other surface and ground water bodies. CECs may be removed from effluent but remain in the treated sewage sludge. Sludge is often used as fertilizer in agricultural fields. Sewage sludge is frequently applied on fields as fertilizers. Recent studies have shown that PFAAs and pharmaceuticals contaminated sludge can not only contaminate groundwater, but also enter crops through absorption via roots (Tanoue et al. 2012). Moreover, the CECs on sewage sludge, which is referred to as biosolids can desorb and leach to contaminate ground water and other. Some of the majorly discussed CECs in published are reviewed in the subsequent sections.

Figure 2.

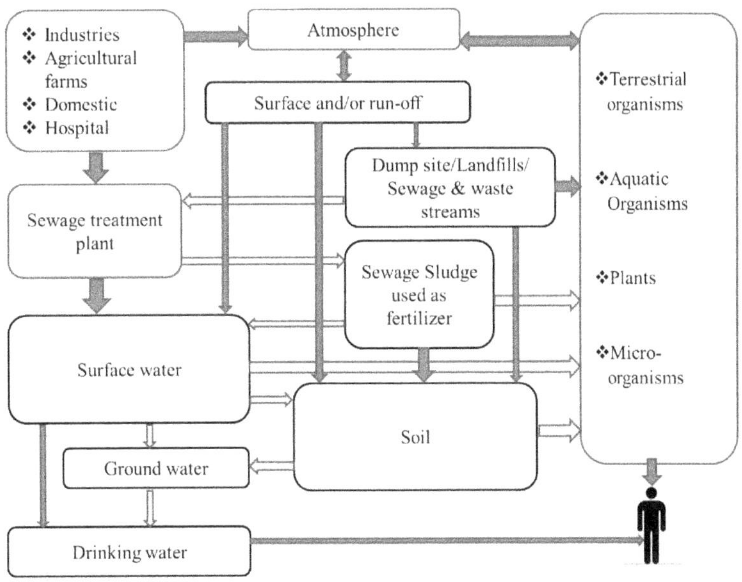

Pharmaceutical Active Compounds

The major source and transfer points for PhACs are ubiquitous. Pharmaceutical active compounds have been detected in environment segments notably in wastewater, sediments, and WWTPs sludge and effluent. Recent studies within Lake Victoria catchment area reported the accumulation of these compounds in wastewater and sewage sludge (Kimosop et al., 2016). In general, WWTPs receiving hospital wastewater had higher levels of detected pharmaceuticals than those receiving influents mainly from domestic sources. In Bugoma Hospital, acetaminophen (472 μg kg^{-1} d.w.), sulfamethoxazole (485.3 μg kg^{-1} d.w.) and propranolol (418 μg kg^{-1} d.w.), were detected at the highest concentrations among the studied WWTPs. Acetaminophen and propranolol are painkillers and are "on counter" drugs, meaning that they can be purchased without physician's recommendation. Sludge produced from Mumias General Hospital and Mumias Hospital Theater contained the highest concentrations of tramadol, with concentrations of 80.1 μg kg^{-1} d.w. and 101 μg kg^{-1} d.w., respectively. Fluconazole was also found to be prevalent in Mumias WWTP with concentrations up to 102 μg kg^{-1} d.w., which is several orders of magnitude higher concentration than in the other WWTPs. The highest individual antibiotic concentrations in wastewater and sludge was ampicillin (0.79 ± 0.07 μg/L and 94 ± 3 ng/g) respectively, while the concentrations of sulfamethoxazole in sewage sludge was (276 ± 12 ng/g), (Kimosop et al 2016). Total discharge loads of the selected antibiotics in the nine WWTPs located around Lake Victoria is 6333.01 mg day^{-1}. There is high likelihood for these compounds to be transferred to agricultural farms through sewage sludge (Jelic et al. 2011).

Perfluorinated Substances

The emission of PFASs into the environment is mainly determined by the use and application of the PFASs containing products and from the precursor molecules. PFAAs are a widely used due to their unique amphophilic properties that find beneficial application in both the commercial market and industry, because of their anti-wetting and surfactant properties. PFASs find application and use in products such as firefighting foams, carpets, leather/apparel, textiles/upholstery, paper and packaging, coatings and coating additives, industrial and household cleaning products, pesticides and insecticides, photographic industry, photolithography and semiconductor, hydraulic fluids and metal plating among other applications. The total PFAAs discharge at Kisumu was 656 mg day^{-1}. The estimated total discharge load of the PFAAs analyzed from five WWTP was 1013 mg day^{-1} (Chrikona et al, 2015). PFOA discharge at Kisumu WWTP was the highest 238 mg day^{-1}. The concentrations of PFHxA ranged from <48 ng kg^{-1} d.w. to 1.54×10^3 ng kg^{-1} d.w., which is comparable to PFHxA concentrations reported in sludge from WWTPs in a previous study from Kenya and Sweden (Chirikona et al. 2015). The concentration of PF-HxA in Mumias Hospital and Mumias Hospital Theater with concentrations of 1.54×10^3 ng kg^{-1} d.w. and 905 ng kg^{-1} d.w., respectively. PFOA was detected in Mumias Hospital Theater at the highest PFAAs concentration of 1.62×10^3 ng kg^{-1} d.w. PFHxA and PFHpA were less frequently detected compared to PFOA and PFDA, which were the dominant homologues in sludge with concentrations from 30 to 1.50 $\times 10^3$ ng kg^{-1} d.w. A report of the levels of Perfluorooctane Sulfonate (PFOS) and Perfluorooctanoic Acid (PFOA) in two fish species namely *Lates niloticus* (Nile perch) and *Oreochromis niloticus* (Nile tilapia) was presented by Orata et al, (2008). The study gave concentrations value of PFOS in Nile perch muscles of up to 10.50 and 35.70 ng/g for liver samples were obtained. Nile tilapia concentration values were of up to 12.40 and 23.70 ng/g for muscles and liver samples respectively.

Pesticides Related Chemicals of Emerging Concern (Herbicides, Insecticides and Fungicides)

Despite their ban or restricted use, of some of the pesticides especially the organochlorine pesticides ((DDT, lindane, aldrin, dieldrin, heptachlor, endrin, endosulfan (both α- and β- isomers and endosulfan sulphate), pesticides are still present in different environmental compartments. A number of studies have been carried out to determine pesticides and their residues in various environmental matrices within the catchment region. The levels obtained for concentrations of pesticides in Basin Rivers and the lake water are in μg/L and ng/L and lower levels. However, levels obtained in sediments and soils are slightly higher in comparison to those obtained for water matrix. All the pesticides analyzed in this study were banned in Kenya in 1986 except for aldrin, dieldrin, lindane, and DDT, the use of which is restricted (Pest Control and Product Board, 1992). There was a suggestion by Getenga et al. (2004) and Abongo et al. (2015) for the ban of organochlorine pesticides in other parts of the Lake Victoria basin. Abongo et al, (2015) further suggested the minimization of residue levels of organophosphates, through restriction of use within the Lake Victoria catchment.

Polychlorinated Phenolic Compounds

A study by Omwoma et al. (2015) sought out to generate baseline concentrations of polychlorinated dibenzo-p-dioxins (PCDDs), polychlorinated dibenzofurans (PCDFs) and dioxin-like polychlorinated biphenyls (dl-PCBs) in surface sediment from selected sites (Kisumu, Homa Bay and Mbita (control) towns) in the Winam Gulf of Lake Victoria, Kenya. The study also determined the influence of human activities on the polychlorinated phenolic compounds concentrations and potential risks to fish eating birds living near the lake. The study reported a total mean concentrations ranging from 17.4 - 812 pg g^{-1} dry wt (∑dl-PCBs), 36.6-813 pg g^{-1} dry wt (∑PCDDs) and 1.45-46.4 pg g^{-1} dry wt (∑PCDFs). The calculated Toxic Equivalents (TEQWHO) ranged from 0.001 - 0.43 (∑dl-PCBs) and 0.09 - 31 (∑PCDD/Fs), with the toxic chemicals emanating from anthropogenic activities, especially near large industrial towns such as Kisumu, Nzoia and Bugoma. The fish landing beaches at Kisumu city were found to be contaminated with respect to dl-PCBs and dioxins, followed by Homa Bay and Mbita. The Relatively high levels of octachlorodibenzo-p-dioxin (OCDD) and octachlorodibenzofuran (OCDF) influenced the TEQ and the ∑PCDFs/∑PCDDs ratios implicating chemical processes as partial sources of the dioxins (Omwoma et al., 2015). The authors concluded that the levels of contaminants obtained in the study showed potential exposure to aquatic organisms and fish eating birds through food chain transfer. Pollution and overfishing were thought to have contributed to stock reduction, specifically Nile perch (*Lates niloticus*), tilapia (*Oreochromis niloticus*) and Dagaa (*Rastrineobola argentea*) in the Lake Victoria.

Adverse Effects of CECs on the Environment and Human Health

The health effects of emerging pollutants depend on a variety of factors, including the level of exposure, duration of exposure and stage of life during exposure. The ecotoxicological and possible adverse health risks of most CECs on human health that is associated with their occurrence in the environment are still unexplained. Studies show that biota can be important environmental 'sinks' for emerging pollutants. For instance, researchers exposed minnow livers to municipal wastewater containing CECs and detected 28 of 32 compounds tested, including pharmaceuticals and personal care products, hormones, and industrial compounds (Vidal-Dorsch et al., 2013). Because some of the CECs are persistent in the environment, the probability that they will enter into the food chain is high. The effect of these emerging chemical pollutants of concern is on all the trophic levels of organisms within the whole food web. These are as a result of inter-dependent of organisms in the food chains of the complex food web, as a result of the complex feeding relationships within an ecosystem. Some of the CECs may undergo processes such as biotransformation or degradation to new compounds and metabolites. However, the properties and fate of metabolites and transformation products are still largely unknown (Fatta et al., 2007). Recent research has shown that some of the CECs and their products of biotransformation are associated with adverse developmental effects in aquatic organisms at environmentally relevant concentrations (Khetan and Collins, 2007). Accumulations of persistent CECs in organism are of particular health concern because they may cause chronic toxicity to affected organisms. The accumulation of persistent CECs such as dioxins and furans in the environment can reach levels that render water resources unfit for human consumption. Table 3 shows some of the CECs and the adverse effect they can cause to human beings. Some of the probable health effects of dioxins and furans include the development of cancer, immune system suppression, reproductive developmental complications and endocrine disruption (Connett, 1998 Tangri (2003) however reported that 90-95% of human exposure to dioxins is from fatty foods, particularly meat and dairy products. This is because dioxins accumulate in fats and oils.

Ecosystem health can be adversely affected by CECs in its environmental matrices. Studies have consistently observed adverse health, behavioral and reproductive effects (Painter et al., 2009, Bisesi et al., 2016, Schultz et al., 2012, Brodin et al., 2013, Crago et al., 2016, Niemuth et al., 2015) in certain organisms exposed to CECs. Consequently, the ecosystem balance will be affected which can adversely affect the ecosystem productivity.

Uncontrolled release of wastewater into aquatic environment may cause adverse health effects on the population of people within the Lake Victoria catchment area. The challenges facing any efforts to control CECs release and removal from various environmental segments are poor urban planning, uncontrolled pollution from agricultural and industrial source practices, barriers to adoption of new technologies of treating the sewage influent, and lack of awareness of the effects of/and new class or pollutants. Introduction of advanced wastewater treatment and water purification technologies have been demonstrated to remove CECs to a significant extend. It is therefore recommended that advanced wastewater treatment techniques be introduced or incorporated in already existing treatment facilities such as WWTPs and drinking water purification plants within the Lake Victoria catchment area. Examples of effective wastewater treatment and water purification technologies are those based on advanced oxidation processes (AOPs) and Ultrafiltration and membrane technology. AOPs are successful to transform toxic organic compounds (e.g. drugs, pesticides, endocrine disruptors etc.) into biodegradable substances. Ultrafiltration membranes and membrane bioreactors can also be applied and it combines advanced clarification through ultrafiltration membranes for a more effective biological degradation of pollution. Enhanced routine monitoring of CECs within the Lake Victoria and other freshwater ecosystems should start. There is need to control /or treatment micro-pollutants at the sources before release to receiving waters, by the adaptation of modern and advanced technologies in wastewater and general waste treatment.

Table 3. Chemicals of emerging concern groups and the adverse effect they can cause to human beings

Chemical of Emerging Concern group	Subgroup	Examples	Main adverse effect and nature of CEC
Pharmaceuticals and Veterinary medicine	synthetic	17-ethinylestradiol (EE2)	EDC
	Painkillers	Ibuprofen, diclofenac, naproxen	EDC
	lipid regulators	Bezafibrate,fenofibric acid,clofibric acid	EDC
	ß-blockers	Propranolol	EDC
	Antibiotics	Trimethroprim, ciprofloxacin,sulfamethoxazole sulfonamides, tetracyclines, chloramphenicol	EDC, Antibiotic resistance in disease causing bacteria complicating treatment of infections
	Antiepileptics	carbamazepine	EDC
	Psychostimulants	caffeine	moderately toxic
	radiocontrast agents	Iopromide	EDC
Animal feed additives	growth enhancement	Arsenic roxarsone	toxic, carcinogenic
Personal care products (PCPs)	sun / UV screens	Methylbenzylidene camphor (4-MBC)	EDC, largely unknown
	antioxidants and preservatives	Parabens (hydroxy benzoic acid)	EDC
	cosmetics, household chemicals	Triclosan	EDC, persistent
	nitro- and polycyclic musks	Tonalide (AHTN)	EDC
	disinfection by-products	N-nitrosodimethyl-amine (NDMA)	carcinogenic
	insect repellants	N,N-diethyl-metatoluamide (DEET)	EDC
Biocides	Herbicides	Alachlor,atrazine,bentazone,2,4 D,diuron	carcinogenic, Toxic, EDC
	Insecticides	Carbaryl, dieldrin, (DDT)	Toxic, EDC
	Fungizides	Vinclozolin	EDC
	Biocides	tributyl tin (TBT) and compounds	EDC
Industrial chemicals	flame retardants, plastics	brominated diphenyl ethers (PBDE)	EDC, carcinogenic
	Surfactants and metabolites,	nonyl phenol (NP), bisphenol A	EDC, Can be toxic to animals, ecosystems, and humans
	Fluoro-surfactants	Perfluoroalkyl acids (PFOS,PFOA)	EDC, carcinogenic, PFOS cause enlargement of liver, induce tumor of the liver
	anti-oxidants (food additives)	butylated hydroxyanisole (BHA)	carcinogenic
	Plasticizers	diethyl phthalate (DEP)	EDC
	coordination complex	EDTA	toxic, mutagenic
	corrosion inhibitors	benzotriazole (BTSA)	toxic
	Coolants	PCBs,	EDC, carcinogenic, mutagenic
	incineration products	dioxins, PAHs	EDC, carcinogenic, mutagenic, enhance HIV replication and mediate carcinogenicity.

REFERENCES

Abong'o, D. A., Wandiga, S. O., & Jumba, I. O. (2018). Occurrence and distribution of organochlorine pesticide residue levels in water, sediment and aquatic weeds in the Nyando River catchment, Lake Victoria, Kenya. *African Journal of Aquatic Science*, *43*(3), 255–270. doi:10.2989/16085914.2018.1490244

Abong'o, D. A., Wandiga, S. O., Jumba, I. O., Van Den, B. P., Nazariwo, B. B., Madadi, V. O., ... Nkedi-Kizza, P. (2015). Organochlorine pesticide residue levels in soil from the Nyando River Catchment, Kenya. *Africa Journal of Physical Sciences*, *2*(1), 18–32.

Arinaitwe K, Kiremire B.T., Muir D.C., Fellin P., Li H., Teixeira C., Mubiru D.N. (2016). Legacy and currently used pesticides in the atmospheric environment of Lake Victoria, East Africa. *Sci Total Environ., 543*(A), 9-18. doi:10.1016/j.scitotenv.2015.10.146

Bisesi, J. H. Jr, Sweet, L. E., van den Hurk, P., & Klaine, S. J. (2016). Effects of an antidepressant mixture on the brain serotonin and predation behavior of hybrid striped bass. *Environmental Toxicology and Chemistry*, *35*(4), 938–945. doi:10.1002/etc.3114 PMID:26076900

Bloutsos, A. A., & Giannopoulos, P. C. (2011). Concentrations of selected toxic elements in airborne particulates of Patras, Greece. *Global NEST Journal*, *13*(2), 109–118.

Bowa, K. O., Lalah, J. O., & Wandiga, S. O. (2009). Spatial and seasonal variations in concentrations of polycyclic aromatic hydrocarbons in water and sediment of Kisumu City Bay of Winam Gulf, Lake Victoria-Kenya. *Bulletin of Environmental Contamination and Toxicology*, *83*, 722–733.

Bowa, K. O., Lalah, J. O., & Wandiga, S. O. (2009). Spatial and seasonal variations in concentrations of polycyclic aromatic hydrocarbons in water and sediment of Kisumu City Bay of Winam Gulf, Lake Victoria-Kenya. *Bulletin of Environmental Contamination and Toxicology*, *83*, 722–733.

Brodin, T., Fick, J., Jonsson, M., & Klaminder, J. (2013). Dilute Concentrations of a Psychiatric Drug Alter Behavior of Fish from Natural Populations. *Science*, *339*(6121), 814–815. doi:10.1126cience.1226850 PMID:23413353

Chirikona, F., Filipovic, M., Ooko, S., & Orata, F. (2015). Perfluoroalkyl acids in selected wastewater treatment plants and their discharge load within the Lake Victoria basin in Kenya. *Environmental Monitoring and Assessment*, *187*(5), 238. doi:10.100710661-015-4425-6 PMID:25861900

Connett, P. (1998). Municipal Waste Incineration: A Poor Solution for the Twenty First Century. *4th Annual International Management Conference*, 1-15.

Crago, J., Bui, C., Grewal, S., & Schlenk, D. (2016). Age-dependent effects in fathead minnows from the anti-diabetic drug metformin. *General and Comparative Endocrinology*, *232*, 185–190. doi:10.1016/j.ygcen.2015.12.030 PMID:26752244

Fatta, K. D., Kalavrouziotis, I. K., Koukoulakis, P. H., & Vasquez, M. I. (2011). The risks associated with the wastewater reuse and xenobiotics in the agroecological environment. *The Science of the Total Environment*, *409*(19), 3555–3563. doi:10.1016/j.scitotenv.2010.03.036 PMID:20435343

Getenga, Z. M., Kengara, F. O., & Wandiga, S. O. (2004). Determination of organochlorine pesticides in soil and water from river Nyando drainage system within Lake Victoria Basin, Kenya. *Bulletin of Environmental Contamination and Toxicology*, *72*(2), 335–342. doi:10.100700128-003-9107-3 PMID:15106770

González-Naranjo, V., Boltes, K., & Biel, M. (2013). Mobility of ibuprofen, a persistent active drug, in soils irrigated with reclaimed water. *Plant, Soil and Environment*, *59*(2), 68–73. doi:10.17221/590/2012-PSE

Halden, R. U. (2015). Epistemology of contaminants of emerging concern and literature meta-analysis. *Journal of Hazardous Materials*, *282*, 2–9. doi:10.1016/j.jhazmat.2014.08.074 PMID:25294779

Jelic, A., Gros, M., Ginebra, A., Cespedes-Sanchez, R., Ventura, F., Petrovic, M., & Barcelo, D. (2011). Occurrence partition and removal of pharmaceutical in sewage water and sludge during wastewater treatment. *Water Research*, *45*(3), 1165–1176. doi:10.1016/j.watres.2010.11.010 PMID:21167546

Kalantzi, O. I., & Alcock, R. A. (2012). Short-chain chlorinated paraffins in biota – levels and effects. *Global NEST Journal*, *14*(1), 66–71.

Khetan, S. K., & Collins, T. J. (2007). Human pharmaceuticals in the aquatic environment: A challenge to green chemistry. *Chemical Reviews*, *107*(6), 2319–2364. doi:10.1021/cr020441w PMID:17530905

Kimosop, S. J., Getenga, Z. M., Orata, F., Okello, V. A., & Cheruiyot, J. K. (2016). Residue levels and discharge loads of antibiotics in wastewater treatment plants (WWTPs), hospital lagoons, and rivers within Lake Victoria Basin, Kenya. *Environmental Monitoring and Assessment*, *188*(9), 532. doi:10.100710661-016-5534-6 PMID:27553946

Kimosop, S. J., Orata, F., & Getenga, Z. (2012). The Influence of Filter Mud Compost and Tithonia diversifolia Leaves on the Dissipation of Diuron in Soils within the Nzoia River Drainage Basin, Kenya. *Bulletin of Environmental Contamination and Toxicology*, *89*(2), 328–333. doi:10.100700128-012-0685-9 PMID:22617949

Kumar, A., Chang, B., & Xagoraraki, I. (2010). Human health risk assessment of pharmaceuticals in water: Issues and challenges ahead. *International Journal of Environmental Research and Public Health*, *7*(11), 3929–3953. doi:10.3390/ijerph7113929 PMID:21139869

Lisouza, F. A., Owuor, P. O., & Lalah, J. O. (2011). Variation in indoor levels of polycyclic aromatic hydrocarbons from burning various biomass types in the traditional grass-roofed households in Western Kenya. *Environ Pollut*, *159*(7), 1810–1815. doi:10.1016/j.envpol.2011.03.032 PMID:21501911

Liu, J., Wang, R., Haung, B., Lin, C., Zhou, J., & Pan, X. (2011). Biological effects and bioaccumulation of steroidal and phenolic endocrine chemicals in highback Crucian carp exposed to wastewater treatment plant effluents. *Journal of Environmental Pollution, 162*, 325-331.

Madadi, V. O. (2005). *Chemodynamics studies and assessment of pesticides residue levels in Lake Victoria catchment area for Rivers Sio and Nzoia* (MSc. Thesis). Department of Chemistry, University of Nairobi.

Manickum, T., John, W., & Terry, S. (2011). Determination of selected steroid estrogens in treated sewage effluent in the Umsuduzi (Duzi) Rier water catchment area. *Hydrology: Current Research*, *2*(3), 45–56.

Mule, S., Nguta, C., Kamau, G., Ongera, G., Wangila, P., Kiplimo, J., & Chirchir, D. (2015). Retention Efficiencies of Halogenated and Non-Halogenated Hydrocarbons in Selected Wetland Ecosystem in Lake Victoria Basin. *Intern. J. of Envir*, *4*(2), 38–52. doi:10.3126/ije.v4i2.12623

Niemuth, N. J., & Klaper, R. D. (2015). Emerging wastewater contaminant metformin causes inters ex and reduced fecundity in fish. *Chemosphere*, *135*, 38–45. doi:10.1016/j.chemosphere.2015.03.060 PMID:25898388

Ogunah, J. A., Owuor, P. O., Kowenje, C. O., & Lalah, J. O. (2016). Adsorption and desorption of chlorsulfuron in agricultural soils of Mara River Basin, Kenya. *AJEA*, *12*(6), 1–10. doi:10.9734/AJEA/2016/26406

Okungu, J. O., Njoka, S., Abuodha, J. O. Z., & Hecky, R. E. (2005). *An Introduction to Lake Victoria Catchment, Water Quality, Physical Limnology and Ecosystem Status (Kenyan Sector).* Lake Victoria Environment Management Project (LVEMP) Kisumu Kenya. Retrieved from http://hdl.handle.net/1834/7140

Omwoma, S., Lalah, J. O., Virani, M., Schramm, K., & Henkelmann, B. (2015). Dioxin-like PCBs and PCDD/Fs in surface sediments near the shore of Winam Gulf, Lake Victoria. *Chemosphere*, *118*, 143–147. doi:10.1016/j.chemosphere.2014.07.062 PMID:25150826

Onunga, D. O., Kowino, I. O., Ngigi, A. N., Aggrey, O., Orata, F., Getenga, Z. M., & Were, H. (2015). Biodegradation of carbofuran in soils within Nzoia River Basin, Kenya. *Journal of Environmental Science and Health. Part. B, Pesticides, Food Contaminants, and Agricultural Wastes*, *50*(6), 387–397. do i:10.1080/03601234.2015.1011965 PMID:25844859

Orata F., Maes A., Werres F., Wilken D. F. (2011). Perflourinated compounds distribution and source identification in sediments of Lake Victoria Gulf Basin. *Soil and Sediment Contamination: An International Journal.* doi:10.1080/15320383.2011.546442

Orata, F., Maes, A., Werres, F., & Wilken, R. D. (2011). Perfluorinated Compounds Distribution and Source Identification in Sediments of Lake Victoria Gulf Basin. *Soil and Sediment Contamination. International Journal (Toronto, Ont.)*, *20*(2), 129–141. doi:10.1080/15320383.2011.546442

Orata F., Quinete N., Maes, A., Werres, F., & Wilken, R. (2008a). Perfluorooctanoic acid and perfluorooctane sulfonate in Nile Perch and tilapia from gulf of Lake Victoria. *African Journal of Pure and Applied Chemistry*, *2*(8), 75-79.

Orata, F., Quinete, N., Werres, F., & Wilken, R. D. (2009). Determination of Perfluorooctanoic acid and Perfluorooctane sulfonate in Lake Victoria Gulf Water. *Bulletin of Environmental Contamination and Toxicology*, *82*(2), 218–222. doi:10.100700128-008-9543-1 PMID:18791652

Orata, F. O. (2018). Conventional Wastewater Treatment Plants as a Discharge and Source Point for Biota Exposure to Micro-pollutants. In R. A. Hauser-Davis & T. E. Parente (Eds.), *Ecotoxicology, Perspectives on Key Issues*. doi:10.1201/b21896-11

Pothitou, P., & Voutsa, D. (2008). Endocrine disrupting compounds in municipal and industrial wastewater treatment plants in Northern Greece. *Chemosphere*, *73*(11), 1716–1723. doi:10.1016/j.chemosphere.2008.09.037 PMID:18954890

Schultz, M. M., Bartell, S. E., & Schoenfuss, H. L. (2012). Effects of Triclosan and Triclocarban, Two Ubiquitous Environmental Contaminants, on Anatomy, Physiology, and Behavior of the Fathead Minnow (Pimephales promelas). *Archives of Environmental Contamination and Toxicology*, *63*(1), 114–124. doi:10.100700244-011-9748-x PMID:22237462

Shala, L., & Foster, G. D. (2010). Surface Water Concentrations and Loading Budgets of Pharmaceuticals and Other Domestic-Use Chemicals in an Urban Watershed (Washington, DC, USA). *Archives of Environmental Contamination and Toxicology*, *58*(3), 551–561. doi:10.100700244-009-9463-z PMID:20098984

Siddique, A., Saied, S., Alam Zaigham, N., Mumtaz, M., Ali Mahar, G., & Mohiuddin, S. (2012). Temporal variability of disinfection by-products concentration in urban public water system. *Global NEST Journal*, *14*(4), 393–398.

Smeti, E., Samios, S., Nikolaou, A., Golfinopoulos, S., & Lekkas, T. D. (2011). Statistical analysis of trihalomethanes in treated-water tanks: Seasonality, local variability and correlations. *Global NEST Journal*, *13*(1), 41–49.

Ssebugere, P., Kiremire, B. T., Henkelmann, B., Bernhöft, S., Wasswa, J., Kasozi, G. N., & Schramm, K. W. (2013, June). PCDD/Fs and dioxin-like PCBs in surface sediments from Lake Victoria, East Africa. *The Science of the Total Environment*, *454–455*(1), 528–533. doi:10.1016/j.scitotenv.2013.03.046 PMID:23567173

Tangri, N. (2003). *Waste Incineration: A Dying Technology: Essential Action for GAIA*. Retrieved from www.no-burn.org

Tanoue, R., Sato, Y., Motoyama, M., Nakagawa, S., Shinohara, R., & Nomiyama, K. (2012). Plant uptake of pharmaceutical chemicals detected in recycled organic manure and reclaimed wastewater. *Journal of Agricultural and Food Chemistry*, *60*(41), 10203–10211. doi:10.1021/jf303142t PMID:23003104

Thomaidis, N. S., Asimakopoulos, A. G., & Bletsou, A. A. (2012). Emerging contaminants: A tutorial mini-review. *Global NEST Journal*, *14*(1), 72–79.

Twesigye, C. K., Simon, M. O., Getenga, Z. M., Mwakalila, S. S., & Nakiranda, J. K. (2011). The Impact of Land Use Activities on Vegetation Cover and Water Quality in the Lake Victoria Watershed. *The Open Environmental Engineering Journal*, *4*(1), 66–77. doi:10.2174/1874829501104010066

Vidal-Dorsch, D. E., Colli-Dula, R. C., Bay, S. M., Greenstein, D. J., Wiborg, L., Petschauer, D., & Denslow, N. D. (2013). Gene Expression of Fathead Minnows (Pimephales Promelas) Exposed to Two Types of Treated Municipal Wastewater Effluents. *Environmental Science & Technology*, *47*(19), 11268–11277. doi:10.1021/es401942n PMID:23919544 vom Saal, F. S., Nagel, S. C., Coe, B. L., Angle, B. M., & Taylor, J. A. (2012). The estrogenic endocrine disrupting chemical bisphenol-A (BPA) and obesity. *Journal of Molecular and Cellular Endocrinology*, *354*(1-2), 74–88. doi:10.1016/j.mce.2012.01.001 PMID:22249005

Xu, N., Xu, Y. F., Xu, S., Li, J., & Tao, H. C. (2012). Removal of estrogens in municipal wastewater treatment plants: A Chinese perspective. *Environmental Pollution*, *165*(0), 215–224. doi:10.1016/j.envpol.2011.12.025 PMID:22226508

Chapter 2
Occurrence and Impact of Emerging Contaminants in Nigeria's Freshwater Resources

Sinmi Abosede
Pan Atlantic University, Nigeria

ABSTRACT

The presence of emerging contaminants (ECs) in freshwater presents a major challenge as they pose serious threats to human life and ecosystems. It is essential that water is of good quality before it can be used by communities and benefit the environment. Nigeria is blessed with abundant water resources; however, the country lacks effective and efficient institutions to provide sanitation services to treat domestic waste and to monitor and regulate industries, helping them protect the environment by ensuring they treat their industrial waste before discharging to the water bodies. Various ECs have been detected in the nation's freshwater resources. They pose significant threat to the environment, and they have the potential to harm aquatic life and human health. Nigeria needs a comprehensive and integrated water resources management plan that gives priority to the monitoring, detection, and treatment of emerging contaminants to mitigate against the potential risks that occur when they are present in freshwater resources.

INTRODUCTION

Water is essential to life and it is needed to sustain livelihoods and the environment. This has been manifested by the United Nations Development Agenda, and access to good quality water is at the heart of the attainment of majority of the 17 Sustainable Development Goals (SDGs) particularly, Goals 3, 6 and 14. SDG Goal 3 seeks to "Ensure healthy lives and promote well-being for all at all ages"; SDG Goal 6 seeks to "Ensure availability and sustainable management of water and sanitation for all" and focuses on ensuring safe drinking water, minimizing the release of hazardous chemicals and materials, and improving wastewater management and the safe reuse of wastewater and SDG Goal 14 seeks to "Conserve and sustainably use the oceans, seas and marine resources for sustainable development". The

DOI: 10.4018/978-1-7998-1871-7.ch002

presence of Emerging Contaminants (ECs) in freshwater presents a major challenge to the attainment of these and the other 14 SDG goals, as they pose serious threats to human life and ecosystems.

ECs are contaminants that are found in drinking water or aquatic environments that have not been previously identified as contaminants. The U.S Geological Survey has defined them as 'any synthetic or naturally occurring chemical that is not commonly monitored in the environment but has the potential to enter the environment and cause known or suspected adverse ecological and/or human health effects (Raghav et al., 2013). The group of compounds classified as Emerging Contaminants are very diverse in terms of their toxicity, behavior in the environment, treatment required and monitoring technique used to identify them. They include Pharmaceuticals and Personal Care Products (PPCPs), Endocrine Disrupting Chemicals (EDCs), polychlorinated biphenyls (PCBs), organochlorine pesticides (OCPs), polycyclic aromatic hydrocarbons (PAHs), polychlorinated dibenzofurans (PCDFs), dibenzo-p-dioxins (PCDDs), steroids, hormones, surfactants and surfactant metabolites, flame retardants, pesticides, industrial additives, nanomaterials and gasoline additives (Inam et al., 2015; Stefanakis & Becker, 2016). The use and production of synthetic organic chemicals for use in various industries, households and agriculture has led to an increase in the occurrence and concentration of these compounds, in wastewater effluents and consequently the environment. Most ECs have only recently been recognized as contaminants, as a result of evidence linking them with various adverse effects observed in the environment, aquatic life and human health. They are usually bioactive, bioaccumulative, with a widespread occurrence and persistence in the environment (Stefanakis & Human, 2015). Globally, it is estimated that approximately 2 million tons of wastewater per day are disposed of within receiving waters, these include various chemicals and industrial, human and agricultural waste products (UNESCO, 2003). The presence of ECs in wastewater is a source of concern, as they are used and released in the environment constantly, contaminating surface water and groundwater resources used for potable water. They pose a threat even at low concentrations and make cause chronic toxicity (Carlsson et al., 2009), endocrine disruption in humans and aquatic life (Bolong et al., 2009) and may result in the development of antibiotic resistance genes (Lubick, 2011).

In Africa, the increased use of synthetic organic compounds for domestic, agriculture and industrial use poses a potential threat to surface and groundwater resources across the region (Sorensen et al., 2015). Sewerage coverage in sub-Saharan Africa is less than 30% and over 80% of the wastewaters produced in large cities in the region are discharged untreated to rivers and lakes (Nyenje et al., 2010). This has led to the accumulation of various ECs in the environment. These ECs pose significant threat to the human and aquatic organisms, with a growing demand for freshwater sources in Africa, there is a need to understand the occurrence and impact of ECs in the environment. There have been a number of studies focused on the occurrence and fate of EC in countries such as Italy, Spain, Sweden, Switzerland, Netherlands, UK, Israel, USA, China, Canada and Germany (Caliman & Gavrilescu, 2009). However, very few studies have been conducted in Africa, to understand the risks associated with the accumulation of these contaminants in the environment (Offiong, et al., 2019). Like most sub-Saharan African countries, in Nigeria, there is a lack of understanding of the risks associated with ECs in the environment, this is coupled with a lack of investment in technology required to monitor and detect the presence of these contaminants in the environment. There is a need to understand the occurrence, source and potential risks of ECs in Nigeria's freshwater resources. This chapter will review the occurrence and impact of ECs in freshwater resources in Nigeria.

BACKGROUND

Nigeria at a Glance

Nigeria has a population of 186 million people, with 51.4% of the population living in rural areas (The World Bank, 2016). It is a country blessed with abundant human and natural resources such as water, crude oil, natural gas, tin ore, bauxite, gold, limestone, lead and iron ore. It is Africa's largest producer of oil and the 11th largest producer of oil in the world and is blessed with renewable energy resources like wind, solar, biomass and hydropower. However, these resources have not translated to improved economic and development conditions for the country, as more than 62.6% of the population live below the poverty line and earn less than $2 a day (NBS, 2017). It has a Human Development Index of 0.527 and is ranked 152 out of 188 countries (UNDP, 2016).

Water Resources in Nigeria

Nigeria has a total surface area of about 923 770 km^2, of which about 1.4% is covered with water. It has huge potential water resources, with annual yields estimated at 267 billion cubic meters of surface water and 52 billion cubic meters of ground water, however this is unevenly distributed across the country. The country is drained mainly by the Niger Basin, the Lake Chad Basin, the South-Western littoral basin and the South-eastern littoral basin. The climate of Nigeria varies between semi-arid in the north to tropical and humid in the south. The average rainfall ranges from about 500 mm/year in the North to over 4,000 mm in the south. The Northern part of the country is especially susceptible to the effects of the high variability in rainfall and some parts have experienced a persistent drought over the last 30 years, which has resulted in the reduction of the wetlands in Hadejia-Nguru and Lake Chad. This is in contrast with the situation in the Southern states, where flooding incidents occur frequently.

With approximately 1600 m^3/capita/year (FAO, 2014) of renewable water resources, Nigeria is not considered to be a water poor country. However, based on the Falkenmark indicator, it has been described as a country experiencing economic water scarcity, which means that as a result of poor water resources management and a lack of investment, the water needs of the population are not being met. The three tiers of government (Federal, State and Local) are responsible for water and sanitation in Nigeria, however there are several agencies and ministries that also have responsibilities relating to water resources management. Other sub-sectors relating to water resources include hydro-electric power generation, flood/erosion control, inland transportation, fisheries/aquaculture, livestock, and mining are undertaken by Federal Ministries of Power, Environment, Transport, Agriculture & Rural Development, Mining & Steel Development, respectively (JICA, 2014). The Federal Ministry of Water Resources (FMWR) is responsible for developing the national water policies, water resources management and approving development projects and is supported by various agencies and parastatals such as the National Water Resource Institute, Nigeria Hydrological Service Agency, Gurara Water Management Authority, Nigeria Integrated Water Management Commission and 12 River Basin Development Agencies (RBDA). The RBDAs are mainly responsible for the provision of bulk water supply for irrigation and urban water supply. The State Water Agencies (SWA) are responsible for the developing and managing water supply facilities with the aim of providing mainly urban and semi-urban water supply and in some cases rural water supply and the Local Government Authorities (LGA) are responsible for the provision of rural water supplies and sanitation in 774 local government areas around the country. The involvement of so many institutions, often leads to confusion and inefficiency in the implementation of programs and projects, as a result of overlapping functions or duplication of tasks.

The sanitation sector is not well developed and the FMWR developed a National Sanitation Policy in 2004 to provide a framework for the sustainable development and management of appropriate sanitation services. The Ministry of Environment for each state is responsible for sanitation. Sanitation access levels are low and at 2015, only 33% of the population had access to a basic sanitation service and 26% practiced open defecation (The World Bank, 2016). Past Government administrations have paid more importance to water supply than sanitation and the sector is badly underdeveloped. Poor sanitation conditions in Nigeria, from contaminated water to unhygienic environments are the primary causes of water related diseases such as cholera, typhoid, diarrhea, guinea worm and schistosomiasis, the mortality rate as a result of these diseases is between 5 – 20% and an estimated 150,00 – 200,000 children under the age of 5, die each year as a result of diarrhea (FMWR, 2004).

Nigeria has a substantial body of legislation to protect the environment. Section 20 of the constitution recognizes the importance of protecting the environment and states that it is an objective of the Nigerian State to improve and protect the air, land, water, forest and wildlife of Nigeria (FGN, 1999). This is further supported by the National Environmental Standards and Regulation Enforcement Agency (NESREA) Act of 2007. It contains all the laws and regulations focused on the protection of the environment and is administered by the Federal Environmental Protection Agency (FEPA), a parastatal of the Ministry of the Environment and provides authority to ensure the compliance with environmental laws, pollution control and prevention through monitory and regulatory measures; it empowers the agency to make and review environmental regulations, effluent discharge limits, prevention of the discharge of harmful substances and prohibits the discharge of hazardous substances into the environment, by imposing a fine of and an imprisonment term of 5 years (ELRI, 2011). The Water Resources Act of 2004 aims to the promote the optimum planning, development of the nation's water resources (FGN, 2004), it makes provision for the authorities to implement pollution prevention plans to protect aquatic life and makes polluters of water resources liable to a fine or an imprisonment term of 6 months.

Currently, in spite of the numerous legislations and regulations, there is little evidence that industrial and domestic pollution is being adequately managed in the country, as there is no enforcement by the relevant agencies. In addition, the various industries have shown little readiness for compliance to the existing regulations, there is an absence of the legal requirement for self-reporting and recording compliance information by corporations, this has made non-compliance easier and increased the difficulty of enforcement (Ahmad, 2003).

Water Pollution and the Presence of Emerging Contaminants in Nigeria Freshwater Resources

In Nigeria there are 57 million people without access to an improved water source and 70 million people without access to improved sanitation (FMWR, 2016). Inadequate access to water and sanitation facilities, increases the risk of contracting common illnesses such as diarrhea and cholera and neglected tropical diseases, such as soil-transmitted parasitic helminth infections, schistosomiasis and trachoma. These diseases are endemic across Nigeria, affecting up to 50% of the population in some regions (The World Bank, 2017). Nigeria lacks a comprehensive strategy on sanitation as a whole, including solid waste disposal, domestic and industrial wastewater disposal and wastewater treatment. Only 5.6% of the population make use of sewerage systems and approximately 46 million people practice open defecation (The

World Bank, 2017). The sanitation sector in Nigeria is unregulated and no one monitors or controls the discharge of harmful compounds to natural water courses in the country. There are hardly any sewerage systems and majority of the wastewater generated by households and industries, is discharged directly untreated to surface waters or the soil, contaminating both surface water and the groundwater. This results in poor water quality as a result of the presence of toxic chemicals, metals, organic pollutants, minerals and bacteria in the water. This poses a significant threat to the quality of surface water and groundwater in the country, which might result in various adverse effects to human and aquatic life. 80% of the nation's industry is concentrated in 4 main states namely - Lagos, Rivers, Kano and Kaduna and the challenges and risks will be more pronounced in these urban areas, due to higher density of contaminant sources and a greater concentration of anthropogenic activity (Sorensen et al., 2015). It has been estimated that poor sanitation practices in the country costs the country $3M every year (WSP, 2012).

Pollution of freshwater resources occurs when inorganic and organic contaminants are released directly or indirectly without proper treatment. These contaminants can interfere with the proper function of the aquatic ecosystems, harming aquatic organisms and the animals and humans that rely on the water resources for various uses. In Nigeria, due to the absence of sewerage systems, domestic and agriculture waste contribute to the pollution of the nation's water resources, however industrial pollution is a particular cause for concern, as many of the industries discharge liquid effluent, solid waste and hazardous waste directly into the nation's water bodies (Okereke, et al. 2016). Very few industries in Nigeria treat their effluent before discharging to water bodies (Taiwo et al., 2012). The negative impact of the activities of oil companies in the Niger-Delta, on the environment and livelihood of the local communities is well documented. The constant oil spills has led to water scarcity, severe water pollution and disrupted the socio-economic activities of people in the region and this has resulted in the alteration of habitats, loss of biodiversity, degradation of forests and depletion of aquatic organisms (UNDP, 2006). However, the majority of the pollution incidences have not been properly documented over the years and detailed analysis of the extent and effects of the pollution caused by oil spills, in the region are not readily available (Ite et al., 2018).

Polycyclic aromatic hydrocarbons (PAHs) are naturally occurring compounds found in coal and tar and anthropogenic compounds generated as a result of the incomplete combustion of organic material such as coal, wood, garbage, and gasoline (Anderson & Hillwalker, 2008). They consist of two or more fused benzene rings arranged in various configurations (Kim, Jahan, Kabir, & Brown, 2013). Examples of PAHs include naphthalene, pyrene, benzopyrene, anthracene, phenanthrene, etc. PAHs are insoluble in water but have an affinity for organic carbon and tend to sorb onto organic sediments in aquatic waters and bioaccumulate in certain aquatic organisms that lack the capacity to effectively biodegrade the compounds. The bioaccumulation of PAHs in aquatic organisms, poses a threat to humans, who consume these organisms. Nduka & Orisakwe (2010) investigated the presence of PAHs in some selected rivers in Niger Delta, South Nigeria. Six PAHs were detected in samples from Anieze River in Port Harcourt, they include; acenaphthene, benzanthracene, benzoperylene, benzofluoranthene, dibenzoanthracene and chrysene, with concentrations ranging from 0.004 mg/l to 0.064 mg/l. Chrysene and fluorine were found in Orash River in Delta State, at concentrations of 0.017 mg/l and fluorene 0.109 mg/l, respectively and Dibenzoanthracene was found in the Ifie-Kporo Creek in Delta State at a concentration of 4.350 mg/l.

Tongo et al., (2017) evaluated the levels and distribution of PAHs in the surface water, sediment and fish samples from Ovia River, Southern Nigeria and it was found that naphalene, acenaphthylene and fluoranthene were the dominant contaminants in all 3 samples, with concentrations ranging from 30.8 -90.7 μg/l. The levels of 16 different PAHs were assessed in aqueous samples, sediments, fish, crab and shrimps collected from the Lagos Lagoon in South West Nigeria (Alani et al., 2012). It was found that the different PAHs bioaccumulated to various degrees in the different aquatic organisms, with concentrations ranging from 44.6 ng/g to 625 ng/g.

Arukwe et al., (2012) reported the presence of various ECs in aquatic waters in Owerri, in South East Nigeria. The ECs were introduced into the environment through surface run-off arising from the leaching of solid waste dumping sites. Samples were collected from the sediment and run-off of the solid waste dumping sites. Phthalate, an integral component of plastic products, was detected as the dominant compound group in all sediment samples and the run-off water sample, while nonylphenols, nonylphenol-polyethoxylates, ethyl methoxy cinnamate (EHMC) and Bisphenol-A (BPA) were determined in the sediment samples. Concentrations ranged from 3 - 3263 ng/L in the run-off samples and 3 - 31,805 μg/kg in the sediment samples.

Perfluorochemicals (PFCs) are man-made contaminants, which are persistent in the environment. They are found in wastewater generated from the textile, paper and leather industry. Two PFCs –perfluorooctanesulfonate (PFOS) and perfluorooctanoate (PFOA) were detected in the surface water, pore water and sediments from major rivers in Nigeria (Ololade, 2014; Ololade et al., 2018). PFOS occurred at higher concentrations than PFOA, with concentrations ranging from 3.9 ng/L in surface water, 10.9 to 20.4 ng/L in pore water and 5.1 to 10.4 ng/g in the sediment.

Organochlorine pesticides are widely used in Nigeria's agricultural sector for the control of pests, in an effort to increase agricultural productivity. They have been banned in many countries for many years because of their persistence, toxicity and bioaccumulation in the environment (Shen et al., 2017). There are laws restricting their use in Nigeria, however, due to a lack of regulation, many old, non-patented and toxic pesticides are used extensively in Nigeria (Nnamonu & Onekutu, 2015). Organochlorine pesticides have been detected in Ikpoba River, a major River in the heart of Benin City, a Nigerian urban town (Ogbeide et al., 2019). The detected pesticides in the surface water samples included hexachlorocyclohexane, chlordane, diphenyltrichloroethane, while aldrin, chlordane, dichlorodiphenyltrichloroethane, dieldrin, endosulfan, endrin, endrin aldehyde, heptachlor and heptachlor were detected in sediment samples. Results showed varying pesticide concentrations in both sediment and water samples with hexachlorocyclohexane having the highest concentration of 0.24 ng/L in water and dieldrin having the highest concentration of 0.99 μg/kg in the sediment. Researchers have also detected the presence of organochlorine pesticide residues in water and fish obtained from rivers in Edo State, South West Nigeria. Lindane, Aldrin and dichlorodiphenytrichloroethanes were detected in the water samples, with concentrations ranging from 0.2879–0.8143 ppb, while Lindane, Aldrin and dichlorodiphenytrichloroethanes were also detected in bottom, middle and top feeders fish with concentrations ranging from 0.003 – 0.063 μg/g.

The levels of some ECs including EDCs and PPCPs were assessed in the Ikpa River Basin (Inam et al., 2015). Seven antibiotic drugs (acetamidophenol, chloramphenicol, ciprofloxacin, erythromycin, lincomycin HCl, roxythromycin, and sulfamethoxazole), three bactericides/antimicrobial agents (sulfathiazole, triclosan and triclocarban), an antiepileptic drug (carbamazepine), an analgesic drug (diclofenac sodium), a resin precursor (bisphenol A), a sunscreen product (oxybenzone), a hormone (equilin), an insect repellent (DEET), and a stimulant (caffeine) were detected in surface water samples from the River Basin, nearby landfill leachate and hospital effluent. The detected concentrations ranged from 0.3 – 218.7 ng/l, with bisphenol A, chloramphenicol, erythromycin, triclocarban and triclosan having the highest concentrations in all samples. The concentration levels of the herbicide, Paraquat dichloride, in the surface water, sediment and dominant fish species of the Warri River Basin was assessed (Ikpesu, 2015). The concentration of the herbicide in the surface water samples ranged from 0.02 μg/L to 3.14

μg/L, while its concentration in sediment samples ranged from 1.64 – 10.06 μg/g. The concentration in the fish samples ranged from 0.88 – 10.870 μg/g.

EDCs such as 4-t-octyphenol, 4-nonylphenol and Bisphenol-A (BPA), were detected in the sediments and waters of River Ogun and River Ibeche in South West Nigeria (Oketola & Fagbemigun, 2013). 4-t-octyphenol and 4-nonylphenol were detected in the water samples with concentrations ranging from 43.9 – 79.4 ng/L, while all 3 compounds were detected in the sediment samples from the rivers with concentrations ranging from 0.4 - 79.4 ng/g.

Effect of Emerging Contaminants on Aquatic Life and Human Health

Various ECs have been identified in the natural water bodies of Nigeria's water system. These contaminants pose significant threat to the environment, aquatic life and human health. ECs can enter the environment through point sources from the discharge of domestic and industrial wastewater and landfill leachates and non-point sources from run-off from agricultural farms, roadways and groundwater contamination (Bolong et al., 2009). The presence of ECs in the environment, poses a potential threat to aquatic life and humans. The estrogenic ethynylestradiol (EE2) is a component of oral contraceptives, which has been linked to the reproductive failure of aquatic organisms. Nash et al., (2004) demonstrated that the exposure of zebrafish to EE2 resulted in deterioration in breeding dynamics and reproduction capacity of the fish. Another study in Ontario, demonstrated that low concentrations of EE2 had an adverse impact on the development of reproductive organs and hatch success of amphibians from a lake in Ontario, Canada (Park & Kidd, 2005). Oral contraceptives containing synthetic estrogen have been shown to induce feminization of male fish, diminishing reproduction and depletion of fish population (Mills et al., 2015). Natural estrogens, from plant sources, such as phytoestrogens has been linked to reproductive failures in animals and it was observed that sheep grazing on certain strains of clover in New Zealand experienced reproductive dysfunction (Adams, 1998).

Exposure to perfluoroalkyl compounds (PFCs) have been linked to male infertility. Experimental evidence from a study on 212 males from the Veneto region in North East Italy, showed that exposure to PFCs interfered with hormonal pathways, potentially leading to male infertility (Di Nisio et al., 2018). It has been suggested that various ECs such as alkylphenols (APs), bisphenol A (BPA), parabens, perfluoroalkyl substances (PFASs), phthalates, polybrominated diphenyl ethers (PBDEs), synthetic musks and triclosan can alter the structure of the mammary gland, its development and function, thereby increasing the risk of breast cancer (Siddique et al., 2016). Other researchers have also suggested that the presence of nonylphenol and BPAs in the environment interferes with animal reproduction and has an adverse effect on the development of ovary and sperm (Kimura et al., 2006; Lawson et al., 2011; McClusky et al., 2006).

Researchers found alterations in the development of the reproductive and endocrine systems of wildlife populations in Lake Apopka, Florida, exposed to organochlorine pesticides such as dichlorodiphenyltrichloroethane (DDT) (Guillette Jr et al., 1994). Gibbs, Bryan, & Pascoe (1991), found that marine gastropods (Nucella lapillus) exposed to tributylins (leached from paints and PVC pipes), experienced sterilization of females, population decline and extinction and the development of male sexual characteristics in females. It has been suggested that prenatal exposure to polychlorinated biphenyls (PCBs) and polychlorinated dibenzofurans (PCDFs) results in neurobehavioral effects such as poor gross motor function and poor infant visual recognition memory (Jacobson & Jacobson, 1997). The presence of antibiotics in water bodies causes resistance among bacterial pathogens and alters the microbial community structure, which has grave consequences for the higher food chain (Daughton & Ternes, 1999). Other researchers have observed the effect of perchlorates on the optimum function of the thyroid gland by competitively inhibiting the transport of iodide through the body (Logan, 2001; Urbansky, 2000).

SOLUTIONS AND RECOMMENDATIONS

To prevent harm to the environment, aquatic life and human health, it is imperative that the Nigerian Government invests in appropriate infrastructure and technology for the treatment of contaminated effluent. In addition to increasing the number of conventional wastewater treatment plants around the country, consisting of basic primary and secondary treatment processes, to remove solids and dissolved contaminants from wastewater, it is essential that focus is also given to the installation of tertiary treatment plants, equipped with the appropriate technology that have been proven to efficiently remove ECs from wastewater. These include Advanced Oxidation Processes, Activated Carbon Adsorption, Membrane Separation Processes, etc.

FURTHER RESEARCH

Additional research should be conducted to identify ECs that pose the greatest risk for human health, the environment and aquatic life. Knowledge regarding the occurrence and risks of ECs is relatively limited in Nigeria; this places the population at a disadvantage, as they face the risk of exposure to these ECs. The Government should place emphasis on educating and sensitizing the population of the potential risks associated with the presence of ECs in the environment. Appropriate technology should be adopted to monitor, detect and remove a wide range of ECs from contaminated effluents. This will require working with research institutions, who can use pilot plant tests to determine the suitability of a particular process to remove particular type of ECs from contaminated ECs.

There is also a need for the development of models, to predict the fate, transport and transformation of ECs in the environment (Stefanakis & Human, 2015). This will be useful to identify the ECs as they occur in the environment, so that appropriate action can be taken to minimize their effect on the environment, aquatic life and human health.

CONCLUSION

Nigeria is blessed with abundant water resources, however, the presence of emerging contaminants and other contaminants in Nigeria's water resources presents a challenge that will affect the nation's ability to meet the United Nation's SDG goals, particularly SDG goals 3, 6 and 14. The available water resources should be fit for consumption, before the people can obtain maximum benefit from it. Nigeria needs to improve its approach to monitoring and regulating the water quality of freshwater bodies. Government agencies should set up periodic monitoring of the nation's water resources to ensure that pollution incidents are identified, before they can cause significant damage to the environment and harm the subsequent users of the water. It needs to adopt an integrated approach to water resources management, a key component of which should include the development of a water quality monitoring network and robust water quality database. Reliable and adequate data of water quality parameters need to be collected and monitored periodically. The present water quality monitoring status in Nigeria involves monitoring only groundwater once every year by each states' water board using FEPA standards (Zejiao & Ekiye Ebiare, 2010). The country also has to increase its spending on infrastructure for sanitation, solid waste management and general pollution control. This will reduce the levels of pollutants that are discharged to the water bodies. In addition, government should ensure enforcement of the various environmental laws and make industrial operators responsible for the waste they generate. They should also engage all stakeholders and embark on a nation-wide campaign sensitizing people about the dangers of discharging untreated waste to the water bodies, educating them about the adverse effects, polluted water has on the environment and human and animal health.

REFERENCES

Adams, N. R. (1998). Natural and anthropogenic environmental oestrogens: The scientific basis for risk assessment. *Pure and Applied Chemistry*, *70*(9), 1855–1862. doi:10.1351/pac199870091855

Ahmad, A. (2003). Policing Industrial Pollution in Nigeria. In *International Environmental Law and Policy in Africa* (pp. 131–150). Springer. doi:10.1007/978-94-017-0135-8_7

Alani, R., Drouillard, K., Kehinde, O., & Alo, B. (2012). Bioaccumulation of polycyclic aromatic hydrocarbons in fish and invertebrates of Lagos Lagoon, Nigeria. *Journal of Emerging Trends in Engineering and Applied Sciences*, *3*(2), 287–296.

Anderson, K. A., & Hillwalker, W. E. (2008). Bioavailability. Encyclopedia of Ecology, 348–357.

Arukwe, A., Eggen, T., & Möder, M. (2012). Solid waste deposits as a significant source of contaminants of emerging concern to the aquatic and terrestrial environments—A developing country case study from Owerri, Nigeria. *The Science of the Total Environment*, *438*, 94–102. doi:10.1016/j.scitotenv.2012.08.039 PMID:22975307

Bolong, N., Ismail, A. F., Salim, M. R., & Matsuura, T. (2009). A review of the effects of emerging contaminants in wastewater and options for their removal. *Desalination*, *239*(1–3), 229–246. doi:10.1016/j.desal.2008.03.020

Caliman, F. A., & Gavrilescu, M. (2009). Pharmaceuticals, personal care products and endocrine disrupting agents in the environment - A review. *Clean - Soil, Air. Water (Basel).*

Carlsson, G., Örn, S., & Larsson, D. G. J. (2009). Effluent from bulk drug production is toxic to aquatic vertebrates. *Environmental Toxicology and Chemistry*, *28*(12), 2656. doi:10.1897/08-524.1 PMID:19186889

Daughton, C. G., & Ternes, T. A. (1999). Pharmaceuticals and personal care products in the environment: Agents of subtle change? *Environmental Health Perspectives*, *107*(suppl 6), 907–938. doi:10.1289/ehp.99107s6907 PMID:10592150

Di Nisio, A., Sabovic, I., Valente, U., Tescari, S., Rocca, M. S., Guidolin, D., ... Plebani, M. (2018). Endocrine disruption of androgenic activity by perfluoroalkyl substances: Clinical and experimental evidence. *The Journal of Clinical Endocrinology and Metabolism*, *104*(4), 1259–1271. doi:10.1210/jc.2018-01855 PMID:30403786

ELRI. (2011). A synopsis of laws and regulations on *environment*. ELRI.

FAO. (n.d.). *AQUASTAT*. Retrieved from http://www.fao.org/nr/water/aquastat/data/query/results.html

FGN. (1999). *Constitution of the Federal Republic of Nigeria*. FGN.

FGN. (2004). *Water Resources Act*. FGN.

FMWR. (2004). *National Water and Sanitation Policy*. FMWR.

FMWR. (2016). *Partnership for Expanded Water Supply, Sanitation & Hygiene*. PEWASH.

Gibbs, P. E., Bryan, G. W., & Pascoe, P. L. (1991). TBT-induced imposex in the dogwhelk, Nucella lapillus: Geographical uniformity of the response and effects. *Marine Environmental Research*, *32*(1–4), 79–87. doi:10.1016/0141-1136(91)90035-7

Guillette, L. J. Jr, Gross, T. S., Masson, G. R., Matter, J. M., Percival, H. F., & Woodward, A. R. (1994). Developmental abnormalities of the gonad and abnormal sex hormone concentrations in juvenile alligators from contaminated and control lakes in Florida. *Environmental Health Perspectives*, *102*(8), 680–688. doi:10.1289/ehp.94102680 PMID:7895709

Ikpesu, T. O. (2015). Assessment of occurrence and concentrations of paraquat dichloride in water, sediments and fish from Warri River Basin, Niger Delta, Nigeria. *Environmental Science and Pollution Research International*, *22*(11), 8517–8525. doi:10.100711356-014-3993-2 PMID:25548014

Inam, E., Offiong, N.-A., Kang, S., Yang, P., & Essien, J. (2015). Assessment of the Occurrence and Risks of Emerging Organic Pollutants (EOPs) in Ikpa River Basin Freshwater Ecosystem, Niger Delta-Nigeria. *Bulletin of Environmental Contamination and Toxicology*, *95*(5), 624–631. doi:10.100700128-015-1639-9 PMID:26341253

Ite, A. E., Harry, T. A., Obadimu, C. O., Asuaiko, E. R., & Inim, I. J. (2018). Petroleum hydrocarbons contamination of surface water and groundwater in the Niger Delta region of Nigeria. *Journal of Environment Pollution and Human Health*, *6*(2), 51–61. doi:10.12691/jephh-6-2-2

Jacobson, J. L., & Jacobson, S. W. (1997). Evidence for PCBs as neurodevelopmental toxicants in humans. *Neurotoxicology*, *18*(2), 415–424. PMID:9291491

JICA. (2014). *Federal Republic of Nigeria the Project for Review and Update of Nigeria - National Water Resources Master Plan 2013 - MASTER PLAN PROGRESS REPORT - 2 SUMMARY*. JICA.

Kim, K.-H., Jahan, S. A., Kabir, E., & Brown, R. J. C. (2013). A review of airborne polycyclic aromatic hydrocarbons (PAHs) and their human health effects. *Environment International*, *60*, 71–80. doi:10.1016/j.envint.2013.07.019 PMID:24013021

Kimura, N., Kimura, T., Suzuki, M., & Totsukawa, K. (2006). Effect of gestational exposure to nonylphenol on the development and fertility of mouse offspring. *The Journal of Reproduction and Development, 52*(6), 789–795. doi:10.1262/jrd.18007 PMID:17038804

Lawson, C., Gieske, M., Murdoch, B., Ye, P., Li, Y., Hassold, T., & Hunt, P. A. (2011). Gene expression in the fetal mouse ovary is altered by exposure to low doses of bisphenol A. *Biology of Reproduction, 84*(1), 79–86. doi:10.1095/biolreprod.110.084814 PMID:20739668

Logan, B. E. (2001). Peer reviewed: Assessing the outlook for perchlorate remediation. *Environmental Science & Technology, 35*(23), 482A–487A. doi:10.1021/es012564o PMID:11770780

Lubick, N. (2011). Tools for tracking antibiotic resistance. *Environmental Health Perspectives, 119*(5), A214–A217. doi:10.1289/ehp.119-a214 PMID:21531665

McClusky, L. M., De Jager, C., & Bornman, M. S. (2006). Stage-related increase in the proportion of apoptotic germ cells and altered frequencies of stages in the spermatogenic cycle following gestational, lactational, and direct exposure of male rats to p-nonylphenol. *Toxicological Sciences, 95*(1), 249–256. doi:10.1093/toxsci/kfl141 PMID:17065434

Mills, M. R., Arias-Salazar, K., Baynes, A., Shen, L. Q., Churchley, J., Beresford, N., ... Jobling, S. (2015). Removal of ecotoxicity of 17α-ethinylestradiol using TAML/peroxide water treatment. *Scientific Reports, 5*(1), 10511. doi:10.1038rep10511 PMID:26068117

Nash, J. P., Kime, D. E., Van der Ven, L. T. M., Wester, P. W., Brion, F., Maack, G., ... Tyler, C. R. (2004). Long-Term Exposure to Environmental Concentrations of the Pharmaceutical Ethynylestradiol Causes Reproductive Failure in Fish. *Environmental Health Perspectives, 112*(17), 1725–1733. doi:10.1289/ehp.7209 PMID:15579420

NBS. (2017). *Nigeria SDGs Indicators Baseline Report 2016 | UNDP in Nigeria.* National Bureau of Statistics.

Nduka, J. K., & Orisakwe, O. E. (2010). Water Quality Issues in the Niger Delta of Nigeria: Polyaromatic and Straight Chain Hydrocarbons in Some Selected Surface Waters. *Water Quality, Exposure, and Health, 2*(2), 65–74. doi:10.100712403-010-0024-5

Nnamonu, L. A., & Onekutu, A. (2015). Green pesticides in Nigeria: An overview. *Journal of Biology. Agriculture and Healthcare, 5*(9), 48–62.

Nyenje, P. M., Foppen, J. W., Uhlenbrook, S., Kulabako, R., & Muwanga, A. (2010). Eutrophication and nutrient release in urban areas of sub-Saharan Africa - A review. *The Science of the Total Environment, 408*(3), 447–455. doi:10.1016/j.scitotenv.2009.10.020 PMID:19889445

Offiong, N.-A. O., Inam, E. J., & Edet, J. B. (2019). Preliminary Review of Sources, Fate, Analytical Challenges and Regulatory Status of Emerging Organic Contaminants in Aquatic Environments in Selected African Countries. *Chemistry Africa*, 1–13.

Ogbeide, O., Uhunamure, G., Okundaye, F., & Ejeomo, C. (2019). First report on probabilistic risk assessment of pesticide residues in a riverine ecosystem in South-South Nigeria. *Chemosphere, 231*, 546–561. doi:10.1016/j.chemosphere.2019.05.105 PMID:31151015

Okereke, J. N., Ogidi, O. I., & Obasi, K. O. (2016). International Journal of Advanced Research in Biological Sciences Environmental and Health Impact of Industrial Wastewater Effluents in Nigeria-A Review. *Int. J. Adv. Res. Biol. Sci*, *3*(6), 55–67.

Oketola, A. A., & Fagbemigun, T. K. (2013). Determination of nonylphenol, octylphenol and bisphenol-A in water and sediments of two major rivers in Lagos, Nigeria. *Journal of Environmental Protection*, *4*(07), 38–45. doi:10.4236/jep.2013.47A005

Ololade, I. A. (2014). Spatial distribution of perfluorooctane sulfonate (PFOS) in major rivers in south-west Nigeria. *Toxicological and Environmental Chemistry*, *96*(9), 1356–1365. doi:10.1080/02772248.2015.1028409

Ololade, I. A., Oladoja, N. A., Ololade, O. O., Oloye, F. F., Adeola, A. O., Alabi, A. B., ... Owolabi, M. B. (2018). Geographical distribution of perfluorooctanesulfonate and perfluorooctanoate in selected rivers from Nigeria. *Journal of Environmental Chemical Engineering*, *6*(4), 4061–4069. doi:10.1016/j.jece.2018.06.020

Park, B. J., & Kidd, K. (2005). Effects of the synthetic estrogen ethinylestradiol on early life stages of mink frogs and green frogs in the wild and in situ. *Environmental Toxicology and Chemistry: An International Journal*, *24*(8), 2027–2036. doi:10.1897/04-227R.1 PMID:16152976

Raghav, M., Eden, S., Mitchell, K., & Witte, B. (2013). *Contaminants of Emerging Concern in Wastewater.* Water Resources Research Center.

Shen, B., Wu, J., & Zhao, Z. (2017). Organochlorine pesticides and polycyclic aromatic hydrocarbons in water and sediment of the Bosten Lake, Northwest China. *Journal of Arid Land*, *9*(2), 287–298. doi:10.100740333-017-0008-4

Siddique, S., Kubwabo, C., & Harris, S. A. (2016). A review of the role of emerging environmental contaminants in the development of breast cancer in women. *Emerging Contaminants*, *2*(4), 204–219. doi:10.1016/j.emcon.2016.12.003

Sorensen, J. P. R., Lapworth, D. J., Nkhuwa, D. C. W., Stuart, M. E., Gooddy, D. C., Bell, R. A., ... Pedley, S. (2015). Emerging contaminants in urban groundwater sources in Africa. *Water Research*, *72*, 51–63. doi:10.1016/j.watres.2014.08.002 PMID:25172215

Stefanakis, A., & Becker, J. (2016). A Review of Emerging Contaminants in Water. In *Impact if Water Pollution on Human Health and Environmental Sustainability* (pp. 55–80). IGI Global. doi:10.4018/978-1-4666-9559-7.ch003

Stefanakis, A., & Human, J. B. (2015). *A Review of Emerging Contaminants in Water: Classification, Sources, and Potential Risks. In Impact of Water Pollution on Human Health and Environmental Sustainability*. IGI Global.

Taiwo, A. M., Olujimi, O. O., Bamgbose, O., & Arowolo, T. A. (2012). Surface Water Quality Monitoring in Nigeria: Situational Analysis and Future Management Strategy. Water Quality Monitoring and Assessment.

The World Bank. (2016). *Health Nutrition and Population Statistics | DataBank*. Retrieved March 22, 2018, from http://databank.worldbank.org/data/reports.aspx?source=health-nutrition-and-population-statistics#

The World Bank. (2017). A Wake Up Call: Nigeria Water Supply, Sanitation, and Hygiene Poverty Diagnostic. *WASH Poverty Diagnostic* (Vol. 431). Author.

Tongo, I., Ezemonye, L., & Akpeh, K. (2017). Distribution, characterization, and human health risk assessment of polycyclic aromatic hydrocarbons (PAHs) in Ovia River, Southern Nigeria. *Environmental Monitoring and Assessment*, *189*(6), 247. doi:10.100710661-017-5931-5 PMID:28466449

UNDP. (2006). *Niger Delta Human Development Report*. Abuja. Retrieved from http://hdr.undp.org/en/content/human-development-report

UNDP. (2016). *Human Development Report 2016 - Human Development for Everyone*. New York. Retrieved from http://hdr.undp.org/sites/default/files/2016_human_development_report.pdf

UNESCO. (2003). *Water for People, Water for Life. The United Nations World Water Development Report*. UNESCO. doi:10.1017/CBO9781107415324.004

Urbansky, E. T. (2000). Perchlorates in the Environment. In H. Rosenkranz (Ed.), Environmental Science Research (Vol. 440, p. 299). New York: Kluwer Academic/Plenum Publishers. doi:10.1007/978-1-4615-4303-9

WSP. (2012). *Economic Impacts of Poor Sanitation in Africa*. Retrieved from https://www.wsp.org/sites/wsp.org/files/publications/WSP-ESI-Nigeria-brochure.pdf

Zejiao, L., & Ebiare, E. (2010). Water quality monitoring in Nigeria; Case Study of Nigeria's industrial cities. *Journal of American Science, 6*(4). Retrieved from http://www.americanscience.org/journalseditor@americanscience.org

ADDITIONAL READING

Bai, X., Lutz, A., Carroll, R., Keteles, K., Dahlin, K., Murphy, M., & Nguyen, D. (2018). Occurrence, distribution, and seasonality of emerging contaminants in urban watersheds. *Chemosphere*, *200*, 133–142. doi:10.1016/j.chemosphere.2018.02.106 PMID:29477762

Bandala, E. R., & Goonetilleke, A. (2016). *Emerging Contaminants in urban stormwater: challenges and perspectives for sustainable water use*. Retrieved from https://sustainabledevelopment.un.org/content/documents/980043_Bandala et al._Emerging Contaminants in urban stormwater_challenges and perspectives for sustainable water use.pdf

Galindo-Miranda, J. M., Guízar-González, C., Becerril-Bravo, E. J., Moeller-Chávez, G., León-Becerril, E., & Vallejo-Rodríguez, R. (2019). Occurrence of emerging contaminants in environmental surface waters and their analytical methodology – a review. *Water Supply*, *19*(7), 1871–1884. doi:10.2166/ws.2019.087

Lapworth, D. J., Baran, N., Stuart, M. E., & Ward, R. S. (2012). Emerging organic contaminants in groundwater: A review of sources, fate and occurrence. *Environmental Pollution, 163*, 287–303. doi:10.1016/j.envpol.2011.12.034 PMID:22306910

Lei, M., Zhang, L., Lei, J., Zong, L., Li, J., Wu, Z., & Wang, Z. (2015). Overview of emerging contaminants and associated human health effects. *BioMed Research International*. PMID:26713315

Richardson, S. D., & Ternes, T. A. (2011). Water Analysis: Emerging Contaminants and Current Issues. *Analytical Chemistry, 83*(12), 4614–4648. doi:10.1021/ac200915r PMID:21668018

Sauvé, S., & Desrosiers, M. (2014). A review of what is an emerging contaminant. *Chemistry Central Journal, 8*(1), 15. doi:10.1186/1752-153X-8-15 PMID:24572188

KEY TERMS AND DEFINITIONS

Economic Water Scarcity: Water scarcity caused by poor water resources management and a lack of investment in water infrastructure, where the needs of the population are not met.

Emerging Contaminants: Any synthetic or naturally occurring chemical that is not commonly monitored in the environment but has the potential to enter the environment and cause known or suspected adverse ecological and or human health effects.

Endocrine Disrupting Compound: These are chemicals that can interfere with the endocrine or hormonal systems of humans and animals.

Organochlorine Pesticide: These are chlorinated hydrocarbons which are used for pesticide control.

Perfluorochemicals (PFCs): These are man-made contaminants, which are persistent in the environment. They are found in wastewater generated from the textile, paper and leather industry.

Pharmaceutical and Personal Care Products (PPCP): These are substances used by individuals for various personal health and cosmetic reasons and by the agricultural sector for boosting the growth of animals and for disease control.

Polycyclic Aromatic Hydrocarbons (PAHs): These are naturally occurring found in coal and tar and anthropogenic compounds generated as a result of the incomplete combustion of organic material such as coal, wood, garbage, and gasoline.

Sustainable Development Goals (SDGs): These are a collection of global goals that are a universal call to action to end poverty, protect the planet and ensure that all people enjoy peace and prosperity.

Chapter 3
Microplastics as Emerging Contaminants:
Occurrence, Toxicology, and Analysis

Bowa O. Kwach
Maseno University, Kenya

Victor Odhiambo Shikuku
Kaimosi Friends University College, Kenya

ABSTRACT

The ever-increasing production of plastics and concomitant poor plastic waste disposal systems explain the recent rising concerns over the occurrence of microplastics in freshwater resources. Microplastics are presently recognized as emerging contaminants owing to the increasing reports on their occurrence in the environment and the associated toxicological effects. This chapter discusses the recent trends in the monitoring of microplastics in freshwater resources, the toxicological effects of microplastics, and the sampling and analysis techniques available for detection and quantification. The challenges in analysis and comparison of various studies and future prospects have also been highlighted.

INTRODUCTION

In 2016, the world production of plastic, mostly used for packaging, exceeded the 320 million tons' mark. This rise in production was accompanied by an unprecedented exponential increase in the occurrence of plastic wastes in different environmental compartments with 5-13 million tons reported to leak into the oceans annually (World Economic Forum, 2016). Inappropriate disposal of plastic wastes accounts for the major fraction of the plastics in both soil and aquatic environments (Jambeck et al., 2015) followed by their breakdown through various mechanisms such as photo-oxidative processes and mechanical forces among others (da Costa et al., 2017). Figure 1 is an illustration of inappropriate disposal of plastic wastes. The aforementioned processes breakdown those larger plastic residues into microplastics (with particles less than 5mm) and nanoplastics (with particles between 1 and 100 nm).

DOI: 10.4018/978-1-7998-1871-7.ch003

The presence of microplastics in the environment is a subject of increasing concern and microplastics are considered as contaminants of emerging concern (CEC). To this effect, microplastic pollution has been recognized as one of the leading environmental issues facing the mankind today (Lambert and Wagner, 2018). Besides compromising the aesthetic value of water, microplastics have been reported to have potential toxicity on biota and humans (Wright et al., 2013; von Moos et al., 2012). Additionally, microplastics present a pathway for introduction of plastic additives and persistent organic pollutants (POPs) into the aquatic environment (Andrady, 2011; Brennecke et al., 2016). Some of the additives are listed in Table 1. The health problems associated with POPs in the food chain are well documented and the details will not be reiterated here (Beckingham and Ghosh, 2017). Therefore, sea-foods from contaminated waters are a potential source of entrance of microplastics and related toxic compounds into the food chain (GESAMP, 2015). As such, the development of analytical tools for detection and quantification of micro and nano-plastics is a crucial step towards efficient monitoring of the environmental fate of these plastics. This chapter summarizes the recent reports and reviews on the occurrence of microplatics in aquatic environments, published data relating to the toxicity of microplastics and the trends in analytical tools for detection and quantification of microplastics are also brought to perspective. The challenges and future opportunities are also highlighted.

Occurrence of Microplastics In the Aquatic Environment

Microplastics are water-insoluble solid polymer particles with dimensions of less than 5mm (Thompson et al., 2009). Those that are less than 1mm in size are informally known as nano-plastics. Though the larger plastics mainly find their way into the environment through inappropriate waste disposal, microplastics on the other hand are primarily generated from fragmentation of the larger plastics. The rate of fragmentation of the larger plastics in the natural environment is, however, unknown and presents an open field of scientific inquiry (Lambert and Wagner, 2018). The paucity of data on the dynamics and fate of microplastics presents a challenge in the assessment of the health risks posed to the aquatic life and man (Wright et al., 2013). Despite these limitations, there has been agitation and attempts to regulate the usage of microplastic for public safety.

Figure 1. Poor plastic waste disposal

Table 1. Additives used in plastics

Additive	Role
Plasticizers	Induce pliability
Flame retardants	Makes the material less flammable
Pigments	Gives color
Surfactants	For modification of surface properties
Antioxidants	Improves durability

Several studies have reported the quantities of microplastics in diverse water resources such as, oceans, seas, drinking water, surface water, groundwater and wastewater (Peng et al., 2017; Li et al., 2018). Mason et al. (2016) reported the occurrence of microplastics in 17 wastewater treatment facilities in U.S.A. On average, the amount of microplastics in wastewater was 0.05 microparticles per liter translating to a discharge of up to 4 million particles per day per facility. Fibers and fragments constituted the largest fraction of the types of particles in the effluents. The study revealed that municipal wastewater treatment facilities are leading point sources of loading microplastics into the waterways. Similar findings were reported by Dris et al. (2015) with regard to the presence of microplastics in untreated and treated effluents from wastewater treatment facilities in Greater Paris. While the treatment process reduces the loading of the microplastics in the influent wastewater by up to 95%, the levels in the treated effluents were significantly high and are a pollution source for the recipient surface waters. Table 2 is a summary of some results on microplastics from wastewater treatment facilities.

These studies unequivocally attest that microplastics are generally incalcitrant to the conventional wastewater treatment processes and are thus released into aquatic environments. Noteworthy, there is also limited data on occurrence of microplastics in African water resources despite the known ban and efforts to control the use of plastic-based materials in some African countries. This could be attributed to the lack of or limited access to technological infrastructure for analysis of microplastics.

Table 2. Microplastics measurements in wastewater treatment plants

Location	Mean abundance	Maximum abundance	Reference
East Bay Municipal WWTP, USA	–	169 p/m^3	(Dyachenko et al., 2017)
WWTP, Finland	–	Influent: 9×10^5 p/m^3 Effluent: 3.5×10^3 p/m^3	(Talvitie et al., 2017)
WWTP, Glasgow	Influent: 1.57×10^4 p/m^3 Effluent: 250 p/m^3	–	(Murphy et al., 2016)
12 WWTPs in Saxony, Germany	–	9×10^3 p/m^3	(Mintenig et al., 2017)
WWTP, St. Petersburg	Influent: 1.6×10^5 p/m^3 Effluent:7×10^3 p/m^3	–	(Talvitie, 2014)
Långeviksverket, Sweden	Influent: 1.5×10^4 p/m^3; Effluent: 8.25 p/m^3	–	(Magnusson and Norén, 2014)
7 WWTPs, Netherland	Influent: 7.3×10^4 p/m^3 Effluent: 5.2×10^4 p/m^3	Influent: 5.66×10^5p/m^3 Effluent: 9.1×10^4 p/m^3	(Leslie et al., 2017)

Recently, Orb (2017) reported the presence of microplastics in 83% of drinking water samples obtained from five continents. These findings raise serious questions concerning the level of safety of drinking water and also underscore the role of non-governmental organizations (NGOs) and the media in creating public awareness and disseminating scientific output on this environmental issue. Literature is replete with publications on microplastics in the marine environments with reports dating back as early as 1970s but with a handful of data on freshwater systems. As Lambert and Wagner (2018) points out, less than 4% of publications works on 'microplastics' dealt with freshwater. However, studies on the occurrence of microplastics in 'freshwaters' is gaining momentum. Zhang and co-workers (2015) reported 192–13,617 microplastic particles km^{-2} in surface water at the Three Gorges Dam, China. These considerably high concentrations were attributed to poor waste management and the absence of wastewater treatment facilities in the peripheral smaller towns. Elsewhere, Mani et al. (2015) reported a mean of 892,777 microplastic particles km^{-2} in the Rhine river, Germany. Other reports include those by Dris et al. (2017) for occurrence of microplastics in surface waters of France and Khan et al. (2017) covering the inland African water resources. With increasing concerns and access to sophisticated equipment, publications on the occurrence and fate of microplastics in freshwater is expected to increase in the coming years. The transport, fate, degradation and persistence of microplastics in freshwater systems have been reviewed elsewhere (Lambert and Wagner, 2018).

Toxicity of Microplastics And Associated Chemicals

Microplastic, when ingested, may have various biological effects on aquatic organisms. Direct exposure experiments have revealed significant biological effects such as; weight loss, reduced feeding activity, increased phagocytic activity, and transference to the lysosomal (storage) system in some invertebrates (Lusher, 2015). In addition, Rochman et al. (2016) suggested that microplastics ranging from 2–5 mm could take longer to pass from the stomachs of organisms and could be retained in the digestive system, potentially increasing the exposure time to adsorbed toxins.

Indirectly, microplastics may be carriers of chemicals that are sorbed onto their surface from their environment (e.g., PCBs or Dichlorodiphenyldichloroethylene, DDEs), or chemicals that are added to the plastic (e.g., plasticizers) in the plastic production process (Mato *et al.*, 2001; Talsness *et al.*, 2009). Some of these classes of compounds are shown in Figure 2. There is potential for both of these types of chemicals to be transferred to organisms. Many plastics are known to adsorb both inorganic and organic chemicals, and indeed are comprised of some compounds linked to endocrine disruption like PCBs, pesticides, metals (Teuten *et al.*, 2007; Velzeboer *et al.*, 2014). These compounds have been found to leach off of the MPs when ingested by aquatic biota, however, those tightly bound do not leach off (Gouin *et al.*, 2011; Koelmans *et al.*, 2016; Zarfl and Matthies, 2010).

Marine studies investigating transport of hydrophobic contaminants (e.g., phenanthrene) by plastic have found that contaminants sorb to plastics more easily than they do to some natural sediments and that microplastics can consequently transfer contaminants to organisms (Teuten *et al.*, 2007). For example, plastic was shown to facilitate the transport of contaminants to the sediment-dwelling lugworm, *A. marina* and to the amphidromous Medaka fish, *O. latipes* (Teuten *et al.*, 2007)

Marine field studies confirm the presence of sorbed environmental contaminants on microplastics (Mato *et al.*, 2001), and laboratory evidence suggests that sorbed contaminants can be transferred to marine fish and invertebrates (Besseling *et al.*, 2012; Browne *et al.*, 2013). Since chemicals are present in water entering treatment plants, in treated effluent, and in drinking water (Morasch *et al.*, 2010; Brausch and

Figure 2. Some organic contaminants that adsorb on microplastic

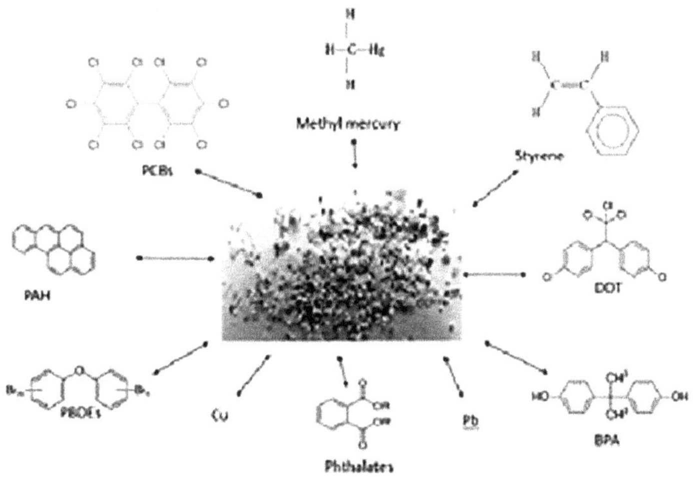

Rand, 2011), there could be concern that freshwater systems close to industrial and population centers may have both a greater microplastic presence, and greater concentrations of chemicals and contaminants, and that biota in these regions may therefore experience greater exposure. Such concerns are valid, but more research is needed, as interactions (chemical sorption/ desorption to plastic and transfer to biota) are complex and not yet fully predictable. Chemical transfer depends on the plastic, the contaminant, the surrounding environment, and the organism that ingests the plastic. For example, the sorption capacity varies between plastics, e.g., polyethylene sorbs greater concentrations of contaminants than other polymers, and the release of contaminants from plastics is facilitated by increased temperature and low pH equivalent, resembling conditions in a warm-blooded animal (Bakir *et al.*, 2014).

Other than affecting the distribution of chemicals in the environment, microplastics may directly or indirectly affect abiotic qualities of the environment. Microplastic accumulation in pelagic and benthic habitats might alter light penetration into the water column or change sediment characteristics (Arthur and Baker, 2011) such as grain size, pore size, and sediment binding capacity to chemicals (Simpson *et al.*, 2005).

ANALYSIS OF MICROPLASTICS

Sampling Methods

One of the challenges facing the assessment of microplastics in the environment is the lack of standardized sample collection, pre-treatment and analysis techniques. This may result in underestimation of the actual amounts of microplastics in a given sample and limits objective comparison of results obtained from other studies carried out using different methods.

The sampling methods reported in literature include volume-reduced and bulk sampling (Hidalgo-Ruz et al., 2012). As the name indicates, the volume-reduced technique entails the reduction of the volume of sample collected during sampling. Here, neuston plankton net and manta trawl are the most often used techniques for the collection of surface water. On the other hand, in bulk sampling, there is no volume reduction. The choice of the surface water is because the microplastics are less dense than water and therefore have the tendency to float and can thus be collected using the manta trawl (Rocha-Santos and Duarte, 2015). The volume of water passing through the filter must be recorded and can be computed using a flow meter (Free et al., 2014). The mesh size, which controls the amount and content of microplastics sampled, must be controlled for accuracy and comparison purposes (Eerkes-Medrano et al., 2015). That is, sampling of the same surface water with mesh of differing sizes would lead to variance in the content of microplastics reported. For instance, an 80-µm mesh resulted in microplastics content 100,000 fold higher than when a 450- µm mesh was used (Lozano and Mouat, 2009). The variance cannot be gainsaid. Mesh sizes of 333 and 500 µm have been reported in literature, with the former as the widely used mesh size (Arthur et al., 2009). The impact of mesh size on the validity of the results cannot be overstated.

Besides the volume-reduced and bulk sampling methods, another sampling technique is the hand-net sampling method (Moore et al., 2011). In this method, a hand bucket is used to collect a given volume of surface water at a certain depth followed by the hand-net filtration. A comparative study of the different sampling methods on the occurrence of microplastics in seawater was in the sequence hand-net, bulk water and manta trawl, in order of increasing suitability (Song et al., 2014). Additionally, sampling method should factor in the type of environment. That is, sampling methods for microplastics in a marine environment are not necessarily the same for freshwater environments. Table 3 is a summary of sampling methods and processes for microplastics in different environmental matrices (Silva et al., 2018).

Figure 3 is an illustration an Avani microplastics sampling trawl net while Figure 4 demonstrates its application during sampling.

It is essential that all sampling equipment be thoroughly cleaned before and between sampling to avoid or minimize cross-contamination of the samples from external sources. The sources of contamination may include clothing gears and atmospheric fallouts. As in other analytical procedures, procedural blanks can be used to assess possible effects of cross-contamination during sampling.

Table 3. Sampling methods and processes for microplastics

Matrix	Process
Water	
Surface water	Grab sampling/use of trawl
Mid-water level	Use of bongo nets
Sediments	
Bottom samples	Use of box corer
Surface samples	Use of non-plastic spades
Seabed samples	Use of bottom trawl
Biological tissues	Dissection, egestion, and regurgitation

Figure 3. A microplastics sampling trawl net

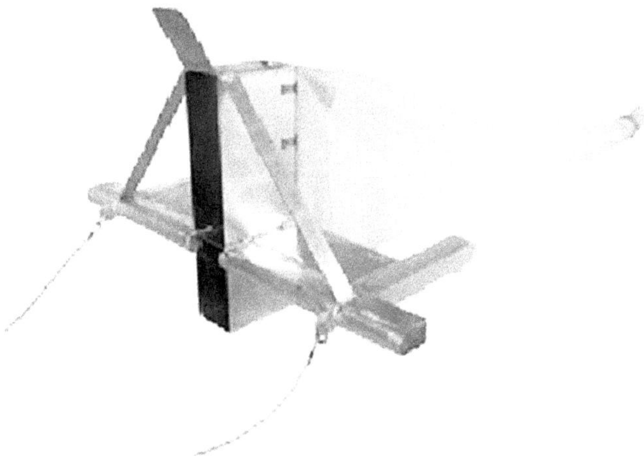

Sample Pre-Treatment Methods

Like other typical analytical procedures, analysis of microplastics involves sample pre-treatment (extraction and purification) and quantification. The pre-treatment procedure depends on the sample matric, whether water, sediment or biological tissues. For water samples, separation of the microplastics from the liquid phase is usually achieved through density flotation by adding NaCl or NaI salt, and floatation, filtration by means of size fractionation or sieving through size exclusion (Rocha-Santos and Duarte, 2015). In an attempt to standardize sampling and sample analysis of microplastics in water, a technical report was published by the US National Oceanic and Atmospheric Administration (NOAA) (Masura et al., 2015). The described protocol for analysis of microplastic, ranging between 0.3-5 mm, in marine waters is described in Figure 5.

Figure 4. Application of trawl net for sampling

Identification and Quantification

Presently, there are no specialized techniques for detection and quantification of microplastics except the already established techniques used for other analytes. The methods include chemical composition analyses, optical and spectroscopic techniques.

Optical Techniques

The preliminary analysis of the sample is usually visual observation, performed either by using the naked eyes examination or by optical microscopy (Song et al., 2015). Evidently, the use of the naked eye observation suffers many inherent limitations including frequent false positives and negatives, besides limited reproducibility. On the other hand, optical microscopy not only eliminates the aforementioned limitations but brings additional information such as textural properties, color, shape and structural features necessary for identification purposes and for distinguishing the microplastics from other sample constituents (Shim et al., 2017). In the analysis of microplastic pollution in three gorges reservoir in China, Zhang et al. (2017) reported the use of optical microscopy to distinguish between sheet, fragments and foam. Similar approach was reported by Sutton et al. (2016) for analysis of microplastics in wastewater. Though visualization is a relatively cheap and simple technique, studies show that it should be coupled with other techniques such as FT-IR and Raman spectroscopies for effective enumeration and classification of microplastics.

Scanning Electron Spectroscopy

Scanning electron microscopy (SEM) is an important tool in the identification of microplastics due to its high resolution images. This is particularly important in distinguishing microplastics from other components of the sample matrix (Cooper and Corcoran, 2010). Additionally, when coupled with EDS, the determination of the elemental composition is enabled, further allowing the discrimination of the

Figure 5. Microplastics analysis procedure proposed by NOAA

carbon-containing plastics from their inorganic counterparts (Shim et al., 2017; Dehghani et al., 2017). Despite the obvious superiority of SEM over naked eye examination, the ability to identify microplastic markers by chemical composition or by their shapes, it also suffers some inherent limitations. Firstly, SEM-EDS is costly and requires several sample preparation steps that reduces the throughput, that is, the number of samples that can be analyzed at a given time. Secondly, color cannot be used as the sole criterion for identification of the microplastics hence the technique is unsuitable for other types of microplastics (Shim et al., 2017). The later limitation may induce discrepancies in quantification of microplastics in a water sample. Lastly, EDS may not distinguish between additives of plastic polymers and adsorbed debris on microplastic surface (Dehghani et al., 2017). As such, SEM-EDS analysis requires supplementation with other techniques such as FTIR spectroscopy.

IR and Raman Spectroscopy

These two spectroscopic techniques are the most commonly used for identification of microplastics in water. Unlike SEM-EDS, FTIR and Raman spectroscopies require little sample preparation, a small amount of the sample is required for analysis and can effectively distinguish between plastic materials and other sample components. FTIR is useful for microplastics of size above 10 μm while the Raman spectroscopy can determine microplastics as low as less than 1 μm (Ribeiro-Claro et al., 2017). FTIR is useful for qualitative analysis where the functional groups of the microplastic polymers are compared with the spectra of standard known plastics. On the other hand, though Raman spectroscopy can be used to identify much small-sized plastics, unlike FTIR, one challenge with the technique is sample degradation by UV exposure (Lenz et al., 2015). To overcome this problem, and achieve more reliable qualitative analysis, the analysis of spectra of degraded polymers at different stages should be included in reference databases (Ribeiro-Claro et al., 2017). Nevertheless, relative to FTIR, Raman spectroscopy better response of non-polar, symmetric bonds. In contrast, FTIR provides better identification of polar bonds (Lenz et al., 2015).

Unlike chemical quantification, quantification of plastics is relatively difficult and highly susceptible to overestimation or underestimation due to their diversity in terms of type, size, color, and morphology combined with the lack of homogeneity within environmental samples. These dynamics affect each step of the analytical process, from sampling to quantification. Furthermore, the lack of standard methods for sampling implies there is wide room for discrepancies and difficulty in comparing different studies with different sampling and quantification approaches.

CHALLENGES AND FUTURE

Like most chemicals in the environment, microplastics also undergo chemical and biological driven transformations in water systems which may lead to sizeable amounts remaining undetected and therefore unreported (Hidalgo-Ruz et al., 2012). Such underestimations have untold consequences on the aquatic life forms and humans as well. Furthermore, such alterations of chemical forms the microplastics also complicate the dependence on spectroscopic techniques for accurate identification and quantification. Additionally, the absence of standardized methods of sampling and analysis further compound the effort for systematic data on occurrence, fate and transport of microplastics in water systems and other environmental compartments. The differences in the Chemistry of environmental samples (soil, water, biological samples) and the sizes of the microplastics also necessitate the use of multiple analytical techniques. Therefore, the need for robust and standardized analytical methods for microplastics detection and quantification cannot be overemphasized.

CONCLUSION

This chapter demonstrates that microplastics are not only emerging contaminants of water resources but they also serve as means of entry for other emerging chemicals of concern into the water systems. Though there are reports on their occurrence in freshwater systems, there is paucity of data especially for water resources in emerging economies. The harmful effects of microplastics are also not clearly understood and opens fields for further research. Nevertheless, published reports on their toxicity have been highlighted. Furthermore, it is also demonstrated that there is need for standardization of sampling, detection and quantification techniques for microplastics monitoring. Relevant environmental agencies need to develop and enforce policies related to microplastics production, use, disposal and monitoring.

REFERENCES

Andrady, A. L. (2011). Microplastics in the marine environment. *Marine Pollution Bulletin*, *62*(8), 1596–1605. doi:10.1016/j.marpolbul.2011.05.030 PMID:21742351

Arthur, C., & Baker, J. (Eds.). (2011). *Proceedings of the Second Research Workshop on Microplastic Debris*. NOAA Technical Memorandum NOS-OR&R-39.

Arthur, C., Baker, J., & Bamford, H. (2009). *Proceedings of the International Research Workshop on the Occurrence, Effects, and Fate of Microplastic Marine Debris*. Academic Press.

Bakir, A., Rowland, S. J., & Thompson, R. C. (2014). Enhanced desorption of persistent organic pollutants from microplastics under simulated physiological conditions. *Environ. Pollut.*, *185*, 16–23. doi:10.1016/j.envpol.2013.10.007 PMID:24212067

Barnes, D. K. A., Galgani, F., Thompson, R. C., & Barlaz, M. (2009). Accumulation and fragmentation of plastic debris in global environments. *Philosophical Transactions of the Royal Society of London. Series B, Biological Sciences*, *364*(1526), 1985–1998. doi:10.1098/rstb.2008.0205 PMID:19528051

Beckingham, B., & Ghosh, U. (2017). Differential bioavailability of polychlorinated biphenyls associated with environmental particles: Microplastic in comparison to wood, coal and biochar. *Environ. Pollut.*, *220*, 150–158. doi:10.1016/j.envpol.2016.09.033 PMID:27650963

Besseling, E., Wegner, A., Foekema, E. M., van den Heuvel-Greve, M. J., & Koelmans, A. A. (2012). Effects of microplastic on fitness and PCB bioaccumulation by the lugworm Arenicola marina (L.). *Environmental Science & Technology*, *47*(1), 593–600. doi:10.1021/es302763x PMID:23181424

Brausch, J. M., & Rand, G. M. (2011). A review of personal care products in the aquatic environment: Environmental concentrations and toxicity. *Chemosphere*, *82*(11), 1518–1532. doi:10.1016/j.chemosphere.2010.11.018 PMID:21185057

Brennecke, D., Duarte, B., Paiva, F., Caçador, I., & Canning-Clode, J. (2016). Microplastics as vector for heavy metal contamination from the marine environment. *Estuarine, Coastal and Shelf Science*, *178*, 189–195. doi:10.1016/j.ecss.2015.12.003

Browne, M. A., Niven, S. J., Galloway, T. S., Rowland, S. J., & Thompson, R. C. (2013). Microplastic moves pollutants and additives to worms, reducing functions linked to health and biodiversity. *Current Bio.*, *23*(23), 2388–2392. doi:10.1016/j.cub.2013.10.012 PMID:24309271

Carpenter, E. J., Anderson, S. J., Harvey, G. R., Miklas, H. P., & Peck, B. B. (1972). Polystyrene spherules in coastal waters. *Sci.*, *178*(4062), 749–750. doi:10.1126cience.178.4062.749 PMID:4628343

Carpenter, E. J., & Smith, K. L. Jr. (1972). Plastics on the Sargasso Sea surface. *Sci.*, *175*(4027), 1240–1241. doi:10.1126cience.175.4027.1240 PMID:5061243

Cooper, D. A., & Corcoran, P. L. (2010). Effects of mechanical and chemical processes on the degradation of plastic beach debris on the island of Kauai, Hawaii. *Marine Pollution Bulletin*, *60*(5), 650–654. doi:10.1016/j.marpolbul.2009.12.026 PMID:20106491

da Costa, J. P., Duarte, A. C., & Rocha-Santos, T. A. P. (2017). Chapter 1-microplastics - occurrence, fate and behavior in the environment. *Comprehensive Analytical Chemistry*, *75*, 1–24. doi:10.1016/bs.coac.2016.10.004

Dehghani, S., Moore, F., & Akhbarizadeh, R. (2017). Microplastic pollution in deposited urban dust, Tehran metropolis. Iran. *Environmental Science and Pollution Research International*, *24*(25), 20360–20371. doi:10.100711356-017-9674-1 PMID:28707239

Dyachenko, A., Mitchell, J., & Arsem, N. (2017). Extraction and identification of microplastic particles from secondary wastewater treatment plant (WWTP) effluent. *Analytical Methods*, *9*(9), 1412–1418. doi:10.1039/C6AY02397E

Eerkes-Medrano, D., Thompson, R. C., & Aldridge, D. C. (2015). Microplastics in freshwater systems: A review of the emerging threats, identification of knowledge gaps and prioritisation of research needs. *Water Research*, *75*, 63–82. doi:10.1016/j.watres.2015.02.012 PMID:25746963

Free, C. M., Jensen, O. P., Mason, S. A., Eriksen, M., Williamson, N. J., & Boldgiv, B. (2014). High-levels of microplastic pollution in a large, remote, mountain lake. *Marine Pollution Bulletin*, *85*(1), 156–163. doi:10.1016/j.marpolbul.2014.06.001 PMID:24973278

GESAMP. (2015). *Sources, Fate and Effects of Microplastics in the Marine Environment: a Global Assessment (IMO/FAO/UNESCO-IOC/UNIDO/WMO/IAEA/UN/UNEP/UNDP Joint Group of Experts on the Scientific Aspects of Marine Environmental Protection).* Rep. Stud. GESAMP No. 90 Available from: http://ec.europa.eu/environment/marine/goodenvironmental-status/descriptor10/pdf/GESAMP_microplastics%20full%20study.pdf

Hidalgo-Ruz, V., Gutow, L., Thompson, R. C., & Thiel, M. (2012). Microplastics in the marine environment: A review of the methods used for identification and quantification. *Environmental Science & Technology*, *46*(6), 3060–3075. doi:10.1021/es2031505 PMID:22321064

Jambeck, J. R., Geyer, R., Wilcox, C., Siegler, T. R., Perryman, M., Andrady, A., ... Law, K. L. (2015). Plastic waste inputs from land into the ocean. *Sci. Magna.*, *347*, 768–771. PMID:25678662

Koelmans, A. A., Besseling, A., & Foekema, E. M. (2014). Leaching of plastic additives to marine organisms. *Environ. Pollut.*, *187*, 49–54. doi:10.1016/j.envpol.2013.12.013 PMID:24440692

Lenz, R., Enders, K., Stedmon, C. A., Mackenzie, D. M. A., & Nielsen, T. G. (2015). A critical assessment of visual identification of marine microplastic using Raman spectroscopy for analysis improvement. *Marine Pollution Bulletin*, *100*(1), 82–91. doi:10.1016/j.marpolbul.2015.09.026 PMID:26455785

Lozano, R., & Mouat, J. (2009). *Marine litter in the North-East Atlantic region: assessment and priorities for response*. KIMO Int.

Lusher, A. L., Burke, A., O'Connor, I., & Officer, R. (2014). Microplastic pollution in the Northeast Atlantic Ocean: Validated and opportunistic sampling. *Marine Pollution Bulletin*, *88*(1-2), 325–333. doi:10.1016/j.marpolbul.2014.08.023 PMID:25224764

Masura, J., Baker, J., Foster, G., & Arthur, C. (2015). *Laboratory Methods for the Analysis of Microplastics in the Marine Environment: Recommendations for Quantifying Synthetic Particles in Waters and Sediments*. NOAA Technical Memorandum NOS-OR&R-48. Available from: https://marinedebris.noaa.gov/sites/default/files/publications-files/noaa_microplastics_methods_manual.pdf

Mato, Y., Isobe, T., Takada, H., Kanehiro, H., Ohtake, C., & Kaminuma, T. (2001). Plastic resin pellets as a transport medium for toxic chemicals in the marine environment. *Environmental Science & Technology*, *35*(2), 318–324. doi:10.1021/es0010498 PMID:11347604

Moore, C., Lattin, G., & Zellers, A. (2011). Quantity and type of plastic debris flowing from two urban rivers to coastal waters and beaches of Southern California. Rev. Gestao Costeira Integr.-. *J. Integrated Coast. Zone Manag*, *11*(1).

Morasch, B., Bonvin, F., Reiser, H., Grandjean, D., de Alencastro, L.F., & Perazzolo, C., Che vre, N., Kohn, T. (2010). Occurrence and fate of micropollutants in the Vidy Bay of Lake Geneva, Switzerland. Part II: Micropollutant removal between wastewater and raw drinking water. *Environmental Toxicology and Chemistry*, *29*(8), 1658–1668. PMID:20821617

Napper, I. E., & Thompson, R. C. (2016). Release of synthetic microplastic plastic fibres from domestic washing machines: Effects of fabric type and washing conditions. *Marine Pollution Bulletin*, *112*(1-2), 39–45. doi:10.1016/j.marpolbul.2016.09.025 PMID:27686821

Peng, J., Wang, J., & Cai, L. (2017). Current understanding of microplastics in the environment: Occurrence, fate, risks, and what we should do. *Integrated Environmental Assessment and Management*, *13*(3), 476–482. doi:10.1002/ieam.1912 PMID:28440924

Qiu, Q., Tan, Z., Wang, J., Peng, J., Li, M., & Zhan, Z. (2016). Extraction, enumeration and identification methods for monitoring microplastics in the environment. *Estuarine, Coastal and Shelf Science*, *176*, 102–109. doi:10.1016/j.ecss.2016.04.012

Ribeiro-Claro, P., Nolasco, M. M., & Araújo, C. (2017). Chapter 5-characterization of microplastics by Raman spectroscopy. *Comprehensive Analytical Chemistry*, *75*, 119–151. doi:10.1016/bs.coac.2016.10.001

Rocha-Santos, T., & Duarte, A. C. (2015). A critical overview of the analytical approaches to the occurrence, the fate and the behavior of microplastics in the environment. *Trends in Analytical Chemistry*, *65*, 47–53. doi:10.1016/j.trac.2014.10.011

Rochman, C. M., Cook, A. M., & Koelmans, A. A. (2016). Plastic debris and policy: Using current scientific understanding to invoke positive change. *Environmental Toxicology and Chemistry*, *35*(7), 1617–1626. doi:10.1002/etc.3408 PMID:27331654

Shim, W. J., Hong, S. H., & Eo, S. (2017). Identification methods in microplastic analysis: A review. *Analytical Methods*, *9*(9), 1384–1391. doi:10.1039/C6AY02558G

Silva, B., Bastos, S., Justino, I. L., da Costa, P., Duarte, C., & Teresa, A. P. (2018). Microplastics in the environment: Challenges in analytical chemistry -A review. *Analytica Chimica Acta*, *1017*, 1–19. doi:10.1016/j.aca.2018.02.043 PMID:29534790

Simpson, S. L., Batley, G. E., Chariton, A. A., Stauber, J. L., King, C. K., Chapman, J. C., ... Maher, W. A. (2005). *Handbook for Sediment Quality Assessment*. Bangor, NSW: CSIRO.

Song, Y. K., Hong, S. H., Jang, M., Han, G. M., Rani, M., Lee, J., & Shim, W. J. (2015). A comparison of microscopic and spectroscopic identification methods for analysis of microplastics in environmental samples. *Marine Pollution Bulletin*, *93*(1-2), 202–209. doi:10.1016/j.marpolbul.2015.01.015 PMID:25682567

Song, Y. K., Hong, S. H., Jang, M., Kang, J. H., Kwon, O. Y., Han, G. M., & Shim, W. J. (2014). Large accumulation of micro-sized synthetic polymer particles in the sea surface microlayer. *Environmental Science & Technology*, *48*(16), 9014–9021. doi:10.1021/es501757s PMID:25059595

Sutton, R., Mason, S. A., Stanek, S. K., Willis-Norton, E., Wren, I. F., & Box, C. (2016). Microplastic contamination in the san Francisco Bay, California, USA. *Marine Pollution Bulletin*, *109*(1), 230–235. doi:10.1016/j.marpolbul.2016.05.077 PMID:27289280

Talsness, C. E., Andrade, A. J. M., Kuriyama, S. N., Taylor, J. A., & vom Saal, F. S. (2009). Components of plastic: Experimental studies in animals and relevance for human health. *Philos. Trans. R. Soc. B*, *364*(1526), 2079–2096. doi:10.1098/rstb.2008.0281 PMID:19528057

Teuten, E. L., Rowland, S. J., Galloway, T. S., & Thompson, R. C. (2007). Potential for plastics to transport hydrophobic contaminants. *Environmental Science & Technology*, *41*(22), 7759–7764. doi:10.1021/es071737s PMID:18075085

Thompson, R. C., Moore, C. J., vom Saal, F. S., & Swan, S. H. (2009). Plastics, the environment and human health: Current consensus and future trends. *Philosophical Transactions of the Royal Society of London. Series B, Biological Sciences*, *364*(1526), 2153–2166. doi:10.1098/rstb.2009.0053 PMID:19528062

Velzeboer, I., Kwadijk, C., & Kolemans, A. A. (2014). Strong sorption of PCBs to nanoplastics, microplastics, carbon nanotubes, and fullerenes. *Environmental Science & Technology*, *48*(9), 4869–4876. doi:10.1021/es405721v PMID:24689832

Von Moos, N., Burkhardt-Holm, P., & Köhler, A. (2012). Uptake and effects of microplastics on cells and tissue of the blue mussel Mytilus edulis L. after an experimental exposure. *Environmental Science & Technology*, *46*(20), 11327–11335. doi:10.1021/es302332w PMID:22963286

World Economic Forum. (2016). *The New Plastics Economy - Rethinking the Future of Plastics, 2016.* Available from: http://www.ellenmacarthurfoundation.org/publications

Wright, S. L., Thompson, R. C., & Galloway, T. S. (2013). The physical impacts of microplastics on marine organisms: A review. *Environ. Pollut.*, *178*, 483–492. doi:10.1016/j.envpol.2013.02.031 PMID:23545014

Zarfl, C., & Matthies, M. (2010). Are marine plastic particles transport vectors for organic pollutants to the Arctic? *Marine Pollution Bulletin*, *60*(10), 1810–1814. doi:10.1016/j.marpolbul.2010.05.026 PMID:20579675

Zhang, K., Xiong, X., Hu, H., Wu, C., Bi, Y., Wu, Y., ... Liu, J. (2017). Occurrence and characteristics of microplastic pollution in Xiangxi Bay of three gorges reservoir, China. *Environmental Science & Technology*, *51*(7), 3794–3801. doi:10.1021/acs.est.7b00369 PMID:28298079

Chapter 4

Polycyclic Aromatic Hydrocarbon Compounds as Emerging Water Pollutants:

Toxicological Aspects of Phenanthrene on Aquatic Animals

Krishna Bhuyan
Department of Life Science and Bioinformatics, Assam University, Silchar, India

Anirudha Giri
Department of Life Science and Bioinformatics, Assam University, Silchar, India

ABSTRACT

Aquatic ecosystems are pivotal resources that nurture diverse life forms apart from providing different ecosystem services. Global pollution, directly and indirectly, depletes the quality and standards of these resources and hampers the animals residing there. The incomplete combustion of all sorts of organic substances found in nature produces and release an emerging group of contaminants known as polycyclic aromatic hydrocarbons (PAHs). There are over a hundred different kinds of PAHs known and 16 amongst them are regarded as priority pollutants including phenanthrene (PHE). PHE is abundantly found in the aquatic environment and poses a higher risk to animals. It causes a vast array of toxicities in aquatic animals including genotoxicity, cardiotoxicity, transgenerational toxicity, neurotoxicity, developmental toxicity, and potentially induces oxidative stress and behavioral alterations. However, many areas of PHE toxicity in aquatic organisms are yet to be properly understood and management measures are yet to be initiated.

DOI: 10.4018/978-1-7998-1871-7.ch004

INTRODUCTION

In recent times, the worldwide developmental drive continuously increases our dependence on fossil fuels as a primary source of energy. Fossil fuels and all other organic materials upon incomplete burning produce and release a complex cluster of emerging pollutants in the environment known as polycyclic aromatic hydrocarbons (PAHs) (Llamas et al., 2017; Sun et al., 2012). Moreover, in the modern era, human activities have been multiplying the waste and effluents generation in the environment containing different PAHs. All sorts of contaminants including PAHs ultimately find their way into the aquatic ecosystem and thereby deplete the quality and standards of the aquatic habitats directly or indirectly (ATSDR, 1995). PAHs are environmentally concerned pollutants with hundreds different kinds and are found in nature with unique chemical structures, prevalence and physicochemical properties (Zelinkova and Wenzl, 2015). Amongst them, 16 compounds have been identified as priority pollutants by USEPA (ATSDR, 2005) because of their persistent presence, potential toxic properties and probable human exposure (Keith, 2015; Zelinkova and Wenzl, 2015). PAHs are widespread in all the three compartments of the environment such as air, water, and soil (Korsh et al., 2015; Maliszewska-Kordybach et al., 2009). Due to their persistent nature in the environment, the bottom sediments of aquatic ecosystems are also contaminated (Perra et al., 2011). A substantial amount of PAHs are also detected in foods that are subjected to drying, grilling, frying, roasting and baking (Mottier et al., 2000). PAHs remain in the environment as mixtures and some of them are recognized as probable carcinogens (IARC, 2010). These compounds can potentially react with other pollutants such as ozone, sulfur dioxide and nitrogen oxides yielding potentially more mutagenic and carcinogenic diones, sulfonic acids and nitro- and dinitro- PAHs respectively (Wild and Jones et al., 1995). Due to their increasing abundance in the environment and toxic potential, PAHs are being regarded as emerging pollutants of concern. The database of Norman Network provides a list of 1036 emerging pollutants which includes most of the PAHs and PAH metabolites including phenanthrene (PHE) (Norman-Network, 2016).

PHE is one of the 16 priority PAH compounds. It is widely abundant in PAH mixtures and is often the most dominant PAH in aquatic ecosystems (Bhagat et al., 2016; Khan et al., 2013; Machado et al., 2014; Sun et al., 2015a). WHO (1983) describes PHE as a major contributor to the total PAH content of the environment. In recent years, the presence of PHE in surface water has been detected worldwide (Callen et al., 2013; Qiao et al., 2006) and their levels are increasing due to source multiplication (Bigus et al., 2014; Lorgeoux et al., 2016). PHE is found to be abundantly higher around urban catchment areas and oil extraction sites (Loughery et al., 2018). Moreover, it has been reported to be toxic to many aquatic test animals (Bhagat et al., 2016; Brette et al., 2017; Gauthier et al., 2016; Khan et al., 2013; Machado et al., 2014; Morais et al., 2014; Nam et al., 2015; Prosser et al., 2011; Torreiro-Melo et al., 2015; Wu et al., 2015; Zhang et al., 2013). However, many aspects of the toxic properties of PHE and its metabolites are unexplored. The toxicological effects of PHE on many target animals such as amphibians are very scarce. Therefore, it is an area of concern among aquatic ecotoxicologists to understand and monitor the potential threats emanating from increasing concentrations of PAH in general which is crucial for the proper preservation of nature and natural resources.

To the best of our knowledge the literature regarding the toxicity of PHE is very limited. Most of the studies on PHE are either reviewed broadly in context to PAH or focused on the biodegradation part. Therefore, the primary objective of the present chapter is to review the current toxicological information of PHE along with its properties, sources, fate and degradation. The toxicological data on PHE published on or after 2014 in different aquatic test animals were reviewed.

BACKGROUND

Physico-Chemical Nature of Phenanthrene

The term "Phenanthrene" is derived from "phenyl" and "anthracene" (Kafilzadeh et al., 2015) with the chemical formula of $C_{14}H_{10}$. It is categorized among the low molecular weight PAHs (178.22) with three benzene ring structure (USEPA, 1987). Its physical appearance is colorless crystalline solid (USEPA, 1987) with melting and boiling points of 100°C and 340°C respectively (Budavari et al., 1989). The volumetric mass density of PHE is 1.18 g/cm^3 and the vapor pressure is 0.091Pascal (Budavari et al., 1989). Like all other PAHs, PHE is water-insoluble (Pearlman et al., 1984) and is highly lipophilic (IPCS, 1998). The octanol/water partition coefficient of phenanthrene is 4.45-4.57 (Meador et al., 1995; USEPA, 1987). Many PAHs including PHE possess chemically and biologically reactive "bay" and "K"- region epoxides which are metabolically formed. These epoxides are considered as suspected carcinogens (Fawell and Hunt, 1988). Like other known carcinogens, PHE is also used as a model substrate to study the metabolism of the "bay" and "K"- regions (Samanta et al., 1999). The physicochemical parameters of phenanthrene are depicted in Figure 1.

Sources and Fate of Phenanthrene

PHE is found ubiquitously in the environment often as principal components of PAH mixtures. PAHs are of pyrogenic, petrogenic and biological origin (Abdel-Shafy and Mansour, 2016). The natural sources of PAHs include volcanic eruptions and forest fires. PHE is the principal component of crude oil (Hook et al., 2018; Piazza et al., 2016; Tsibart et al., 2014) and is primarily produced in the environment through the incomplete combustion of fossil fuels (Loughery et al., 2018). USEPA (1987) listed some common sources of PHE that include combustion of coal, oil and wood, vehicular emissions, municipal incinerators, iron and steel works, coke, synfuel, oil shale and aluminum plants. PHE is ubiquitously found in aquatic systems and could be detected in wastewater, surface water, tap water and dried lake sediments. A substantial amount of PHE has been detected in many kinds of seafood that were collected from contaminated waters as well as in smoked and charcoal-broiled foods (USEPA, 1988; IARC, 1983).

Upon release, PHE can be directly inhaled and dermally absorbed by the exposed animals (Varanasi, 1989). The released PHE mix with the atmosphere and is subjected to short and long-range transport before settling in an aquatic ecosystem through wet and dry deposition, surface runoff, groundwater runoff and runoff from industrial areas (ATSDR, 1995) as shown in Figure 2. The transport of PAHs including PHE completely depends on the physicochemical properties of the compound. PHE being water-insoluble with an octanol-water partition coefficient (log Kow) of 4.35 adsorbs and settles on the

Figure 1. Physico-chemical parameters of phenanthrene

Name	Chemical formula	Structure	Physical appearance	Melting point	Boiling point	Mass density	Vapor pressure	Log K_{ow}
Phenanthrene (85-01-8)	$C_{14}H_{10}$	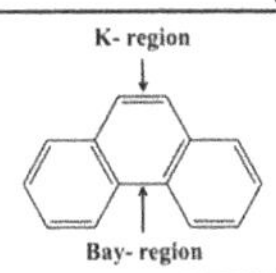	Colorless crystalline solid^a	100°C^b	340°C^c	1.18 g/cm3^d	0.091 pascal^e	4.45-4.57^f

\# U.S. EPA, 1987^{a,f}; Budavari et al., 1989^{b,c,d,e}; Meador et al., 1995^f

Figure 2. Schematic representation of sources and transport of phenanthrene

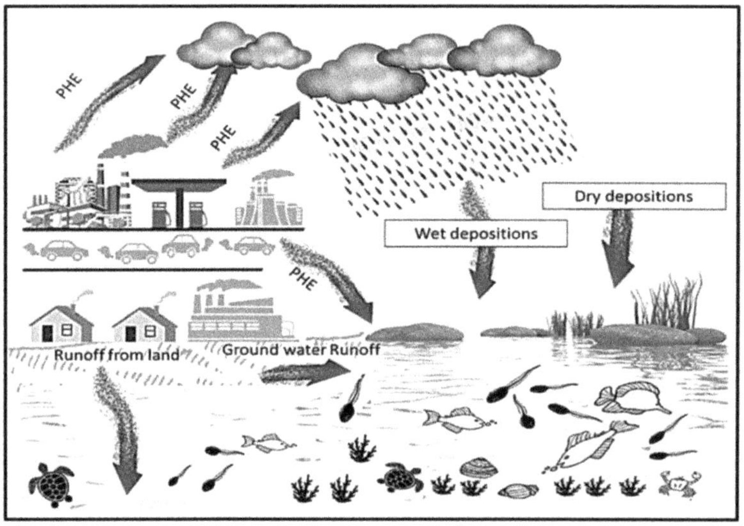

bottom sediments or remains suspended in the water column. Due to its high log Kow, PHE can be bioaccumulated in the food chains of both terrestrial and aquatic ecosystems (Lee and McCain, 2015). The lipophilic properties of PHE facilitate its passage through the fatty layers of the animals. Accumulation of PAHs in fatty tissues is subjected to the formation of micelles in the intestinal lumen (Jandacek and Genuis, 2013; Porter et al., 2007). In the long term, sediment trapped of PHE and lipophilic accumulation possesses a higher risk of PHE toxicity to aquatic biota (Jee et al., 2004). The organisms possessing aryl hydrocarbon hydroxylase (AHH) enzyme systems effectively metabolize PHE and decrease their bioaccumulation in the tissues. In the environment, PHE gets degraded by photolytic effects or by certain microbes such as fungi and bacteria (ATSDR, 1995).

PHENANTHRENE TOXICITY IN AQUATIC MODEL ANIMALS

PHE is reported to be very toxic to aquatic organisms (Bhagat et al., 2016; Morais et al., 2014; Richardi et al., 2018). Considerable research has been carried out to study the toxicological aspects of water abundant PAH like PHE on many aquatic test systems and some key researches from the year 2014-2018 has been depicted in Table 1. PHE has been reported to be genotoxic (Bhagat et al., 2016; Machado et al., 2014; Morais et al., 2014; Van Houten et al., 2018), cardiotoxic (Brette et al., 2017; Cypher et al., 2017; Zhang et al., 2013), neurotoxic (Jee and Kang, 2004; Nam et al., 2015), teratogenic (Wassenberg and Di Giulio, 2004), inducer of oxidative stress (Afifi et al., 2017; Bhagat et al., 2016; Machado et al., 2014). PHE has also been reported to alter behavioral traits (Torreiro-Melo et al., 2015; Gauthier et al., 2016) and is a potent endocrine disruptor (De Campos et al., 2018). All these studies revealed the toxic properties of PHE in both aquatic invertebrates and vertebrates. The following sections provide a detailed account of the latest research dealing with toxicities of PHE in both invertebrates and vertebrates.

Acute Toxicity

Acute toxicity is the primary measure to assess the toxicity of any contaminant. The acute toxicity study provides unique median lethal concentrations (LC_{50}) for contaminants on specific test systems. The LC_{50} values of PHE have been documented in this section are obtained from different research findings from the year 2014 onwards on aquatic model animals. Mu et al. (2014) determined the acute toxicity of PHE on marine medaka (*Oryzias melastigma*) and the 25-day LC_{50} was found to be >756 μg/L. Likewise, the LC_{50} values of PHE on zebrafish (120 hpf) is reported to be 310 μg/L (Vergauwen et al., 2015). In the same species, the UV enhanced 96h LC_{50} value was lower and reported to be 271 μg/L (Willis and Oris, 2014). In another study, the 96h LC_{50} value of PHE in adult zebrafish was found to be as high as 920μg/L (Kim et al., 2018).

Wu et al. (2015) reported the LC_{50} values of PHE on different native and non-native aquatic organisms. The reported LC_{50} values of PHE in fishes such as *Rhodens sinensis, Pseudorasbora parva, Misgurnus anguillicaudatus* were 2.55, 0.55, and 3.68 μg/L respectively. However, in planktonic crustacean (*Daphnia magna*) and benthic crustacean (*Macrobrachium nipponens*) the LC_{50} values are reported to be 0.28 and 1.08 μg/L respectively. The LC_{50} values of PHE in other groups of animals like the annelid (*Limnodrilus hoffmeisteri*), insect (*Chironomus plumosus*) and anuran amphibian (*Rana limnocharis*) were 0.799, 0.462, and 0.631 μg/L respectively (Wu et al., 2015). Moreover, a comparative study by Zhao et al. (2016) in the embryo and larvae of red sea bream (*Pagrosomus major*) observed that the 48h LC_{50} of PHE were 1.97 and 1.15 mg/L respectively. Other studies have reported LC_{50} values of PHE to be 0.49 mg/L in the marine zooplankton *Artemia salina* (Lu et al., 2018) and 96 h LC_{50} value to be 1.41 mg/L in the fish *Clarias gariepinus* (Sogbanmu et al., 2018). Therefore, the LC_{50} value of PHE not only varies in different species of aquatic organisms, but also may vary within the same species in different geographical locations suggesting their unique sensitivity towards contaminants like PHE.

Chronic and Sub-lethal Toxicity

Toxicity on Aquatic Invertebrates

Aquatic invertebrates inhabit both marine and freshwater ecosystems and provide ample ecosystem services. They are key balancers of the aquatic food chains and serve the ecosystem as a food source for higher trophic animals (Covich et al., 1999). They are sensitive to pollutants present in the environment and serve as reliable indicators of water quality and ecosystem health (Reynoldson and Metcalfe-Smith, 1992; Slooff, 1983). PHE is capable of causing different toxicities in aquatic invertebrates. Toxicological evaluation of PHE has been mostly carried out in invertebrates such as mollusks and crustaceans. However, little is known about the toxicity of PHE in echinoderms and annelids.

PHE affects oxidative stress parameters by altering the activity of Glutathione-S-transferase, catalase and malondialdehyde (MDA) levels in tropical gastropod *Morula granulate*. Tail DNA was considerably increased in the exposed erythrocytes in comet assay (Bhagat et al., 2016). Zhang et al. (2014) reported enhancement of toxic properties of PHE due to the influence of different suspended sediment concentrations (0, 1, 3, 5 g L^{-1}) during 96h exposure in *Daphnia magna*. The study revealed that PHE (0 – 0.8 mg L^{-1}) could altered superoxide dismutase, increased immobilization of *D. magna* which was further exacerbated with the addition of suspended sediments. The study emphasized the importance of the suspended sediment particles while assessing the water quality and in formulating the water quality standards. Dos Reis et al. (2015) used native oyster *Crassostrea brasiliana* to assess the molecular and histological responses of PHE at concentrations of 100 and 1000 μg L^{-1}. They found that the exposure of PHE for 10 days caused tubular atrophy of digestive gland and increased mucous cells in the mantle. Cytochrome P450 (CYP) is one of the most established biomarkers of environmental contamination (Dorrington et al., 2012) and Cytochrome *P450 2AU1* (*CYP2AU*1) gene transcript is the most important amongst all other transcripts as the levels are activated upon PHE exposure. The levels of *CYP2AU1* is reported to be significantly elevated upon PHE exposure in the mantle region. It is suggested that for evaluating the histological alterations and *CYP2AU1* transcript, mantle should be considered as a target organ in *Crassostrea brasiliana* (Dos Reis et al., 2015). Using transcriptomic analysis, Luchmann et al. (2015) studied the key metabolic pathways in xenobiotic metabolism in *Crassostrea brasiliana* and suggested that amongst numerous gene transcripts, some biomarker genes such as CYP1, CYP2, CYP17, HSC71, GRP78, HSP70-12A, and HSP70-12B were up regulated following PHE exposure.

PHE has been reported to induce antioxidant biomarkers like ethoxyresorufin-O-deethylase, glutathione S-transferase, superoxide dismutase, and glutathione and aryl hydrocarbon hydroxylase in gills of *Sinonovacula constricta* (Li et al., 2016). Activation of cellular detoxification mechanisms has also been reported in *Sinonovacula constricta* following PHE exposure. In a different study, Piazza et al. (2016) analyzed the effects of 24h exposure of 50 and 200 μg/L of PHE on 13 different genes related to cytochrome P450 (CYP), glutathione S-transferase (GST), sulfotransferase (SULT), flavin-containing monooxygenase and fatty acid-binding proteins in gill of scallops *Nodipecten nodosus*. PHE increased the transcriptional levels of *CYP2UI-like, CYP2D20-like, CYP3A11-like, GSTomega-like, SULT1B1-like* genes. Besides, glutathione reductase (GR), glutathione peroxidase (GPX), and glucose-6-phosphate dehydrogenase (G6PDH) activities were increased by PHE. However, depuration decreased the toxicity and restored the GR, GPX activities and all up-regulated biotransformation genes except for *SULT1B1-like*. PHE caused abnormal mantle and shell formation in pacific oyster *Crassostrea gigas* (Nogueira et al., 2017). They further reported that exposure to 2.0 μg/L PHE lowered the calcium levels of oyster shells. The genes such as *CYP30C1*, *Cg-Tal*, *Cg-Tyr*, *Calmodulin* were up-regulated and *CYP2AU2*, *Ferritin*, *Nacrein*, and *Insulin-Like* were found to be down-regulated after PHE exposure (Nogueira et al., 2017).

PHE is a water-insoluble compound and can adsorb to sediments resulting in sublethal effects to benthic organisms (Marini and Frapiccini, 2014). Richardi et al. (2018) assessed the biological responses of *Chironomus sancticaroli* to PHE. They found that PHE exposure delayed growth and molting as well as increased the alpha and beta esterases. PHE exposure even caused brush border disruption, gastric caeca regression and lumen area reduction. Besides, larval nuclear alteration in the trophocytes was also observed following PHE exposure. Overmans et al. (2018) studied PHE toxicity on Scleractinian coral (*Acropora tenuis*) and the effect of UVA exposure on its toxicity. Interestingly, the report showed that UVA exposure did not show any effect on PHE toxicity. However, PHE could reduce the superoxide dismutase (SOD) activity and also down-regulated the expression of tumor suppressor protein p53.

Toxicity on Aquatic Vertebrates

Recent studies in aquatic vertebrates almost exclusively focused on fish as a model for toxicity assessment of PHE. Toxicity assessment of PHE on other model aquatic animals like amphibians and reptiles is negligible. PAHs such as PHE are toxic to fishes and their toxicity is expressed by different mechanisms such as aryl hydrocarbon receptor (AhR) binding, CYP1A up-regulation and induction of phase I biotransformation. The toxicity mechanisms also include non-polar narcosis and enzyme

Table 1. Lethal and sub-lethal studies of phenanthrene on aquatic animals from 2014- 2018

Year	Model animals with age/size in length /stage	Concentrations	Toxicity	References
2014	*Mytilus galloprovincialis* (4–5 cm)	0.01, 0.1, 1.0 and 10 μg L^{-1}	• Induced oxidative stress. • Increased tail DNA, Olive moment (OM) and Tail moment (TM).	*Dailianis et al., 2014*
2014	*Daphnia magna* (6–24h old)	0–0.8 mg L^{-1}	• Increased immobilization. • Superoxide dismutase increased in higher concentration.	*Zhang et al., 2014*
2014	*Poecilia vivipara* (29.48 ± 5.27 mm)	10, 20 and 200 μg L^{-1}	• Induced oxidative stress.	*Machado et al., 2014*
2014	*Danio rerio* (5- 7 day)	50 μg L^{-1}	• Photoinduced mortality.	*Willis and Oris, 2014*
2014	*Oryzias melastigma* (Fertilized eggs)	100, 200, 400, and 800 μg L^{-1}	• Developmental malformation observed. • Blue sac disease (BSD) indexes gradually increased and Induced CYP1A protein expression.	*Mu et al., 2014*
2015	*Crassostrea brasiliana*	100 and 1000 μg L^{-1}	• Tubular atrophy in digestive diverticula. • Increased number of mucous cells in the mantle and induce *CYP2AU1* signal in gills, mantle, digestive diverticula, and intestine.	*Dos Reis et al., 2015*
2015	*Danio rerio* (120 hpf)	0,49,87,150, 217, and 362 μg L^{-1}	• Reduced swimming activity. • Edema of yolk and pericardium. • Malformation of yolk, eye, and mouth and tail curvature observed.	*Vergauwen et al., 2015*
2015	*Dicentrarchus labrax* (85.2±8.5 mm)	-	• Induced hepatic histopathological changes. • Oxidative stress and DNA damage	*Martins et al., 2015*
2015	*Oryzias melastigma* (1 month old)	0.06, 0.6, 6 and 60 μg L^{-1}	• Cardiotoxicity. • Reproductive toxicity.	*Sun et al., 2015a*
2016	*Morula granulate* (2 to 3 cm)	10, 25, 50, and 100 /L	• DNA damage. • Oxidative stress.	*Bhagat et al.,2016*
2016	*Sinonovacula constricta* (14.62 ± 2.26 cm)	0.5 and 4.5 μg L^{-1}	• Oxidative stress.	*Li et al., 2016*
2016	*Nodipecten nodosus* (6.0 - 8.0 cm)	50 and 200 μg L^{-1}	• Increased transcriptional levels of CYP2UI-like, CYP2D20-like, CYP3A11-like, GSTomega-like, SULT1B1-like genes. • GR. GPX and G6PDH activities increased.	*Piazza et al., 2016*
2016	*Clarias gariepinus* (Juveniles)	6.2 and 76 g L^{-1}	• In treated diploid glycogen contents elevated. • In triploid, mRNA level of fushi tarazu-factor 1 (ftz-f1). Increased plasma alkaline phosphatase (ALP) and lactate dehydrogenase (LDH). • In both, the incidence of histopathological lesions in the liver and gills observed with elevated tryptophan hydroxylase2 (tph2).	*Karami et al., 2016*
2017	*Oncorhynchus mykiss* (429 ± 78g)	0.001, 0.01, 0.1, 1.0, 10, 100 μM	• Reduced bile salt-activated lipase (BAL) activity in desalted luminal extracts.	*De Gelder et al., 2017*
2017	*Danio rerio* (24 hpf)	0, 1, 100, and 1000 μg L^{-1}	• Cardiotoxicity.	*Cypher et al., 2017*
2017	*Crassostrea gigas* (Shell length of 10-17 cm)	0.02 and 2.0 μg L^{-1}	• Decreased the frequency of normal development and shell size • Calcium levels decreased in oyster shells. • *CYP30C1*, *Cg-Tal*, *Cg-Tyr*, *Calmodulin* were up-regulated • *CYP2AU2*, *Ferritin*, *Nacrein*, and *Insulin-Like* were down-regulated • Minor decrease in normal larval development.	*Nogueira et al., 2017*
2017	*Danio rerio* (6 dpf)	100,500 and 1000 μg L^{-1}	• Decreased spare respiratory capacity and maximal respiration. • Increased non-mitochondrial respiration.	*Raftery et al., 2017*
2017	*Opsanus beta*	IP-injected (5 μg PHE in 2 μL peanut oil)	• Plasma cortisol levels were significantly higher.	*Reddam et al., 2017*

continued on following page

Table 1. Continued

Year	Model animals with age/size in length /stage	Concentrations	Toxicity	References
2017	*Acanthopagrus latus*	IP-injected (0, 2, 20 and 40 mg kg^{-1})	• Decrease WBC, C3/ C4 levels, • Lysosomal-membrane stability and lysozyme activity. • Significantly increased Cortisol level. • The size and number of Mitochondrial membrane potentials (MMCs) higher.	*Shirmohammadi et al., 2017a*
2017	*Acanthopagrus latus* (15.8 ± 0.2 cm)	IP-injected (2, 20, or 40 mg kg^{-1})	• Oxidative stress in head, kidney, and spleen.	*Shirmohammadi et al., 2017b*
2017	*Gobiocypris rarus* (Six-month-old)	8.9, 82.3, or 510.0 μg/L	• Hepato-toxicity • Apoptosis with increasing Caspase 9 and caspase 3 enzyme activity. • Significant decrease in the MMP.	*Hong et al., 2017*
2018	*Pimephales promelas* (adult)	202 μg L^{-1}	• Condition factor was reduced. • Reproductive toxicity.	*Loughery et al, 2018 b*
2018	*Epinephelus marginatus* (Juveniles)	0.1 mg L^{-1} and 1 mg L^{-1} (*in vivo*) 8.91 mg L^{-1} (*in-vitro*)	• Increase of the area of the hepatocytes *in vivo*. • Increased number of melanomacrophagic centers and hemosiderosis in the spleen and reduced 11-KT levels *in vitro*.	*De Campos et al., 2018*
2018	*Danio rerio*	50, 35.6, 11.2, 5, and 1.00 μM	• Individual phenanthrene did not alter cardiovascular parameters.	*Geier et al, 2018*
2018	*Pimephales promelas* (18-24 months of age)	29.8, 389 or 943 μg L^{-1}	• Increased proportions of the cortical alveolar stage in the ovaries. • No observed effect on 17β-estradiol production, hormone synthesis, and signaling pathways.	*Loughery et al, 2018 a*
2018	*Anoplopoma fimbria* (30-70 mm)	0.05,0.5,0.2 and 2.0 mg L^{-1}	• Mortality of the individuals.	*McConville et al., 2018*
2018	*Chironomus sancticaroli* (larvae)	0.12, 0.78,1.01 and 1.21 mg L^{-1}	• Malpighian tubules showed brush border reduction brush border disruption, gastric caeca regression, and lumen area reduction. • Increased Alpha and Beta esterase. • The fat body showed nuclear alteration in the trophocytes. • Salivary glands were subject to cytoplasm vacuolation.	*Richardi et al., 2018*
2018	*Acropora tenuis*	900,450, 225, 112.5, 56.3, 28.1 and 14.1 μg L^{-1}	• Caused oxidative stress gene expression of p53 was down-regulated.	*Overmans et al., 2018*

ethoxyresorufin-ortho-deethylase (Boehler et al. 2018). According to Brette et al. (2017), PHE has the key moiety for disrupting the physiology of heart muscle cells and may be the major cause of worldwide vertebrate cardiotoxicity.

Mu et al. (2014) studied the embryotoxicity of PHE in the marine medaka (*Oryzias melastigma*). PHE caused developmental deformities and mortality in exposed fish embryos. PHE significantly affected the marine medaka embryos by acting on the peripheral vascular system, yolk sac and embryonic incubation period. The early embryonic development was hampered by PHE exposure leading to a series of malformations collectively called as blue sac disease (BSD). With increasing PHE concentration, BSD index gradually increased in the exposed group (Mu et al., 2014).

PHE considerably alters oxidative stress and modulates cellular antioxidant potential in different fish species such as estuarine guppy *Poecilia vivipara* (Machado et al., 2014), European sea bass *Dicentrarchus labrax* (Martins et al., 2016) and yellowfin seabream *Acanthopagrus latus* (Shirmohammadi et al., 2017a). In estuarine guppy, PHE decreased antioxidant defenses against reactive oxygen species in the gills and muscle cells and reduced SOD activity in the liver (Machado et al., 2014). Other than oxidative stress parameters, effects of PHE on DNA integrity and damage were also studied in *Dicentrarchus labrax* using comet assay (Martins et al., 2016). In this study, PHE is reported to induce DNA strand breaks without clastogenic/aneugenic lesions and also induced hepatic histopathological changes. However, PHE did not show any significant oxidative stress alterations in embryos of marine medaka (Mu et al., 2014). Sun et al. (2015a) used Marine medaka for evaluating the reproductive and transgenerational toxicities of PHE. The study primarily focused on the female embryos which were exposed for 80 days with PHE concentration 0.06, 0.6, 6 and 60μg/L. In the brain, PHE significantly down regulated the mRNA levels of salmon-type gonadotropin-releasing hormone, luteinizing hormone and follicle-stimulating hormone. In the ovary, exposure of PHE suppressed cytochrome P450 aromatase gene CYP19A and estrogen receptor α while in the liver it down-regulated vitellogenin VTG1 and VTG2.

Vergauwen et al. (2015) used a passive dosing format using silicone O-rings in standard 24-well polystyrene plates to assess acute toxicity in Zebrafish embryo. PHE at concentrations 49, 87, 150, 217 and 362μg/L caused reduced swimming activity and mortality of the embryos. It also forms edema in yolk and pericardial regions and even caused deformities in the eye, mouth as well as in tail. Zebrafish was predominantly used as a test organism in other studies to assess the toxicity of PHE and found to be toxic to mitochondrial functions (Raftery et al., 2017); caused cardiotoxicity by decreasing heart rate, cardiac output and arterial red blood cell velocity in normoxia and hypoxic conditions (Cypher et al. 2017). It also decreased both arterial and venous velocity of red blood cells in caudal vessels with the formation of cardiac edema and lordosis (Cypher et al., 2017). On the other hand, Geier et al. (2018) did not observe any significant effect of PHE on any cardiovascular parameters in the same species.

Hong et al. (2017) analyzed the apoptotic mechanisms of PHE toxicity in the fish *Gobiocypris rarus* and found that PHE exposure induced increased liver lesions and hepatocyte proportions in the liver. Besides, activity of caspase 9 and caspase 3 enzymes were affected with significant decrease in the mitochondrial membrane potential. Loughery et al. (2018) studied the effect of sub-chronic exposure to PHE in fathead minnows *Pimephales promelas* at morphometric as well as transcriptional level. PHE increased the cortical alveolar stage in the ovaries and decreased the proportion of cortical alveolar oocytes. Female fish showed decreased gonadosomatic index and proportions of vitellogenic oocytes. Conversely, it increased the proportion of perinuclear spermatogonia and reduced the condition factor K (Loughery et al., 2018).

In African catfish (*Clarias gariepinus*), PHE concentrations of 6.2 and 76μg/L elevated the glycogen contents in triploid individuals. The highest tested concentration of PHE (76μg/L) had increased mRNA level of fushi tarazu-factor 1 (ftz-f1), plasma alkaline phosphatase and lactate dehydrogenase. In addition, expression of tryptophan hydroxylase 2 (tph2) was found to be elevated in both diploid and triploid individuals (Karami et al., 2016). De Gelder et al. (2017) observed an alteration in lipase activity in desalted luminal extracts of rainbow trout (*Oncorhynchus mykiss*) and reported lower bile salt-activated lipase (BAL) activity following PHE exposure apart from affecting immune function in aquatic organisms. PHE has also been reported to suppress immune function in yellowfin seabream *Acanthopagrus latus* following PHE exposure (Shirmohammadi et al., 2017b). PHE induced alterations in hormonal function like induction of cortical stress have been reported in *Acanthopagrus latus* (Shirmohammadi et al., 2017b) and Gulf toadfish *Opsanus beta* (Reddam et al., 2017). However, PHE did not induce significant endocrine disruption in protogynous dusky grouper (*Epinephelus marginatus*) either *in vivo* or *in vitro* (De Campos et al., 2018).

PHENANTHRENE DEGRADATION AND REMEDIATION

PHE caused serious toxic effects on aquatic as well as terrestrial animals including the human beings. Therefore, it becomes a matter of global concern to eliminate these pollutants from the environment. Elimination in terms of degradation and bioremediations are the important aspects of removal or transformations of PAHs from contaminated environments.

Photodegradation is a well-recognized method for removal of PAH in the environment. It basically includes both pathways of photodimerisation and photooxidation. PAH compounds may get directly degraded through photoionization or by transferring energy and electrons from the excited PAH through cation radical intermediates via reactions with singlet and triplet PAH state to molecular oxygen (Chen et. al., 2009; Fasnacht and Blough, 2003; Miller and Olejnik 2001). The photocatalytic materials such as titanium oxide, zinc oxide, silicon dioxide were also used extensively for the efficient PAH degradation (Jiang et al., 2018; Woo et al., 2009; Wen et al., 2003).

The physico-chemical methods include use of nonionic surfactants and suitable adsorbents (Liu and Chen, 2015; Rao et al., 2017). A number of adsorbents increase the adsorption of PAHs viz., activated carbon (Lamichhane et al., 2016), biochar (Oleszczuk et al., 2012; Yang et. al., 2018), graphene oxide (Han et. al., 2018; Mu et. al., 2019) laccase carrying electro spun nanofibers (Dai et. al., 2013) and magnetic nanomaterial (Ríos and Zougagh, 2016; Shanker et. al., 2017). Other techniques like ultrasound, electric beams and electric discharge have also been employed in a number of studies for degradation of PAH compounds (Dailianis et al., 2014; Little et al., 2002; Wang et al., 2019).

From the environmental point of view, bioremediation is recognized as sustainable and eco-friendly technique for degradation of environmental contaminants including organic compounds (Cerniglia, 1989; Kronenberg et al.; 2017). The bioremediation processes include microbial degradation (Feng et al., 2017; Schwarz et al., 2018), plant-animal-microbe interactions (Deng and Zeng, 2017) and phytoremdiation (Peng et al., 2014; Pulford and Watson 2003). Phytoremediation has emerged as a prominent technology involving different plants and associated microrganisms for the active removal of PHE from contamination sites (Ouvrard et al., 2014; Peng et al., 2014). Phytoremediation of PHE is reported to be associated with endophytes (Khan et al., 2014; Liu et al., 2014a); transgenic plants (Peng et al., 2014); and some tress species such as willow and poplar (Sun et al., 2015b). Submerged plants such as *Vallisneria spiralis and Hydrilla verticillata* are reported to remediate substantial amount of PHE from contaminated water (He and Chi, 2017; Liu et al., 2014b).

Microbial bioremediation used different bacterial, fungal and algal species that efficiently remediate the PHE. Bacterial genera such as *Bacillus* (Ferreira et al., 2016; Liu et al., 2019; Rabodonirina et al., 2019), *Pseudomonas* (Kumari et al., 2018; Rabodonirina et al., 2019), *Alcanivorax* (Rodrigues et al., 2017; Schwarz et al., 2018), *Marinobacter* (Schwarz et al., 2018), *Enterococcus*(Schwarz et al., 2018), *Flavobacterium* (Rodrigues et al., 2017), *Streptomyces* (Chaudhary et al, 2011) are efficient for PAHs remediation. Likewise, different fungal species such as *Ganoderma lucidum* (Agarwal et al., 2018; Ting et al. 2011), *Trametes versicolor* (Han et al. 2004), *Trametes hirsute* (Hidayat et al., 2018), *Pleurotus ostreatus* (Tirado-Torres et al. 2016), *Pleurotus eryngii* (Wu et al. 2016) have also been used for bioremediation of PHE with encouraging outcomes. In addition, Schwarz et al. (2018) reported three new genera of fungi namely *Crypotococcus*, *Cladosporium* and *Tremellales* to be highly effective in PHE biodegradation.

A number of different algal species have been successfully tested for bioremediation of PHE (Hong et al. 2008; Zhang et al., 2019). According to Zhang et al. (2019), the green tide algae *Ulva prolifera* showed high adsorptive removal efficiency with maximum adsorption capacity of PHE 1.97 μg g−1. Under photosynthetic conditions, an algal-bacterial consortium viz., *Pseudomonas migulae* and *Chlorella sorokiniana* were able to effectively biodegrade PHE dissolved in silicone oil or tetradecane in the absence of exogenous oxygen supply (Muñoz et al., 2003). The emerging technologies discussed so far are mostly described under controlled laboratory conditions. However, for large-scale field applications, further modifications and development are necessary in future.

RECOMMENDATIONS AND CONCLUSIONS

Studies evaluating the potential toxicity of PHE in aquatic organisms have reported positive toxicity data in most of the tested parameters. However, most of these studies have focused on either mollusks or model fish species. The development of water quality guidelines requires toxicological data in different species across taxonomic groups. Therefore, future toxicological studies on PHE need to address this issue by adopting a taxonomically inclusive approach. Most of the studies discussed here neither report acute toxicity (LC_{50}) data nor defined NOEL concentrations which are important for inter-laboratory comparisons as well as regulatory purposes. Last but not the least, the potential toxicity of PHE to genetic material, reproduction and behavioral aspects such as swimming behavior in aquatic animals is largely understudied and require greater attention

In conclusion, it can be stated that sufficient evidence is available indicating that an increasing concentration of PHE in aquatic ecosystems is a potential threat to the organisms inhabiting these ecosystems. Therefore, greater attention should be given to understand the consequences of this potential threat and initiate corrective measures. Different approaches have been tested for possible remediation of PHE using both biotic and abiotic methods. The bioremediation of PHE appears to be a promising prospect in this direction and should be pursued further for field level applications.

ACKNOWLEDGMENT

The authors thankfully acknowledge Assam University, Silchar, India, for providing necessary infrastructure and Department of Biotechnology, Government of India for financial support.

REFERENCES

Abdel-Shafy, H. I., & Mansour, M. S. M. (2016). A review on polycyclic aromatic hydrocarbons: Source, environmental impact, the effect on human health and remediation. *Egyptian Journal of Petroleum*, *25*(1), 107–123. doi:10.1016/j.ejpe.2015.03.011

Afifi, M., Alkaladi, A., Zinada, O. A. A., & Couderchet, M. (2017). Alteration in antioxidant genes expression in some fish caught from Jeddah and Yanbu coast as a bio-indicator of oil hydrocarbons pollution. *Saudi Journal of Biological Sciences*, *24*(7), 1580–1587. doi:10.1016/j.sjbs.2015.06.014 PMID:30294228

Agrawal, N., Verma, P., & Shahi, S. K. (2018). Degradation of polycyclic aromatic hydrocarbons (phenanthrene and pyrene) by the ligninolytic fungi *Ganoderma lucidum* isolated from the hardwood stump. *Bioresources and Bioprocessing*, *5*(1), 11. doi:10.118640643-018-0197-5

ATSDR. (1995). *Toxicological profile for otto fuels II, U.S. Department of Health and Human Services*. Atlanta, GA: U.S. Department of Health and Human Services, Public Health Service.

ATSDR. (2005). *Toxicology profile for polyaromatic hydrocarbons. ATSDR's Toxicological Profiles on CD-ROM*. Boca Raton, FL: CRC Press.

Bhagat, J., Sarkar, A., & Ingole, B. S. (2016). DNA Damage and Oxidative Stress in Marine Gastropod *Morula granulata* Exposed to Phenanthrene. *Water, Air, and Soil Pollution*, *227*(4), 114. doi:10.100711270-016-2815-1

Bigus, P., Tobiszewski, M., & Namieśnik, J. (2014). Historical records of organic pollutants in sediment cores. *Marine Pollution Bulletin*, *78*(1-2), 26–42. doi:10.1016/j.marpolbul.2013.11.008 PMID:24300286

Boehler, S., Lörracher, A. K., Schubert, J., & Braunbeck, T. (2018). Comparative live-imaging of in vivo EROD (ethoxyresorufin- O -deethylase) induction in zebrafish (Danio rerio) and fathead minnow (Pimephales promelas) embryos after exposure to PAHs and river sediment extracts. *The Science of the Total Environment*, *621*, 827–838. doi:10.1016/j.scitotenv.2017.11.101 PMID:29202294

Brette, F., Shiels, H. A., Galli, G. L. J., Cros, C., Incardona, J. P., Scholz, N. L., & Block, B. A. (2017). A Novel Cardiotoxic Mechanism for a Pervasive Global Pollutant. *Scientific Reports*, *7*(1), 41476. doi:10.1038rep41476 PMID:28139666

Budavari, S., O'Neil, M. J., Smith, A., & Heckran, P. E. (1989). The Merck Index. In *An encyclopedia of chemicals, drugs, and biologicals* (11th ed.). Rahway, NJ: Merck.

Callén, M. S., López, J. M., Iturmendi, A., & Mastral, A. M. (2013). Nature and sources of particle associated polycyclic aromatic hydrocarbons (PAH) in the atmospheric environment of an urban area. *Environmental Pollution*, *183*, 166–174. doi:10.1016/j.envpol.2012.11.009 PMID:23245526

Cerniglia, C. E., & Heitkamp, M. A. (1989). Microbial degradation of polycyclic aromatic hydrocarbons(PAH) in the aquatic environment. In U. Varanasi (Ed.), *Metabolism of polycyclic aromatic hydrocarbons in the aquatic environment* (pp. 41–68). Boca Raton, FL: CRC Press, Inc.

Chaudhary, P., Sharma, R., Singh, S. B., & Nain, L. (2011). Bioremediation of PAH by *Streptomyces* sp. *Bulletin of Environmental Contamination and Toxicology*, *86*(3), 268–271. doi:10.100700128-011-0211-5 PMID:21301805

Chen, Y., Hu, C., Hu, X., & Qu, J. (2009). Indirect Photodegradation of Amine Drugs in Aqueous Solution under Simulated Sunlight. *Environmental Science & Technology*, *43*(8), 2760–2765. doi:10.1021/es803325j PMID:19475946

Covich, A. P., Palmer, M. A., & Crowl, T. A. (1999). The Role of Benthic Invertebrate Species in Freshwater Ecosystems. *Bioscience*, *49*(2), 119. doi:10.2307/1313537

Cypher, A. D., Consiglio, J., & Bagatto, B. (2017). Hypoxia exacerbates the cardiotoxic effect of the polycyclic aromatic hydrocarbon, phenanthrene in *Danio rerio. Chemosphere*, *183*, 574–581. doi:10.1016/j.chemosphere.2017.05.109 PMID:28570901

Dai, Y., Niu, J., Yin, L., Xu, J., & Xu, J. (2013). Laccase-carrying electrospun fibrous membrane for the removal of polycyclic aromatic hydrocarbons from contaminated water. *Separation and Purification Technology*, *104*, 1–8. doi:10.1016/j.seppur.2012.11.013

Dailianis, S., Tsarpali, V., Melas, K., Karapanagioti, H. K., & Manariotis, I. D. (2014). Aqueous phenanthrene toxicity after high-frequency ultrasound degradation. *Aquatic Toxicology (Amsterdam, Netherlands)*, *147*, 32–40. doi:10.1016/j.aquatox.2013.12.006 PMID:24373890

De Campos, M. F., Lo Nostro, F. L., Da Cuña, R. H., & Moreira, R. G. (2018). Endocrine disruption of phenanthrene in the protogynous dusky grouper *Epinephelus marginatus* (Serranidae: Perciformes). *General and Comparative Endocrinology*, *257*, 255–263. doi:10.1016/j.ygcen.2017.06.020 PMID:28652135

De Gelder, S., Sæle, Ø., de Veen, B. T. H., Vos, J., Flik, G., Berntssen, M. H. G., & Klaren, P. H. M. (2017). The polycyclic aromatic hydrocarbons benzo[a]pyrene and phenanthrene inhibit intestinal lipase activity in rainbow trout (*Oncorhynchus mykiss*). *Comparative Biochemistry and Physiology Part C: Toxicology & Pharmacology*, *198*, 1–8. doi:10.1016/j.cbpc.2017.04.008 PMID:28476648

Deng, S., & Zeng, D. (2017). Removal of phenanthrene in contaminated soil by combination of alfalfa, white-rot fungus, and earthworms. *Environmental Science and Pollution Research International*, *24*(8), 7565–7571. doi:10.100711356-017-8466-y PMID:28116628

Dorrington, T., Zanette, J., Zacchi, F. L., Stegeman, J. J., & Bainy, A. C. D. (2012). Basal and 3 methylcholanthrene-induced expression of cytochrome P450 1A, 1B and 1C genes in the Brazilian guppy, *Poecilia vivipara. Aquatic Toxicology (Amsterdam, Netherlands)*, *124-125*, 106–113. doi:10.1016/j.aquatox.2012.08.006 PMID:22940225

Dos Reis, I. M. M., Mattos, J. J., Garcez, R. C., Zacchi, F. L., Miguelão, T., Flores-Nunes, F., ... Bainy, A. C. D. (2015). Histological responses and localization of the cytochrome P450 (CYP2AU1) in *Crassostrea brasiliana* exposed to phenanthrene. *Aquatic Toxicology (Amsterdam, Netherlands)*, *169*, 79–89. doi:10.1016/j.aquatox.2015.10.011 PMID:26519834

Fasnacht, M. P., & Blough, N. V. (2003). Mechanisms of the Aqueous Photodegradation of Polycyclic Aromatic Hydrocarbons. *Environmental Science & Technology*, *37*(24), 5767–5772. doi:10.1021/es034389c PMID:14717193

Fawell, J. K., & Hunt, S. (1988). The polycyclic aromatic hydrocarbons. In J. K. Fawell & S. Hunt (Eds.), *Environmental toxicology: organic pollutants* (pp. 241–269). doi:10.1002/jat.2550090414

Feng, N. X., Yu, J., Zhao, H. M., Cheng, Y. T., Mo, C. H., Cai, Q. Y., ... Wong, M. H. (2017). Efficient phytoremediation of organic contaminants in soils using plant–endophyte partnerships. *The Science of the Total Environment*, *583*, 352–368. doi:10.1016/j.scitotenv.2017.01.075 PMID:28117167

Ferreira, L., Rosales, E., Danko, A. S., Sanromán, M. A., & Pazos, M. M. (2016). *Bacillus thuringiensis* a promising bacterium for degrading emerging pollutants. *Process Safety and Environmental Protection, 101*, 19–26. doi:10.1016/j.psep.2015.05.003

Gauthier, P. T., Norwood, W. P., Prepas, E. E., & Pyle, G. G. (2016). Behavioural alterations from exposure to Cu, phenanthrene, and Cu-phenanthrene mixtures: Linking behaviour to acute toxic mechanisms in the aquatic amphipod, *Hyalella azteca*. *Aquatic Toxicology (Amsterdam, Netherlands)*, *170*, 377–383. doi:10.1016/j.aquatox.2015.10.019 PMID:26596825

Geier, M. C., James Minick, D., Truong, L., Tilton, S., Pande, P., Anderson, K. A., ... Tanguay, R. L. (2018). Systematic developmental neurotoxicity assessment of a representative PAH Superfund mixture using zebrafish. *Toxicology and Applied Pharmacology*, *354*, 115–125. doi:10.1016/j.taap.2018.03.029 PMID:29630969

Han, B., Li, Y., Qian, B., He, Y., Peng, L., & Yu, H. (2018). Adsorption and determination of polycyclic aromatic hydrocarbons in water through the aggregation of graphene oxide. *Open Chemistry*, *16*(1), 716–725. doi:10.1515/chem-2018-0078

Han, M. J., Choi, H. T., & Song, H. G. (2004). Degradation of phenanthrene by Trametes versicolor and its laccase. *Journal of Microbiology (Seoul, Korea)*, *42*(2), 94–98. PMID:15357301

He, Y., & Chi, J. (2017). Pilot-scale demonstration of phytoremediation of PAH-contaminated sediments by *Hydrilla verticillata* and *Vallisneria spiralis*. *Environmental Technology*, *40*(5), 605–613. doi:10.1080/09593330.2017.1398783 PMID:29076392

Hidayat, A., & Yanto, D. H. Y. (2018). Biodegradation and metabolic pathway of phenanthrene by a new tropical fungus, *Trametes hirsuta* D7. *Journal of Environmental Chemical Engineering*, *6*(2), 2454–2460. doi:10.1016/j.jece.2018.03.051

Hong, X., Qin, J., Chen, R., Yuan, L., Zha, J., Huang, C., ... Wang, Z. (2017). Phenanthrene-Induced Apoptosis and Its Underlying Mechanism. *Environmental Science & Technology*, *51*(24), 14397–14405. doi:10.1021/acs.est.7b04045 PMID:29161501

Hong, Y. W., Yuan, D. X., Lin, Q.-M., & Yang, T. L. (2008). Accumulation and biodegradation of phenanthrene and fluoranthene by the algae enriched from a mangrove aquatic ecosystem. *Marine Pollution Bulletin*, *56*(8), 1400–1405. doi:10.1016/j.marpolbul.2008.05.003 PMID:18597790

Hook, S. E., Mondon, J., Revill, A. T., Greenfield, P. A., Stephenson, S. A., Strzlecki, J., ... Barrett, S. (2018). Monitoring sublethal changes in fish physiology following exposure to a light, unweathered crude oil. *Aquatic Toxicology (Amsterdam, Netherlands)*, *204*, 27–45. doi:10.1016/j.aquatox.2018.08.013 PMID:30173120

IARC. (1983). IARC Monographs on the Evaluation of Carcinogenic Risks to Humans. *Polynuclear aromatic compounds, Part 1, chemical, environmental and experimental data, 32*, 1–453.

IARC. (2010). IARC Monographs on the Evaluation of Carcinogenic Risks to Humans. *Some non-heterocyclic polycyclic aromatic hydrocarbons and some related exposures, 92*, 1–853.

IPCS. (1998). *Selected Non-heterocyclic Polycyclic Aromatic Hydrocarbons, Environmental Health Criteria 202*. Geneva: International Programme on Chemical Safety, World Health Organization.

Jandacek, R. J., & Genuis, S. J. (2013). An Assessment of the Intestinal Lumen as a Site for Intervention in Reducing Body Burdens of Organochlorine Compounds. *The Scientific World Journal*, *2013*, 1–10. doi:10.1155/2013/205621 PMID:23476122

Jee, J. H., & Kang, J. C. (2004). Effect of phenanthrene on haematological parameters in olive flounder, *Paralichthys olivaceus* (Temminch et Schlegel). *Aquaculture Research*, *35*(14), 1310–1317. doi:10.1111/j.1365-2109.2004.01152.x

Jee, J. H., Kim, S. G., & Kang, J. C. (2004). Effects of phenanthrene on growth and basic physiological functions of the olive flounder, *Paralichtys olivaceus*. *Journal of Experimental Marine Biology and Ecology*, *304*(1), 123–136. doi:10.1016/j.jembe.2003.12.001

Jiang, L., Yuan, X., Zeng, G., Liang, J., Chen, X., Yu, H., ... Xiong, T. (2018). In-situ synthesis of direct solid-state dual Z-scheme WO 3 /g-C 3 N 4 /Bi 2 O 3 photocatalyst for the degradation of refractory pollutant. *Applied Catalysis B: Environmental*, *227*, 376–385. doi:10.1016/j.apcatb.2018.01.042

Kafilzadeh, F., Khosrobak, A., & Jamali, H. (2015). Isolation and Identification of Phenanthrene Degrading Bacteria from the Soil around Oil Company of Andimeshk and Investigation of Their Growth Kinetics. *Polycyclic Aromatic Compounds*, *36*(1), 58–71. doi:10.1080/10406638.2015.1037006

Karami, A., Romano, N., Hamzah, H., Simpson, S. L., & Yap, C. K. (2016). Acute phenanthrene toxicity to juvenile diploid and triploid African catfish (*Clarias gariepinus*): Molecular, biochemical, and histopathological alterations. *Environmental Pollution*, *212*, 155–165. doi:10.1016/j.envpol.2016.01.055 PMID:26845363

Keith, L. H. (2015). The Source of U.S. EPA's Sixteen PAH Priority Pollutants. *Polycyclic Aromatic Compounds*, *35*(2-4), 147–160. doi:10.1080/10406638.2014.892886

Khan, E. A., Bertotto, L., Dale, K., Lille-Langøy, R., Yadetie, F., Karlsen, O. A., ... Chen, Y. (2013). A battery of bioassays for the evaluation of phenanthrene biotoxicity in soil. *Archives of Environmental Contamination and Toxicology*, *65*(1), 47–55. doi:10.100700244-013-9879-3 PMID:23440446

Khan, Z., Roman, D., Kintz, T., delas Alas, M., Yap, R., & Doty, S. (2014). Degradation, Phytoprotection and Phytoremediation of Phenanthrene by Endophyte Pseudomonas putida, PD1. *Environmental Science & Technology*, *48*(20), 12221–12228. doi:10.1021/es503880t PMID:25275224

Kim, K., Jeon, H. J., Choi, S. D., Tsang, D. C. W., Oleszczuk, P., Ok, Y. S., ... Lee, S. E. (2018). Combined toxicity of endosulfan and phenanthrene mixtures and induced molecular changes in adult Zebrafish (*Danio rerio*). *Chemosphere*, *194*, 30–41. doi:10.1016/j.chemosphere.2017.11.128 PMID:29197246

Korsh, J., Shen, A., Aliano, K., & Davenport, T. (2015). Polycyclic Aromatic Hydrocarbons and Breast Cancer: A Review of the Literature. *Breast Care (Basel, Switzerland)*, *10*(5), 316–318. doi:10.1159/000436956 PMID:26688678

Kronenberg, M., Trably, E., Bernet, N., & Patureau, D. (2017). Biodegradation of polycyclic aromatic hydrocarbons: Using microbial bioelectrochemical systems to overcome an impasse. *Environmental Pollution*, *231*, 509–523. doi:10.1016/j.envpol.2017.08.048 PMID:28841503

Kumari, S., Regar, R. K., & Manickam, N. (2018). Improved polycyclic aromatic hydrocarbon degradation in a crude oil by individual and a consortium of bacteria. *Bioresource Technology*, *254*, 174–179. doi:10.1016/j.biortech.2018.01.075 PMID:29413920

Lamichhane, S., Bal Krishna, K. C., & Sarukkalige, R. (2017). Surfactant-enhanced remediation of polycyclic aromatic hydrocarbons: A review. *Journal of Environmental Management*, *199*, 46–61. doi:10.1016/j.jenvman.2017.05.037 PMID:28527375

Lamichhane, S., Krishna, K. C. B., & Sarukkalige, R. (2016). Polycyclic aromatic hydrocarbons (PAHs) removal by sorption: A review. *Chemosphere*, *148*, 336–353. doi:10.1016/j.chemosphere.2016.01.036 PMID:26820781

Lee, S. H., & McCain, M. C. (2015). Wildlife Toxicity Assessment for Phenanthrene. In M. A. Williams, G. Reddy, M. J. Quinn Jr, & M. S. Johnson (Eds.), *Wildlife Toxicity Assessments for Chemicals of Military Concern* (pp. 591–598). Elsevier Inc. doi:10.1016/B978-0-12-800020-5.00031-4

Li, L., Jiang, M., & Shen, X. (2016). Variability in antioxidant/detoxification enzymes of *Sinonovacula constricta* exposed to benzo[a]pyrene and phenanthrene. *Marine Pollution Bulletin*, *109*(1), 507–511. doi:10.1016/j.marpolbul.2016.05.013 PMID:27216044

Little, C., Hepher, M. J., & El-Sharif, M. (2002). The sono-degradation of phenanthrene in an aqueous environment. *Ultrasonics*, *40*(1-8), 667–674. doi:10.10160041-624x(02)00196-8

Liu, H., Meng, F., Tong, Y., & Chi, J. (2014). Effect of plant density on phytoremediation of polycyclic aromatic hydrocarbons contaminated sediments with *Vallisneria spiralis*. *Ecological Engineering*, *73*, 380–385. doi:10.1016/j.ecoleng.2014.09.084

Liu, J., & Chen, W. (2015). Remediation of phenanthrene contaminated soils by nonionic-anionic surfactant washing coupled with activated carbon adsorption. *Water Science and Technology*, *72*(9), 1552–1560. doi:10.2166/wst.2015.357 PMID:26524446

Liu, S. H., Zeng, Z. T., Niu, Q. Y., Xiao, R., Zeng, G. M., Liu, Y., ... Tao, J.-J. (2019). Influence of immobilization on phenanthrene degradation by *Bacillus* sp. P1 in the presence of Cd (II). *The Science of the Total Environment*, *655*, 1279–1287. doi:10.1016/j.scitotenv.2018.11.272 PMID:30577120

Llamas, A., Al-Lal, A. M., García-Martínez, M. J., Ortega, M. F., Llamas, J. F., Lapuerta, M., & Canoira, L. (2017). Polycyclic Aromatic Hydrocarbons (PAHs) produced in the combustion of fatty acid alkyl esters from different feedstocks: Quantification, statistical analysis and mechanisms of formation. *The Science of the Total Environment*, *586*, 446–456. doi:10.1016/j.scitotenv.2017.01.180 PMID:28215803

Lorgeoux, C., Moilleron, R., Gasperi, J., Ayrault, S., Bonté, P., Lefèvre, I., & Tassin, B. (2016). Temporal trends of persistent organic pollutants in dated sediment cores: Chemical fingerprinting of the anthropogenic impacts in the Seine River basin, Paris. *The Science of the Total Environment*, *541*, 1355–1363. doi:10.1016/j.scitotenv.2015.09.147 PMID:26479909

Loughery, J. R., Kidd, K. A., Mercer, A., & Martyniuk, C. J. (2018). Part B:Morphometric and transcriptomic responses to sub-chronic exposure to the polycyclic aromatic hydrocarbon phenanthrene in the fathead minnow (Pimephales promelas). *Aquatic Toxicology (Amsterdam, Netherlands)*, *199*, 77–89. doi:10.1016/j.aquatox.2018.03.026 PMID:29621673

Lu, J., Tian, S., Lv, X., Chen, Z., Chen, B., Zhu, X., & Cai, Z. (2018). TiO 2 nanoparticles in the marine environment: Impact on the toxicity of phenanthrene and Cd 2+ to marine zooplankton *Artemia salina*. *The Science of the Total Environment*, *615*, 375–380. doi:10.1016/j.scitotenv.2017.09.292 PMID:28988072

Lüchmann, K. H., Clark, M. S., Bainy, A. C. D., Gilbert, J. A., Craft, J. A., Chipman, J. K., ... Schroeder, D. C. (2015). Key metabolic pathways involved in xenobiotic biotransformation and stress responses revealed by transcriptomics of the mangrove oyster *Crassostrea brasiliana*. *Aquatic Toxicology (Amsterdam, Netherlands)*, *166*, 10–20. doi:10.1016/j.aquatox.2015.06.012 PMID:26186662

Machado, A. A., Hoff, M. L., Klein, R. D., Cordeiro, G. J., Lencina Avila, J. M., Costa, P. G., & Bianchini, A. (2014). Oxidative stress and DNA damage responses to phenanthrene exposure in the estuarine guppy *Poecilia vivipara*. *Marine Environmental Research*, *98*, 96–105. doi:10.1016/j.marenvres.2014.03.013 PMID:24722053

Maliszewska-Kordybach, B., Smreczak, B., & Klimkowicz-Pawlas, A. (2009). Effects of anthropopressure and soil properties on the accumulation of polycyclic aromatic hydrocarbons in the upper layer of soils in selected regions of Poland. *Applied Geochemistry*, *24*(10), 1918–1926. doi:10.1016/j.apgeochem.2009.07.005

Marini, M., & Frapiccini, E. (2014). Do lagoon area sediments act as traps for polycyclic aromatic hydrocarbons? *Chemosphere*, *111*, 80–88. doi:10.1016/j.chemosphere.2014.03.037 PMID:24997903

Martins, M., Ferreira, A. M., Costa, M. H., & Costa, P. M. (2015). Comparing the genotoxicity of a potentially carcinogenic and a noncarcinogenic PAH, singly, and in binary combination, on peripheral blood cells of the European sea bass. *Environmental Toxicology*, *31*(11), 1307–1318. doi:10.1002/tox.22135 PMID:25728603

Meador, J. P., Stein, J. E., Reichert, W. L., & Varanasi, U. (1995). Bioaccumulation of polycyclic aromatic hydrocarbons by marine organisms. *Reviews of Environmental Contamination and Toxicology*, *143*, 79–165. doi:10.1007/978-1-4612-2542-3_4 PMID:7501868

Miller, J. S., & Olejnik, D. (2001). Photolysis of polycyclic aromatic hydrocarbons in water. *Water Research*, *35*(1), 233–243. doi:10.1016/S0043-1354(00)00230-X PMID:11257878

Morais, G. dos S., Pesenti, E. C., Cestari, M. M., & Navarro-Silva, M. A. (2014). Genotoxic effect of Phenanthrene on *Chironomus sancticaroli* (Diptera: Chironomidae). *Zoologia*, *31*(4), 323–328. doi:10.1590/S1984-46702014000400003

Mottier, P., Parisod, V., & Turesky, R. J. (2000). Quantitative Determination of Polycyclic Aromatic Hydrocarbons in Barbecued Meat Sausages by Gas Chromatography Coupled to Mass Spectrometry. *Journal of Agricultural and Food Chemistry*, *48*(4), 1160–1166. doi:10.1021/jf991205y PMID:10775366

Mu, J., Wang, J., Jin, F., Wang, X., & Hong, H. (2014). Comparative embryotoxicity of phenanthrene and alkyl-phenanthrene to marine medaka (*Oryzias melastigma*). *Marine Pollution Bulletin*, *85*(2), 505–515. doi:10.1016/j.marpolbul.2014.01.040 PMID:24559736

Mu, L., Zhou, Q., Zhao, Y., Liu, X., & Hu, X. (2019). Graphene Oxide Quantum Dots Stimulate Indigenous Bacteria to Remove Oil Contamination. *Journal of Hazardous Materials*, *366*, 694–702. doi:10.1016/j.jhazmat.2018.12.044 PMID:30583239

Munoz, R., Guieysse, B., & Mattiasson, B. (2003). Phenanthrene biodegradation by an algal bacterial consortium in two-phase partitioning bioreactors. *Applied Microbiology and Biotechnology*, *61*(3), 261–267. doi:10.100700253-003-1231-9 PMID:12698286

Nam, T. H., Jeon, H. J., Mo, H., Cho, K., Ok, Y. S., & Lee, S. E. (2015). Determination of biomarkers for polycyclic aromatic hydrocarbons (PAHs) toxicity to earthworm (*Eisenia fetida*). *Environmental Geochemistry and Health*, *37*(6), 943–951. doi:10.100710653-015-9706-z PMID:25920560

Nogueira, D. J., Mattos, J. J., Dybas, P. R., Flores-Nunes, F., Sasaki, S. T., Taniguchi, S., ... Bainy, A. C. D. (2017). Effects of phenanthrene on early development of the Pacific oyster *Crassostrea gigas* (Thunberg, 1789). *Aquatic Toxicology (Amsterdam, Netherlands)*, *191*, 50–61. doi:10.1016/j.aquatox.2017.07.022 PMID:28800408

NORMAN-Network. (2016). *Emerging Substances*. Retrieved from http://www.norman-network.net/sites/default/files/files/Emerging_substances_list_Feb_16/NORMAN%20list_2016_FINAL.XLSX

Oleszczuk, P., Hale, S. E., Lehmann, J., & Cornelissen, G. (2012). Activated carbon and biochar amendments decrease pore-water concentrations of polycyclic aromatic hydrocarbons (PAHs) in sewage sludge. *Bioresource Technology*, *111*, 84–91. doi:10.1016/j.biortech.2012.02.030 PMID:22391590

Ouvrard, S., Leglize, P., & Morel, J. L. (2014). PAH phytoremediation: Rhizodegradation or rhizoattenuation? *International Journal of Phytoremediation*, *16*(1), 46–61. doi:10.1080/15226514.2012.759527 PMID:24912214

Overmans, S., Nordborg, M., Díaz-Rúa, R., Brinkman, D. L., Negri, A. P., & Agustí, S. (2018). Phototoxic effects of PAH and UVA exposure on molecular responses and developmental success in coral larvae. *Aquatic Toxicology (Amsterdam, Netherlands)*, *198*, 165–174. doi:10.1016/j.aquatox.2018.03.008 PMID:29550714

Pearlman, R. S., Yalkowsky, S. H., & Banerjee, S. (1984). Water solubilities of polynuclear aromatic and heteroaromatic compounds. *Journal of Physical and Chemical Reference Data*, *103*(2), 555–562. doi:10.1063/1.555712

Peng, R.-H., Fu, X.-Y., Zhao, W., Tian, Y.-S., Zhu, B., Han, H.-J., ... Yao, Q.-H. (2014). Phytoremediation of Phenanthrene by Transgenic Plants Transformed with a Naphthalene Dioxygenase System from *Pseudomonas*. *Environmental Science & Technology*, *48*(21), 12824–12832. doi:10.1021/es5015357 PMID:25299803

Perra, G., Pozo, K., Guerranti, C., Lazzeri, D., Volpi, V., Corsolini, S., & Focardi, S. (2011). Levels and spatial distribution of polycyclic aromatic hydrocarbons (PAHs) in superficial sediment from 15 Italian marine protected areas (MPA). *Marine Pollution Bulletin*, *62*(4), 874–877. doi:10.1016/j.marpolbul.2011.01.023 PMID:21310440

Piazza, R. S., Trevisan, R., Flores-Nunes, F., Toledo-Silva, G., Wendt, N., Mattos, J. J., ... Bainy, A. C. D. (2016). Exposure to phenanthrene and depuration: Changes on gene transcription, enzymatic activity and lipid peroxidation in gill of scallops *Nodipecten nodosus*. *Aquatic Toxicology (Amsterdam, Netherlands)*, *177*, 146–155. doi:10.1016/j.aquatox.2016.05.025 PMID:27286572

Porter, C. J. H., Trevaskis, N. L., & Charman, W. N. (2007). Lipids and lipid-based formulations: Optimizing the oral delivery of lipophilic drugs. *Nature Reviews. Drug Discovery*, *6*(3), 231–248. doi:10.1038/nrd2197 PMID:17330072

Prosser, C. M., Unger, M. A., & Vogelbein, W. K. (2011). Multistressor interactions in the zebrafish (*Danio rerio*): Concurrent phenanthrene exposure and *Mycobacterium marinum* infection. *Aquatic Toxicology (Amsterdam, Netherlands)*, *102*(3-4), 177–185. doi:10.1016/j.aquatox.2011.01.011 PMID:21356180

Pulford, I., & Watson, C. (2003). Phytoremediation of heavy metal-contaminated land by trees—A review. *Environment International*, *29*(4), 529–540. doi:10.1016/S0160-4120(02)00152-6 PMID:12705950

Qiao, M., Wang, C., Huang, S., Wang, D., & Wang, Z. (2006). Composition, sources, and potential toxicological significance of PAHs in the surface sediments of the Meiliang Bay, Taihu Lake, China. *Environment International*, *32*(1), 28–33. doi:10.1016/j.envint.2005.04.005 PMID:15996733

Rabodonirina, S., Rasolomampianina, R., Krier, F., Drider, D., Merhaby, D., Net, S., & Ouddane, B. (2019). Degradation of fluorene and phenanthrene in PAHs-contaminated soil using *Pseudomonas* and *Bacillus* strains isolated from oil spill sites. *Journal of Environmental Management*, *232*, 1–7. doi:10.1016/j.jenvman.2018.11.005 PMID:30453222

Raftery, T. D., Jayasundara, N., & Di Giulio, R. T. (2017). A bioenergetics assay for studying the effects of environmental stressors on mitochondrial function in vivo in zebrafish larvae. *Comparative Biochemistry and Physiology Part C: Toxicology & Pharmacology*, *192*, 23–32. doi:10.1016/j.cbpc.2016.12.001 PMID:27939721

Rao, M. A., Di Rauso Simeone, G., Scelza, R., & Conte, P. (2017). Biochar based remediation of water and soil contaminated by phenanthrene and pentachlorophenol. *Chemosphere*, *186*, 193–201. doi:10.1016/j.chemosphere.2017.07.125 PMID:28778017

Reddam, A., Mager, E. M., Grosell, M., & McDonald, M. D. (2017). The impact of acute PAH exposure on the toadfish glucocorticoid stress response. *Aquatic Toxicology (Amsterdam, Netherlands)*, *192*, 89–96. doi:10.1016/j.aquatox.2017.08.014 PMID:28942071

Reynoldson, T. B., & Metcalfe-Smith, J. L. (1992). An overview of the assessment of aquatic ecosystem health using benthic invertebrates. *Journal of Aquatic Ecosystem Health*, *1*(4), 295–308. doi:10.1007/BF00044171

Richardi, V. S., Vicentini, M., Morais, G. S., Rebechi, D., da Silva, T. A., Fávaro, L. F., & Navarro-Silva, M. A. (2018). Effects of phenanthrene on different levels of biological organization in larvae of the sediment-dwelling invertebrate *Chironomus sancticaroli* (Diptera:Chironomidae). *Environmental Pollution*, *242*, 277–287. doi:10.1016/j.envpol.2018.06.091 PMID:29990935

Ríos, Á., & Zougagh, M. (2016). Recent advances in magnetic nano materials for improving analytical processes. *Trends in Analytical Chemistry*, *84*, 72–83. doi:10.1016/j.trac.2016.03.001

Rodrigues, E. M., Morais, D. K., Pylro, V. S., Redmile-Gordon, M., de Oliveira, J. A., Roesch, L. F. W., ... Tótola, M. R. (2017). Aliphatic Hydrocarbon Enhances Phenanthrene Degradation by Autochthonous Prokaryotic Communities from a Pristine Seawater. *Microbial Ecology*, *75*(3), 688–700. doi:10.100700248-017-1078-8 PMID:28971238

Samanta, S. K., Chakraborti, A. K., & Jain, R. K. (1999). Degradation of phenanthrene by different bacteria: Evidence for novel transformation sequences involving the formation of 1-naphthol. *Applied Microbiology and Biotechnology*, *53*(1), 98–107. doi:10.1007002530051621 PMID:10645629

Schwarz, A., Adetutu, E. M., Juhasz, A. L., Aburto-Medina, A., Ball, A. S., & Shahsavari, E. (2018). Microbial Degradation of Phenanthrene in Pristine and Contaminated Sandy Soils. *Microbial Ecology*, *75*(4), 888–902. doi:10.100700248-017-1094-8 PMID:29080101

Shanker, U., Jassal, V., & Rani, M. (2017). Degradation of toxic PAHs in water and soil using potassium zinc hexacyanoferrate nanocubes. *Journal of Environmental Management*, *204*, 337–348. doi:10.1016/j.jenvman.2017.09.015 PMID:28910732

Shirmohammadi, M., Chupani, L., & Salamat, N. (2017b). Responses of immune organs after single-dose exposure to phenanthrene in yellowfin seabream (*Acanthopagrus latus*): CYP1A induction and oxidative stress. *Chemosphere*, *186*, 686–694. doi:10.1016/j.chemosphere.2017.08.051 PMID:28826128

Shirmohammadi, M., Salamat, N., Ronagh, M. T., Movahedinia, A., & Hamidian, G. (2017a). Assessment of immune status of yellowfin seabream (*Acanthopagrus latus*) during short term exposure to phenanthrene. *Comparative Biochemistry and Physiology Part C: Toxicology & Pharmacology*, *195*, 78–90. doi:10.1016/j.cbpc.2017.02.012 PMID:28257924

Slooff, W. (1983). Benthic macroinvertebrates and water quality assessment:Some toxicological considerations. *Aquatic Toxicology (Amsterdam, Netherlands)*, *4*(1), 73–82. doi:10.1016/0166-445X(83)90062-0

Sogbanmu, T., Osibona, A., Oguntunde, O., & Otitoloju, A. (2018). Biomarkers of toxicity in *Clarias gariepinus* exposed to sublethal concentrations of polycyclic aromatic hydrocarbons. *African Journal of Aquatic Science*, *43*(3), 1–12. doi:10.2989/16085914.2018.1491825

Sun, J., Ni, H., & Zeng, H. (2012). Ecological risk assessment of parent and halogenated polycyclic aromatic hydrocarbons in surface sediments from an urban river in south China. *Environmental Toxicology and Chemistry*, *31*(8), 1867–1873. doi:10.1002/etc.1890 PMID:22619089

Sun, L., Zuo, Z., Chen, M., Chen, Y., & Wang, C. (2015a). Reproductive and transgenerational toxicities of phenanthrene on female marine medaka (*Oryzias melastigma*). *Aquatic Toxicology (Amsterdam, Netherlands)*, *162*, 109–116. doi:10.1016/j.aquatox.2015.03.013 PMID:25805703

Sun, Y. Y., Xu, H. X., Li, J. H., Shi, X. Q., Wu, J. C., Ji, R., & Guo, H. Y. (2015b). Phytoremediation of soils contaminated with phenanthrene and cadmium by growing willow (Salix×Aureo-PendulaCL "j1011"). *International Journal of Phytoremediation*, *18*(2), 150–156. doi:10.1080/15226514.2015.1073668 PMID:26247604

Ting, W. T. E., Yuan, S. Y., Wu, S. D., & Chang, B. V. (2011). Biodegradation of phenanthrene and pyrene by *Ganoderma lucidum. International Biodeterioration & Biodegradation*, *65*(1), 238–242. doi:10.1016/j.ibiod.2010.11.007

Tirado-Torres, D., Gayosso-Canales, M., Marmolejo-Santillan, Y., Romo-Gomez, C., & Acevedo-Sandoval, O. (2016). Removal of polycyclic aromatic hydrocarbons by *Pleurotus ostreatus* sp. ATCC38540 in liquid medium. *Academia Journal of Scientific Research*, *4*(10), 376–379.

Torreiro-Melo, A. G. A. G., Silva, J. S., Bianchini, A., Zanardi-Lamardo, E., & de Carvalho, P. S. M. (2015). Bioconcentration of phenanthrene and metabolites in bile and behavioral alterations in the tropical estuarine guppy *Poecilia vivipara. Chemosphere*, *132*, 17–23. doi:10.1016/j.chemosphere.2014.12.079 PMID:25770832

Tsibart, A., Gennadiev, A., Koshovskii, T., & Watts, A. (2014). Polycyclic aromatic hydrocarbons in post-fire soils of drained peatlands in western Meshchera (Moscow region, Russia). *Solid Earth*, *5*(2), 1305–1317. doi:10.5194e-5-1305-2014

USEPA (United States Environmental Protection Agency). (1987). Health Effects Assessment for Dimethylphenols. EPA/600.8-88/031. Environmental Criteria Assessment Office.

USEPA (United States Environmental Protection Agency). (1988). *Interim sediment criteria values for nonpolar hydrophobic organic contaminants*. Washington, DC: Criteria and Standards Division.

Van Houten, B., Santa-Gonzalez, G. A., & Camargo, M. (2018). DNA repair after oxidative stress: Current challenges. *Current Opinion in Toxicology*, *7*, 9–16. doi:10.1016/j.cotox.2017.10.009 PMID:29159324

Varanasi, U., Stein, J. E., & Nishimoto, M. (1989). Biotransformation and disposition of PAH in fish. In U. Varanasi (Ed.), *Metabolism of Polycyclic Aromatic Hydrocarbons in the Aquatic Environment* (pp. 93–151). Boca Raton, FL: CRC Press.

Vergauwen, L., Schmidt, S. N., Stinckens, E., Maho, W., Blust, R., Mayer, P., ... Knapen, D. (2015). A high throughput passive dosing format for the Fish Embryo Acute Toxicity test. *Chemosphere*, *139*, 9–17. doi:10.1016/j.chemosphere.2015.05.041 PMID:26026258

Wang, L., Hua, X., Zhang, L., Song, N., Dong, D., & Guo, Z. (2019). Influence of organic carbon fractions of freshwater biofilms on the sorption for phenanthrene and ofloxacin: The important role of aliphatic carbons. *The Science of the Total Environment*, *685*, 818–826. doi:10.1016/j.scitotenv.2019.06.203 PMID:31238285

Wassenberg, D. M., & Di Giulio, R. T. (2004). Synergistic Embryotoxicity of Polycyclic Aromatic Hydrocarbon Aryl Hydrocarbon Receptor Agonists with Cytochrome P4501A Inhibitors in *Fundulus heteroclitus. Environmental Health Perspectives*, *112*(17), 1658–1664. doi:10.1289/ehp.7168 PMID:15579409

Wen, S., Zhao, J., Sheng, G., Fu, J., & Peng, P. (2003). Photocatalytic reactions of pyrene at TiO2/water interfaces. *Chemosphere*, *50*(1), 111–119. doi:10.1016/S0045-6535(02)00420-4 PMID:12656236

WHO. (1983). International Agency for Research on Cancer (IARC) Monographs on the Evaluation of Carcinogenic Risks to Humans: Polynuclear Aromatic Compounds, Part 1, Chemical, Environmental and Experimental Data. IARC.

Wild, S. R., & Jones, K. C. (1995). Polynuclear aromatic hydrocarbons in the United Kingdom environment: A preliminary source inventory and budget. *Environmental Pollution*, *88*(1), 91–108. doi:10.1016/0269-7491(95)91052-M PMID:15091573

Willis, A. M., & Oris, J. T. (2014). Acute photo-induced toxicity and toxicokinetics of single compounds and mixtures of polycyclic aromatic hydrocarbons in zebrafish. *Environmental Toxicology and Chemistry*, *33*(9), 2028–2037. doi:10.1002/etc.2648 PMID:24889326

Woo, O. T., Chung, W. K., Wong, K. H., Chow, A. T., & Wong, P. K. (2009). Photocatalytic oxidation of polycyclic aromatic hydrocarbons: Intermediates identification and toxicity testing. *Journal of Hazardous Materials*, *168*(2-3), 1192–1199. doi:10.1016/j.jhazmat.2009.02.170 PMID:19361920

Wu, J. Y., Yan, Z. G., Liu, Z. T., Liu, J., Liang, F., Wang, X. N., & Wang, W. L. (2015). Development of water quality criteria for phenanthrene and comparison of the sensitivity between native and non-native species. *Environmental Pollution*, *196*, 141–146. doi:10.1016/j.envpol.2014.09.024 PMID:25463707

Wu, M., Xu, Y., Ding, W., Li, Y., & Xu, H. (2016). Mycoremediation of manganese and phenanthrene by *Pleurotus eryngii mycelium* enhanced by Tween 80 and saponin. *Applied Microbiology and Biotechnology*, *100*(16), 7249–7261. doi:10.100700253-016-7551-3 PMID:27102128

Yang, X., Chen, Z., Wu, Q., & Xu, M. (2018). Enhanced phenanthrene degradation in river sediments using a combination of biochar and nitrate. *The Science of the Total Environment*, *619-620*, 600–605. doi:10.1016/j.scitotenv.2017.11.130 PMID:29156278

Zelinkova, Z., & Wenzl, T. (2015). The Occurrence of 16 EPA PAHs in Food – A Review. *Polycyclic Aromatic Compounds*, *35*(2-4), 248–284. doi:10.1080/10406638.2014.918550 PMID:26681897

Zhang, C., Lu, J., & Wu, J. (2019). Adsorptive removal of polycyclic aromatic hydrocarbons by detritus of green tide algae deposited in coastal sediment. *The Science of the Total Environment*, *670*, 320–327. doi:10.1016/j.scitotenv.2019.03.296 PMID:30904645

Zhang, X., Xia, X., Dong, J., Bao, Y., & Li, H. (2014). Enhancement of toxic effects of phenanthrene to *Daphnia magna* due to the presence of suspended sediment. *Chemosphere*, *104*, 162–169. doi:10.1016/j.chemosphere.2013.10.086 PMID:24275150

Zhang, Y., Huang, L., Zuo, Z., Chen, Y., Wang, C. (2013). Phenanthrene exposure causes cardiac arrhythmia in embryonic zebrafish via perturbing calcium handling. *Aquatic Toxicology, 142– 143*, 26– 32.

Zhao, Y., Wang, X., Lin, X., Zhao, S., & Lin, J. (2016). Comparative developmental toxicity of eight typical organic pollutants to red sea bream (*Pagrosomus major*) embryos and larvae. *Environmental Science and Pollution Research International*, *24*(10), 9067–9078. doi:10.100711356-016-6282-4 PMID:27130341

ADDITIONAL READING

Abdel-Shafy, H. I., & Mansour, M. S. M. (2016). A review on polycyclic aromatic hydrocarbons: Source, environmental impact, the effect on human health and remediation. *Egyptian Journal of Petroleum*, *25*(1), 107–123. doi:10.1016/j.ejpe.2015.03.011

ATSDR. (1995). *Toxicological profile for otto fuels II, U.S. Department of Health and Human Services, Atlanta (GA), Agency for Toxic Substances and Disease Registry (ATSDR). Toxicological profile for polycyclic aromatic hydrocarbons*. Atlanta, GA: U.S. Department of Health and Human Services, Public Health Service.

Fasnacht, M. P., & Blough, N. V. (2003). Mechanisms of the Aqueous Photodegradation of Polycyclic Aromatic Hydrocarbons. *Environmental Science & Technology*, *37*(24), 5767–5772. doi:10.1021/es034389c PMID:14717193

IARC (1983). IARC Monographs on the Evaluation of Carcinogenic Risks to Humans. Polynuclear aromatic compounds, Part 1, chemical, environmental and experimental data. 32, 1–453.

IARC (2010). IARC Monographs on the Evaluation of Carcinogenic Risks to Humans. Some non-heterocyclic polycyclic aromatic hydrocarbons and some related exposures, 92, 1–853.

Torreiro-Melo, A. G. A. G., Silva, J. S., Bianchini, A., Zanardi-Lamardo, E., & de Carvalho, P. S. M. (2015). Bioconcentration of phenanthrene and metabolites in bile and behavioral alterations in the tropical estuarine guppy *Poecilia vivipara. Chemosphere*, *132*, 17–23. doi:10.1016/j.chemosphere.2014.12.079 PMID:25770832

USEPA (United States Environmental Protection Agency). (1987). Health Effects Assessment for Dimethylphenols. EPA/600.8-88/031. Environmental Criteria Assessment Office, Cincinnati, OH.

USEPA (United States Environmental Protection Agency). (1988). *Interim sediment criteria values for nonpolar hydrophobic organic contaminants*. Washington, District of Columbia: Criteria and Standards Division.

KEY TERMS AND DEFINITIONS

DNA Damage: It is an alteration or modification in DNA structure that changes its coding properties and affects its normal functioning during transcription or replication.

LC_{50}: LC_{50} stands for the "50% Lethal concentration." It is the measure of toxicity which is defined as median lethal concentration at which 50% of the test organisms died during the observation period.

Oxidative Stress: It is the state of oxidative damage when the production of free radicals overwhelms the power of antioxidants in the body.

Toxicity: It is the specific degree of a substance or a mixer of substances to being toxic or poisonous to an organism.

Chapter 5
Analytical Perspectives on Emerging Organic Contaminants in the Aquatic Ecosystem

Hossein Miraji
The University of Dodoma, Tanzania

Othman Chande Othman
University of Dar es Salaam, Tanzania

Faustin Ngassapa
University of Dar es Salaam, Tanzania

Mureithi Eunice
University of Dar es Salaam, Tanzania

ABSTRACT

This chapter introduces readers to the background of emerging contaminants by defining emerging contaminants and telling their history and their corresponding effects. It describes the dynamic properties of emerging contaminants such as advection and dispersion, chemistry, and their reactivity behavior. Lastly, it tells the analytical methodologies on sample preparation such as solid phase extraction and solid-phase micro-extraction, detection and quantification of organic ECs, and it proposes future perspectives of emerging contaminants.

DOI: 10.4018/978-1-7998-1871-7.ch005

BACKGROUND OF EMERGING CONTAMINANTS

Definition

Emerging contaminants (ECs) represent a group of synthetic, natural and some microbial contaminants either naturally occurring or introduced in the environmental matrices, profoundly in the water bodies (Richardson & Ternes, 2017) and little in the atmosphere (Barroso et al., 2019). Several scholars have defined ECs with different watchwords, yet the meaning remains intact (US EPA, 2018). Some essential phases defining ECs are presented in Figure 1. These include; emerging contaminants being chemical contaminants that had not previously detected, or emerging contaminants are contaminants suspected to cause human health risks but not yet fully understood (Richardson, Exposure, & Agency, 2006).

Emerging contaminants are also defined as contaminants that cannot be removed by conventional methods for water and wastewater treatment (Klamerth et al., 2010). In addition, emerging contaminants are contaminants that are continuously detected in the aquatic systems at trace levels thus requiring modern technology for their detection (Pal et al., 2014). Emerging contaminants have also been defined as a group of contaminants lacking monitoring regulations. Being the case, ECs have recently gained scientific attention since they are the least globally investigated contaminants, they are suspected to causes ecological risks and they do not fall under routine laboratory analyses (Miraji et al., 2018). Despite being neglected contaminants, their discovery goes back to 1960s.

The History of Emerging Contaminants

Major chemical and industrial revolutions are reflected back to the outcome of World War II. In order to meet economical demands, organic chemicals were synthesized in a manner that they can last longer meanwhile maintain their efficacy. The main purpose of pesticides were combating against harmful pests while their long term environmental effects were yet considered. As a result, these chemicals were washed out into the surface and groundwater bodies where they accumulated in the soils, sediments, water, aquatic and terrestrial organisms. Thus, in 1960 Hueper associated these organic chemicals with human cancerous risks (National Academy of Sciences, 1999). In 1962, Rachel Carson wrote a book titled Silent Spring that awakens U.S public awareness and the interest for combating against disadvantaged organic chemicals such as DDT (Carson, 1962). More than hundreds of synthetic organic chemical traces that were observed in the rivers in the 1970s were alleged being mutagenic, carcinogenic and teratogenic.

Figure 1. Definitions of Emerging Contaminants

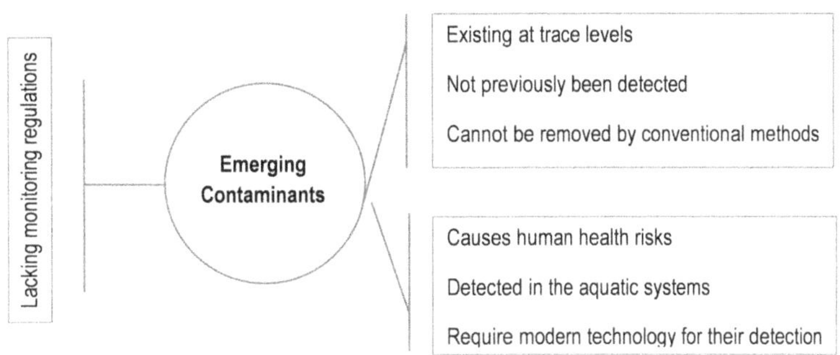

The 1974 US safe drinking water act (SDWA) and the 1976 regulation for six widely used pesticides were implemented, yet the synergic effect and in-situ generation of water disinfection byproducts were not anticipated. Again, the 1979 SDWA accounted for and listed trihalomethanes as an organic chemical need regulations. More inclusion of these chemical contaminants under US EPA regulations was achieved by the 1986 and 1996 SDWA (Raucher, 1996).

The historical trends of ECs presented in the Figure 2 depicts that, after the 1960s public awareness on the effects of chemical contaminants, US EPA became the frontier in fighting, monitoring and regulating chemicals with public health risks. US EPA then banned the use of DDT in 1972 (Roberts, 2006) and aldrin (Schafer & Kegley, 2002) due to their environmental persistence and public health effects. During this period, the US EPA started shortlisting contaminants without environmental monitoring regulations, which were termed as contaminant candidate list. The purposes of this list were to enable global contribution and documentation of names of all known chemical contaminants without monitoring regulations so that they are given special attention. Since 1998, the list is periodically updated in order to include any new contaminant requiring public attention (EPA, 1998, 2005, US EPA, 2009, 2015; 2018). A contextual ground of continuous publication of contaminant candidate list is upon inclusion of new generated emerging contaminants. Unattended and delayed responses on the environmental existence of emerging contaminants threaten the present and future ecological sustainability. The effects of ECs occur once an organism is expoisure to them.

Effects of Emerging Contaminants

The current contaminant candidate list of emerging contaminants comprises of 100 chemicals and 12 microbial contaminants (US EPA, 2015a). These contaminants have a wide range of classes such as pharmaceuticals, cosmetics, water disinfection byproducts, endocrine disrupting compounds, pesticides, surfactants, algal toxins and enteroviruses (Jose et al., 2012). Pharmaceuticals are chemical compounds that at very small quantity affects and alter the physiology of exposed organisms (Gogoi et al., 2017). Some human and animal drugs are persistence in the environment and are not removable by conventional wastewater treatment processes. Continuous exposure of microbes in the drug residuals alters their genetic makeup resulting in the resistance genes against antibiotic treatment (Wang et al., 2015). Six antibiotic

Figure 2. Historical Trends of Emerging Contaminants

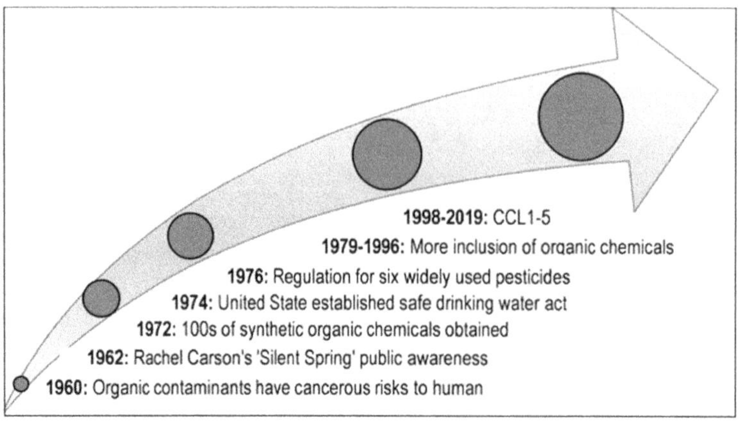

resistance genes against amoxicillin were reported by Tehrani, which presents a real ecological threat like untreatable diseases and suppression of human body immunity (Tahrani et al., 2015).

Cosmetics such as sunscreens, beauty creams, skin lotion, hair sprays, dyes and shampoo are examples of body care products. They contain fatal chemical additives such as octyl dimethyl-p-aminobenzoic acid, ethylhexyl meth-oxycinnamate, octyl methoxycinnamate, benzophenone-3, octocrylene, butyl methoxydibenzoylmethane, terephthalylidine dicamphor sulfonic acid, ethylhexyl triazone, phenylbenzimidazolesulfonic acid, ethylhexyl salicylate, 4-methylbenzylidene camphor, and 1-(4-tert-butylphenyl)-3-(4-methoxyphenyl)-1,3-propanedione (Richardson, 2006). These chemicals find their ways to the environment directly from swimming pools, washing and bathing in the water bodies or indirectly from wastewater treatment facilities. Triclosan and triclocarban are antimicrobial when applied on the skin, their accumulation in the environment results in toxicity effects of non-target organisms. Personal care products contents are suspected to have endocrine disruption property (Peng et al., 2016) and have the estrogenic property (Gogoi et al., 2017).

Endocrine-disrupting compounds are human synthetic chemicals available in the pesticides, food additives, metal additives and in human cosmetics. They naturally mimic hormones such as androgens, estrogens, thyroid hormones and production of overstimulation. They can get into the human body through ingestion of food, skin adsorption, pregnancy and breast-feeding. EDC are fatal as they cause cancerous tumor, birth defects, developmental disorders, interferes the endocrine system of fishes and wildlife, metabolism, reproduction, immunocompetence and general behavior (Iwanowicz et al., 2019), and hormonal-dependency malignancy such as prostate cancer (Hu et al., 2016).

Algal toxins effects result from a waste chemical released by algae associated with bloom, which cannot simply removed by conventional water treatment techniques. Algal toxins are neural toxins, causes liver diseases and strongly associated with gastrointestinal infections (Committee Report, 1999). Enteroviruses are of picornaviridae family among the smallest RNA viruses with about 71 different types. They are transmitted orally, respiration or through fecal contaminations resulting in effects such as gastroenteritis outbreaks and hepatitis E (Committee Report, 1999).

Water disinfection byproducts (WDB) are synthetic chemical useful in rocket fuel, antioxidant and additive for softeners, which may get in the environment through pesticides applications, industrial discharges and wastewater treatment plants. WDB such as nitrosodimethylamine (NDMA) is naturally formed when traces chlorine used for water treatment reacts with natural humic acids in the water to form WDB. These chemicals are family of potent carcinogenic causative agents of tumors in the liver, kidney, respiratory track and blood vessels in the mice and rats (US EPA, 2014).

SPATIAL OCCURRENCE OF EMERGING CONTAMINANTS

Emerging contaminants originate from the human, animal, agricultural, clinical, military, municipal and domestic consumables. All sources of ECs and practices are on the terrestrial however, domestics washouts, latrine leachates, out-of-use drugs, body excrete and wastewater treatment plants channels them to the water bodies. As they are mostly lipophilic they adsorb on the surface of sediments, thus at some points, sediments become sinks. Simply, ECs are in the slinging equilibrium between terrestrial and marine environment. Distribution of organic emerging contaminants in the terrestrial environment is the least investigated area of ECs compared to marine pollution (Darnerud et al., 2001; Headley & McMartin, 2004; Styszko et al., 2015).

Mobility of ECs in the Aquatic Environment

The movement of soluble and suspended materials from uphill towards lower lands is a natural phenomenon that requires external forces. These forces may be physical movements such as land and glacier slides, transportation of sands, sludge and distribution of water from treatment plants to consumers.

They may also be (Chakraborty & Kumar, 2016), land washing, domestic and latrine wastewaters (Pastore, 2015), municipal and clinical drainages, which are all channeled into streams and river whose destinations are lakes, oceans, seas and river basins. At some points, saturated sediments become stationary sources of these contaminants (Thomas, Dye, Schlabach, & Langford, 2007). Organic matters such as grasses and suspended planktons absorb and move with the moving water as well (Subedi & Kannan, 2014).

Figure 3 summarizes the transport behavior of chemical contaminant in the water. Once ECs in the water bodies they can be adsorbed on the sediment and suspended matters, remain suspended in the water, leach into the ground, physically removed though pumping out water, evaporated or keep on mobility and dilution to the infinity. Phytoplankton ecosystem is directly affected by the ECs cycle since feeding on the least member in the food chain to the higher ranks is a complex web that get into the human being, while excretion and dead individuals return the circle to the starting point again (Jaiswal, Kumar, & Yadav, 2011).

Chemistry of Some Organic Emerging Contaminants

The general overview of ECs might be thought of to be a single contaminant or group of contaminants with similar properties. In fact, their chemistry, intended uses, mode of action and toxicity level broadly differ. As an environmental contaminant, solubility, reactivity, and persistence of selected classes are of interest.

Figure 3. Transport Behavior of Emerging Contaminants in the Water

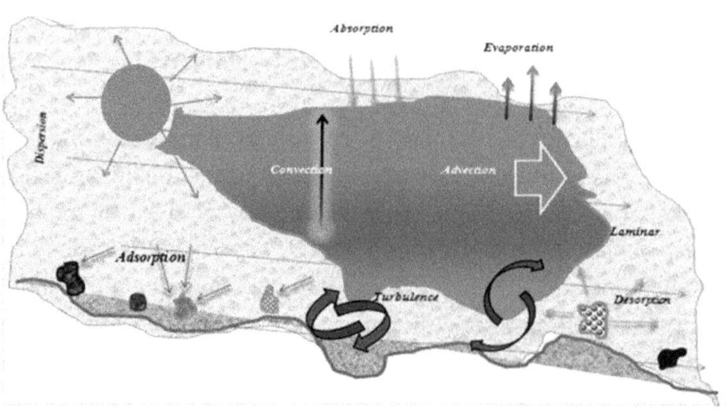

Solubility

The insight to ECs is a broad spectrum; there are inorganic contaminants such as Co, ClO_3^-, Ge, Mn, Te and molybdenum, and organics such as formaldehyde and NDMA, which are water and other inorganic solvents soluble (Beita-Sandí, Ersan, Uzun, & Karanfil, 2016; Chemical Book, 2016; Golfinopoulos, Nikolaou, & Lekkas, 2003). Contrarily, industrial solvents such as aniline, herbicides such as acrolein, insecticide such as disulfoton, estrogenic hormones such as equilenin, veterinary and human pharmaceuticals such as estrone and nonylphenol used in the lubricating oil additives, resins, plasticizers and antioxidants for plastic and rubber are organic chemicals thus they are hydrophilic in nature (US EPA, 2015b). Usually, the survival of microbes relies on the physical and chemical characteristics of the media rather than solubility rules. Despite the hydrophobicity and lipophilicity of these contaminants, water which is inorganic universal solvent is very abundant, buffer-like solvent and responsible for most natural reactions on the planet. In this case, whether soluble or insoluble, yet ECs prevails in the water, but mostly bioaccumulated (Huerta et al., 2015; Yan-long et al., 2016) in the biological systems due to their lipophilic nature and therefore undergoes bioconcentration (H. Hu et al., 2016).

Reaction and Persistence

Environmental reactions of ECs are the least reported, nevertheless, two processes are anticipated, where contaminants may react with the existing complex matrices or metabolism process (Stuart, Lapworth, Crane, & Hart, 2012). Metals in the environment never exist as a free pure element; rather they may form stable oxides, salts and complexes that apply to chlorate ions too.

Some ECs such as perfluorinated surfactants, molinate and human drugs such as erythromycin and tetracycline have a stable, unreactive and polar chemical structure making them merely inert when in the environment.

Persistence means the half-life of a chemical is more than 2 days in the air, 182 days in the water, 365 days in the sediments and 182 days in the soil. Thus, having chemically inert contaminants means they will reside in the respective environmental matrix for a long period. Once this happens, continuous accumulation occurs, in turn, threatens the ecosystem due to probable toxicity that might happen. Yet, the toxicity of contaminants varies among organisms based on the type and root of contaminant (Darnerud et al., 2001), dosage (Zoeller & Vandenberg, 2015) and the overall weight of the exposed organisms. The smaller the body, the more it is exposed thus the severe the effect and vice versa.

Figure 4.

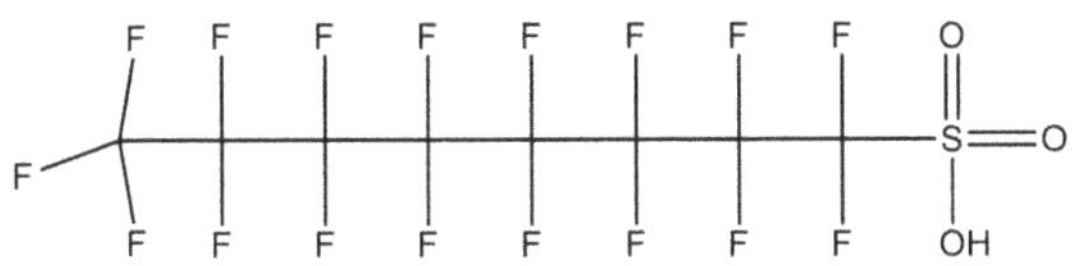

Perfluorooctane Sulfonic acid

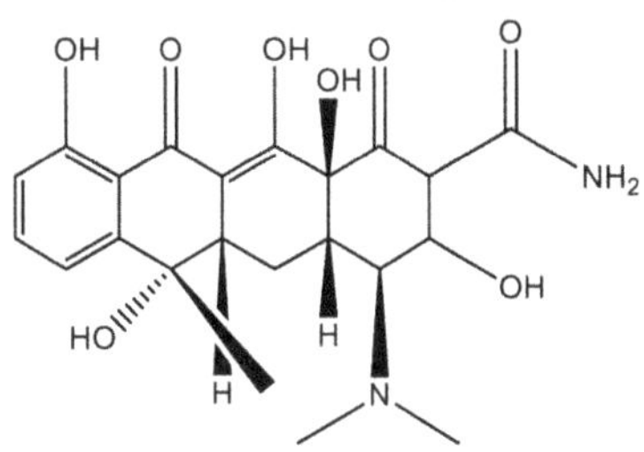

Tetracycline

ANALYTICAL APPROACHES ON EMERGING ORGANIC CONTAMINANTS

Analytical intervention on the existing of ECs involves diagnosis on the existence, sampling (Mills, 2015), sample handling, analysis and interpretation of the results. A diagnostic tool is a primary knowledge on suspecting probability of existing of these contaminants based on the theoretical and practical experiences. The suspected environment must have some important indicators of the existence of ECs such WWSP, municipal, domestic and clinical wastes and uncontrolled damping are inclusive in the decision (Bound & Voulvoulis, 2005; Brown, de Banate, & Rother, 2010; Tran, Hu, Li, & Ong, 2014).

Technological Developments in Sample Preparation

Liquid-liquid partition is also known as solvent extraction is a separation technique involving solubility principles. Two immiscible solvents especially water and organic solvents are mixed together preferably in the separation funnel to allow the dissolution of aqueous soluble constituents into the water phase and dissolution of organic soluble constituents in the organic phase. All hydrophilic chemical constituents such as inorganic salts remain in the lower aqueous phase while hydrophobic constituents of the mixtures such as organic ECs reside in the upper organic phase. The main purpose of solvent extraction is to separate in most cases organic constituents from a matrix, purify and bring required analytes into a measurable state. In particular, it is a simple technique in terms of training and adaptation, although not environmental friend resulting excessive use of organic solvents. SPE techniques are sensitive especially on the number of chemicals absorbed since once saturation is reached it can no longer absorb (Golfinopoulos et al., 2003; Iwanowicz et al., 2019; Zhang et al., 2019).

Supercritical fluid extraction (SFE) is the extraction technique of one component from another by using supercritical fluids as the extraction solvent. A supercritical fluid is a fluid between the typical gas and liquid state. However, this technique is not applicable in the case of emerging contaminants as they can be destroyed under heat (Haggarty & Burgess, 2017).

Solid-phase extraction (SPE) is a sample preparatory technique involving interaction between solid stationery against a mobile liquid phase. During this physical and chemical interactions, constituents of the mobile phase anchor to the stationary phase where they can be later removed via elution. This approach may be normal or reverse phase depends on the relationship between chemical composition of the stationary and constituent of interest in the mobile phase. This technique isolates cleanup and concentrates analytes of interest before qualitative or quantitative analysis (Agüera et al., 2005).

Solid-phase micro-extraction (SPME) is one of SPE technique using a fiber coated with extraction polymer (liquid) or sorbents (solid) such as mixture of polyacrylate, polyethylene glycol, divinylbenzene, carboxy, and polydimethylsiloxane that can extract volatile and non-volatile chemical materials from fluid phase. After completion of extraction, the SPME fiber is placed in the separation column of the GC/MS or appropriate instrument for desorption followed by analysis. It is quick, simple, and effective up to parts per trillion and it is solvent-free technique (Orata, 2012). A semipermeable membrane device (SPMD) is a type of passive sampling device used for extraction of organic chemicals from aquatic environments. Passive sampling is an online technique involves application of specific designed device for selective accumulation of environmental contaminants over a certain period. SPMD relies on the kinetic and potential effects of water over time, does not need sample grabbing and has been effective for ECs analysis (Curran, Rabin, Prada, & Furton, 2005).

Syringe and disc-type hydrophilic-lipophilic balance (HLB) cartridges are SPE forms of extraction technique constitute of styrene and divinyl copolymers designed for reverse phase extraction of analytes, which work under broad range of pH. They require preconditioning, loading of water sample for extraction of organic contaminants, washing of extracts, and then followed by elution with an organic solvent. It is an effective technique, has high recovery and mostly adapted compared to other extraction techniques (Agüera et al., 2005; Weigel, Kallenborn, & Hühnerfuss, 2004).

Detection and Quantification of Organic ECs

Extraction of merging contaminants is an important process for separation before qualitative and quantitative analysis. Analysis of cleanly extracted analytes is achieved by using GC, GC/MS, GC/MS/MS, HPLC, HPLC/MS or HPLC/MS/MS. Either of the instrument of choice, separation through the interaction between a stationary phase and mobile phase containing extracted ECs before analysis is inevitable (Lidia et al., 2015; H. Zhang, Bayen, & Kelly, 2015).

Separation of these components in the gas chromatography is achieved through stainless steel or glass packed columns or capillary columns whose inner lining consist of polar (polyethyleneglycol) or non-polar (squalene) stationary phase. Polarity affinities between volatile vaporized sample from the injection port with stationary phase in the column retain organic compounds at different strength that through carrier gas (Helium) each isolated component emerge in the detector at different time known as retention time. A detector gives qualitative and/or quantitative data depends on the need. A GC may be coupled with two columns in order to enhance separation and resolution power, making it a multi-dimensional gas chromatography (2D GC). Gas chromatography may be coupled with single mass spectrometry detector (GC/MS) for determination of molecular mass of a molecule. Two mass detector (GC/MS/MS) also called Gas Chromatography-Tandem Mass Spectrometry are designed with high sensitivity EI source, ultimate sensitivity, selectivity and with large mass range up to 1050 m/z for enhancement of analysis (Unger et al., 2010).

High-Performance Liquid Chromatography (HPLC) employs a system of pumps pressurizing liquid solvent that carries a mixture sample to the separation column packed with solid adsorbents. Polarity difference between constituents of the sample mixture and the stationary phase results in the retardation of some chemical species leading to the separation of each fraction. Each fraction is detected and quantified at different retention times in the detector. Similar to GC, HPLC can be coupled with the second column (2D model), single mass analyzer (HPLC/MS) or even two mass analyzers (HPLC/MS/MS) for improved performance. HPLC is more preferred than GC technique since polarity and volatility effects of ECs have least affects to HPLC (Klamerth et al., 2010).

Derivatization is a chemical process involving the transformation of chemical species with unrequired properties to a new product with improved properties that can be analyzed by a GC/MS. Polar chemical species such as carbonyls, mines, thiols, phosphates and hydroxyl groups in this reaction are converted to non-polar esters and others in order to make them volatile. Derivatization increase volatility, improve selectivity, efficiency, stability and enhance detectability of analytes. For example, silylation with N-(tert-butyldimethylsilyl)-N-methyltrifluoroacetamide (MTBSTFA) is performed before analysis (Aznar et at al., 2014; Weigel et al., 2004).

Silylation is the prevalent type of derivatization used in GC-MS analysis for hydroxyl and other compounds. Dry residues of wastewaters are dissolved in 50 μL of MTBSTFA. Samples are then heated at 65° C for 40 min using digital dry bath thermo-block Labnet and subsequently analyzed in GC-MS system

(Nosek et al., 2014), (Aznar et al., 2014), (Aznar et al., 2016). Using (MTBSTFA + 1% t-BDMCS) i.e. addition of t-BDMCS catalyzes reactions of hindered alcohols and amines. Silylation reduces polarity and hydrogen bonding effects of compounds (Orata, 2012). The alternative derivatizing agent is MSTFA where derivatization is done by heating the extracts with 100 µl N-methyl-N-trimethylsilyltrifl uoro-acetamide (MSTFA) for 20 min at 60 °C, then followed by GC-MS analysis (Macherey-Nagel, 2012).

FUTURE PERSPECTIVES OF EMERGING CONTAMINANTS

The world population is continuously increasing, but the number of existing natural resources is depleting while the industrial sector grows and therefore the increase in environmental degradation through pollution. Continuous production and disposal of the existing chemicals and yet to be known chemicals may have synergic effect on the environment to product unknown chemical products with unknown effects. Global occurrence of ECs is no longer a deniable phenomenon that regulatory and responsible institutions have to come out with regulations that will accommodate both ECs and other pollutants that may be anticipated to occur in the environment. As ECs are the global issue, there are essential resources such as availability analytical manuals, standards, and portable instrumentation could be given priority in terms of quick availability and reduced prices as support to researchers on addressing ECs. The availability of universal columns implanted or with plugin system of reversible derivatizing materials could facilitate the process un-doughtily.

REFERENCES

Agüera, A., Perez Estrada, L. A., Ferrer, I., Thurman, E. M., Malato, S., & Fernandez-Alba, R. (2005). Application of time-of-flight mass spectrometry to the analysis of phototransformation products of diclofenac in water under natural sunlight. *Journal of Mass Spectrometry*, *40*(7), 908–915. doi:10.1002/jms.867 PMID:15934037

Aznar, R., Sánchez-Brunete, C., Albero, B., Rodríguez, J. A., & Tadeo, J. L. (2014). Occurrence and analysis of selected pharmaceutical compounds in soil from Spanish agricultural fields. *Environmental Science and Pollution Research International*, *21*(6), 4772–4782. doi:10.100711356-013-2438-7 PMID:24363053

Barroso, P. J., Santos, J. L., Martín, J., Aparicio, I., & Alonso, E. (2019). Emerging contaminants in the atmosphere: Analysis, occurrence and future challenges. *Critical Reviews in Environmental Science and Technology*, *49*(2), 104–171. doi:10.1080/10643389.2018.1540761

Beita-Sandí, W., Ersan, M. S., Uzun, H., & Karanfil, T. (2016). Removal of N-nitrosodimethylamine precursors with powdered activated carbon adsorption. *Water Research*, *88*, 711–718. doi:10.1016/j.watres.2015.10.062 PMID:26584342

Book, C. (2016). *Germanium CAS#_ 7440-56-4*. Retrieved from http://www.chemicalbook.com/ProductChemicalPropertiesCB7733835_EN.htm#MSDSA

Bound, J. P., & Voulvoulis, N. (2005). Household disposal of pharmaceuticals as a pathway for aquatic contamination in the United Kingdom. *Environmental Health Perspectives*, *113*(12), 1705–1711. doi:10.1289/ehp.8315 PMID:16330351

Brown, R. J., de Banate, M. A., & Rother, K. I. (2010). Artificial sweeteners: a systematic review of metabolic effects in youth. *International Journal of Pediatric Obesity, 5*, 305–312.

Carson, R. (1962). *Silent Spring* (1st ed.). Houghton Mifflin Company.

Chakraborty, S., & Kumar, R. N. (2016). Assessment of groundwater quality at MSW landfill site using standard and AHP based water quality index : A case study from Ranchi, Jharkhand, India. *Environmental Monitoring and Assessment*, *188*(335), 18. PMID:27155859

Committee Report. (1999). Emerging Pathogens: Viruses, protozoa and Algal Toxins. *AWWA, 91*(9), 110–121.

Curran, A. M., Rabin, S. I., Prada, P. A., & Furton, K. G. (2005). Comparison of the Volatile Organic Compounds Present in Human Odor Using SPME-GC/MS. *Journal of Chemical Ecology*, *31*(7), 1607–1619. doi:10.100710886-005-5801-4 PMID:16222796

Darnerud, P. O., Eriksen, G. S., Jóhannesson, T., Larsen, P. B., & Viluksela, M. (2001). Polybrominated diphenyl ethers: occurrence, dietary exposure, and toxicology. *Environmental Health Perspectives, 109*, 49–68.

Gogoi, A., Mazumder, P., Tyagi, V. K., Chaminda, G. G. T., An, A. K., & Kumar, M. (2017). Occurrence and Fate of Emerging Contaminants in Water Environment: A Review. *Groundwater for Sustainable Development*, 37.

Golfinopoulos, S. K., Nikolaou, A. D., & Lekkas, T. D. (2003). The occurrence of disinfection by-products in the drinking water of Athens, Greece. *Environmental Science and Pollution Research International*, *10*(6), 368–372. doi:10.1065/espr2002.06.124 PMID:14690026

Haggarty, J., & Burgess, K. E. (2017). Recent advances in liquid and gas chromatography methodology for extending coverage of the metabolome. *Current Opinion in Biotechnology*, *43*, 77–85. doi:10.1016/j.copbio.2016.09.006 PMID:27771607

Headley, J. V., & McMartin, D. W. (2004). A review of the occurrence and fate of naphthenic acids in aquatic environments. *Journal of Environmental Science and Health. Part A, Toxic/Hazardous Substances & Environmental Engineering*, *39*(8), 1989–2010. doi:10.1081/ESE-120039370 PMID:15332664

Hu, D., Hu, W., Xie, L., Li, Y., Birch, L., & Prins, G. S. (2016). Actions of Estrogenic Endocrine Disrupting Chemicals on Human Prostate Stem/Progenitor Cells and Prostate Carcinogenesis. *The Open Biotechnology Journal*, *10*(77), 76–97. doi:10.2174/1874070701610010076

Hu, H., Deng, Y., Fan, Y., Zhang, P., Sun, H., Gan, Z., & Yao, Y. (2016). Effects of artificial sweeteners on metal bioconcentration and toxicity on a green algae Scenedesmus obliquus. *Chemosphere*, *150*, 285–293. doi:10.1016/j.chemosphere.2016.02.043 PMID:26915590

Huerta, B., Rodriguez-mozaz, S., Nannou, C., Nakis, L., Ruhí, A., Acuña, V., & Barcelo, D. (2015). Determination of a broad spectrum of pharmaceuticals and endocrine disruptors in the biofilm from a wastewater treatment plant-impacted river. *The Science of the Total Environment*. PMID:26087856

Iwanowicz, L. R., Pinkney, A. E., Guy, C. P., Major, A. M., Munney, K., Blazer, V. S., ... Smith, D. R. (2019). Temporal evaluation of estrogenic endocrine disruption markers in smallmouth bass (Micropterus dolomieu) reveals seasonal variability in intersex. *The Science of the Total Environment*, *646*, 245–256. doi:10.1016/j.scitotenv.2018.07.167 PMID:30055487

Jaiswal, D. K., Kumar, A., & Yadav, R. R. (2011). Analytical Solution to the One-Dimensional Advection-Diffusion Equation with Temporally Dependent Coefficients. *Journal of Water Resource and Protection*, *2011*(January), 76–84. doi:10.4236/jwarp.2011.31009

Jose, M., Ramos, G., Agüera, A., & Fernández-alba, A. R. (2012). Occurrence and persistence of organic emerging contaminants and priority pollutants in five sewage treatment plants of Spain: Two years pilot survey monitoring. *Environmental Pollution*, *164*, 263–273. PMID:22387188

Klamerth, N., Malato, S., Maldonado, M. I., Agüera, A., & Fernández-Alba, A. R. (2010). Application of photo-fenton as a tertiary treatment of emerging contaminants in municipal wastewater. *Environmental Science & Technology*, *44*(5), 1792–1798. doi:10.1021/es903455p PMID:20131796

Lidia, C., Kaklamanos, G., Ioana, A., Adriana, A., & Linda, M. (2015). High sensitive multiresidue analysis of pharmaceuticals and antifungals in surface water using U-HPLC-Q-Exactive Orbitrap HRMS. Application to the Danube river basin on the Romanian territory. *The Science of the Total Environment*, *532*, 501–511. doi:10.1016/j.scitotenv.2015.06.010 PMID:26100729

Mills, G. (2015). *Active and Passive Sampling for Pollutants of Emerging Concern - Sheffield 4th March 2015. In Emerging Contaminants in Waters and Soils, Practical Considerations: Sampling*. Analysis and Consequences.

Miraji, Chande, & Ngassapa. (2018). The Evolutionary Engagement of Analytical Techniques in the Eco-toxicity of Emerging Contaminants. *International Journal for Research & Development in Technology*, *3*(3), 42–48.

National Academy of Sciences. (1999). Historical Overview of Drinking Water Contaminants and Public Water Utilities. In Identifying Future Drinking Water Contaminants (p. 260). Author.

Orata, F. (2012). *Derivatization Reactions and Reagents for Gas Chromatography Analysis*. Advanced Gas Chromatography. doi:10.5772/33098

Pal, A., He, Y., Jekel, M., Reinhard, M., & Gin, K. Y. (2014). Emerging contaminants of public health significance as water quality indicator compounds in the urban water cycle Untreated water Sewer system. *Environment International*, *71*, 46–62. doi:10.1016/j.envint.2014.05.025 PMID:24972248

Pastore, M. C. (2015). Reworking the relation between sanitation and the city in Dar es Salaam, Tanzania. *Environment and Urbanization*, *27*(2), 473–488. doi:10.1177/0956247815592285

Peng, X., Xiong, S., Ou, W., Wang, Z., Tan, J., Jin, J., ... Fan, Y. (2016). Persistence, temporal and spatial profiles of ultraviolet absorbents and phenolic personal care products in riverine and estuarine sediment of the Pearl River catchment, China. *Journal of Hazardous Materials*, *8*. doi:10.1016/j.jhazmat.2016.05.020 PMID:27209124

Raucher, R. S. (1996). Public Health and Regulatory Considerations of the Safe Drinking Water Act. *Annual Review of Public Health*, *17*(1), 179–202. doi:10.1146/annurev.pu.17.050196.001143 PMID:8724223

Richardson, S. D. (2006). Environmental Mass Spectrometry: Emerging Contaminants and Current Issues. *Analytical Chemistry*, *78*(12), 4021–4046. doi:10.1021/ac060682u PMID:16771539

Richardson, S. D., Exposure, N., & Agency, U. S. E. P. (2006). Environmental Mass Spectrometry : Emerging Contaminants and Current Issues. *Analytical Chemistry*, *78*(12), 4021–4046. doi:10.1021/ac060682u PMID:16771539

Richardson, S. D., & Ternes, T. A. (2017). Water Analysis: Emerging Contaminants and Current Issues. *America Chemical Society*, *90*, 398–428. PMID:29112806

Roberts, D. (2006). DDT returns. *Nature Publishing Group*, *12*(8), 870–871. PMID:16892021

Schafer, A. K. S., & Kegley, S. E. (2002). Persistent Toxic Chemicals in the US Food Supply Persistent toxic chemicals in the US food supply. *Journal of Epidemiology and Community Health*, *56*(11), 813–817. doi:10.1136/jech.56.11.813 PMID:12388566

Stuart, M., Lapworth, D., Crane, E., & Hart, A. (2012). *Review of risk from potential emerging contaminants in UK groundwater*. Academic Press.

Styszko, K., Nosek, K., Motak, M., & Bester, K. (2015). Preliminary selection of clay minerals for the removal of pharmaceuticals, bisphenol A and triclosan in acidic and neutral aqueous solutions. *Comptes Rendus. Chimie*, *18*(10), 1134–1142. doi:10.1016/j.crci.2015.05.015

Subedi, B., & Kannan, K. (2014). Fate of artificial sweeteners in wastewater treatment plants in New York State, U.S.A. *Environmental Science & Technology*, *48*(23), 13668–13674. doi:10.1021/es504769c PMID:25365516

Tahrani, L., Sou, L., Mehri, I., Najjari, A., Hassan, A., Van Loco, J., & Ben, H. (2015). Isolation and characterization of antibiotic-resistant bacteria from pharmaceutical industrial wastewaters. *Microbial Pathogenesis Journal*, *89*, 54–61. doi:10.1016/j.micpath.2015.09.001 PMID:26343496

Thomas, K. V., Dye, C., Schlabach, M., & Langford, K. H. (2007). Source to sink tracking of selected human pharmaceuticals from two Oslo city hospitals and wastewater treatment works. *Journal of Environmental Monitoring*, *9*(12), 1410–1418. doi:10.1039/b709745j PMID:18049781

Tran, N. H., Hu, J., Li, J., & Ong, S. L. (2014). Suitability of artificial sweeteners as indicators of raw wastewater contamination in surface water and groundwater. *Water Research*, *48*, 443–456. doi:10.1016/j.watres.2013.09.053 PMID:24156949

Unger, K. K., Ditz, R., Machtejevas, E., & Skudas, R. (2010). Liquid Chromatography-Its Development and Key Role in Life Science Applications. *Liquid Chromatography*, *49*(13), 2300–2312. PMID:20232421

US EPA. (1998). *Contaminant Candidate List 1 - CCL 1*. Retrieved November 6, 2015, from http://www2.epa.gov/ccl/contaminant-candidate-list-1-ccl-1

US EPA. (2005). *Contaminant Candidate List 2 - CCL 2*. Retrieved from http://www2.epa.gov/ccl/contaminant-candidate-list-2-ccl-2

US EPA. (2009). *Contaminant Candidate List 3 - CCL 3*. Retrieved from http://www2.epa.gov/ccl/contaminant-candidate-list-3-ccl-3

US EPA. (2014). *Technical Fact Sheet – N-nitroso-dimethylamine*. Retrieved from https://www.epa.gov/sites/production/files/2014-03/documents/ffrrofactsheet_contaminant_ndma_january2014_final.pdf

US EPA. (2015a). *Draft Contaminant Candidate List 4-CCL 4*. Retrieved from http://www2.epa.gov/ccl/draft-contaminant-candidate-list-4-ccl-4

US EPA. (2015b). *Fact Sheet: Drinking Water Contaminant Candidate List 4 (CCL 4)– Draft*. Retrieved from http://www2.epa.gov/ccl

US EPA. (2018). *Contaminant Candidate List 5 (CCL 5)*. Retrieved May 17, 2019, from https://www.epa.gov/ccl/contaminant-candidate-list-5-ccl-5

Wang, J., Mao, D., Mu, Q., & Luo, Y. (2015). Fate and proliferation of typical antibiotic resistance genes in five full-scale pharmaceutical wastewater treatment plants. *The Science of the Total Environment*, *526*, 366–373. doi:10.1016/j.scitotenv.2015.05.046 PMID:25991498

Weigel, S., Kallenborn, R., & Hühnerfuss, H. (2004). Simultaneous solid-phase extraction of acidic, neutral and basic pharmaceuticals from aqueous samples at ambient (neutral) pH and their determination by gas chromatography-mass spectrometry. *Journal of Chromatography. A*, *1023*(2), 183–195. doi:10.1016/j.chroma.2003.10.036 PMID:14753684

Yan-long, W., Jun-hui, C., Li-yuan, G. A. O., Shuai, W., Xiao-ling, Z., Cheng-jun, S. U. N., & Xiao-ru, W. (2016). Determination of Eight Typical Lipophilic Algae Toxins in Particles Suspended in Seawater by Ultra Performance Liquid Chromatography-Tandem Mass Spectrometry. *Chinese Journal of Analytical Chemistry*, *44*(3), 335–341. doi:10.1016/S1872-2040(16)60911-8

Zhang, D., Bond, T., Krasner, S. W., Chu, W., Pan, Y., Xu, B., & Yin, D. (2019). Trace determination and occurrence of eight chlorophenylacetonitriles: An emerging class of aromatic nitrogenous disinfection byproducts in drinking water. *Chemosphere*, *220*, 858–865. doi:10.1016/j.chemosphere.2018.12.127

Zhang, H., Bayen, S., & Kelly, B. C. (2015). Multi-residue analysis of legacy POPs and emerging organic contaminants in Singapore's coastal waters using gas chromatography – triple quadrupole tandem mass spectrometry. *The Science of the Total Environment*, *523*(1), 219–232. doi:10.1016/j.scitotenv.2015.04.012 PMID:25863512

Zoeller, R. T., & Vandenberg, L. N. (2015). Assessing dose–response relationships for endocrine disrupting chemicals (EDCs): A focus on non-monotonicity. *Environmental Health*, *14*(42), 1–5. PMID:25971795

Chapter 6
Technological Developments in the Determination of Emerging Contaminants in Water

Magali Kemmerich
https://orcid.org/0000-0002-8374-4106
Universidade Federal de Pelotas, Brazil

ABSTRACT

Emerging contaminants (ECs), which constitute a group of chemicals, such as personal care products, food additives, and endocrine disruptors, are not commonly monitored and have the potential to cause adverse effects on the environment and humans. In water, even at low concentrations, they pose risks to environmental health. Several technologies have been developed to determine these compounds; the most common ones are liquid-liquid extraction (LLE), solid-phase extraction (SPE), solid-phase microextraction (SPME), stir-bar sorptive extraction (SBSE), and dispersive liquid-liquid microextraction (DLLME) with analysis by gas and liquid chromatography (GC and LC) coupled to mass spectrometry (MS). This chapter analyzes these methodologies and highlights their possibilities and limitations and compiles the most recent advances in this field.

INTRODUCTION

Emerging contaminants (ECs) in environmental matrices, mainly in water, have been reported in the literature, since they pose significant risks to environmental health. As a result, many methodologies have been proposed in order to determine concentrations of ECs, such as some drugs, personal care products (PCPs), food additives and endocrine disruptors. However, there has been a shortage of bibliography that organizes new proposals for technologies which can determine ECs in water and meet students', researchers' and industrial chemical analysts' needs. Therefore, this chapter aims to provide an overview of modern and technical analytical tools for determining ECs and to introduce enough theory so that they can be understood. Possibilities and limitations of these technologies are highlighted and the most recent advances in this field are compiled.

DOI: 10.4018/978-1-7998-1871-7.ch006

EMERGING CONTAMINANTS IN WATER SAMPLES

A wide range of synthetic organic compounds has been used by society for different purposes, such as food production and preservation, industrial manufacturing processes and human and animal healthcare. Many chemicals, such as organic and inorganic compounds, pathogens and nanoparticles (NPs), have been identified as contaminants in water sources. ECs can be found in different groups of chemicals, such as pharmaceuticals and personal care products (PPCPs), endocrine-disrupting chemicals (EDCs) and polybrominated diphenylethers (PBDEs). Recent research has revealed the occurrence of ECs in sewage effluents, surface and ground waters, precipitation and drinking waters at trace levels. Direct analysis of ECs in water is difficult because, in some samples, they are found at low concentrations and are associated with the components of the matrix, i. e., organic matter. Therefore, sample preparation has currently moved towards environmental friendliness, low cost, miniaturization, automation and simplicity. The following section discusses the application of different sample preparation techniques and analytical method analysis of ECs.

SAMPLE PREPARATION

Solid-Phase Extraction (SPE)

Fundamentals

Solid-Phase Extraction (SPE) is the most common technique in preparation of water samples, in which analytes are found at levels from µg/L to ng/L. SPE can be used for four purposes: analyte concentration, analyte isolation, interferent isolation or sample storage. The first is the most common one, in which analytes are retained in the solid phase for further elution (Jardim, 2010).

SPE is a sorption extraction technique which was introduced in the early 70's to account for drawbacks of liquid-liquid extraction (LLE). Separation of analytes occurs by liquid-solid separation, based on the separation mechanisms of low-pressure liquid chromatography (Vidal & Riekkola & Canals, 2012).

Methodology

The most common SPE devices are syringe cartridges, which consist of 1 to 6 mL polypropylene or glass tube containing about 50 to 500 mg solid phase, 40 to 60 µm particle size, attached to the tube by two 20 µm pore size filters. The choice of a cartridge depends mainly on sample volume, concentrations and physicochemical properties of analytes (Buszewski & Szultka, 2012). In general, the procedure of isolating/concentrating an analyte in SPE involves the following 4 steps (Figure 1).

The conditioning step aims to activate the sorbent and to elute some impurity found in the cartridge. The best option for conditioning is a solvent whose characteristics are similar to the ones of the solvent in which the sample is dissolved. It is important that the sorbent should not dry at this stage, so as not to form preferential pathways that compromise extraction. Variable sample volumes usually range from 100 to 1,000 mL. Sample pH may be critical to reach adequate retention of the analyte in the sorbent. In addition, speed of application can be critical in some cases. Ideally, this step should be slow, with flow rate below 2 mL min^{-1}. The clean-up step is critical to eliminate compounds from the matrix that may

Figure 1. Steps for extraction in the solid phase in the concentration mode of the compounds of interest: (a) Conditioning, (b) Percolation of the sample, (c) Clean-up and (d) Elution of the analyte

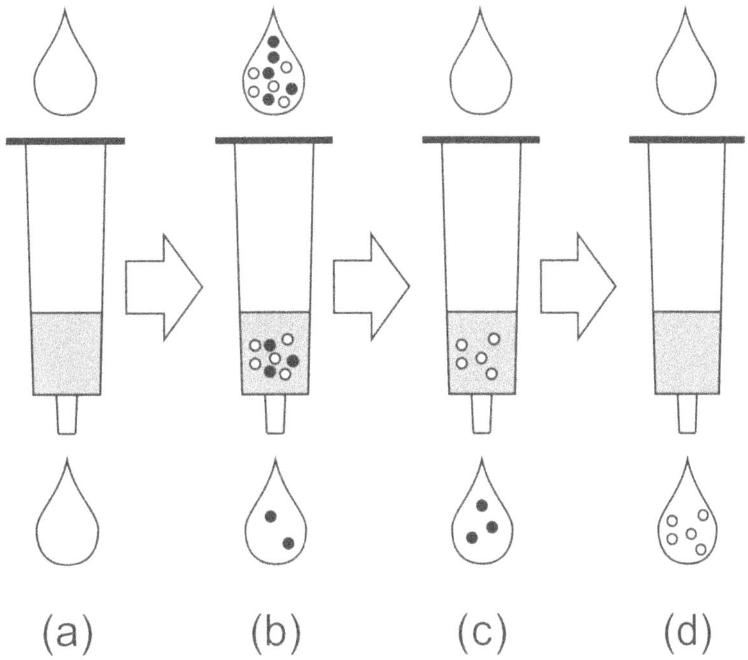

interfere with the analytical procedure. The clean-up solution usually contains little organic solvent, has low saline concentration or presents an ideal pH for interferent elution (Jardim, 2010).

In analyte elution, a small volume of eluent should be used in order to concentrate analytes found in the sample. The eluent should elute the analyte but should not allow elution of interferents that had been eliminated in the previous step. Acetonitrile, methanol, hexane and mixtures of acetonitrile-methanol and methanol-acetone are some solvents which have been currently used. The use of two or more aliquots of eluent increases extraction efficiency.

Sorbent materials

Most commercially available sorbents are based on organic groups chemically bonded to silica. Sorbents based on C18 and C8 are used for extracting apolar and low-polarity analytes. However, residual silanol groups may have negative influence on the separation of polar analytes, which may be irreversibly retained (Nováková & Vlcková, 2009; Buszewski & Szultka, 2012).

Table 1 shows recent application of SPE to the determination of ECs. Besides C18 sorbents, others have been studied and employed, mainly Strata-X and HLB. Sorbents based on hydrophilic polymers have the advantages of being stable over a wide pH range from 1 to 14 and highly acceptable by the sample. A novel copolymer of N-vinylpyrrolidone-divinylbenzene, commercially available from Waters, with hydrophilic-lipophilic balance (HLB), is Oasis® HLB. This sorbent has good retention in a broad spectrum of analytes, such as lipophilic, hydrophobic, acidic, basic and neutral ones (Buszewski & Szultka, 2012).

Possibilities and Limitations

SPE has replaced LLE in some official methods of drinking water analysis due to its low consumption of solvents, clean-up, absence of emulsions, low waste volumes, broad selectivity, easy automation (on-line and off-line), analyte concentration capacity and commercial availability of various types of cartridges (volumes and sorbents), besides high throughput (Sousa-Silva et al., 2015).

SPE disadvantages include high sampling time, high costs of cartridges and manifolds, carryover in automated systems and the need for sample pre-treatments, such as filtration and centrifugation. Moreover, in some cases, SPE-sorption efficiency may not be enough to achieve the required sensitivity (Sousa-Silva et al., 2015; Nováková & Vlcková, 2009).

Recent Advances

The synthesis of new sorbents may obtain clean and small solid phases (40 μm diameter), thus reducing sample volumes (≤100 mL). In addition, new materials have been developed, with high sample load-

Table 1. Recent applications of different sorbents to the determination of ECs in water by Solid-Phase Extraction (SPE)

Compound	Analytes	Sample treatment	LOD range	Reference
PPCPs and EDCs	17β-oestradiol, oestrone, 17α-ethynylestradiol, octylphenols, nonylphenols and their monoethoxylates and diethoxylates, bisphenol A, formononetin, biochanin A, daidzein, genistein and phytosterol sitosterol	Filter; adjusted pH (2); SPE (HLB)	0.6-5.5 ng L^{-1}	Rocha et al., 2016
	Nonylphenol, nonylphenol monoethoxylate, nonylphenol diethoxylates, bisphenol A, phytoestrogen and sitosterol	Filter; SPE (ENVI-C18)	< 10 ng L^{-1}	Feng et al., 2016
	15 novel psychoactive substances (NPS), 3 drugs of abuse and 2 antidepressants	SPE (Strata-X-Drug B cartridges)	0.01-1.09 ng L^{-1}	Peng & Gautam & Hall, 2019
	17-α-ethinylestradiol, 17-β-estradiol and estrone	On-line and Off-line SPE (Oasis® HLB)	LOQs: 0.035-0.1 ng L^{-1}	Barreca et al., 2019
PBDEs	Five Hydroxylated Polybrominated Diphenyl Ethers (OH-BDEs) and Triclosan.	SPE (HLB); clean-up: silica column	0.002-0.05 ng L^{-1}	Kerrigan et al., 2015
	Seven alkylphenol ethoxylates isomers, six polybrominated diphenyl ethers congeners, pentabromobiphenyl and hexabromocyclododecane.	Adjusted pH (acid); SPE (Strata-X)	0.01-0.2 μg L^{-1}	Chokwe et al., 2015
	Eight polybrominated diphenyl ethers (PBDEs) congener, six novel brominated flame retardants (NBFRs) and dechlorane plus (DP) (anti-isomer and syn-isomer).	SPE (ENVI)	0.017-0.03 pg L^{-1}	Khan et al., 2016

LOD: Limit of Detection; LOQ: Limit of Quantification.

ing capacities and efficiency to retain more polar analytes. The development of novel functionalized polymers, immunosorbents, molecularly printed polymers and restricted access materials seems to be the new step in the synthesis of novel sorbents (Ramos, 2012).

New materials, such as carbon nanotubes (CNTs), have been introduced. They comprise a wide range of allotropic carbon forms, including graphite nanofibers, nanodiamonds, fullerenes and graphenes. Selectivity of the analyte can occur either by covalent attachment or by other interactions with the functionalized surface of CNTs, which provides excellent extraction performance to a wide variety of analytes (Liu & Shi & Jiang, 2012).

Hyphenation of SPE with LC or GC, in SPE on-line mode, has reduced sample amounts and simplified sample preparation. Both require the use of modern solid phases and selective equipment, such as MS (Barreca et al., 2019).

Another recent development is the use of microfluidics and on-chip analytical systems. In the future, the use of micro- and nanoscale-based techniques will result in rapid and sensitive analytical methods of sample preparation and multi-residue analysis (Buszewski & Szultka, 2012).

Solid-Phase Microextraction (SPME)

Fundamentals

Solid-phase microextraction (SPME) was described and introduced by Arthur and Pawliszyn in the 90's (Arthur & Pawliszyn, 1990) and its use has been increasing due to its speed, easy operation, little sample handling, low number of sources of contamination and high analyte detectability.

SPME is a non-exhaustive technique based on the equilibrium of analytes between the sample matrix and the extraction phase. The extraction phase can be directly exposed either to the sample media (direct immersion, DI) or to its headspace (HS). When the SPME coating is placed directly in contact with the sample, the amount of analyte extracted at equilibrium (n_e) can be described by Equation 1, where K_{fs} is the distribution coefficient of the analyte between the coating and the sample matrix, V_f is the volume of the extraction phase, V_s is the volume of the sample and C_s is the analyte concentration in the sample matrix (Sousa-Silva et al., 2015).

$$n_e = \frac{K_{fs} x V_s x V_f}{K_{fs} x V_f + V_s} x C_s \quad \textit{Equation 1}$$

When the sample volume is large enough, Equation 1 can be simplified and result in Equation 2. This simplification denotes that the amount extracted by the coating is directly proportional to the concentration of the sample and completely independent of sample volume, a fact that enables on-site sampling in rivers and lakes (Sousa-Silva et al., 2015).

$$n_e = K_{fs} x V_f x C_s \quad \textit{Equation 2}$$

Extraction efficiency in SPME techniques can, therefore, be enhanced by increasing the magnitude of K_{fs} (chemistry of the extraction phase) and/or by increasing the volume or active surface area of the extraction phase (geometry of the extraction phase) (Sousa-Silva et al., 2015).

SPME can be coupled to different interfaces. In gas chromatography (GC) applications, the SPME fiber allows for complete automation due to the commercialization of autosamplers dedicated to SPME and its similarity to the common injection syringe used in GC applications. When coupling SPME to liquid chromatography (LC), in-tube SPME must be used; it consists of an extraction phase coated onto the inner walls of fused-silica tubing (Sousa-Silva et al., 2015).

Methodology

The SPME basic device consists of a stick (10 mm or 20 mm long, from 110 to 160 µm in diameter) coated with films up to 100 µm thick sorbent polymers (polydimethylsiloxane - PDMS, PA polyacrylate or polyethylene glycol-CW) or solid adsorbent (Carboxen®-CAr, microparticulate activated carbon and polystyrene-divinylbenzene resin-DVB) dispersed in polymers (CAR-PDMS, CW-DVB). The most traditional carrier has been fused silica fiber (piece of optic fiber).

The SPME technique occurs in two stages (Figure 2). In the SPME extraction step, analytes found in the sample are sorbed by a thin film of sorbent deposited on the support. The section covered with the sorbent material is directly placed either in contact with the sample (direct extraction mode) or in its confining space in the vapor phase - the headspace (headspace extraction mode).

Its extent depends either on partition or on adsorption of the analyte between the extractive phase and the matrix. After extraction, desorption of analytes occurs, so, the fiber is withdrawn from the sample

Figure 2. Use of SPME in extraction (a) SD mode, (b)HS mode and (c) desorption in GC system

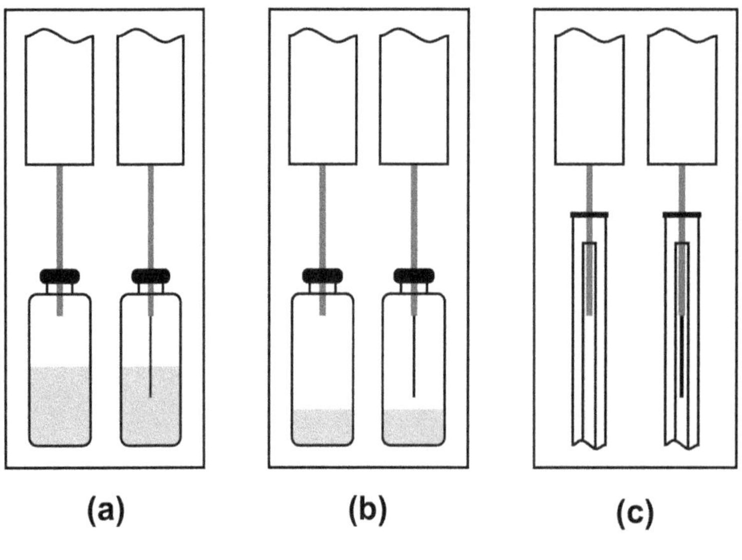

and inserted into the gas chromatography injector (Pawliszyn, 1997). Table 2 shows studies that have employed SPME to determine ECs in the last years.

Possibilities and Limitations

SPME can be easily employed to carry out on-site analyses. Besides, when environmental matrices containing solid particulate are studied, extensive binding of analytes to the solid matter might occur, leading to both decrease in free concentrations of analytes and in method sensitivity. On the other hand, analyses of complex matrices by SPME impose substantial challenges, especially in the case of trace-level analytes, since they can decrease coating lifetime and change extraction properties of the coating (Sousa-Silva et al., 2015).

Recent Advances

Novel extraction phases based on nanomaterials, ionic liquids (IL), molecularly imprinted polymers (MIPs), single- or multi-walled CNTs (SWCNTs/MWCNTs), graphene and metal-organic frameworks (MOFs) were applied to wastewater analysis, increasing the extraction efficiency over traditional techniques and achieving detection limit in picogram order (Liu & Shi & Jiang, 2012).

Different techniques, such as dipping, microwave-assisted, electrochemical deposition and immobilization with epoxy glue, have been proposed to prepare coating. On-site sampling based on SPME not only produces accurate, precise, and fast analytical data, but also decreases the overall possibility of errors (Piri-Moghadam et al., 2016; Sousa-Silva et al., 2015).

Table 2. Applications of solid-phase microextraction (SPME) to the determination of ECs in water

Compound	Analytes	Sample treatment	LOD Range	Reference
PPCPs	Methylparaben, ethylparaben, propylparaben, butylparaben, 2,4,6-trichlorophenol (2,4,6-TCP) and triclosan	Filter; HS-SPME	4-21 pg L^{-1}	Regueiro et al., 2009a
	Phenazone, carbamazepine, ketoprofen, naproxen, bezafibrate and ibuprofen	Filter; SPME-MD	0.05-12 ng L^{-1}	Padrón & Ferrera & Rodríguez, 2009
	Benzophenone, octyl salicylate, homosalate, 3-(4-methylbenzylidene) camphor, 2-hydroxy4-methoxybenzophenone	Adjusted pH (5); SPME	0.5-6.8 ng L^{-1}	Zhang & Lee, 2012
PBDEs	Seven polybrominated diphenyl ethers: BDE-35, BDE-47, BDE-77, BDE-99, BDE-100, BDE-153, BDE-154	Sol-gel SWNTs-TSO-OH-coated; SPME-TD	0.08-0.8 ng L^{-1}	Zhang et al., 2009
	Seven polybrominated diphenyl ethers: BDE-28, BDE-47, BDE-100, BDE-99, BDE-154, BDE-153, BDE-183	SPME with Fe_3O_4-coated bamboo charcoal fibers	0.25-0.62 ng L^{-1}	Zhao et al., 2013

LOD: Limit of Detection; MD: micellar desorption; SWNTs-TSO-OH-coated: polymer-functionalized single-walled carbon nanotubes with hydroxyl-terminated silicone oil; TD: thermal desorption.

Stir-Bar Sorptive Extraction (SBSE)

Fundamentals

Stir-Bar Sorptive Extraction (SBSE) is a sorption method which was proposed in 1999. It casually emerged from SPME, since the inclusion of a polymer phase in a conventional stir bar promoted high sorption capacity, ease of handling, high robustness and sensitivity (Baltussen et al., 1999).

The SBSE device uses polydimethylsiloxane (PDMS) as polymeric phase, a silicone-type sorbent coating. This non-polar polymeric phase promotes hydrophobic interactions with target molecules and the retention mechanism occurs mainly through Van-der-Waals forces, but hydrogen bonds can also occur with the oxygen atoms of the PDMS, depending on the chemical structure of the analytes (Nogueira, 2015).

Efficiency of SBSE yields of an analyte relate to the partitioning between the PDMS phase of the stir bar and the water sample. Behavior is similar to the distribution described by the octanol-water partition coefficients ($K_{o/w}$) during static equilibrium. Thus, $K_{O/W}$ and phase ratio ($\beta = V_W/V_{PDMS}$, in which V_W is the water sample volume and V_{PDMS} is the PDMS volume) are important parameters to predict recovery. Equation 3 might be used to calculate theoretical recovery (%) (Nogueira, 2015).

$$Recovery(\%) = \frac{\frac{K_{o/w}}{2}}{1+\frac{K_{o/w}}{2}} x100 \quad \textit{Equation 3}$$

Methodology

The principle of SBSE operation consists of two fundamental steps: 1st) extraction or enrichment of analytes for the polymer: and 2nd) retroextraction or desorption of analytes from the extractive phase to the chromatographic system. Figure 3 depicts the SBSE analytical device, which consists of a magnetic stir bar incorporated into a glass jacket (10-20 mm in length) typically coated with 24–126 μL (0.3–1.0 mm in thickness) of PDMS (Nogueira, 2015).

Extraction of target analytes occurs from the sample bulk towards the PDMS polymer by an absorption process, using sampling modes of immersion or HS. Subsequently, desorption of solutes from the polymeric phase to the chromatographic system takes place, using the modes of thermal desorption (TD) or liquid desorption (LD) (Nogueira, 2015). Table 3 shows the application of SBSE to the determination of ECs in water samples, in which TD and LD modes are employed.

Possibilities and Limitations

SBSE is an environment-friendly technique, which is reliable in terms of enrichment capacity, outstanding performance, great sensitivity and selectivity for ultra-trace analysis of non-polar to medium-polar organic compounds from complex matrices. In spite of all advantages, this analytical approach is not effective enough to monitor the large group of polar organic compounds (Nogueira, 2015).

Figure 3. Schematic representation of (a) SBSE analytical device, (b) retroextraction in the SBSE-LD mode, (c) extraction by immersion and (d) Bar Adsorptive Microextraction (BAμE)

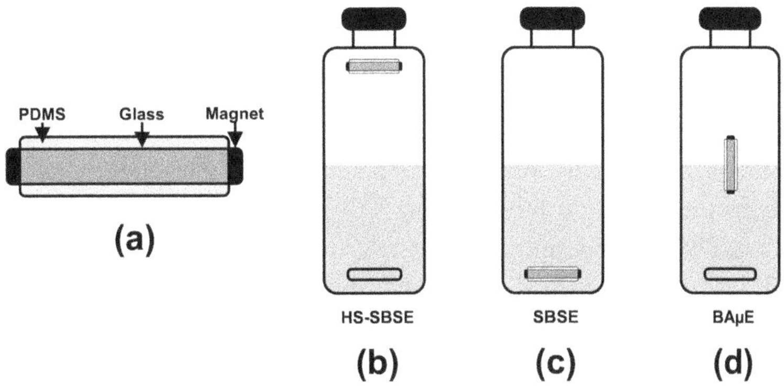

Recent Advances

Since SBSE is not suitable for polar compounds, mainly for the use of PDMS sorbent, alternative coating phases have been developed to achieve better affinity, selectivity and flexibility. Thus, new sorbent phases proposed for SBSE and novel complementary microextraction techniques (e.g. Bar Adsorptive Microextraction, BAμE) have shown to be innovative and have overcome most analytical challenges and limitations (Nogueira, 2015).

Table 3. Applications of stir-bar sorptive extraction (SBSE) to the determination of ECs in water

Compound	Analytes	Sample treatment	LOD Range	Reference
PCPs	Eighteen synthetic musk fragrances	SBSE-TD	LOQs: 5-80 ng L^{-1}	Arbulu et al., 2011
	Four UV filters (2,2-dihydroxy-4-methoxybenzophenone, benzophenone-3, octocrylene, and octyldimethyl-p-aminobenzoic acid) and two antimicrobial agents (triclocarban and triclosan)	Adjusted pH (3); SBSE-LD	2.5 ng L^{-1}	Pedrouzo et al., 2010
EDCs and PBDEs	Ten polybrominated diphenyl ethers and four polybrominated biphenyls	SBSE-TD	1.1-6.0 ng L^{-1}	Prieto et al., 2008a
	Six water and oil repellents, four preservatives, two plasticizers, seven surfactants, a flame retardant, four hormones, fourteen pharmaceutical compounds, an UV-filter and nine pesticides.	SBSE	3.6-45 ng L^{-1}	Aparicio et al., 2017
OCs	102 contaminants: fragrances, UV filters, repellents, endocrine disruptors, biocides, polycyclic aromatic hydrocarbons (PAHs), polychlorinated biphenyls (PCBs), and several types of pesticides	SBSE	< 1 ng L^{-1}	Pintado-Herrera et al., 2014

LOD: Limit of Detection; LOQ: Limit of Quantification.

Liquid-Liquid Extraction (LLE)

Fundamentals

LLE is based on the distribution of a compound between two immiscible liquids, in which this compound has different solubility. Generally, one of the phases is aqueous and the other one is an organic solvent. During extraction, the more hydrophilic compound prefers the aqueous phase, whereas the most hydrophobic one prefers the organic solvent (Dean, 2009).

The distribution ratio of a compound between two immiscible phases is constant and consists of the Law of Nernst (Equation 4), where K_D is the distribution constant, C_{org} is the concentration of the compound in the organic solvent and C_{aq} is the concentration of the compound in the aqueous phase.

$$K_D = \frac{C_{org}}{C_{aq}} \quad \textit{Equation 4}$$

The fraction of the extracted compound (E) is given by Equation 5, where V_{org} is the volume of the organic solvent and V_{aq} is the volume of the aqueous phase.

$$E = \frac{C_{org} x V_{org}}{C_{org} x V_{org} + C_{aq} x V_{aq}} \quad \textit{Equation 5}$$

Some LLE are carried out by using tens or hundreds of milliliters of each phase. In general, to reach high recoveries, two or three new extractions with new portions of solvents are required. To improve K_D extraction, several strategies, such as replacement of the organic solvent, addition of salt to the aqueous phase, which decreases the concentration of the compound in the aqueous phase, and addition of an ion pair reagent, can be used (Dean, 2009; Majors, 2009).

Methodology

Liquid-liquid extraction (LLE) is performed with a selective solvent for an analyte in the sample or for matrix components that must be removed. LLE, is usually performed in a separation funnel or test tube, in three steps (Figure 4).

The first step of the LLE is to add an immiscible organic solvent to the sample. It is subsequently stirred. After a short period of time, both layers separate and the layer of interest is removed. If necessary, extraction should be repeated with the use of new aliquots of organic solvent (Barth et al., 2015). Table 4 shows studies that have applied LLE to the determination of ECs in the last years. Either dichloromethane or a mixture of dichloromethane/hexane is usually employed.

Possibilities and Limitations

LLE is very simple and does not require any complex apparatus, since it has many routine applications, such as most the United States Environmental Protection Agency (EPA) methods applied to wastewater

Figure 4. Simplified diagram of LLE steps: (a) mixing the liquid sample with the organic solvent; (b) stirring the mixture; and (c) separating and removing the extraction solvent for analysis

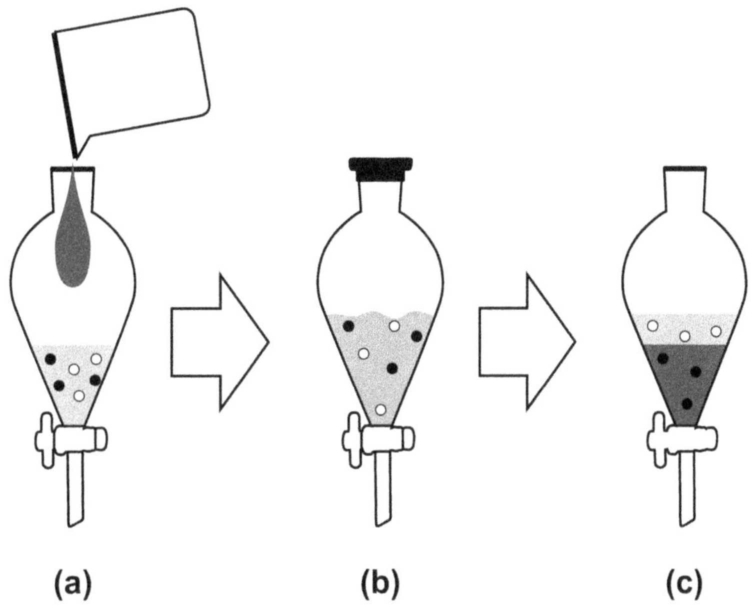

samples. This technique allows the use of a large range of solvents, with different solubility and selectivity (Queiroz & Collins & Jardim, 2001).

Since LLE needs large amounts of organic solvents, it is expensive and hazardous, besides contributing to environmental pollution. In addition, automation and emulsion formation are hard tasks, a fact that can increase extraction time. Its main drawback is the formation of emulsions due to the occurrence of surfactants, mainly in the case of wastewater samples (Piri-Moghadam et al., 2016).

Table 4. Applications of liquid-liquid extraction (LLE) to the determination of ECs in water

Compound	Analytes	Sample treatment	LOD Range	Reference
OPs	Around 150 organic pollutants, including PAHs, octyl/nonyl phenols, PCBs, PBDEs and pesticides (insecticides, herbicides, fungicides and several relevant metabolites)	LLE (DCM, neutral-acid)	LOIs: 0.1-1 μg L^{-1}	Portolés et al., 2011
EDCs and PBDEs	Eight polybrominated diphenyl ethers congeners and sixteen polycyclic aromatic hydrocarbons	Filter; LLE (DCM)	0.008-1.2 ng L^{-1}	Wang et al., 2013
	Nine hydroxylated polybrominated diphenyl ethers (OH-PBDEs), acting as emerging endocrine disruptors	Adjusted pH (3); LLE (HEX-DCM, 1:1, v/v)	0.0039-0.0220 μg L^{-1}	Yu et al., 2015

LOD: Limit of Detection; LOI: Limit of Identification;OPs: organic pollutants; DCM: dichloromethane; HEX: hexane.

Recent Advances

Some techniques of microextraction derived from LLE are liquid-liquid dispersion microextraction (DLLME), liquid-liquid microextraction with microporous membrane (MMLLE) and liquid-liquid microextraction with hollow cylindrical membranes (HF-LPME). In support-assisted liquid-liquid extraction (ELLAS), an aqueous sample is applied to the surface of the matrix inside a cylindrical syringe, which is similar to the one used in SPME. The use of syringes allows easy automation and prevents emulsions formation. In assisted-pressure LLE, organic solvents are used at high temperatures (> 200 °C) and pressures (> 20,000 kPa) in order to keep the solvent liquid (Nováková & Vlcková, 2009).

Dispersive Liquid-Liquid Microextraction (DLLME)

Fundamentals

Dispersive liquid-liquid microextraction (DLLME) was proposed by Rezaee *et al.* for the preconcentration of polycyclic aromatic hydrocarbons (PAHs) in water samples in 2006. Authors defined extraction recovery (EF) in Equation 6, where C_{sed} is the analyte concentration in the sedimented phase and C_0 is the initial analyte concentration in the sample.

$$EF = \frac{C_{sed}}{C_0} \textit{ Equation 6}$$

ER is defined as the ratio of the amount of analyte in the sedimented phase to the initial concentration in the sample in Equation 7, where V_{sed} is the volume of the sedimented phase and V_0 is the volume of the aqueous phase.

$$ER = \frac{V_{sed}}{V_0} xEFx100 \textit{ Equation 7}$$

Relative recoveries (RR) can be calculated from Equation 8, where $C_{founded}$ is the analyte concentration measured from the sample after analyte addition, C_{real} is the native analyte concentration and C_{added} is the amount of analyte that was added to the sample.

$$RR = C_{founded} - \frac{C_{real}}{C_{added}} \textit{ Equation 8}$$

Methodology

Steps of DLLME are shown in Figure 5. The first step is to inject a suitable mixture of the dispersing and extracting solvents into the aqueous sample which contains analytes. Many conventional DLLME typically use from 20 to 100 µL chlorinated solvents as the extraction solvent, from 0.5 to 2 mL disperser solvent and from 5 to 10 mL aqueous sample (Leong et al., 2014).

In the extraction step, the extraction solvent is dispersed in the aqueous phase in very fine droplets. Due to the large surface area between the extraction solvent and the aqueous sample, chemical equilibrium is reached rapidly and extraction time is independent (Rezaee & Yamini & Faraji, 2010). In the centrifugation separation step, the organic phase settles to the bottom of the extractor tube, which is transferred to a vial for further analyte determination.

Addition of salts, such as NaCl, promotes increase in the sedimented phase volume, due to decrease in solubility of the extraction solvent in the presence of the salt. In the literature, many studies can be found with the addition from 0.5 to 30% of NaCl. Also, adjusting the pH to the medium allows analytes to remain in the neutral form and facilitates partitioning of analytes into microdroplets of the extraction solvent (Ahmad et al., 2015).

Table 5 shows studies that have employed DLLME to determine ECs in the last years. DLLME can be coupled to other techniques, such as DLLME based on solidification of floating organic droplet (DLLME-SFO), ionic liquid assisted DLLME (IL-DLLME) and UA-DLLME (Leong et al., 2014).

Possibilities and Limitations

DLLME major advantages include simplicity, minimal use of harmful solvents, rapid extraction and low cost. Despite its advantages, this method also has some difficulties regarding solvent requirements.

Figure 5. Simplified diagram of DLLME stages: (a) water sample, (b) rapid injection of dispersant and extraction solvent mixture into the sample; (c) dispersion of the extracting solvent in the form of micro-droplets: and (d) separation and extraction of extraction solvent for analysis

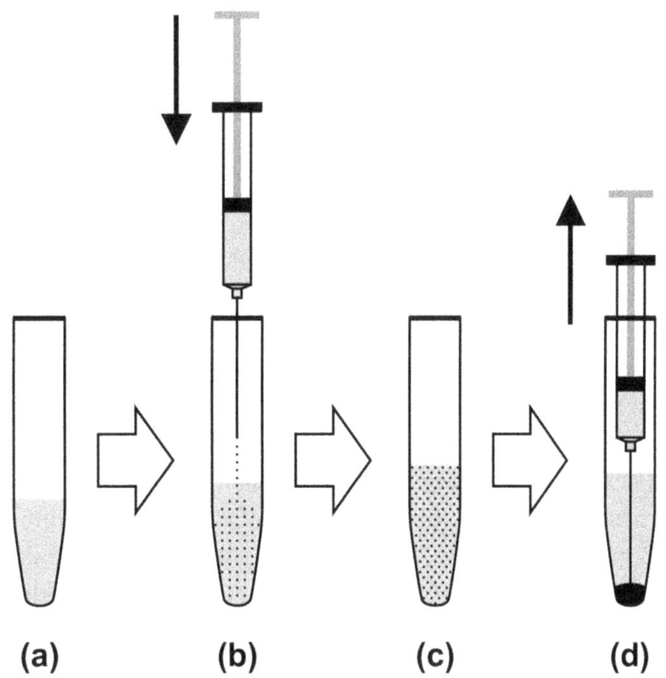

Table 5. Applications of dispersive liquid-liquid microextraction (DLLME) to the determination of ECs in water

Compound	Analytes	Sample treatment	LOD Range	Reference
PCPs	Five salicylate and benzophenone-type ultraviolet (UV) filters: ethylhexyl salicylate (EHS), 3,3,5-trimethyl-cyclohexyl salicylate (HMS), 2-hydroxy-4-methoxybenzophenone (BP-3), 2,4-dihydroxy-benzophenone (BP-1) and 2,2-dihydroxy-4-methoxybenzophenone (BP-8)	UA-DLLME	1-2 ng L^{-1}	Wu et al., 2013
	Eight commonly used UV-filters.	Adjusted pH (2.5); filter; DLLME	10-30 ng L^{-1}	Benede et al., 2014
EDCs	Bisphenol-A, 4-cumylphenol, 4-tertbutylphenol, 4-octylphenol and 4-n-nonylphenol	SDME; DLLME	0.2-1.6 ng mL^{-1}	López-Darias et al., 2010
PBDEs	Thirteen brominated flame retardants and twelve novel brominated flame retardants.	DLLME (TCE)	0.007-0.023 ng L^{-1}	Li et al., 2016

LOD: Limit of Detection.

It must be immiscible in water, form tiny droplets in it and have higher affinity towards the analyte (Ahmad et al., 2015).

Recent Advances

Some interesting DLLME include methods which use special devices; procedures based on solidification of FSO; low-density solvent-based solvent demulsification DLLME procedures; procedures based on the adjustment of solvents mixture density; and procedures based on automation of DLLME by sequential injection analysis (Ahmad et al., 2015).

Other Sample Preparation Techniques

Besides the sample preparation techniques that have already been described in this chapter, there are many other techniques which have been used for extracting ECs from water samples, as shown in Table 6.

Regueiro et al. (2009b) developed ultrasound-assisted emulsification (USAEME), which has been found to improve efficiency by increasing the rate of mass transfer between both immiscible phases. Besides, this method has been successfully developed to deal with complex water samples and reduce the amount of solvent by raising the concentration factor. USAEME is an efficient, simple, rapid and inexpensive alternative to other extraction techniques, such as SPE, SPME and Liquid-Phase Microextraction (LPME) (Leong et al., 2014).

Microextraction by Packed Sorbent (MEPS) is a new automatable microextraction technique that consists of a packed sorbent within a microliter syringe. MEPS, which requires small sample and solvent volumes, besides short extraction time, has been widely used in accurate drug and pharmaceutical determination that is comparable to SBSE (García-Córcoles et al., 2019). In addition, MEPS seems to be a promising approach to couple micro-SPE to GC-MS, although most MEPS applications are associated with HPLC analysis (Moeder et al., 2010).

Table 6. Other sample preparation techniques applied to the determination of ECs in water

Compound	Analytes	Sample treatment	LOD Range	Reference
PCPs	Parabens, triclosan and related phenols	Filter; USAEME	7.4-133 pg mL^{-1}	Regueiro et al., 2009b
	UV filter and polycyclic musk compounds	MEPS (C18)	35-87 ng L^{-1}	Moeder et al., 2010
	Seven UV-filters compounds	Adjusted pH (3); MNPs-based d-SPE	0.2-6 ng L^{-1}	Román et al., 2011
OPs	Polycyclic aromatic hydrocarbons (PAHs), polychlorinated biphenyls (PCBs), phthalate esters (PEs), nonylphenols (NPs), polybrominated biphenyls (PBBs) and polybrominated diphenyl ethers (PBDEs).	MASE-LVI-PTV	0.1-222 ng L^{-1}	Prieto et al., 2008b
PBDEs	Six polychlorinated biphenyls (PCBs) and two polybrominated diphenyl ethers (PBDEs)	PC-HFME	0.04-0.21 ng L^{-1}	Basheer et al., 2007
PPCPs	141 compounds, like as: atenolol, caffeine, cannabidiol, ibuprofen, naproxen, progesterone, triclosan and trimethroprim	POCIS	0.015-2.5 ng L^{-1}	Zenobio et al., 2015

LOD: Limit of Detection.

Dispersive Solid-Phase Extraction (d-SPE) was first described by Anastassiades et al. (2003) as a clean-up step. It employes a SPE sorbent dispersed in a sample solution which contains analytes. In this approach, since all analytes interact equally with sorbent particles, high capacity per amount of sorbent is achieved and traditional SPE problems, such as channeling and blocking of cartridges and disks, are avoided (Román et al., 2011).

Membrane-Assisted Solvent Extraction (MASE) is an extraction technique based on non-porous membranes. It was first described by Hauser and Popp (2001) for organochloride determination in water samples. In this technique, analytes dissolve in the membrane material (i.e., polypropylene) and pass through it from the donor phase to the acceptor one. Since MASE uses low volumes of organic solvent (400 to 1000 μL), along with large volume injection (LVI), it provides good Limit of Detection (LOD) (Prieto et al., 2008).

Basheer et al. (2007) described the Polymer-Coated Hollow Fiber Microextraction (PC-HFME) technique as an on-site sample preparation approach for seawater samples. For the analysis, only the fiber and extracted analytes are transported to the laboratory. Besides, it showed to be a cost-effective and convenient sampling technique and a complementary technique to SPME.

Polar Organic Chemical Integrative Sampler (POCIS) is a passive sampler which enables sample collection and sample extraction to occur at the same time (Alvarez, 1999). POCIS comprises a solid extraction sorbent sandwiched between two microporous membranes whose pore size is 100 nm. The sorbent appears to be more specific for pesticide compounds and some hormones when it is a triphasic mixture or more specific for pharmaceutical compounds when it is the HLB phase (Morin et al., 2012).

INSTRUMENTATION

In order to study occurence, transport and fate of ECs in the environment, different analytical techniques have been used. Besides there are some works employing spectroscopic (Gowen et al., 2012; Li & Wang & Li, 2016), electrochemical (Gupta & Dubey & Malik, 2013; Moraes et al., 2015) and electrophoresis (Zhang & Zhang & Zhang, 2015; Le et al., 2016) techniques, gas and liquid chromatography (GC and LC, respectively) coupled to mass spectrometry (MS or MS/MS) have been the main modern analytical techniques (Wu et al., 2010) for ECs determination. Selection usually depends on physicochemical properties of analytes, for example, less volatile and polar compounds are normally detected by LC, while volatile compounds are analyzed by GC.

Gas Chromatography Coupled to Mass Spectrometry

Fundamentals

The most common technique in gas chromatography is elution, i. e., gas stream passes continuously through the column and when the vaporized sample is introduced rapidly into this gas stream, it is drawn through the column, which contains the stationary phase. Substances found in the sample are separated, and then get to the detector, which generates a signal to the recorder.

Many factors, such as length, internal diameter, temperature, flow rate of the mobile phase, sample volume, injection technique and characteristics of substances, affect the efficiency of a column. During the analysis, the temperature may remain constant or undergo variation, which may be linear or non-linear. During the analysis, the column temperature may be increased in order to decrease retention of high boiling substances. In general, choice of gas (nitrogen, helium or hydrogen) depends on the type of detector, but it can also affect efficiency.

The chromatographic column is a long tube which contains the stationary phase. This tube can be made of several materials, such as glass, fused silica and PTFE, coated with a protective material, which is usually polyimide. With the introduction of capillary columns, which enable very high efficiency to be reached due to the fact that they are longer (10 to 100 m), selectivity of the stationary phase becomes less important (Collins & Braga & Bonato, 2009).

Among detectors that can be used in both gas and liquid chromatography, Mass Spectrometry (MS) has stood out due to its versatility. MS makes quantitative analyses more reliable because the analyte is identified by its characteristic retention time (t_R) and also by the precursor ion and product ion.

The basic principle of MS is the generation of organic or inorganic compound ions and separation through its mass/charge ratio (m/z), followed by detection of the respective m/z and abundance. The mass spectrometer consists of an ion source, a mass analyzer and a detector; the two last operate under high vacuum conditions. The source of ions does not need to be under vacuum (Gross, 2004).

Ionization sources used in GC-MS are electron ionization (EI) and chemical ionization (CI), positive (PCI) and negative (NCI) modes. Depending on the type of ionization mode, MS can offer great selectivity and sensitivity. NCI is the most common one, since it provides higher sensitivity than the one of electron impact ionization (EI) (García-Corcoles et al, 2019).

EI is more popular, since it provides many ions to identify the analyte. In EI, ions are generated in the gas phase and under vacuum, thus, interaction among ions after formation is unlikely to happen. Therefore, the EI mass spectrum is reproducible, with no reactions or collisions occurring inside the ion source.

Mass analyzers separate ions in agreement with the m/z ratio. Choosing the most appropriate mass analyzer depends on certain factors, such as the desired mass range, resolution and available feature. Some examples coupled to gas chromatograpgy are: quadrupole, triple quadrupole, ion trap and time of flight (TOF). Regarding the coupling of MS to GC or LC techniques, Quadrupole (Q) and triple Quadrupole (QqQ) mass analyzers are the most popular ones, due to their simplicity, relatively low price, good linear range, easy understanding and operation.

Quadrupole is usually composed of four metal bars, arranged in pairs, to which a DC and an alternating potential RF are applied. Ions produced at the ionization source focus on the center of the region among the four cylinders and cross the quadrupole axially. Their trajectories depend on the produced electric field where only ions of a given m/z have a stable one and reach the detector (Chiaradia & Collins & Jardim, 2008).

The triple quadrupole (TQ) analyzer consists of three quadrupoles in series; the second quadrupole (Q2 or q) is used as a collision cell, in which fragmentation of selected ions in the first quadrupole (Q1 or MS1) occurs, usually by collision-induced dissociation (CID) and also as a driver of the ions produced at the third quadrupole (Q3 or MS2). All quadrupoles are controlled to transmit ions of a single m/z ratio or a range of m/z ratios to generate more accurate analytical information (Chiaradia & Collins & Jardim, 2008).

Applications

GC-MS or GC-MS/MS must be combined with derivatization to analyze most pharmaceuticals, due to their polarity. Sebok et al. (2008) reported a method of 63 multiclass pollutant analysis that included ibuprofen, gemfibrozil, naproxen, ketoprofen, diclofenac sodium salt and others, by GC-MS after SPE sample preparation. LOQ ranged from 0.92 to 600 ng L^{-1}. The method was successfully applied to the quantification of pollutants found in Hungarian influent and effluent wastewaters and in the Danube River.

Many EDCs require a derivatization step to modify their structure and meet the conditions of GC-MS. Since derivatization increases the complexity of the process and total analysis time, LC has replaced GC-MS for EDC detection in environmental samples (García-Córcoles et al, 2019). In order to determine EDCs, high-resolution GC-MS/MS has been employed by Ikonomou et al. (2008), who reached LODs in ng L^{-1} range. The method allowed the analysis of many organic wastewater contaminants in a complex matrix.

Due to the hydrophobic characteristics of PBDEs, either GC-MS or GC-MS/MS is usually used for their determination, rather than LC (Wu et al, 2010). Prieto et al. (2008) employed GC-MS after SBSE sample preparation for PBDEs determination in water samples. Method LODs ranged from 1.1 to 6.0 ng L^{-1}.

Liquid Chromatography Coupled to Mass Spectrometry

Fundamentals

Since most compounds are not sufficiently volatile to be analyzed by Gas Chromatography, High Performance Liquid Chromatography (HPLC) has been very prominent. In HPLC, the main requirement of the analysis is the solubility of the compounds in the mobile (liquid) phase (Collins & Braga & Bonato, 2009).

HPLC is based on the distribution of analytes and matrix components by polarity between the stationary phase and the mobile (liquid) one. It can be performed either as normal phase chromatography, with a stationary phase which is more polar than the mobile phase, or as reverse phase chromatography, with a stationary phase that is more non-polar than the mobile one (Collins & Braga & Bonato, 2009). The latter is the most employed one.

Different parameters can be optimized in the analysis by liquid chromatography, mainly mobile and stationary phases. In reversed phase chromatography, the most widely used in LC applications, an aqueous phase containing additives which facilitate ionization of analytes is applied. Additives may be formic acid or acetic acid, at concentrations ranging from 0.01% to 0.1% (v/v). Besides, an organic phase, usually methanol or acetonitrile, is used. Separation is carried out in gradient of mobile phase, optimized for each analyte or group of analytes.

In order to obtain higher resolution and faster analyses, Ultra High Performance Liquid Chromatography (UHPLC) has been used. This technique is based on the same principles of separation of HPLC, however, it employs chromatographic columns with reduced dimensions (5-10 cm in length and internal diameters of 1-2.1 mm), filled with FE ≤ 2 μm particles and can reach high linear velocities of mobile phase (Maldaner & Jardim, 2012).

The use of UHPLC with high performance requires certain modifications, such as equipment capable of working at high pressures; rapid and accurate injection system in a small volume range; reduced internal volumes (connections, sampling loop, detector cell and pumps); detectors with high data acquisition rates, appropriate columns and stationary phase (Maldaner & Jardim, 2012).

Among these stationary phases, there are a third generation hybrid silica particles produced and patented by Waters, which consist of hybrid silica (Ethylene Bridged Hybrid, BEH) particles incorporated with a small amount of surface charge. The column which has this stationary phase has a wide usable pH range, thus allowing a versatile and robust separation technology for the development of analytical methods (Fekete et al., 2014).

Advantages of the UHPLC system over HPLC are the small sample volume, increased sensitivity, separation efficiency, easy transfer of a method developed in HPLC to UHPLC and the possibility of analyzing a large number of samples in short time. However, due to reduction in internal diameters and decreased porosity of filters, the UHPLC system is more susceptible to clogging than HPLC. Furthermore, the use of high pressures may alter the order of elution of compounds, displacement or overlapping of chromatographic peaks, besides generating gradients that influence retention and efficiency (Fekete et al., 2014; Maldaner & Jardim, 2012).

When coupling LC with Mass Spectrometry (LC-MS or LC-MS/MS), sources of ions are electrospray ionization (ESI), atmospheric pressure chemical ionization (APCI) and atmospheric pressure photoionization (APPI). Some of the mass analyzers are Q, QqQ, TOF, Q-TOF and – more recently – Orbitrap.

In ESI, which is the most common, the sample is dissolved and pressurized in a stainless steel capillary tube and voltage, usually between 3 and 5 kV, is applied. As a result, the liquid emerges from the capillary at atmospheric pressure in the form of aerosol. As desolvation takes place, assisted by a continuous flow of dry gas (usually N_2), the droplet size is reduced until repulsion force among similar loads becomes higher than the cohesive forces of the liquid phase (surface tension). Thus, coulomb explosion, which generates droplets with sizes equivalent to 10% of the original droplet size, takes place. A series of explosions occurs until analyte ions are produced from these beads, which are transferred into the mass spectrometer by a series of focusing devices. In this process, only the ionization occurs at atmospheric pressure; the mass spectrometer is under vacuum. It is ideal for analysis of more polar molecules with high molecular mass (Chiaradia & Collins & Jardim, 2008).

Applications

LC-MS allows the detection of compounds with extreme polarities (acidic compounds), thermally labile ones (steroids) and those which have high molecular weight (antibiotics) with great sensitivity and selectivity (García-Córcoles et al., 2019).

By comparison with GC-MS/MS, LC-MS/MS is more favorable in the analysis of pharmaceuticals and their metabolites. Besides, UHPLC has been increasingly used for the analysis of pharmaceuticals, due to the fact that shorter run times are needed and smaller amount of solvents are consumed. Batt et al. (2008) developed a method for the analysis of 48 pharmaceuticals and 6 metabolites, by employing SPE and UHPLC-MS/MS analyses. Method LODs ranged from 1.0 to 51 ng L^{-1}. It was applied to effluent and surface water analysis, whose samples had concentrations from 7 to 2950 ng L^{-1}.

For the determination of EDCs, LC-MS/MS has been increasingly applied, although GC-MS has still been employed. ESI and APCI are the most widely used LC-MS interfaces in the detection of EDCs in water samples while QqQ is the most sensitive one. Kuster et al. (2008) developed a method for 21 ECs, including seven estrogens and three progestogens, with SPE followed by LC-MS/MS analysis. Method LODs were below 0.85 ng L^{-1} for estrogens and beow 3.94 ng L^{-1} for progestogens. The method was applied to the analysis of the Llobregat river basin, in Spain. Estrone-3-sulfate, estrone, estriol and progesterone were found in ng L^{-1} range.

SOLUTIONS AND RECOMMENDATIONS

The analysis of ECs requires previous planning, considering its objectives and available materials and equipment. In the case of water monitoring analyses, the major limitations may be the low levels of concentration that need to be achieved and sample preparation techniques, such as SPE, SBSE and DLLME, which also concentrate the compounds of interest. SPE has currently been the most common technique, since it has been widely marketed in different configurations.

Chromatographic analysis, when coupled to mass spectrometry, provides reliable and accurate results. In sample screening, high-resolution equipment, such as Q-TOF and Orbitrap, previously applied to the analysis of ECs, is recommended. Concerning quantification, the QqQ is more advantageous, due to its low cost and highly precise results.

FUTURE RESEARCH DIRECTIONS

Determination of emerging compounds in water is a relatively new subject in the field of Analytical Chemistry. In sample preparation, trends have focused on technique miniaturization, automation, online and low-cost analyses. Therefore, there has been growing interest in the synthesis of new sorbents and liquid extraction phases, besides the use of microfluidics and on-chip analytical systems, so as to achieve better affinity, selectivity and flexibility.

Regarding the analysis, the UHPLC-MS/MS system has increased sensitivity and separation efficiency, allowing the use of reduced limits of quantification and, thus, monitoring water samples with high robustness, accuracy and precision. High resolution Q-TOF and Orbitrap should be further studied in the near future.

CONCLUSION

The wide diversity of physical-chemical properties of ECs, the unknown structures of their metabolites and their existence at trace level in water samples, make the analysis of these compounds an analytical challenge. However, due to studies which show their potential ecotoxicological effects and aquatical environmental analysis results with their increasing occurrence, efforts have been made to develop rapid and sensitive methods for their determination at trace concentrations.

For many years, classical sample preparation techniques, such as LLE and SPE, were employed. Nowadays, many microextraction techniques, such as SPME, SBSE and DLLME, have been developed and applied, thus, solving deficiencies of classical techniques. GC-MS and LC-MS have been widely employed in the analysis of ECs in water samples for quantification of trace levels with high precision and sensitivity. There are still several possibilities of innovation and development of analytical methods for the determination of ECs and their metabolites, a current focus of global Analytical Chemistry.

ACKNOWLEDGMENT

This research was supported by CNPq [grant number 151866/2018-0].

REFERENCES

Ahmad, W., Al-Sibaai, A. A., Bashammakh, A. S., Alwael, H., & El-Shahawi, M. S. (2015). Recent advances in dispersive liquid-liquid microextraction for pesticide analysis. *Trends in Analytical Chemistry*, *72*, 181–192. doi:10.1016/j.trac.2015.04.022

Alvarez, D. A. (1999). *Development of an Integrative Sampling Device for Hydrophilic Organic Contaminants in Aquatic Environments*. Columbia, MO: University of Missouri-Columbia.

Anastassiades, M., Lehotay, S. J., Stajnbaher, D., & Schenck, F. J. (2003). Fast and easy multiresidue method employing acetonitrile extraction/partitioning and "dispersive solid-phase extraction" for the determination of pesticide residues in produce. *Journal of AOAC International*, *86*, 412. PMID:12723926

Aparicio, I., Martín, J., Santos, J. L., Malvar, J. L., & Alonso, E. (2017). Stir bar sorptive extraction and liquid chromatography-tandem mass spectrometry determination of polar and non-polar emerging and priority pollutants in environmental waters. *Journal of Chromatography. A*, *1500*, 43–52. doi:10.1016/j.chroma.2017.04.007 PMID:28416215

Arbulu, M., Sampedro, M. C., Unceta, N., Gomez-Caballero, A., Goicolea, M. A., & Barrio, R. J. (2011). A retention time locked gas chromatography-mass spectrometry method based on stir-bar sorptive extraction and thermal desorption for automated determination of synthetic musk fragrances in natural and wastewaters. *Journal of Chromatography. A*, *1218*(20), 3048–3055. doi:10.1016/j.chroma.2011.03.012 PMID:21444087

Arthur, C. L., & Pawliszyn, J. (1990). Solid phase microextraction with thermal desorption using fused silica optical fibers. *Analytical Chemistry*, *62*(19), 2145–2148. doi:10.1021/ac00218a019

Baltussen, E., Sandra, P., David, F., & Cramers, C. (1999). Stir bar sorptive extraction (SBSE), a novel extraction technique for aqueous samples: Theory and principles. *The Journal of Microcolumn Separations*, *11*(10), 737–747. doi:10.1002/(SICI)1520-667X(1999)11:10<737::AID-MCS7>3.0.CO;2-4

Barreca, S., Busetto, M., Colzani, L., Clerici, L., Daverio, D., Dellavedova, P., ... Ubaldi, V. (2019). Determination of estrogenic endocrine disruptors in water at sub-ng L−1 levels in compliance with Decision 2015/495/EU using offline-online solid phase extraction concentration coupled with high performance liquid chromatography-tandem mass spectrometry. *Microchemical Journal*, *147*, 1186–1191. doi:10.1016/j.microc.2019.04.030

Barth, T., Calixto, L. A., Jabor, V. A. P., & Borges, K. B. (2015). Extração Líquido-Líquido. In Preparo de amostras para análise de compostos orgânicos. Rio de Janeiro: LTC.

Basheer, C., Vetrichelvan, M., Valiyaveettil, S., & Kee Lee, H. (2007). On-site polymer-coated hollow fiber membrane microextraction and gas chromatography-mass spectrometry of polychlorinated biphenyls and polybrominated diphenyl ethers. *Journal of Chromatography. A*, *139*(2), 157–164. doi:10.1016/j.chroma.2006.11.008 PMID:17118376

Batt, A. L., Kostich, M. S., & Lazorchak, J. M. (2008). Analysis of ecologically relevant pharmaceuticals in wastewater and surface water using selective solid-phase extraction and UPLC-MS/MS. *Analytical Chemistry*, *80*(13), 5021–5030. doi:10.1021/ac800066n PMID:18498179

Benede, J. L., Chisvert, A., Salvador, A., Sanchez-Quiles, D., & Tovar-Sanchez, A. (2014). Determination of UV filters in both soluble and particulate fractions of seawaters by dispersive liquid–liquid microextraction followed by gas chromatography-mass spectrometry. *Analytica Chimica Acta*, *812*, 50–58. doi:10.1016/j.aca.2013.12.033 PMID:24491764

Buszewski, B., & Szultka, M. (2012). Past, Present, and Future of Solid Phase Extraction: A Review. *Critical Reviews in Analytical Chemistry*, *42*(3), 198–213. doi:10.1080/07373937.2011.645413

Chiaradia, M. C., Collins, C. H., & Jardim, I. C. S. F. (2008). O estado da arte da cromatografia associada à espectrometria de massas acoplada à espectrometria de massas na análise de compostos tóxicos em alimentos. *Quimica Nova*, *31*(3), 623–636. doi:10.1590/S0100-40422008000300030

Chokwe, T. B., Okonkwo, J. O., Sibali, L. L., & Ncube, E. J. (2015). Alkylphenol ethoxylates and brominated flame retardants in water, fish (Carp) and sediment samples from the Vaal River, South Africa. *Environmental Science and Pollution Research International*, *22*(15), 11922–11929. doi:10.100711356-015-4430-x PMID:25869432

Collins, C.H., Braga, G.L., & Bonato, P.S. (2009). *Fundamentos de Cromatografia* (2nd ed.). Campinas: Unicamp.

Dean, J. R. (2009). Classical Approaches for aqueous extraction. In Extraction Techniques in Analytical Sciences. West Sussex, UK: John Wiley and Sons. doi:10.1002/9780470682494.ch2

Dimpe, K. M., & Nomngongo, P. N. (2016). Current sample preparation methodologies for analysis of emerging pollutants in different environmental matrices. *Trends in Analytical Chemistry*, *82*, 199–207. doi:10.1016/j.trac.2016.05.023

Fekete, S., Schappler, J., Veuthey, J.-L., & Guillarme, D. (2014). Current and future trends in UHPLC. *Trends in Analytical Chemistry*, *63*, 2–13. doi:10.1016/j.trac.2014.08.007

Feng, L., Yang, G., Zhu, L., Xu, J., Xu, X., & Chen, Y. (2016). Distribution and risk assessment of endocrine-disrupting pesticides in drinking water sources from agricultural watershed. *Water, Air, and Soil Pollution*, *227*(1), 23–32. doi:10.100711270-015-2704-z

García-Córcoles, M. T., Rodríguez-Gómez, R., de Alarcón-Gómez, B., Cipa, M., Martín-Pozo, L., Kauffmann, J.-M., & Zafra-Gómez, A. (2019). Chromatographic Methods for the Determination of Emerging Contaminants in Natural Water and Wastewater Samples: A Review. *Critical Reviews in Analytical Chemistry*, *49*(2), 160–186. doi:10.1080/10408347.2018.1496010 PMID:30285467

Gowen, A. A., Tsenkova, R., Bruen, M., & O'donell, C. (2012). Vibrational Spectroscopy for Analysis of Water for Human Use and in Aquatic Ecosystems. *Critical Reviews in Environmental Science and Technology*, *42*(23), 2546–2573. doi:10.1080/10643389.2011.592758

Gross, J. H. (2004). Mass spectrometry. A textbook. Springer.

Gupta, A. K., Dubey, R. S., & Malik, J. K. (2013). Application of Modern Electroanalytical techniques: Recent trend in Pharmaceutical and Drug Analysis. *International Journal of Pharmaceutical Sciences and Research*, *4*, 2450–2458.

Hauser, B., & Popp, J. (2001). Membrane-assisted solvent extraction of organochlorine compounds in combination with large-volume injection/gas chromatography-electron capture detection. *Journal of Separation Science*, *24*(7), 551–560. doi:10.1002/1615-9314(20010801)24:7<551::AID-JSSC551>3.0.CO;2-2

Ikonomou, M. G., Cai, S. S., Fernandez, M. P., Blair, J. D., & Fischer, M. (2008). Ultra-trace analysis of multiple endocrine-disrupting chemicals in municipal and bleached KraftMill effluents using gas chromatography-high-resolution mass spectrometry. *Environmental Toxicology and Chemistry*, *27*(2), 243–251. doi:10.1897/07-136R.1 PMID:18348620

Jardim, I. C. S. F. (2010). Extração em Fase Sólida: Fundamentos Teóricos e Novas Estratégias para Preparação de Fases Sólidas. *Scientia Chromatographica*, *2*, 13–25.

Kerrigan, J. F., Engstrom, D. R., Yee, D., Sueper, C., Erickson, P. R., Grandbois, M., ... Arnold, W. A. (2015). Quantification of hydroxylated polybrominated diphenyl ethers (OH-BDEs), triclosan, and related compounds in freshwater and coastal systems. *PLoS One*, *10*(10), e0138805. doi:10.1371/journal.pone.0138805 PMID:26466159

Khan, M. U., Li, J., Zhang, G., & Naseem, R. (2016). Malik first insight into the levels and distribution of flame retardants in potable water in Pakistan: An underestimated problem with an associated health risk diagnosis. *The Science of the Total Environment*, *565*, 346–359. doi:10.1016/j.scitotenv.2016.04.173 PMID:27177141

Kuster, M., de Alda, M. J. L., Hernando, M. D., Petrovic, M., Martín-Alonso, J., & Barceló, D. (2008). Analysis and occurrence of pharmaceuticals, estrogens, progestogens and polar pesticides in sewage treatment plant effluents, river water and drinking water in the Llobregat River basin (Barcelona, Spain). *Journal of Hydrology (Amsterdam)*, *358*(1-2), 112–123. doi:10.1016/j.jhydrol.2008.05.030

Le, M. D., Duong, H. A., Nguyen, M. H., Sáiz, J., Pham, H. V., & Mai, T. D. (2016). Screening determination of pharmaceutical pollutants in different water matrices using dual-channel capillary electrophoresis coupled with contactless conductivity detection. *Talanta*, *160*, 512–520. doi:10.1016/j.talanta.2016.07.032 PMID:27591645

Leong, M. I., Fuh, M.-R., & Huang, S.-D. (2014). Beyond dispersive liquid-liquid microextraction. *Journal of Chromatography. A*, *1335*, 2–14. doi:10.1016/j.chroma.2014.02.021 PMID:24582396

Li, B., Sun, S. J., Huo, C. Y., Li, W. L., Zhu, N. Z., Qi, H., ... Ma, W. L. (2016). Occurrence and fate of PBDEs and novel brominated flame retardants in a wastewater treatment plant in Harbin, China. *Environmental Science and Pollution Research International*, *23*(19), 19246–19256. doi:10.100711356-016-7142-y PMID:27364485

Li, Z., Wang, J., & Li, D. (2016). Applications of Raman spectroscopy in detection of water quality. *Applied Spectroscopy Reviews*, *51*(4), 333–357. doi:10.1080/05704928.2015.1131711

Liu, Q., Shi, J., & Jiang, G. (2012). Application of graphene in analytical sample preparation. *Trends in Analytical Chemistry*, *37*, 1–11. doi:10.1016/j.trac.2012.03.011

López-Darias, J., Germán-Hernández, M., Pino, V., & Afonso, A. M. (2010). Dispersive liquid–liquid microextraction versus singledrop microextraction for the determination of several endocrine-disrupting phenols from seawaters. *Talanta*, *80*(5), 1611–1618. doi:10.1016/j.talanta.2009.09.057 PMID:20152385

Majors, R. (2009). Sample Preparation. In Introduction to Modern Liquid Chromatography (3rd ed.; pp. 757-808). Hoboken, NJ: John Wiley and Sons.

Maldaner, L., & Jardim, I. C. S. F. (2012). UHPLC – Uma abordagem atual: Desenvolvimentos e desafios recentes. *Scientia Chromatographica*, *4*(3), 197–207. doi:10.4322c.2012.014

Moeder, M., Schrader, S., Winkler, U., & Rodil, R. (2010). At-line microextraction by packed sorbent-gas chromatography-mass spectrometry for the determination of UV filter and polycyclic musk compounds in water samples. *Journal of Chromatography. A*, *1217*(17), 2925–2932. doi:10.1016/j.chroma.2010.02.057 PMID:20334867

Moraes, F. C., Rossi, B., Donatoni, M. C., de Oliveira, K. T., & Pereira, E. C. (2015). Sensitive determination of 17b-estradiol in river water using a graphene based electrochemical sensor. *Analytica Chimica Acta*, *881*, 37–43. doi:10.1016/j.aca.2015.04.043 PMID:26041518

Morin, N., Miège, C., Randon, J., & Coquery, M. (2012). Chemical calibration, performance, validation and applications of the polar organic chemical integrative sampler (POCIS) in aquatic environments. *Trends in Analytical Chemistry*, *36*, 144–175. doi:10.1016/j.trac.2012.01.007

Nogueira, J. M. F. (2015). Stir-bar sorptive extraction: 15 years making sample preparation more environment friendly. *Trends in Analytical Chemistry*, *71*, 214–223. doi:10.1016/j.trac.2015.05.002

Nováková, L., & Vlcková, H. (2009). A review of current trends and advances in modern bio-analytical methods: Chromatography and sample preparation. *Analytica Chimica Acta*, *656*(1-2), 8–35. doi:10.1016/j.aca.2009.10.004 PMID:19932811

Núñez, O., Gallart-Ayala, H., Martins, C. P. B., & Lucci, P. (2012). New trends in fast liquid chromatography for food and environmental analysis. *Journal of Chromatography. A*, *1228*, 298–323. doi:10.1016/j.chroma.2011.10.091 PMID:22153282

Padrón, M. E. T., Ferrera, Z. S., & Rodríguez, J. J. S. (2009). Coupling of solid-phase microextraction with micellar desorption and high performance liquid chromatography for the determination of pharmaceutical residues in environmental liquid samples. *Biomedical Chromatography*, *23*(11), 1175–1185. doi:10.1002/bmc.1240 PMID:19444801

Pal, A., He, Y., Jekel, M., Reinhard, M., & Gin, K. Y.-H. (2014). Emerging contaminants of public health significance as water quality indicator compounds in the urban water cycle. *Environment International*, *71*, 46–62. doi:10.1016/j.envint.2014.05.025 PMID:24972248

Pawliszyn, J. (1997). *Solid phase microextraction: theory and practice*. New York: Wiley-VHC.

Pedrouzo, M., Borrull, F., Marce, R. M., & Pocurull, E. (2010). Stir-barsorptive extraction and ultra-high-performance liquid chromatography-tandem mass spectrometry for simultaneous analysis of UV filters and antimicrobial agents in water samples. *Analytical and Bioanalytical Chemistry*, *397*(7), 2833–2839. doi:10.100700216-010-3743-3 PMID:20428847

Peng, Y., Gautam, L., & Hall, S. W. (2019). The detection of drugs of abuse and pharmaceuticals in drinking water using solid-phase extraction and liquid chromatography-mass spectrometry. *Chemosphere*, *223*, 438–447. doi:10.1016/j.chemosphere.2019.02.040 PMID:30784750

Pintado-Herrera, M. G., González-Mazo, E., & Lara-Martín, P. A. (2014). Atmospheric pressure gas chromatography-time-of-flight-mass spectrometry (APGC-ToF-MS) for the determination of regulated and emerging contaminants in aqueous samples after stir bar sorptive extraction (SBSE). *Analytica Chimica Acta*, *851*, 1–13. doi:10.1016/j.aca.2014.05.030 PMID:25440658

Piri-Moghadam, H., Ahmadi, F., & Pawliszyn, J. (2016). A critical review of solid phase microextraction for analysis of water samples. *Trends in Analytical Chemistry*, *85*, 133–143. doi:10.1016/j.trac.2016.05.029

Portolés, T., Pitarch, E., Lopez, F., & Hernandez, J. F. (2011). Development and validation of a rapid and wide-scope qualitative screening method for detection and identification of organic pollutants in natural water and wastewater by gas chromatography time-of-flight mass spectrometry. *Journal of Chromatography. A*, *1218*(2), 303–315. doi:10.1016/j.chroma.2010.11.010 PMID:21134677

Prieto, A., Basauri, O., Rodil, R., Usobiaga, A., Fernández, L. A., Etxebarria, N., & Zuloaga, O. (2010). Stir-bar sorptive extraction: A view on method optimisation, novel applications, limitations and potential solutions. *Journal of Chromatography. A*, *1217*(16), 2642–2666. doi:10.1016/j.chroma.2009.12.051 PMID:20083248

Prieto, A., Telleria, O., Etxebarria, N., Fernández, L. A., Usobiaga, A., & Zuloaga, O. (2008b). Simultaneous preconcentration of a wide variety of organic pollutants in water samples. Comparison of stir bar sorptive extraction and membrane-assisted solvent extraction. *Journal of Chromatography. A*, *1214*(1-2), 1–10. doi:10.1016/j.chroma.2008.10.060 PMID:18990398

Prieto, A., Zuloaga, O., Usobiaga, A., Etxebarria, N., & Fernandez, L. A. (2008a). Use of experimental design in the optimisation of stir bar sorptive extraction followed by thermal desorption for the determination of brominated flame retardants in water samples. *Analytical and Bioanalytical Chemistry*, *390*(2), 739–748. doi:10.100700216-007-1712-2 PMID:18038264

Queiroz, S. C. N., Collins, C. H., & Jardim, I. C. S. F. (2001). Métodos de Extração e/ou Concentração de Compostos Encontrados em Fluidos Biológicos para Posterior Determinação Cromatográfica. *Quimica Nova*, *24*(1), 68–76. doi:10.1590/S0100-40422001000100013

Ramos, L. (2012). Critical overview of selected contemporary sample preparation techniques. *Journal of Chromatography. A*, *1221*, 84–98. doi:10.1016/j.chroma.2011.11.011 PMID:22137129

Regueiro, J., Becerril, E., Garcia-Jares, C., & Llompart, M. (2009a). Trace analysis of parabens, triclosan and related chlorophenols in water by headspace solid-phase microextraction with in situ derivatization and gas chromatography-tandem mass spectrometry. *Journal of Chromatography. A*, *1216*(23), 4693–4702. doi:10.1016/j.chroma.2009.04.025 PMID:19414177

Regueiro, J., Llompart, M., Psillakis, E., Garcia-Monteagudo, J. C., & Garcia-Jares, C. (2009b). Ultrasound-assisted emulsification-microextraction of phenolic preservatives in water. *Talanta*, *79*(5), 1387–1397. doi:10.1016/j.talanta.2009.06.015 PMID:19635375

Rezaee, M., Assadi, Y., Hosseini, M.-R. M., Aghaee, E., Ahmadi, F., & Berijani, S. (2006). Determination of organic compounds in water using dispersive liquid–liquid microextraction. *Journal of Chromatography. A*, *1116*(1-2), 1–9. doi:10.1016/j.chroma.2006.03.007 PMID:16574135

Rezaee, M., Yamini, Y., & Faraji, M. (2010). Evolution of dispersive liquid–liquid microextraction method. *Journal of Chromatography. A*, *1217*(16), 2342–2357. doi:10.1016/j.chroma.2009.11.088 PMID:20005521

Rocha, M. J., Cruzeiro, C., Reis, M., Pardal, M. A., & Rocha, E. (2016). Pollution by oestrogenic endocrine disruptors and b-sitosterol in a South-Western European River (Mira, Portugal). *Environmental Monitoring and Assessment*, *188*(4), 240–254. doi:10.100710661-016-5236-0 PMID:27004433

Román, I. P., Chisvert, A., & Canals, A. (2011). Dispersive solid-phase extraction based on oleic acid-coated magnetic nanoparticles followed by gas chromatography-mass spectrometry for UV-filter determination in water samples. *Journal of Chromatography. A*, *1218*(18), 2467–2475. doi:10.1016/j.chroma.2011.02.047 PMID:21411104

Sebok, A., Vasanits-Zsigrai, A., Helenkar, A., Zaray, G., & Molnar-Perl, I. (2009). Multiresidue analysis of pollutants as their trimethylsilyl derivatives, by gas chromatography-mass spectrometry. *Journal of Chromatography. A*, *1216*(12), 2288–2301. doi:10.1016/j.chroma.2009.01.056 PMID:19201001

Souza-Silva, E. A., Jiang, R., Rodriguez-Lafuente, A., Gionfriddo, E., & Pawliszyn, J. (2015). A critical review of the state of the art of solid-phase microextraction of complex matrices I. Environmental analysis. *Trends in Analytical Chemistry*, *71*, 224–235. doi:10.1016/j.trac.2015.04.016

Tankiewicz, M., Fenik, J., & Biziuk, M. (2011). Solventless and solvent-minimized sample preparation techniques for determining currently used pesticides in water samples: A review. *Talanta*, *86*, 8–22. doi:10.1016/j.talanta.2011.08.056 PMID:22063507

Vidal, L., Riekkola, M.-L., & Canals, A. (2012). Ionic liquid-modified materials for solid-phase extraction and separation: A review. *Analytica Chimica Acta*, *715*, 19–41. doi:10.1016/j.aca.2011.11.050 PMID:22244164

Wang, X., Xi, B., Huo, S., Sun, W., Pan, H., Zhang, J., ... Liu, H. (2013). Characterization, treatment and releases of PBDEs and PAHs in a typical municipal sewage treatment plant situated beside an Urban River, East China. *Journal of Environmental Sciences (China)*, *25*(7), 1281–1290. doi:10.1016/S1001-0742(12)60201-0 PMID:24218838

Wu, J., Zhang, L., & Yang, Z. (2010). A Review on the Analysis of Emerging Contaminants in Aquatic Environment. *Critical Reviews in Analytical Chemistry*, *40*(4), 234–245. doi:10.1080/10408347.2010.515467

Wu, J. W., Chen, H. C., & Ding, W. H. (2013). Ultrasound-assisted dispersive liquid-liquid microextraction plus simultaneous silylation for rapid determination of salicylate and benzophenone-type ultraviolet filters in aqueous samples. *Journal of Chromatography. A*, *1302*, 20–27. doi:10.1016/j.chroma.2013.06.017 PMID:23831000

Yu, B., Zhang, R., Liu, P., Zhang, Y., Zhang, Y., & Bai, Y. (2015). Determination of nine hydroxylated polybrominated diphenyl ethers in water by precolumn derivatization-gas chromatography mass spectrometry. *Journal of Chromatography. A*, *1419*, 19–25. doi:10.1016/j.chroma.2015.09.081 PMID:26454789

Zenobio, J. E., Sanchez, B. C., Leet, J. K., Archuleta, L. C., & Sepúlveda, M. S. (2015). Presence and effects of pharmaceutical and personal care products on the Baca National Wildlife Refuge, Colorado. *Chemosphere*, *120*, 750–755. doi:10.1016/j.chemosphere.2014.10.050 PMID:25465958

Zhang, H., & Lee, H. K. (2012). Simultaneous determination of ultraviolet filters in aqueous samples by plunger-in-needle solid-phase microextraction with graphene-based sol-gel coating as sorbent coupled with gas chromatography-mass spectrometry. *Analytica Chimica Acta*, *742*, 67–73. doi:10.1016/j.aca.2012.03.016 PMID:22884209

Zhang, W., Sun, Y., Wu, C., Xing, J., & Li, J. (2009). Polymer-functionalized single-walled carbon nanotubes as a novel sol-gel solidphase micro-extraction coated fiber for determination of poly-brominated diphenyl ethers in water samples with gas chromatography-electron capture detection. *Analytical Chemistry*, *81*(8), 2912–2920. doi:10.1021/ac802123s PMID:19364140

Zhang, Z., Zhang, D., & Zhang, X. (2015). Simultaneous determination of pharmaceutical and personal care products in wastewater by capillary electrophoresis with head-column field-amplified sample stacking. *Analytical Methods*, *6*(19), 7978–7983. doi:10.1039/C4AY01562B

Zhao, R.-S., Liu, Y. L., Chen, X.-F., Yuan, J.-P., Bai, A.-Y., & Zhou, J.-B. (2013). Preconcentration and determination of polybrominated diphenyl ethers in environmental water samples by solid-phase microextraction with Fe_3O_4-coated bamboo charcoal fibers prior to gas chromatography-mass spectrometry. *Analytica Chimica Acta*, *769*, 65–71. doi:10.1016/j.aca.2013.01.027 PMID:23498122

ADDITIONAL READING

Dimpe, K. M., & Nomngongo, P. N. (2016). Current sample preparation methodologies for analysis of emerging pollutants in different environmental matrices. *Trends in Analytical Chemistry*, *82*, 199–207. doi:10.1016/j.trac.2016.05.023

Fekete, S., Schappler, J., Veuthey, J.-L., & Guillarme, D. (2014). Current and future trends in UHPLC. *Trends in Analytical Chemistry*, *63*, 2–13. doi:10.1016/j.trac.2014.08.007

García-Córcoles, M. T., Rodríguez-Gómez, R., de Alarcón-Gómez, B., Cipa, M., Martín-Pozo, L., Kauffmann, J.-M., & Zafra-Gómez, A. (2019). Chromatographic Methods for the Determination of Emerging Contaminants in Natural Water and Wastewater Samples: A Review. *Critical Reviews in Analytical Chemistry*, *49*(2), 160–186. doi:10.1080/10408347.2018.1496010 PMID:30285467

Piri-Moghadam, H., Ahmadi, F., & Pawliszyn, J. (2016). A critical review of solid phase microextraction for analysis of water samples. *Trends in Analytical Chemistry*, *85*, 133–143. doi:10.1016/j.trac.2016.05.029

Wu, J., Zhang, L., & Yang, Z. (2010). A Review on the Analysis of Emerging Contaminants in Aquatic Environment. *Critical Reviews in Analytical Chemistry*, *40*(4), 234–245. doi:10.1080/10408347.2010.515467

KEY TERMS AND DEFINITIONS

DLLME: Dispersive liquid-liquid microextraction is a sample preparation technique based on the injection of extraction solvent (immiscible in water), which is assisted by a dispersive solvent (miscible in both water and extraction solvents) within an aqueous solution.

GC-MS: Gas chromatography coupled to mass spectrometry is an analytical method used to identify and quantify different substances in a sample.

LC-MS: Liquid chromatography coupled to mass spectrometry is an analytical method used to identify and quantify different substances in a sample.

LLE: Liquid-liquid extraction is a sample preparation technique based on the separation of a specific component from a heterogeneous mixture of liquids with different solubilities into two different immiscible liquids.

SBSE: Stir-bar sorptive extraction is a sample preparation technique based on the partioning between a PDMS phase of the stir-bar and the water sample.

SPE: Solid-phase extraction is a sample preparation technique based on different interactions of analytes with a solid phase and a liquid one;

SPME: Solid-phase microextraction is a sample preparation technique based on the equilibrium of analytes between the sample matrix and the extraction phase.

Chapter 7
Semiconductor Nanocomposites-Based Photoelectrochemical Aptamer Sensors for Pharmaceuticals Detection

Kevin Otieno Okoth
Tom Mboya University College, Kenya

Ruth Nduta Wanjau
Kenyatta University, Kenya

Maurice Otieno Odago
State University of New York at Oneonta, USA

ABSTRACT

Development of sensors for pharmaceuticals has become very essential. This is due to the need to monitor the release and toxicological effects of pharmaceuticals into the environment. In this work, the authors explored bismuth sulphide (Bi2S3) nanorods and graphene as photoactive material for constructing a photoelectrochemical (PEC) aptasensor for sulfadimethoxine (SDM) detection, exhibiting high sensitivity, stability, and reproducibility. In another experiment, Mo-doped BiVO4 (Mo-BiVO4) and graphene nanocomposites were explored as photoactive material to construct a visible light-driven photoelectrochemical biosensor. Graphene in the nanocomposites was very essential in immobilizing streptomycin aptamer through π-π stacking interaction. Finally, graphene doped CdS (GR-CdS) synthesized via one pot hydrothermal technique and gold nanoparticles (Au NPs) were employed to construct a PEC aptasensor for diclofenac (DCF).

DOI: 10.4018/978-1-7998-1871-7.ch007

INTRODUCTION

Over the past few decades, a number of organic compounds that have been found to resist biodegradation, bioaccumulate in organisms and therefore possess the potential to adversely affect the environment and human health have been identified. Amongst these chemicals, antibiotics play an important role in treatment and prevention of bacterial infection in human and animals as well as serving as growth promoters. Development of reliable analytical techniques has led to knowledge on their toxicities levels. However, monitoring these chemicals still remains a big challenge since most are not covered by standard monitoring techniques. Despite the great efforts aimed at developing reliable analytical techniques to identify and quantify the emerging pollutants in the environment, serious challenges such as existence of a number of isomers, occurance in very low concentrations and existence of the compounds in very complex environments still remain. So far, techniques such as chromatography, spectroscopy, colorimetric methods, enzyme linked immunosorbent assay (ELISA) and electrophoresis have been applied to the determination of these pollutants and exhibited high sensitivity and accuracy. Unfortunately, applications of some of these techniques have been constrained by the sophisticated instruments involved, long preparation time, complicated procedures that require skilled manpower and high detection cost. In addition, some of these techniques are plagued by low detection levels while some electrochemical techniques lack specificity. However, with the ever increasing demand for sensitive and selective detection, great efforts have been made towards development of modern electrochemical and photoelectrochemical sensors that are applicable not only in laboratories but in a variety of other settings.

Fabrication of Photoelectrochemical Sensors

Generally, photoelectrochemical (PEC) reactions refer to processes that involve use of semiconductor materials to transform light irradiation into other forms of energy. In a typical PEC reaction, three processes are basically involved. The first reaction involves generation of electron-hole pairs upon irradiation of the semiconductors. In the second step, the photogenerated charges are separated with holes migrating to the valency band (VB) while the electrons are transported to the conduction band (CB). Lastly, the surface chemical reaction where reduction by photogenerated electrons or (and) oxidation by photogenerated holes occur as shown in Figure 1. For example, photogenerated holes have been suggested to oxidize organic molecules in PEC sensors and photocatalysts while photogenerated electrons have found wide acceptance in water splitting.

It has been widely stated that PEC sensing acquires the advantages of electrochemical analysis such as high sensitivity and low limit of detection, accuracy, precision, linearity in wide range of measurement, short time of analysis, small sample volume, possibility of simultaneous measurements, moderate cost and portability has gained interest in qualitative and quantitative pharmaceutical analysis. However, it possesses greater sensitivity due to the fact the complete separation of the excitation source of light and signal read-out. Thus, to perform a PEC experiment, apart from an electrochemical work station and a cell consisting of three electrodes; working, counter and reference electrodes, an excitation source usually consisting of an irradiation source, a monochromator, and a chopper must accompany. In addition, PEC sensors are also generally cheaper, simpler and easier to miniaturize when compared to optical sensors. As a matter of fact, almost all drugs (X. Wang et al., 2015; Xu et al., 2016; S. Zhou et al., 2014) and even biomolecules such as glucose (J. Zhang et al., 2015) and dopamine (Yan et al., 2015) can be analysed using PEC technique as they can be oxidised by the photogenerated holes with the amount of

Figure 1. Schematic diagram showing the PEC reaction process

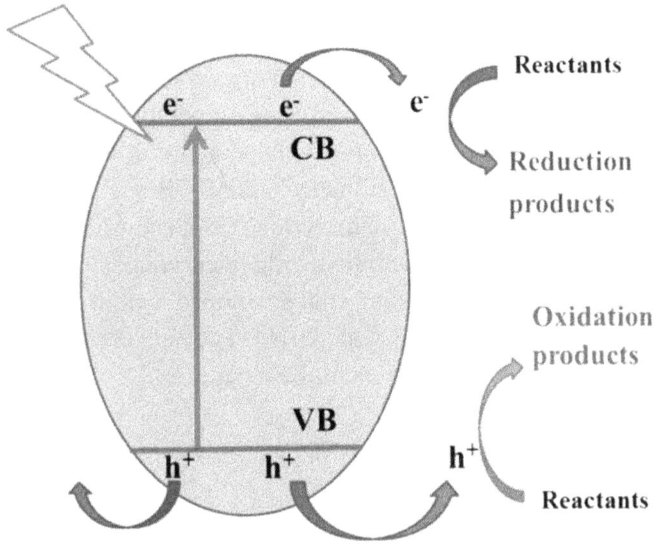

photocurrent corresponding to the amount of biomolecules oxidised. Despite all these attributes, selectivity of the sensors remains a challenge especially whenever biorecognition molecules are not employed in the construction of the sensor. Additionally, choice of photoactive materials for PEC sensors remains a challenge as some of them undergo photocorrosion while others experience high recombination rates of photogenerated carriers, in effect influencing the detection limit.

Inorganic semiconductors such as TiO_2 NPs, ZnO nanorods, and various quantum dots (QDs) including CdS and GQDs, and their composites have been widely employed in fabrication of PEC sensors due to their superior photocatalytic properties. These materials have been used to enhance the photoelectrochemical property of various substrate surfaces including; fluorine doped tin oxide (FTO), indium doped tin oxide (ITO), gold and glassy carbon electrode (GCE) electrodes. To obtain the most suitable material for PEC experiment, a number of factors have been suggested. These include: the semiconductors should have bandgaps that allow absorption of incidence light across the UV-visible region of solar spectrum; the semiconductor surface should minimize the recombination of electron-hole pairs that arise upon photoirradiation; the photocatalyst should possess adequate quality and quantity of photocatalytic reaction sites, and semiconductor should be such that the bottom of the conduction band (CB) falls at a more negative potential than the reduction potential while the top of the valency band (VB) appears at a more positive potential than the oxidation potential, to achieve thermodynamically feasible redox reactions (Tu, Zhou, & Zou, 2013).

Since most semiconductors do not meet most of these requirements, two or more semiconductors with different band gaps have been used to form heterojunctions. Alternatively, dye sensitization, plasmon resonance, polymer addition and construction of carbon-semiconductor composites have been employed so as to improve the charge separation, photocurrent conversion efficiency and lowering the band gaps (Cong, Fan, Wang, Abdel-Halim, & Zhu, 2016; Huo, Xu, Zhang, & Xu, 2015; J. Li et al., 2014; Yixin Liu et al., 2016). To a large extent, graphene and its derivatives have also been widely explored in construction of PEC sensors when incorporated into semiconductors and QDs. These have been attributed to: (a) Graphene possess very high electron transfer ability and thus is able to enhance separation between the electron and holes resulting from photoexcitation as shown in Fig. 2. This in turn results in enhanced photocatalytic performance as more holes are available for photooxidation; (b) The large surface area provided by graphene and its derivatives enhances the available active sites for the composite nanomaterials; (c) graphene can be used to tune the band gap of semiconductors since they enhance the light absorption intensity in addition to enlarging the absorption range, for example, TiO_2 can be red shifted into the visible region; (d) Graphene can also improve the stability of CdS QDs through hybridization process in a composite (Perreault, de Faria, & Elimelech, 2015). By using BiOCl and graphene sheets hybrid, a novel sensor was fabricated for glucose detection (Gopalan, Muthuchamy, & Lee, 2017). The sensor exhibited great selectivity and reproducibility, wide linear range, high sensitivity and was effective in analysis of glucose in human serum. Jiang et al developed a chlorpyrifos sensor based on BiOI and nitrogen doped graphene composite (Jiang et al., 2016). The enhanced activity was attributed to the formation of a p-n junction that ensured efficient charge separation.

Hydrothermal Synthesis of Semiconductors

To date, a number of methods have been proposed for the synthesis of nano-structured materials for application in electroanalysis. Some of these methods include: liquid phase deposition (LPD), chemical vapor deposition (CVD), solvothermal methods, ball milling and electrodeposition. In some cases, more than one method have been utilized in a combination to obtain nanomaterial with improved electrocatalytic activity. It is important to note that hydrothermal technique has been most widely employed in the synthesis of metal oxides and sulphides. This has been attributed to some of its outstanding advantages that include: the process is generally inexpensive and easy control as it makes use of relatively cheap precursor materials; often generate high yield of products, uniformly controlled size and shape of products can be achieved by tuning the process; the process also involves self- purification and thus easier to attain

Figure 2. Schematic illustration of the charge carrier transfer for GR-semiconductor nanocomposite

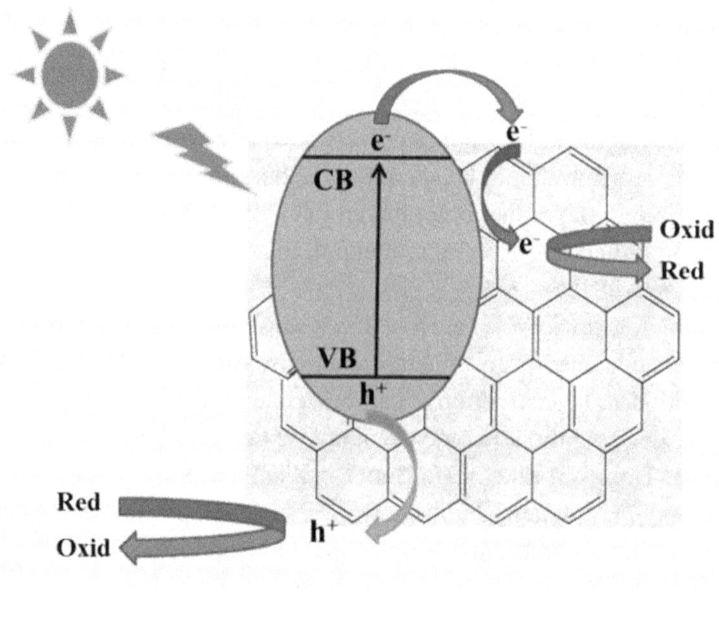

products of high quality; it is environmentally friendly since the reaction happens in a closed system in addition to the use of water as a solvent, and low energy consumption as reactions generally done under 300 °C (Dunne, Starkey, Gimeno-Fabra, & Lester, 2014; Shandilya, Rai, & Singh, 2016). Subsequently, a number of composites have been prepared including Bi_2S_3 nanorods that have been synthesized and applied in PEC sensing (B. Sun et al., 2014; M. Wang et al., 2014), photodetectors (Chen et al., 2015), solar cells (G. Li, Chen, & Gao, 2014), photocatalytic degradation of pollutants (Rauf et al., 2015; B. Weng, Zhang, Zhang, Tang, & Xu, 2015) and supercapacitors (Nie, Lu, Lei, Yang, & Wang, 2015). These applications stem from the high incident photon to electron conversion efficiency, reasonably low band-gap (1.3-1.7 ev) with excellent absorption in the visible light range (B. Sun et al., 2014). Synthesis was made possible using $Bi(NO_3)_3{\cdot}5H_2O$ and Na_2S in a Teflon lined stainless steel autoclave maintained at 120 °C for 24 hours (Okoth, Yan, Liu, & Zhang, 2016). A composite of graphene was then achieved through ultrasonication of suspensions. Bi_2S_3 nanorods were established to be uniformly distributed on the GR nanosheets as seen in SEM, Fig. 3, while XRD patterns of the samples showed that all the diffraction peaks matched well with those of pure Bi_2S_3 orthorhombic phase. Besides, UV-visible spectra were used to study the effects of the amount of graphene on the optical property of the composite and it was quite obvious that increase in GR enhanced absorption of Bi_2S_3 in the visible region which was advantageous for PEC activity.

To tame the deficiencies of TiO_2 and ZnO which include wide band gaps that respond mainly to UV irradiation that could potentially damage biomolecules, monoclinic scheelite $BiVO_4$ has been used to design photocatalysts. It possesses a band gap of about 2.4 eV and has very strong absorption in the visible region in addition to other unique properties such as ferroelasticity, chemical stability, ionic conductivity, non-toxicity (W. Zhao et al., 2016). Due to these properties, it has received much attention in fabrication of visible light photocatalytic degradation (Yao, Iwai, & Ye, 2008), PEC sensing (L. N. Wang et al., 2015), photocatalyst for dye decoloration, (Shan et al., 2016; Yin et al., 2013) as well as solar water oxidation (Ma, Pendlebury, Reynal, Le Formal, & Durrant, 2014). Despite these great attributes, $BiVO_4$ experiences low quantum yield and fast recombination of photogenerated charge carriers that have limited its practical applications. To date, a number of approaches have been taken to suppress recombination of photogenerated electron-hole pairs including coupling with other semiconductors such as WO_3 (Rao et al., 2014), g-C_3N_4 (C. J. Li et al., 2014), TiO_2 (Singh et al., 2016), MoS_2 (H. L. Li et al., 2015) and BiOI (Huang, He, Du, Chu, & Zhang, 2015). In addition, doping with elements such as W (Pattengale, Ludwig, & Huang, 2016) and Mo (Nair, Perkins, Lin, & Law, 2016; Okoth, Yan, & Zhang, 2017) has been explored. Moreover, improved separation of charges have been achieved by forming graphene/$BiVO_4$ composites since graphene is able to enhance transfer electrons (H. S. Park, Ha, Ruoff, & Bard, 2014).

To prepare monoclinic Mo-$BiVO_4$, hydrothermal synthesis using a mixture of $Na_2MoO_4{\cdot}2H_2O$, $Bi(NO_3)_3$ and NH_4VO_3 was applied for 24 h at a temperature of 200 °C. The products that were obtained were washed and dried and later mixed with graphene that had been prepared via the modified hummers' method. From Fig. 4, surface characterization was done and it is evident especially from EDS analysis and XPS spectrum that Mo, Bi, O and C were available and fairly distributed within GR-Mo-$BiVO_4$. However, it was noted that Mo doping did not obviously affect the main structural features of $BiVO_4$. On the other hand, the DRS spectra showed that both Mo-$BiVO_4$ and $BiVO_4$ were photo-excited under visible light irradiation while incorporation of GR to form GR-Mo-$BiVO_4$ resulted into a remarkable increase in absorption intensity due to the induced broad absorption by graphene in the visible region.

Design of CdS nanostructures such as nanorods, nanowires, nanoflowers, nanosheets and zero dimensional nanoparticles have been well documented due to their various properties. Hydrothermal

Figure 3. SEM images of (A) Bi_2S_3, (B) GR and (C) GR-Bi_2S_3. (C) UV-visible absorption spectra of (a) Bi_2S_3 and GR-Bi_2S_3 composites with different weight ratios of Bi_2S_3 to GR; (b) 100:1, (c) 20:1, (d) 11:1 and (e) 9:1 and (D) XRD patterns of GO, GR, GR-Bi_2S_3 and Bi_2S_3.

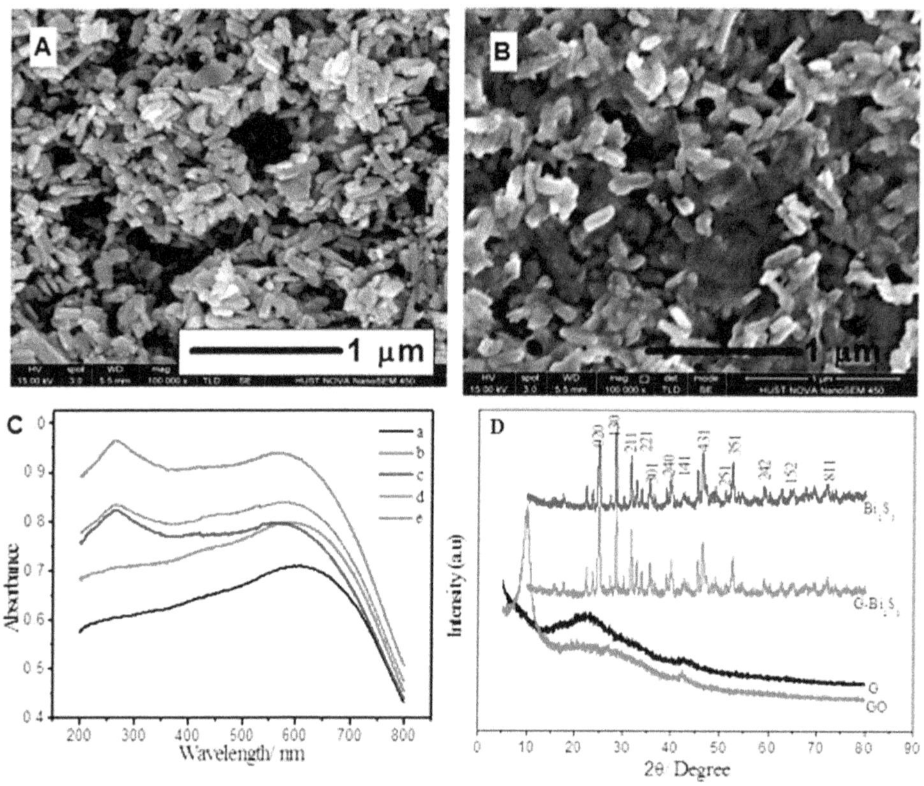

reaction is by far, the most widely reported technique for CdS preparation. In addition to reaction conditions, other factors such as source of Sulphur, addition of surfactants and Cd^{2+} source have been reported to impact the morphology of the nanostructures. Specifically, L-cysteine(G. Yu et al., 2016), thiourea (Kuang et al., 2016), thioacetamide (S. Wang et al., 2014), DMSO (Han, Yang, Zhang, & Xu, 2014) and Sulphur powder (Bera, Kundu, & Patra, 2015) have all been explored as the major sources of S^{2-} in the various synthesis and application. As a matter of fact, the semiconductor that has been widely explored to develop photocatalysts due to its desirable band gap of 2.4 ev, in addition to high absorption coefficient of visible light. Unfortunately, CdS has limitations that include photocorrosion under strong illumination as well as high recombination rate of photogenerated charge carriers. So far, lots of literature have reported various initiatives such as coupling with other semiconductors with matched energy levels such as TiO_2 and ZnO; incorporation carbon nanostructures such as graphene and doping with noble metals, all aimed at improving the separation of photoexcited holes and electrons and hence photocatalytic activity. A combination of synthesis of a composite of CdS and graphene and doping with Au NPs strategies was explored to fabricate a PEC biosensor (Okoth, Yan, Feng, & Zhang, 2018). Typically, GR-CdS nanocomposite was prepared via one pot hydrothermal synthesis with GO suspension, cadmium acetate and thioacetamide serving as the starting materials. Cysteamine stabilized Au NPs were obtained from a process that involved $NaBH_4$ reduction of $HAuCl_4$. Characterization was done

Figure 4. SEM images of (A) $BiVO_4$, (B) Mo-$BiVO_4$ and (C) GR-Mo-$BiVO_4$, EDS mapping, EDS and XPS spectra of GR-Mo-$BiVO_4$.

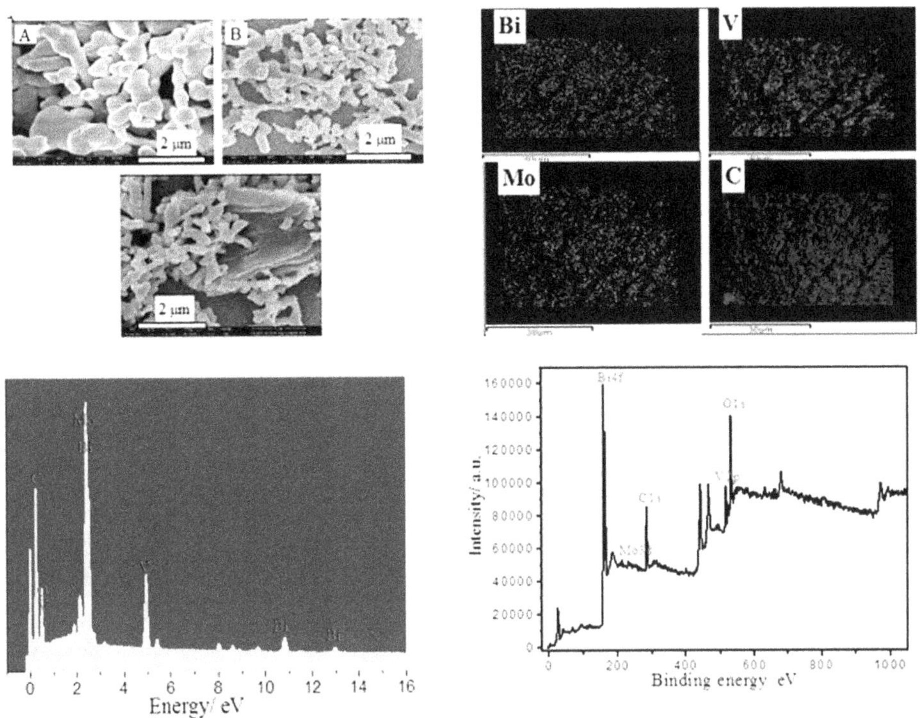

through XRD and FTIR analysis as is exhibited in Fig. 5 and both confirmed the reduction of GO to GR in the hydrothermal reaction. Since the occurance of Au NPs could not be established using SEM, the influence of these nanomaterials on GR-CdS activity were determined by PEC. It is quite clear that photocurrent intensity increased with increase in the amount of Au NPs till 12 μl due to SPR effect. In addition, further increase in Au NPs was observed to result in lower photoresponse due to increased film thickness that hindered the transfer of the photogenerated electrons.

PEC APTAMER SENSORS

Aptamers

A biosensor is a device through which specific quantitative analytical information can be obtained from a biorecognition element that is in direct spatial contact with a transducer. The main components of biosensors usually include: the biorecognition elements, the detection devices (transducer) and displays usually consisting of a signal amplifier and the data processor. Biosensors have also been constructed based DNA or RNA aptamers as bioreceptors and these are generally referred to as aptasensors. Aptamers are synthetic functional oligonucleotide receptors engineered with a high affinity and specificity towards diverse targets. By in vitro evolutionary selection method referred to as systematic evolution of ligands by exponential enrichment (SELEX), aptamers have been selected against a wide variety of targets. The

Figure 5. XRD patterns (A) and FTIR spectra (B) of GO and GR-CdS. (C) UV- absorption spectra of (a) GR-CdS and (b) Au. Inset is the PL spectrum of GR-CdS and (D) The influence of Au NPs amount on the photocurrent response of Au/GR-CdS modified electrodes in 0.1 M Na_2SO_4 at a bias potential of 0.2 V.

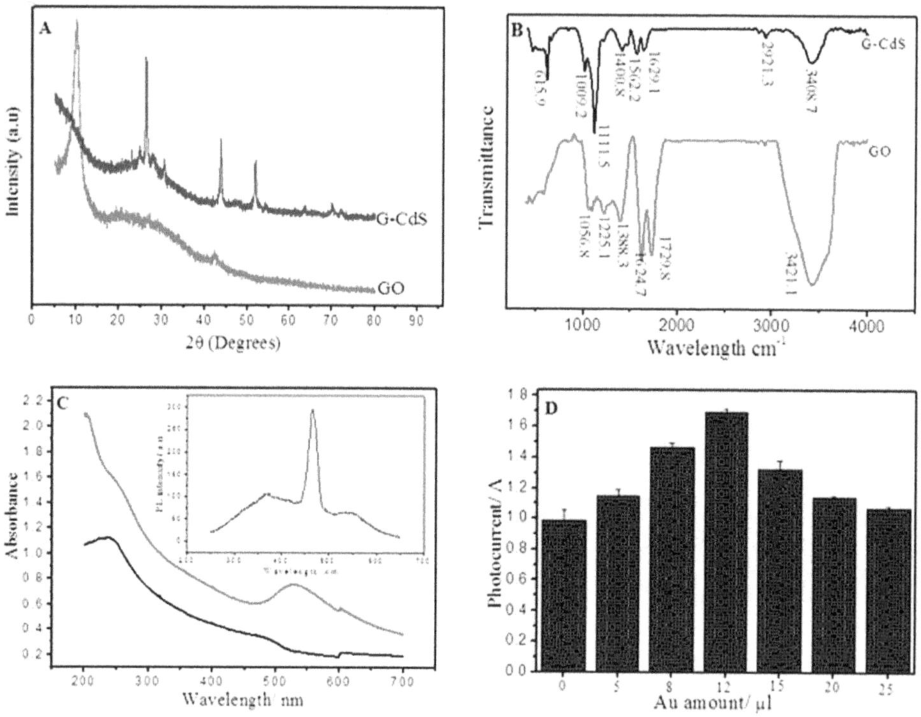

SELEX process may be divided into three steps namely binding, partitioning and amplification which are iteratively repeated to obtain sequences with improved binding ability towards the target of interest. So far, there exists a number of high affinity aptamers that have been synthesized for metal ions (M. Zhao, Fan, Chen, Shi, & Zhu, 2015), cells (Hashkavayi, Raoof, Ojani, & Kavoosian, 2017), organic environmental pollutants (P. Yu et al., 2016; L. Zhou, Wang, Li, & Li, 2014), proteins (Q. Q. Wang et al., 2015; X. Weng & Neethirajan, 2016), antibiotics (Danesh, Ramezani, Emrani, Abnous, & Taghdisi, 2016; S. Liu et al., 2017), and small molecules (Yang, Wang, Xiang, Yuan, & Chai, 2014) analysis. Due to improvement in technology, aptamers can be synthesized in a shorter time, at lower costs and can be easily modified with different functional groups so that they fit the various immobilization strategies for practical applications. Moreover, aptamers are thermally stable and reusable and are thus gaining some ground against antibodies in fabrication of biosensors.

Aptamer Selection Methods

In the SELEX method, the process is initiated by incubating a target of interest with a random library, usually consisting of 10^{13}-10^{16} single stranded DNA or RNA sequences. Each sequence usually adopts a unique subset of 3-dimensional structural combination of base pairing, stacking, sugar packing and non-canonical intramolecular interactions. Due to structural complexity established, the probability of selecting an aptamer that can strongly and specifically bind to the target of interest is increased. This is usually followed by iterative cycles of absorption, recovery of bound DNA/RNA and amplification (Iliuk, Hu, & Tao, 2011). At this stage, separation that imparts the purity and selectivity of the aptamer is done by capillary electrophoresis. The bound aptamers are then eluted and amplified by PCR to generate new pools for the next selection cycle usually with more stringent conditions (Ilgu & Nilsen-Hamilton, 2016). Actually, 8-15 cycles are often performed to obtain nucleic acids having the highest binding affinity for the target. Subsequently, the resultant sequences are cloned and sequenced, then their binding affinity and Gibbs energy calculated so that aptamers with the highest affinity and specificity for target analyte is selected (Iliuk et al., 2011; Meirinho, Dias, Peres, & Rodrigues, 2016). The SELEX process can be summarized as shown in Fig. 6. Through this technique, SRT aptamer with sequence 5'-TAGGGAATTCGTCGAC-GGATCCGGGGTCTGTTCTGCTTTGTTCTGTCGGGTCGTCTGCAGGTCGACGCATGCGCCG-3' and sulfadimethoxine (SDM) aptamer with sequence 5´-NH_2-(GAGGGCAACGAGTGTTTATAGA-3´) among others have been synthesized. The aptamers were obtained from Shanghai Sangon Biotech Co. Ltd. (Shanghai, China) and kept in a refrigerator below 4 °C. Though capillary electrophoresis method of separation has been credited with enhanced selection efficiency due to fast selection ability based on electrophoretic mobility, its application is limited when it comes to neutral species that do not possess electrophoretic mobility shift.

An alternative method is the bead-based combinatorial library selection method, where a combinatorial oligonucleotide library is synthesized using pool method on noncleavable beads followed by exposure to the desired target. After that, the bead-based library is incubated with the fluorescently tagged protein and manually picked under the fluorescence. Alternatively, the target could be labelled with biotin so that the bound aptamer beads are separated by magnetic attraction using streptavidin-coated magnetic nanoparticles. Just as aforementioned, the last step usually involves polymerase chain reaction (PCR) amplification of the sequence of the selected aptamers. One of the most outstanding features of this method is the short selection time since it does not involve iteractive selection, and PCR amplification cycles synonymous with capillary electrophoresis related SELEX. In addition, the bead based selection method allows diverse chemical modifications to be introduced into the oligonucleotides which could improve the specificity of the selected aptamer (Thiviyanathan & Gorenstein, 2012).

Figure 6. Schematic illustration of the SELEX aptamers selection process

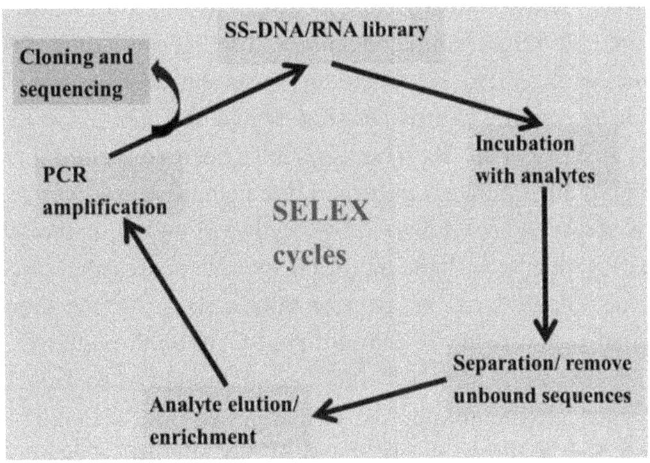

Aptamer Immobilization Techniques

The immobilization of the aptamer plays a very important part in the biosensor performance since the technique used to immobilize the aptamer onto the electrode could impact on the stability, affinity and response time, sensitivity, flexibility of use and the specificity of the aptamer towards the target. In fact, the mode of immobilization could affect the orientation of the aptamer that could subsequently affect accessibility to the target analyte, the selectivity as well as the stability of the bound aptamer onto the electrode. Therefore, in determining the most appropriate immobilization technique, crucial factors such as the physicochemical properties of the intended target, the kind of electrode surface, and the experimental conditions of the aptasensor are often considered (Meirinho et al., 2016; Xiaojuan Zhang & Yadavalli, 2011). Generally, three immobilization strategies have been employed in aptamer immobilization as show in Fig. 7 and they include: physisorption (physical adsorption), chemisorption (chemical adsorption) and covalent attachment through functional groups.

Physical adsorption of aptamers has been a widely applied immobilization technique that makes use of electrostatic interaction especially the π-π stacking interaction between the aptamers and carbon nanocomposites (J.-W. Park, Tatavarty, Kim, Jung, & Gu, 2012). The main advantage of this technique especially when used with graphene nanocomposites is that it does not alter the intrinsic properties of the material. By exploring the π-π stacking interaction between GQDs and chloramphenicol (CAP) aptamer, Liu et al were able to design a novel PEC aptasensor that was sensitively used to determine CAP in the range 10-250 nM (Yong Liu, Yan, Okoth, & Zhang, 2015). Through direct physisorption of *Staphyilococcus aureus* aptamer on SWCNT, real time potentiometric monitoring of *Staphyilococcus aureus* in skin was possible (Zelada-Guillen, Luis Sebastian-Avila, Blondeau, Riu, & Xavier Rius, 2012). However, physisorptiom method suffers from detachment of aptamers from electrode surface and may therefore not be efficient in the construction of reusable aptamer sensors (Meirinho et al., 2016). In addition, fewer binding sites have been claimed to result to the lower sensitivities for aptasensors fabricated through π stacking when compared to those prepared via covalent bonding (Zelada-Guillen et al., 2012).

In chemisorption, the conjugation of aptamers on gold electrode or gold nanoparticles is normally achieved through interaction of thiol from thiol-labelled aptamers and gold surface (Meirinho et al., 2016). This method is effective since it makes use of the strong interaction between the thiol groups (-SH) and the gold surface to form strong covalent bonds. Furthermore, Sulphur and gold atoms are able to form stable and flexible aptamer layers leading to immobilization of a large amount of aptamers that could generate improved sensitivity. By using a composite of ZnO and Au NPs, a PEC aptasensor was fabricated by attachment of thiol terminated aptamer for determination of BPA, achieving a detection limit of 0.5 nM (Qiao et al., 2016). Alternatively, based on strong interactions between aptamers and avidin, streptavidin or neutravidin (Ahmad Raston, Nguyen, & Gu, 2016; Ilkhani, Sarparast, Noori, Bathaie, & Mousavi, 2015; Wang et al., 2016) sensors have been fabricated.

In as far as covalent immobilization is concerned, the technique involves modification of the electrode with functional groups such as; amine (-NH_2), hydroxyl (-OH), carboxylic acid (-COOH), and aldehyde (-CHO) which then interact with the modified aptamer possessing the corresponding chemical functional group to form an ordered aptamer layers on the electrode surface. Generally, surfaces such as CdS are first functionalized with carboxylic acid groups (-COOH) through EDC and NHS. The amine modified aptamer are then covalently attached onto the functionalized surface followed by blocking of the remaining sites using monoethanolamine (MEA) (Zang, Lei, Hao, & Ju, 2014). Chitosan with plenty of amino groups has also been explored with (Shandost-fard, Salimi, Sharifi, & Korani, 2013; Shen,

Figure 7. Various techniques of aptamers immobilization

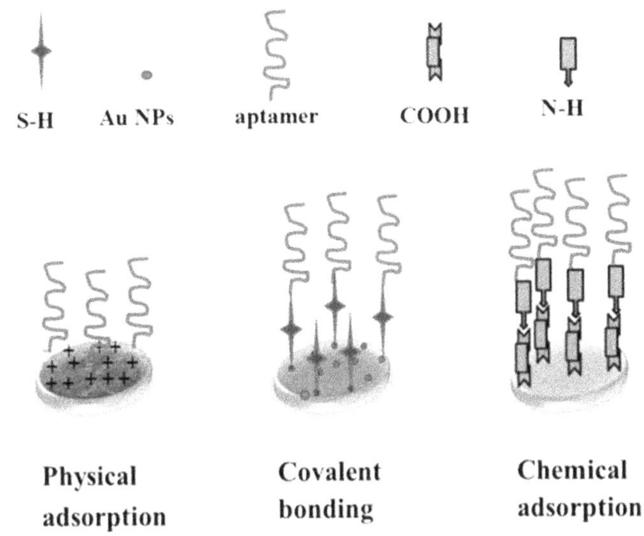

Guo, Sun, & Wang, 2014) or without (Erdem, Eksin, & Muti, 2014) glutaraldehyde as a cross linking agent to covalently immobilize various aptamers. In a recent study, a thiolated cocaine aptamer was covalently attached onto a conducting polymer (polythiophene-*g*-polyalanine) via Sulfosuccinimidyl 4-[N-maleimidomethyl]cyclohexane-1-carboxylate (Sulfo-SMCC) chemistry, then used in electrochemical analysis (Bozokalfa et al., 2016).

PEC Aptamer Sensing

Recently, PEC aptamer sensing has increasingly gained attention as an effective method of analysis. Due to some properties associated with aptamers as bioreceptors in the construction of aptasensors, aptamer based sensors have been found to offer some advantages over bioreceptors based biosensors. These include: chemical stability under a wide range of buffer conditions, high selectivity due to their high affinity towards specific analytes, resistance to harsh treatments without loss of bioactivity, reversible natural denaturation, adaptability to various targets and existence of an array immobilization techniques (Ilgu & Nilsen-Hamilton, 2016; Kim, Raston, & Gu, 2016). Moreover, the small size of aptamers allows their efficient immobilization at high densities in different surfaces that could obviously impact the sensitivity of the resulting sensor. It is worth noting that very few ions, biomolecules and drugs aptamers have been developed due to the complicated nature of the SELEX technique and so very few aptasensors have be constructed. Additionally, some of these aptamers are expensive and also require stringent storage conditions to ensure stability and optimal activity. The principle of PEC aptasensors is based on photooxidation or reduction of biomolecules to enable photogenerated electrons transfer between the analyte and the aptamers under light irradiation to generate the PEC response (Zeng, Tu, Li, Bao, & Dai, 2014). To be able to achieve detection, a specific binding of the target molecule to the biorecognition element immobilized on a suitable semiconductor modified electrode occurs and this produces either discrete or continuous signals which are proportional to the amount of the analytes. Subsequently, the signals are transformed into digital format that can be recognized and interpreted. The above processes are summarized in Fig. 8. From this figure, there could either be decline in photoresponse due to steric hindrance or increase in photoresponse due to conformation change/ oxidation upon reaction with a target.

Application of Aptamers in PEC Sensing of Selected Pharmaceuticals

As has been stated before, the design, fabrication and production, miniaturization and automation of biosensors for application in real samples could be more easily achieved with aptamers than with biorecognition molecules. In these fabrications, a decrease in electrical signals due to enhanced steric hindrance or increased photocurrent response upon interaction of the aptamer with the target molecules have usually been observed. For instance, in a typical fabrication of a PEC sensor for sulfadimethoxine (SDM), a composite of graphene-Bi_2S_3 prepared by mixing suspensions of graphene (GR) and Bi_2S_3 were drop-casted on FTO to serve as a working electrode. SDM is a low-cost broad-spectrum antibiotic that has been widely utilized in the veterinary treatment of coccidiosis and other bacterial infections (Song, Jeong, Jeon, Jo, & Ban, 2012), and thus found its way into the environment through agricultural application of contaminated manure. It has been established that low concentration of SDM in aquatic environment could be toxic both to the target organisms and green algae hence the need for routine analysis. Upon irradiation of the modified electrode, a considerably high photocurrent response was realized as shown in Fig. 9, and this was accounted for by the high incident photon to electron conversion efficiency of Bi_2S_3, as well as the fast charge transfer rate synonymous with GR. As mentioned earlier, covalent mode of immobilization of aptamers possess the advantage of stability aptamer layers. With the help of glutaraldehyde and chitosan, the amino terminated aptamers were immobilized on the electrode. Evidently, immobilization of aptamers led to a steep decline in the photocurrent intensity attributed to steric hindrance that impeded the electrons transfer. These results did not only confirm the successful fabrication of the aptasensor but also the fact that covalently bonded aptamers could provide better results when compared to those of physisorbed aptamers. Moreover, they correlate well with the results attained using the EIS analysis. For instance, results from EIS analysis showed that the net increase in R_{et} value

Figure 8. Schematic illustration on the PEC aptasensing process

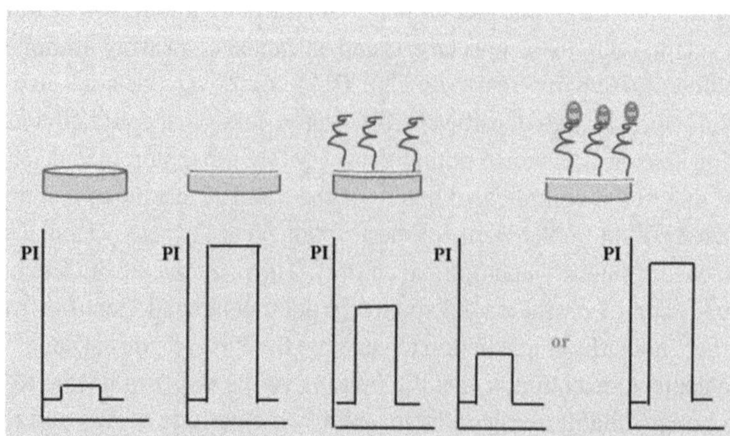

for aptamers that were covalently bonded were approximately three times higher than that for physically adsorbed aptamer thus the amino groups plenty in chitosan were more effective in the immobilization of the amine-terminated aptamer.

The fabricated sensor was then employed in the determination of various concentrations of SDM with a detection limit (3S/N) of 0.55 nM being achieved. To examine the effectiveness of the proposed sensor, selectivity and reproducibility were explored and were found to be satisfactory especially due to the fact that aptamers have tendencies of binding to specific molecules. Furthermore, the working of the proposed sensor was explored in a veterinary drug formulation. Similarly, the drug was determined using the established HPLC technique and the results as shown in Table 1 indicates reliability of the proposed technique. The Bi_2S_3 nanostructures have very narrow band width that enabled them to absorb light in the visible region and were able to generate photoelectrons and holes on light illumination. However, with the high electrons and holes recombination rate, graphene was incorporated to accept photogenerated electrons from the conduction band (CB) of Bi_2S_3 and subsequently accelerate their transfer to the electrode. This made the suppression of charge recombination possible with the resultant photogenerated holes being transported to the valence band (VB) of Bi_2S_3 where they were used to oxidize SDM molecules as shown in Fig. 9.

Another drug, streptomycin (SRT) which belongs to aminoglycoside family of antibiotics has been widely employed as a veterinary drug for treatment of bacterial infections. In addition, it has been employed as a pesticide to arrest the growth of bacteria, fungi and algae in vegetables, fruits, ornamental plants, seeds and aquaria (N. D. Zhou et al., 2013). Uncontrolled application of SRT could be of great concern as this could result into the presence of the antibiotic residues in foodstuffs and water bodies which could pose health risks ranging from nausea and fever to more severe ailments such as photodermatitis

Figure 9. Photocurrent (A) and EIS (B) responses of different electrodes: (a) FTO, (b) GR-Bi_2S_3, (c) CS/GR-Bi_2S_3 (d) aptamer/CS/GR-Bi_2S_3, (e) SDM-aptamer/CS/GR-Bi_2S_3. Schematic illustration for PEC aptasensing of SDM on APT/CS/GR-Bi_2S_3/FTO electrode (C) while (D) represents the photocurrent responses of the proposed aptasensor towards (a-f) 1, 10, 20, 50, 80 and 100 nM SDM.

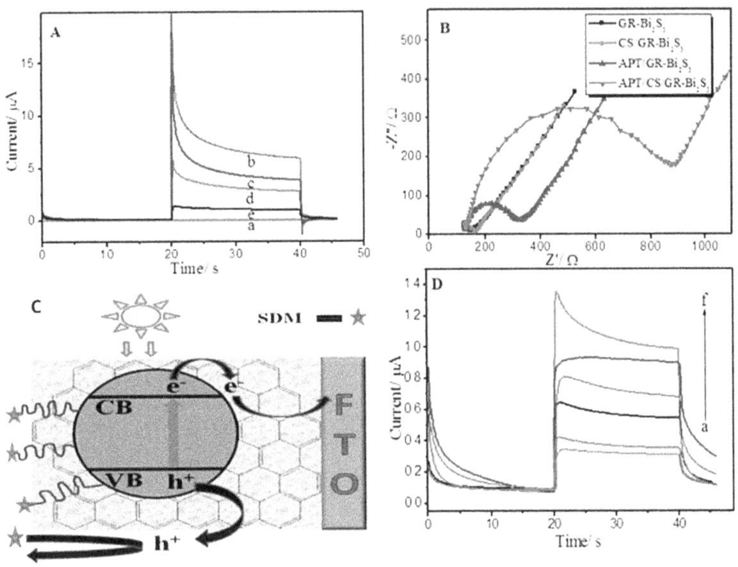

Table 1. Assay of SDM in pharmaceutical formulations (n=3)

Drug	Amount labeled (g)	Determined by PEC (g)	Determined by HPLC (g)
SDM	8.3	8.389± 0.0682	8.458 ± 0.001

and anaphylaxis (B. Q. Liu et al., 2013). Consequently, a simple, sensitive but effective technique for monitoring trace amounts of SRT was proposed. In the fabrication of the sensor, hybrid material prepared by directly mixing graphene (2 g L^{-1}) and Mo-$BiVO_4$ (2 g L^{-1}) to form GR-Mo-$BiVO_4$ was drop casted on an FTO glass. This was then followed by immobilization of SRT aptamer via physical adsorption.

Photoelectrochemical and photoluminescence studies performed suggested that Mo doping immensely improved the capacity of $BiVO_4$ to photogenerate charge carriers as shown in Fig. 10. The Mo atoms replaced the partial sites of vanadium resulting into a less distorted structure at Bi center which enhanced the lifetime of the electrons thus improved PEC activity for $BiVO_4$. Furthermore, incorporation of GR into the composite improved the transportation of photogenerated electrons and inhibiting the recombination of electron-hole pairs as shown in in photoluminescence results thus GR-Mo-$BiVO_4$ was quite useful for PEC sensing. However, immobilized aptamers hindered the transfer of electrons suggesting that of π-π bonds were formed between graphene and the aptamer molecules. Once all the experimental conditions were optimized, the proposed sensor was used for quantitative determination of SRT in the range 0.1 to 100 nM with a detection limit of 0.0481 nM and this was much lower than those found with other techniques as shown in Table 2. Furthermore, the aptasensor was established to be fairly selective when determined in the presence of other antibiotics while the applicability was confirmed using commercial drugs. On another hand, the stability of the aptasensors was assessed by determining the extent of photocurrent change in100 nM SRT after the as-prepared sensors were stored at 4 °C for 3 weeks. The results revealed that the aptasensor retained about 93% of its photocurrent, confirming long term stability.

Table 2. Comparison of selected analytical techniques for streptomycin determination

Method	Linear range / M	LOD / M	Reference
Fluorescence	2×10^{-6}- 7.0×10^{-5}	2.0×10^{-7}	(Lan et al., 2015)
Fluorescent aptasensor	6.0×10^{-8}-2.0×10^{-6}	5.45×10^{-8}	(Taghdisi, Danesh, Nameghi, Ramezani, & Abnous, 2016)
Fluorescent quenching aptasensor	-	4.76×10^{-8}	(Emrani et al., 2016)
Electrochemical aptasensor	3.0×10^{-8}- 1.5×10^{-6}	1.14×10^{-8}	(Danesh et al., 2016)
Electrochemical immunosensor	3.43×10^{-11}- 3.43×10^{-8}	3.43×10^{-12}	(B. Q. Liu et al., 2011)
Electrochemical molecularly imprinted sensor	3.43×10^{-11}- 1.37×10^{-8}	6.86×10^{-12}	(B. Q. Liu et al., 2013)
Colorimetric aptasensor	-	7.31×10^{-8}	(Emrani et al., 2016)
Colorimetry	3.0×10^{-8}-3.0×10^{-7}	3.0×10^{-9}	(Xiaofang Zhang et al., 2013)
Colorimetric sensor	8.0×10^{-8}-4.8×10^{-7}	2.0×10^{-9}	(J. Y. Sun et al., 2011)
Electrochemical quartz crystal nanobalance	2.08×10^{-10}- 6.86×10^{-9}	2.06×10^{-10}	(Mishra, Sharma, & Bhand, 2015)
Photoelectrochemical aptasensor	1.0×10^{-10}- 1.0×10^{-7}	4.81×10^{-11}	This work

An anti-inflammatory drug, diclofenac (DCF) that is usually administered to patients after surgery or those with acute and chronic inflammation was also analyzed. This was vital as prolonged use of this drug could result into life threatening heart or circulation problems in man as well inducing renal lesions in gills of fish, essentially affecting the health of aquatic organisms (Arvand, Gholizadeh, & Zanjanchi, 2012). In brief, cysteamine stabilized Au NPs were drop casted onto GR-CdS modified FTO electrode. PEC studies conducted as exhibited in Fig. 11 showed that GR-CdS generated a high photoresponse ascribed to the strong absorption of visible light ability by CdS in addition to excellent electrical conductivity of graphene that ensured efficient transfer of photogenerated electrons. Further increase in photocurrent intensity was noted when Au NPs were deposited due to surface plasmon resonance that increased the energy transfer from Au surface to CdS, improving the photogeneration of holes and electrons. Moreover, the Au NPs were beneficial to the formation of bonds with the thiol terminated aptamer. The resultant electrode was coated with an aptamer solution and used as the working electrode. After incubation with DCF, a significant enhancement of photoresponse was observed, confirming that DCF molecules were captured of by the aptamers and these were oxidized by the photogenerated holes hence the abundance of photogenerated electron that got to the electrode. Under optimized conditions, the PEC response of the proposed sensor was found to linearly increase with increase in DCF concentration in the wide detection range of 1-150 nM. Moreover, an accptable detection limit estimated to be 0.78 nM was obtained, demonstrating great sensitivity. The device exhibited excellent selectivity, stability and furthermore, great promise was noted when applied in real sample analysis.

Figure 10. (A) Photocurrent responses and (B) Photoluminescence spectra of (a) $BiVO_4$, (b) Mo-$BiVO_4$ and (c) GR-Mo-$BiVO_4$. Photoresponses of the aptasensor towards (a-f) 0.1,1, 10, 20, 50 and 100 nM SRT(C) and (D), towards 100 nM of SRT in the presence of interferences.

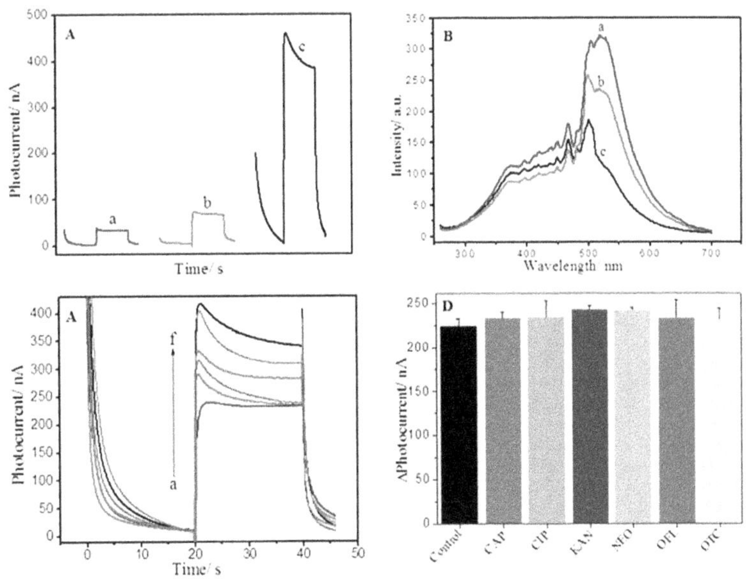

CONCLUSION

In summary, tremendous work has been done to construct novel aptamer sensors by making use of various functionalized nanomaterials with excellent PEC activity and aptamers as recognition units. Various non-toxic semi-conductor nanocomposites have been prepared via the simple hydrothermal technique and applied in PEC sensing. By combining an appropriate amount of GR and Bi_2S_3 nanorods through sonication, an excellent nanocomposite was developed that exhibited improved PEC activity. Besides, GR was also used to construct a composite with Mo-$BiVO_4$. In such studies, GR not only enhanced the PEC activity by improving electron transfer rate, but also provided a platform to immobilize SRT aptamer. The materials were thereby used to fabricate sensitive and selective PEC aptamer sensors for SDM and SRT. In another fabrication process, CdS nanomaterial was grown on GR in a hydrothermal preparation. Such a composite provided the advantages of admirable photoactivity of CdS in the visible light region and fast transfer of electrons of GR. The Au NPs deposited on GR-CdS further enhanced the sensitivity of the sensor, aided the immobilization of SH- terminated aptamer with the resultant electrode being reliably used in analysis of DCF in environmental sample. This work therefore offers insights into some recent achievements in the development of sensitive and selective PEC biosensors for pharmaceuticals. From such accomplishments, more nanomaterials could be explored in PEC sensing by tuning their

Figure 11. (A) Photocurrent responses in 0.1 M Na_2SO_4 at an applied potential of 0.2 V and (B) Nyquist plots of the various electrodes in 0.1 M KCl containing 5.0 mM $K_3Fe(CN)_6/K_4Fe(CN)_6$. Photocurrent responses of the aptasensor to various concentrations of DCF (C) and corresponding calibration curve (D).

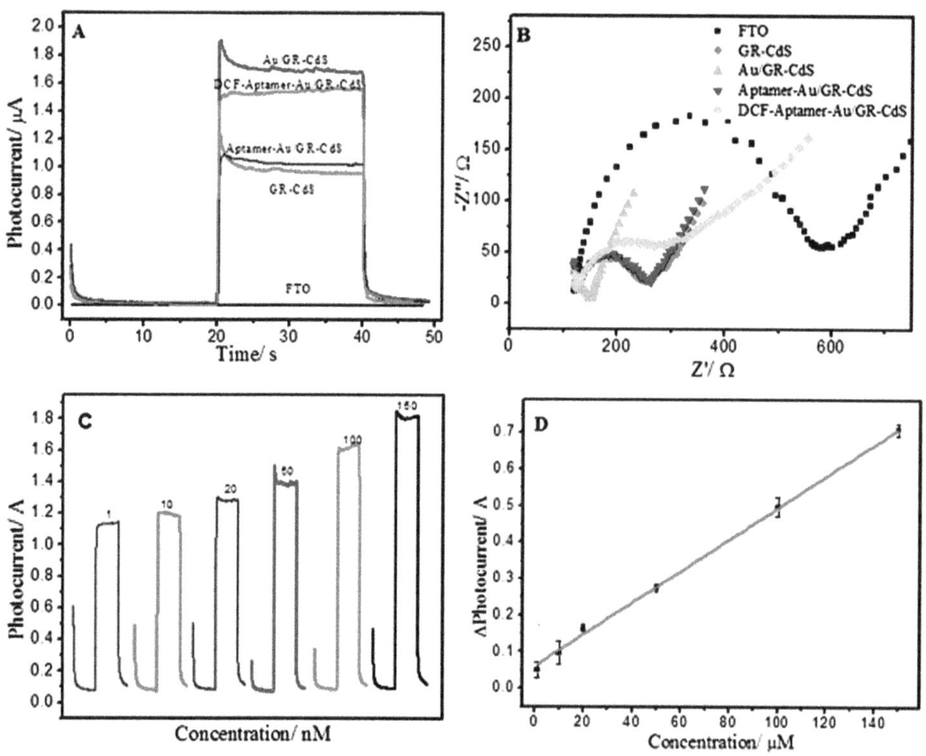

properties to make them responsive in the visible region while a variety of aptamers could be designed to make determination of more analytes possible.

REFERENCES

Ahmad Raston, N. H., Nguyen, V.-T., & Gu, M. B. (2016). A new lateral flow strip assay (LFSA) using a pair of aptamers for the detection of Vaspin. *Biosensors & Bioelectronics*. doi:10.1016/j.bios.2016.11.061 PMID:27916536

Arvand, M., Gholizadeh, T. M., & Zanjanchi, M. A. (2012). MWCNTs/$Cu(OH)_2$ nanoparticles/IL nanocomposite modified glassy carbon electrode as a voltammetric sensor for determination of the non-steroidal anti-inflammatory drug diclofenac. *Materials Science & Engineering C-Materials for Biological Applications*, *32*(6), 1682–1689. doi:10.1016/j.msec.2012.04.066 PMID:24364977

Bera, R., Kundu, S., & Patra, A. (2015). 2D Hybrid Nanostructure of Reduced Graphene Oxide-CdS Nanosheet for Enhanced Photocatalysis. *ACS Applied Materials & Interfaces*, *7*(24), 13251–13259. doi:10.1021/acsami.5b03800 PMID:26029992

Bozokalfa, G., Akbulut, H., Demir, B., Guler, E., Gumus, Z. P., Odaci Demirkol, D., ... Yagci, Y. (2016). Polypeptide Functional Surface for the Aptamer Immobilization: Electrochemical Cocaine Biosensing. *Analytical Chemistry*, *88*(7), 4161–4167. doi:10.1021/acs.analchem.6b00760 PMID:26928030

Chen, G., Yu, Y., Zheng, K., Ding, T., Wang, W., Jiang, Y., & Yang, Q. (2015). Fabrication of Ultrathin Bi_2S_3 Nanosheets for High-Performance, Flexible, Visible-NIR Photodetectors. *Small*, *11*(24), 2848–2855. doi:10.1002mll.201403508 PMID:25703599

Cong, X., Fan, G.-C., Wang, X., Abdel-Halim, E. S., & Zhu, J.-J. (2016). Enhanced photoelectrochemical aptasensing platform amplified through the sensitization effect of CdTe@CdS core-shell quantum dots coupled with exonuclease-I assisted target recycling. *Journal of Materials Chemistry. B, Materials for Biology and Medicine*, *4*(36), 6117–6124. doi:10.1039/C6TB01807F

Danesh, N. M., Ramezani, M., Emrani, A. S., Abnous, K., & Taghdisi, S. M. (2016). A novel electrochemical aptasensor based on arch-shape structure of aptamer-complimentary strand conjugate and exonuclease I for sensitive detection of streptomycin. *Biosensors & Bioelectronics*, *75*, 123–128. doi:10.1016/j.bios.2015.08.017 PMID:26302000

Dunne, P. W., Starkey, C. L., Gimeno-Fabra, M., & Lester, E. H. (2014). The rapid size- and shape-controlled continuous hydrothermal synthesis of metal sulphide nanomaterials. *Nanoscale*, *6*(4), 2406–2418. doi:10.1039/C3NR05749F PMID:24435800

Emrani, A. S., Danesh, N. M., Lavaee, P., Ramezani, M., Abnous, K., & Taghdisi, S. M. (2016). Colorimetric and fluorescence quenching aptasensors for detection of streptomycin in blood serum and milk based on double-stranded DNA and gold nanoparticles. *Food Chemistry*, *190*, 115–121. doi:10.1016/j.foodchem.2015.05.079 PMID:26212949

Erdem, A., Eksin, E., & Muti, M. (2014). Chitosan-graphene oxide based aptasensor for the impedimetric detection of lysozyme. *Colloids and Surfaces. B, Biointerfaces*, *115*, 205–211. doi:10.1016/j.colsurfb.2013.11.037 PMID:24362059

Gopalan, A. I., Muthuchamy, N., & Lee, K. P. (2017). A novel bismuth oxychloride-graphene hybrid nanosheets based non-enzymatic photoelectrochemical glucose sensing platform for high performances. *Biosensors & Bioelectronics*, *89*, 352–360. doi:10.1016/j.bios.2016.07.017 PMID:27453439

Han, C., Yang, M.-Q., Zhang, N., & Xu, Y.-J. (2014). Enhancing the visible light photocatalytic performance of ternary CdS-(graphene-Pd) nanocomposites via a facile interfacial mediator and co-catalyst strategy. *Journal of Materials Chemistry. A, Materials for Energy and Sustainability*, *2*(45), 19156–19166. doi:10.1039/C4TA04151H

Hashkavayi, A. B., Raoof, J. B., Ojani, R., & Kavoosian, S. (2017). Ultrasensitive electrochemical aptasensor based on sandwich architecture for selective label-free detection of colorectal cancer (CT26) cells. *Biosensors & Bioelectronics*, *92*, 630–637. doi:10.1016/j.bios.2016.10.042 PMID:27829554

Huang, H. W., He, Y., Du, X., Chu, P. K., & Zhang, Y. H. (2015). A General and Facile Approach to Heterostructured Core/Shell $BiVO_4$/BiOI p-n Junction: Room-Temperature in Situ Assembly and Highly Boosted Visible-Light Photocatalysis. *ACS Sustainable Chemistry & Engineering*, *3*(12), 3262–3273. doi:10.1021/acssuschemeng.5b01038

Huo, H., Xu, Z., Zhang, T., & Xu, C. (2015). Ni/CdS/TiO_2 nanotube array heterostructures for high performance photoelectrochemical biosensing. *Journal of Materials Chemistry. A, Materials for Energy and Sustainability*, *3*(11), 5882–5888. doi:10.1039/C4TA07190E

Ilgu, M., & Nilsen-Hamilton, M. (2016). Aptamers in analytics. *Analyst (London)*, *141*(5), 1551–1568. doi:10.1039/C5AN01824B PMID:26864075

Iliuk, A. B., Hu, L., & Tao, W. A. (2011). Aptamer in Bioanalytical Applications. *Analytical Chemistry*, *83*(12), 4440–4452. doi:10.1021/ac201057w PMID:21524128

Ilkhani, H., Sarparast, M., Noori, A., Bathaie, S. Z., & Mousavi, M. F. (2015). Electrochemical aptamer/antibody based sandwich immunosensor for the detection of EGFR, a cancer biomarker, using gold nanoparticles as a signaling probe. *Biosensors & Bioelectronics*, *74*, 491–497. doi:10.1016/j.bios.2015.06.063 PMID:26176209

Jiang, D., Du, X., Chen, D., Li, Y., Hao, N., Qian, J., ... Wang, K. (2016). Facile wet chemical method for fabricating p-type BiOBr/n-type nitrogen doped graphene composites: Efficient visible-excited charge separation, and high-performance photoelectrochemical sensing. *Carbon*, *102*, 10–17. doi:10.1016/j.carbon.2016.02.027

Kim, Y. S., Raston, N. H. A., & Gu, M. B. (2016). Aptamer-based nanobiosensors. *Biosensors & Bioelectronics*, *76*, 2–19. doi:10.1016/j.bios.2015.06.040 PMID:26139320

Kuang, P. Y., Zheng, P. X., Liu, Z. Q., Lei, J. L., Wu, H., Li, N., & Ma, T. Y. (2016). Embedding Au Quantum Dots in Rimous Cadmium Sulfide Nanospheres for Enhanced Photocatalytic Hydrogen Evolution. *Small*, *12*(48), 6735–6744. doi:10.1002mll.201602870 PMID:27709776

Lan, M. H., Liu, W. M., Ge, J. C., Wu, J. S., Sun, J. Y., Zhang, W. J., & Wang, P. F. (2015). A selective fluorescent and colorimetric dual-responses chemosensor for streptomycin based on polythiophene derivative. *Spectrochimica Acta Part a-Molecular and Biomolecular Spectroscopy, 136*, 871-874. doi:10.1016/j.saa.2014.09.106

Li, C. J., Wang, S. P., Wang, T., Wei, Y. J., Zhang, P., & Gong, J. L. (2014). Monoclinic Porous $BiVO_4$ Networks Decorated by Discrete g-C_3N_4 Nano-Islands with Tunable Coverage for Highly Efficient Photocatalysis. *Small, 10*(14), 2783–2790. doi:10.1002mll.201400506 PMID:24740883

Li, G., Chen, X., & Gao, G. (2014). Bi_2S_3 microspheres grown on graphene sheets as low-cost counter-electrode materials for dye-sensitized solar cells. *Nanoscale, 6*(6), 3283–3288. doi:10.1039/C3NR06093D PMID:24509629

Li, H. L., Yu, K., Lei, X., Guo, B. J., Fu, H., & Zhu, Z. Q. (2015). Hydrothermal Synthesis of Novel MoS_2/$BiVO_4$ Hetero-Nanoflowers with Enhanced Photocatalytic Activity and a Mechanism Investigation. *The Journal of Physical Chemistry C, 119*(39), 22681–22689. doi:10.1021/acs.jpcc.5b06729

Li, J., Tu, W., Li, H., Han, M., Lan, Y., Dai, Z., & Bao, J. (2014). In Situ-Generated Nano-Gold Plasmon-Enhanced Photoelectrochemical Aptasensing Based on Carboxylated Perylene-Functionalized Graphene. *Analytical Chemistry, 86*(2), 1306–1312. doi:10.1021/ac404121c PMID:24377281

Liu, B. Q., Tang, D. P., Zhang, B., Que, X. H., Yang, H. H., & Chen, G. N. (2013). Au(III)-promoted magnetic molecularly imprinted polymer nanospheres for electrochemical determination of streptomycin residues in food. *Biosensors & Bioelectronics, 41*, 551–556. doi:10.1016/j.bios.2012.09.021 PMID:23058661

Liu, B. Q., Zhang, B., Cui, Y. L., Chen, H. F., Gao, Z. Q., & Tang, D. P. (2011). Multifunctional Gold-Silica Nanostructures for Ultrasensitive Electrochemical Immunoassay of Streptomycin Residues. *ACS Applied Materials & Interfaces, 3*(12), 4668–4676. doi:10.1021/am201087r PMID:22059488

Liu, S., Wang, Y., Xu, W., Leng, X., Wang, H., Guo, Y., & Huang, J. (2017). A novel sandwich-type electrochemical aptasensor based on GR-3D Au and aptamer-AuNPs-HRP for sensitive detection of oxytetracycline. *Biosensors & Bioelectronics, 88*, 181–187. doi:10.1016/j.bios.2016.08.019 PMID:27544787

Liu, Y., Ma, H., Zhang, Y., Pang, X., Fan, D., Wu, D., & Wei, Q. (2016). Visible light photoelectrochemical aptasensor for adenosine detection based on CdS/PPy/g-C_3N_4 nanocomposites. *Biosensors & Bioelectronics, 86*, 439–445. doi:10.1016/j.bios.2016.06.089 PMID:27424261

Liu, Y., Yan, K., Okoth, O. K., & Zhang, J. (2015). A label-free photoelectrochemical aptasensor based on nitrogen-doped graphene quantum dots for chloramphenicol determination. *Biosensors & Bioelectronics, 74*, 1016–1021. doi:10.1016/j.bios.2015.07.067 PMID:26264269

Ma, Y. M., Pendlebury, S. R., Reynal, A., Le Formal, F., & Durrant, J. R. (2014). Dynamics of photogenerated holes in undoped $BiVO_4$ photoanodes for solar water oxidation. *Chemical Science (Cambridge), 5*(8), 2964–2973. doi:10.1039/C4SC00469H

Meirinho, S. G., Dias, L. G., Peres, A. M., & Rodrigues, L. R. (2016). Voltammetric aptasensors for protein disease biomarkers detection: A review. *Biotechnology Advances, 34*(5), 941–953. doi:10.1016/j.biotechadv.2016.05.006 PMID:27235188

Mishra, G. K., Sharma, A., & Bhand, S. (2015). Ultrasensitive detection of streptomycin using flow injection analysis-electrochemical quartz crystal nanobalance (FIA-EQCN) biosensor. *Biosensors & Bioelectronics*, *67*, 532–539. doi:10.1016/j.bios.2014.09.033 PMID:25266252

Nair, V., Perkins, C. L., Lin, Q. Y., & Law, M. (2016). Textured nanoporous Mo:$BiVO_4$ photoanodes with high charge transport and charge transfer quantum efficiencies for oxygen evolution. *Energy & Environmental Science*, *9*(4), 1412–1429. doi:10.1039/C6EE00129G

Nie, G., Lu, X., Lei, J., Yang, L., & Wang, C. (2015). Facile and controlled synthesis of bismuth sulfide nanorods-reduced graphene oxide composites with enhanced supercapacitor performance. *Electrochimica Acta*, *154*, 24–30. doi:10.1016/j.electacta.2014.12.090

Okoth, O. K., Yan, K., Feng, J., & Zhang, J. (2018). Label-free photoelectrochemical aptasensing of diclofenac based on gold nanoparticles and graphene-doped CdS. *Sensors and Actuators. B, Chemical*, *256*, 334–341. doi:10.1016/j.snb.2017.10.089

Okoth, O. K., Yan, K., Liu, Y., & Zhang, J. (2016). Graphene-doped Bi_2S_3 nanorods as visible-light photoelectrochemical aptasensing platform for sulfadimethoxine detection. *Biosensors & Bioelectronics*, *86*, 636–642. doi:10.1016/j.bios.2016.07.037 PMID:27471154

Okoth, O. K., Yan, K., & Zhang, J. (2017). Mo-doped BiVO4 and graphene nanocomposites with enhanced photoelectrochemical performance for aptasensing of streptomycin. *Carbon*, *120*, 194–202. doi:10.1016/j.carbon.2017.04.079

Park, H. S., Ha, H.-W., Ruoff, R. S., & Bard, A. J. (2014). On the improvement of photoelectrochemical performance and finite element analysis of reduced graphene oxide-$BiVO_4$ composite electrodes. *Journal of Electroanalytical Chemistry*, *716*, 8–15. doi:10.1016/j.jelechem.2013.08.036

Park, J.-W., Tatavarty, R., Kim, D. W., Jung, H.-T., & Gu, M. B. (2012). Immobilization-free screening of aptamers assisted by graphene oxide. *Chemical Communications*, *48*(15), 2071–2073. doi:10.1039/C2CC16473F PMID:22143382

Pattengale, B., Ludwig, J., & Huang, J. (2016). Atomic Insight into the W-Doping Effect on Carrier Dynamics and Photoelectrochemical Properties of $BiVO_4$ Photoanodes. *The Journal of Physical Chemistry C*, *120*(3), 1421–1427. doi:10.1021/acs.jpcc.5b11451

Perreault, F., de Faria, A. F., & Elimelech, M. (2015). Environmental applications of graphene-based nanomaterials. *Chemical Society Reviews*, *44*(16), 5861–5896. doi:10.1039/C5CS00021A PMID:25812036

Qiao, Y., Li, J., Li, H., Fang, H., Fan, D., & Wang, W. (2016). A label-free photoelectrochemical aptasensor for bisphenol A based on surface plasmon resonance of gold nanoparticle-sensitized ZnO nanopencils. *Biosensors & Bioelectronics*, *86*, 315–320. doi:10.1016/j.bios.2016.06.062 PMID:27387262

Rao, P. M., Cai, L. L., Liu, C., Cho, I. S., Lee, C. H., Weisse, J. M., ... Zheng, X. L. (2014). Simultaneously Efficient Light Absorption and Charge Separation in WO_3/$BiVO_4$ Core/Shell Nanowire Photoanode for Photoelectrochemical Water Oxidation. *Nano Letters*, *14*(2), 1099–1105. doi:10.1021/nl500022z PMID:24437363

Rauf, A., Sher Shah, M. S. A., Choi, G. H., Humayoun, U. B., Yoon, D. H., Bae, J. W., ... Yoo, P. J. (2015). Facile Synthesis of Hierarchically Structured Bi_2S_3/Bi_2WO_6 Photocatalysts for Highly Efficient Reduction of Cr(VI). *ACS Sustainable Chemistry & Engineering*, *3*(11), 2847–2855. doi:10.1021/acssuschemeng.5b00783

Shan, L., Liu, Y., Ma, C., Dong, L., Liu, L., & Wu, Z. (2016). Enhanced Photocatalytic Performance in Ag^+-Induced $BiVO_4/\text{-}Bi_2O_3$ Heterojunctions. *European Journal of Inorganic Chemistry*, *2016*(2), 232–239. doi:10.1002/ejic.201500936

Shandilya, M., Rai, R., & Singh, J. (2016). Review: Hydrothermal technology for smart materials. *Advances in Applied Ceramics*, *115*(6), 354–376. doi:10.1080/17436753.2016.1157131

Shandost-fard, F., Salimi, A., Sharifi, E., & Korani, A. (2013). Fabrication of a highly sensitive adenosine aptasensor based on covalent attachment of aptamer onto chitosan-carbon nanotubes-ionic liquid nanocomposite. *Biosensors & Bioelectronics*, *48*, 100–107. doi:10.1016/j.bios.2013.03.060 PMID:23660341

Shen, G., Guo, Y., Sun, X., & Wang, X. (2014). Electrochemical Aptasensor Based on Prussian Blue-Chitosan-Glutaraldehyde for the Sensitive Determination of Tetracycline. *Nano-Micro Letters*, *6*(2), 143–152. doi:10.1007/BF03353778

Singh, A. P., Kodan, N., Mehta, B. R., Held, A., Mayrhofer, L., & Moseler, M. (2016). Band Edge Engineering in $BiVO_4/TiO_2$ Heterostructure: Enhanced Photoelectrochemical Performance through Improved Charge Transfer. *ACS Catalysis*, *6*(8), 5311–5318. doi:10.1021/acscatal.6b00956

Song, K.-M., Jeong, E., Jeon, W., Jo, H., & Ban, C. (2012). A coordination polymer nanobelt (CPNB)-based aptasensor for sulfadimethoxine. *Biosensors & Bioelectronics*, *33*(1), 113–119. doi:10.1016/j.bios.2011.12.034 PMID:22244734

Sun, B., Qiao, F., Chen, L., Zhao, Z., Yin, H., & Ai, S. (2014). Effective signal-on photoelectrochemical immunoassay of subgroup J avian leukosis virus based on Bi_2S_3 nanorods as photosensitizer and in situ generated ascorbic acid for electron donating. *Biosensors & Bioelectronics*, *54*, 237–243. doi:10.1016/j.bios.2013.11.021 PMID:24287410

Sun, J. Y., Ge, J. C., Liu, W. M., Fan, Z. Y., Zhang, H. Y., & Wang, P. F. (2011). Highly sensitive and selective colorimetric visualization of streptomycin in raw milk using Au nanoparticles supramolecular assembly. *Chemical Communications*, *47*(35), 9888–9890. doi:10.1039/c1cc12910d PMID:21811731

Taghdisi, S. M., Danesh, N. M., Nameghi, M. A., Ramezani, M., & Abnous, K. (2016). A label-free fluorescent aptasensor for selective and sensitive detection of streptomycin in milk and blood serum. *Food Chemistry*, *203*, 145–149. doi:10.1016/j.foodchem.2016.02.017 PMID:26948599

Thiviyanathan, V., & Gorenstein, D. G. (2012). Aptamers and the next generation of diagnostic reagents. *Proteomics. Clinical Applications*, *6*(11-12), 563–573. doi:10.1002/prca.201200042 PMID:23090891

Tu, W., Zhou, Y., & Zou, Z. (2013). Versatile Graphene-Promoting Photocatalytic Performance of Semiconductors: Basic Principles, Synthesis, Solar Energy Conversion, and Environmental Applications. *Advanced Functional Materials*, *23*(40), 4996–5008. doi:10.1002/adfm.201203547

Wang, L. N., Han, D. X., Ni, S., Ma, W. G., Wang, W., & Niu, L. (2015). Photoelectrochemical device based on Mo-doped $BiVO_4$ enables smart analysis of the global antioxidant capacity in food. *Chemical Science (Cambridge)*, *6*(11), 6632–6638. doi:10.1039/C5SC02277K PMID:29435212

Wang, M., Yin, H., Shen, N., Xu, Z., Sun, B., & Ai, S. (2014). Signal-on photoelectrochemical biosensor for microRNA detection based on Bi_2S_3 nanorods and enzymatic amplification. *Biosensors & Bioelectronics*, *53*, 232–237. doi:10.1016/j.bios.2013.09.069 PMID:24141112

Wang, P., Wan, Y., Deng, S., Yang, S., Su, Y., Fan, C., ... Zuo, X. (2016). Aptamer-initiated on-particle template-independent enzymatic polymerization (aptamer-OTEP) for electrochemical analysis of tumor biomarkers. *Biosensors & Bioelectronics*, *86*, 536–541. doi:10.1016/j.bios.2016.07.025 PMID:27448543

Wang, Q. Q., Zhou, Z. X., Zhai, Y. L., Zhang, L. L., Hong, W., Zhang, Z. Q., & Dong, S. J. (2015). Label-free aptamer biosensor for thrombin detection based on functionalized graphene nanocomposites. *Talanta*, *141*, 247–252. doi:10.1016/j.talanta.2015.04.012 PMID:25966410

Wang, S., Li, J., Zhou, X., Zheng, C., Ning, J., Zhong, Y., & Hu, Y. (2014). Facile preparation of 2D sandwich-like CdS nanoparticles/nitrogen-doped reduced graphene oxide hybrid nanosheets with enhanced photoelectrochemical properties. *Journal of Materials Chemistry. A, Materials for Energy and Sustainability*, *2*(46), 19815–19821. doi:10.1039/C4TA04624B

Wang, X., Yan, T., Li, Y., Liu, Y., Du, B., Ma, H., & Wei, Q. (2015). A competitive photoelectrochemical immunosensor based on a CdS-induced signal amplification strategy for the ultrasensitive detection of dexamethasone. *Scientific Reports*, *5*. doi:10.1038rep17945 PMID:26648409

Weng, B., Zhang, X., Zhang, N., Tang, Z. R., & Xu, Y. J. (2015). Two-dimensional MoS_2 nanosheet-coated Bi_2S_3 discoids: Synthesis, formation mechanism, and photocatalytic application. *Langmuir*, *31*(14), 4314–4322. doi:10.1021/la504549y PMID:25625414

Weng, X., & Neethirajan, S. (2016). A microfluidic biosensor using graphene oxide and aptamer-functionalized quantum dots for peanut allergen detection. *Biosensors & Bioelectronics*, *85*, 649–656. doi:10.1016/j.bios.2016.05.072 PMID:27240012

Xu, L., Li, H., Yan, P., Xia, J., Qiu, J., Xu, Q., ... Yuan, S. (2016). Graphitic carbon nitride/BiOCl composites for sensitive photoelectrochemical detection of ciprofloxacin. *Journal of Colloid and Interface Science*, *483*, 241–248. doi:10.1016/j.jcis.2016.08.015 PMID:27552431

Yan, Y., Liu, Q., Du, X., Qian, J., Mao, H., & Wang, K. (2015). Visible light photoelectrochemical sensor for ultrasensitive determination of dopamine based on synergistic effect of graphene quantum dots and TiO2 nanoparticles. *Analytica Chimica Acta*, *853*, 258–264. doi:10.1016/j.aca.2014.10.021 PMID:25467467

Yang, C., Wang, Q., Xiang, Y., Yuan, R., & Chai, Y. (2014). Target-induced strand release and thionine-decorated gold nanoparticle amplification labels for sensitive electrochemical aptamer-based sensing of small molecules. *Sensors and Actuators. B, Chemical*, *197*, 149–154. doi:10.1016/j.snb.2014.02.036

Yao, W., Iwai, H., & Ye, J. (2008). Effects of molybdenum substitution on the photocatalytic behavior of $BiVO_4$. *Dalton Transactions (Cambridge, England)*, (11): 1426–1430. doi:10.1039/b713338c PMID:18322621

Yin, C., Zhu, S. M., Chen, Z. X., Zhang, W., Gu, J. J., & Zhang, D. (2013). One step fabrication of C-doped $BiVO_4$ with hierarchical structures for a high-performance photocatalyst under visible light irradiation. *Journal of Materials Chemistry. A, Materials for Energy and Sustainability*, *1*(29), 8367–8378. doi:10.1039/c3ta11833a

Yu, G., Wang, X., Cao, J., Wu, S., Yan, W., & Liu, G. (2016). Plasmonic Au nanoparticles embedding enhances the activity and stability of CdS for photocatalytic hydrogen evolution. *Chemical Communications*, *52*(11), 2394–2397. doi:10.1039/C5CC10066F PMID:26732587

Yu, P., Liu, Y., Zhang, X., Zhou, J., Xiong, E., Li, X., & Chen, J. (2016). A novel electrochemical aptasensor for bisphenol A assay based on triple-signaling strategy. *Biosensors & Bioelectronics*, *79*, 22–28. doi:10.1016/j.bios.2015.12.007 PMID:26686919

Zang, Y., Lei, J., Hao, Q., & Ju, H. (2014). "Signal-On" Photoelectrochemical Sensing Strategy Based on Target-Dependent Aptamer Conformational Conversion for Selective Detection of Lead(II) Ion. *ACS Applied Materials & Interfaces*, *6*(18), 15991–15997. doi:10.1021/am503804g PMID:25170538

Zelada-Guillen, G. A., Luis Sebastian-Avila, J., Blondeau, P., Riu, J., & Xavier Rius, F. (2012). Label-free detection of Staphylococcus aureus in skin using real-time potentiometric biosensors based on carbon nanotubes and aptamers. *Biosensors & Bioelectronics*, *31*(1), 226–232. doi:10.1016/j.bios.2011.10.021 PMID:22154169

Zeng, X., Tu, W., Li, J., Bao, J., & Dai, Z. (2014). Photoelectrochemical Biosensor Using Enzyme-Catalyzed in Situ Propagation of CdS Quantum Dots on Graphene Oxide. *ACS Applied Materials & Interfaces*, *6*(18), 16197–16203. doi:10.1021/am5043164 PMID:25154012

Zhang, J., Tu, L., Zhao, S., Liu, G., Wang, Y., & Yue, Z. (2015). Fluorescent gold nanoclusters based photoelectrochemical sensors for detection of H2O2 and glucose. *Biosensors & Bioelectronics*, *67*, 296–302. doi:10.1016/j.bios.2014.08.037 PMID:25190086

Zhang, X., & Yadavalli, V. K. (2011). Surface immobilization of DNA aptamers for biosensing and protein interaction analysis. *Biosensors & Bioelectronics*, *26*(7), 3142–3147. doi:10.1016/j.bios.2010.12.012 PMID:21227676

Zhang, X., Zhang, Y., Zhao, H., He, Y., Li, X., & Yuan, Z. (2013). Highly sensitive and selective colorimetric sensing of antibiotics in milk. *Analytica Chimica Acta*, *778*, 63–69. doi:10.1016/j.aca.2013.03.059 PMID:23639400

Zhao, M., Fan, G.-C., Chen, J.-J., Shi, J.-J., & Zhu, J.-J. (2015). Highly Sensitive and Selective Photoelectrochemical Biosensor for Hg^{2+} Detection Based on Dual Signal Amplification by Exciton Energy Transfer Coupled with Sensitization Effect. *Analytical Chemistry*, *87*(24), 12340–12347. doi:10.1021/acs.analchem.5b03721 PMID:26599580

Zhao, W., Liu, Y., Wei, Z. B., Yang, S. G., He, H., & Sun, C. (2016). Fabrication of a novel p-n heterojunction photocatalyst n-$BiVO_4$@p-MoS_2 with core-shell structure and its excellent visible-light photocatalytic reduction and oxidation activities. *Applied Catalysis B: Environmental*, *185*, 242–252. doi:10.1016/j.apcatb.2015.12.023

Zhou, L., Wang, J., Li, D., & Li, Y. (2014). An electrochemical aptasensor based on gold nanoparticles dotted graphene modified glassy carbon electrode for label-free detection of bisphenol A in milk samples. *Food Chemistry*, *162*, 34–40. doi:10.1016/j.foodchem.2014.04.058 PMID:24874354

Zhou, N. D., Wang, J. Y., Zhang, J., Li, C., Tian, Y. P., & Wang, J. (2013). Selection and identification of streptomycin-specific single-stranded DNA aptamers and the application in the detection of streptomycin in honey. *Talanta*, *108*, 109–116. doi:10.1016/j.talanta.2013.01.064 PMID:23601877

Zhou, S., Kong, Y., Shen, Q., Ren, X., Zhang, J. R., & Zhu, J. J. (2014). Chronic myeloid leukemia drug evaluation using a multisignal amplified photoelectrochemical sensing platform. *Analytical Chemistry*, *86*(23), 11680–11689. doi:10.1021/ac502969x PMID:25372503

Chapter 8
Electrochemical Treatment of Wastewater:
An Emerging Technology for Emerging Pollutants

Isaiah Adesola Oke
https://orcid.org/0000-0002-7082-7682
Obafemi Awolowo University, Nigeria

Lukman Salihu
University of Hafr Al-Batin, Saudi Arabia

Aladesanmi Temitope A.
https://orcid.org/0000-0003-2046-8738
Obafemi Awolowo University, Nigeria

Fehintola Ezekiel Oluwaseun
Adeyemi Colloege of Education, Nigeria

Amoko S. Justinah
Adeyemi Colloege of Education, Nigeria

Hammed O. Oloyede
Adeyemi Colloege of Education, Nigeria

ABSTRACT

This chapter presents an overview of over 529 articles on designs, models, laboratory setups, and applications of electrochemical processes from 1973 to 2017 with particular attention paid to the removal of emerging environmental pollutants. The chapter demonstrates that electrochemical and advanced oxidation processes are efficient despite the economic implications of the technologies. The electrodes in

DOI: 10.4018/978-1-7998-1871-7.ch008

use arranged from monopolar to bipolar mode, which varies from the electroplating baths, recalcitrant organic contaminants, and eluates of an ion-exchange unit and the number of electrodes in a stack to a variant of rotating cathode cell. Application of the process can be in the form of a static anode and a rotating disk cathode. The narrow spacing between the electrodes in the pump cells allow the entrance of the effluent and effective wastewater treatment. It was concluded that electrochemical treatment techniques have a variety of laboratory setups and a wider range of applications.

INTRODUCTION

Antibiotics are molecules that kill, or stop the growth of, microorganisms, including both bacteria and fungi. Antibiotics that kill bacteria are called "bactericidal". Antibiotics that stop the growth of bacteria are called "bacteriostatic". Antibiotics affect four major components and functions of cells as follows:

a) inhibition of bacteria's cell wall synthesis; this class of antibiotics includes vancomycin and β-lactam antibiotics such as penicillins, cephalosporins and carbapenems,
b) inhibition of protein synthesis including tetracyclines, aminoglycosides, macrolides and chloramphenicol ;
c) Cell membrane; and
d) DNA synthesis inhibitors such as fluoroquinolones and sulfonamides that inhibit folic acid synthesis.

Antibiotics contributed over years in bacterial infections control. These drugs have an important role in the rise of life expectancy. Despite all the developments in antibiotic industry, infectious diseases remain the second cause of death worldwide; this is due to the development of antibiotics resistance organisms, which decreased current antibiotics effectiveness (Salma and Rafik, 2015). Generally, there are three main ways for antibiotics to enter the environment as follows (Derakhshan et al. 2016):

a) Through the excretion, industries and the entrance to the sewage network and eventually surface water or groundwater or soil. The amounts of any antibiotic's metabolism in the human body depend on its chemical structure and also its operation in the body; and
b) Through natural resources. Some different antibiotics such as beta lactams, streptomycin, amino glycosides etc. are produced by bacteria in soil.
c) Through the discard of expired and unused antibiotics which may be carried through the health services, sewage or household waste. In such cases, the antibiotics are carried to landfill through the discard of waste.

The human body cells react to antibiotics in very low systemic levels. Antibiotics exist in drinking water or food can increase the levels of these compounds in body. These compounds can reach the body tissues and cause different responses from body. Low concentrations of antibiotics can act as a vaccine for bacteria and make them resistant to the antibiotics used in the treatment of diseases. The bacteria resistance can happen due to the existence of antibiotics in hospitals and health centers wastewater, sewage, soil etc. On open land, raw water and wastewater containing antibiotic, bacteria and resistant bacteria would be used for irrigation, while excess sludge is used as fertilizer. In this action, the resistant bacteria directly enter the food chain. In addition, concentrations of antibiotics less than the required rate for diseases treatment have a significant role in bacteria resistance, which can even transmit to the bacteria genetics. Literature has revealed that the chronic effects of antibiotics are more in the environment (wastewater treatment systems, surface water, etc) and hence the need for pollution reduction and the removal of these compounds from water (Derakhshan et al. 2016).

BACKGROUND

Environmental pollution is a serious problem with which human is faced and due to the population growth, the importance of environmental pollution control and its exponential growth prevention is needed more than ever. Problems caused by the presence of persistent pollutants in the environment emerging pollutants require investigation to identify various techniques to treat, reduce or remove these pollutants from environment (Derakhshan et al. 2016). Pollutants in water and wastewaters can be categorized into two (conventional and emerging pollutants).

a) Conventional pollutants are physical, chemical and biological. These are pollutants includes compounds that have been found in the environment, wastewater and drinking water and are regulated.
b) Emerging pollutants includes compounds that have been found in the environment (antibiotics inclusive), wastewater and drinking water relatively recently and are presently unregulated. These are pharmaceuticals and personal care products (PPCPs), have been detected at low levels in environment and surface water, leading to concerns that these compounds may have an impact on human health and aquatic life. There are more than 1036 emerging pollutants, their metabolites and transformation products, are listed as present in the European aquatic environment (Geissen et al.,2015 ; NORMAN, 2019). There are several studies made countrywide overviews of emerging organic contaminants (EOCs) including PPCPs in the groundwater of Italy (Meffe and Bustamante, 2014), Spain (Jurado et al., 2012) and UK (Stuart et al., 2012), providing useful information on the presence, sources, and potential risks to the environment in their countries. Qian et al., (2015) reviews the occurrence of frequently detected PPCPs, including antibiotics, anti-inflammatories, lipid-regulators, carbamazepine, caffeine, and N,N-diethyl-m-toluamide in groundwater, with special concern to the progress made over the past three years. High concentration of sulfonamides was reported in the groundwater down gradient of a landfill site (Holm et al., 2012). Sulphonamides in the groundwater were detected in five more studies conducted in Switzerland (Morasch 2013, Lapwoth et al., 2012), Spain (Cabezaa et al., 2012, Lopez – Serna et al, 2013), USA (Schaider et al., 2014), China (Peng et al., 2014) and Africa (Ebele et al., 2017). Figure 1, Figure 2, and Figure 3 present types emerging pollutants, concept for management of emerging pollutants and multi-scale assessment of the ecological and human health risks.

Water and wastewater treatments are carried out by the conventional wastewater treatment techniques (Zhen et al. 2015). Conventional water and wastewater treatment technologies commonly used are as follows (EPA, 2010):

Figure 1. Types of emerging pollutants

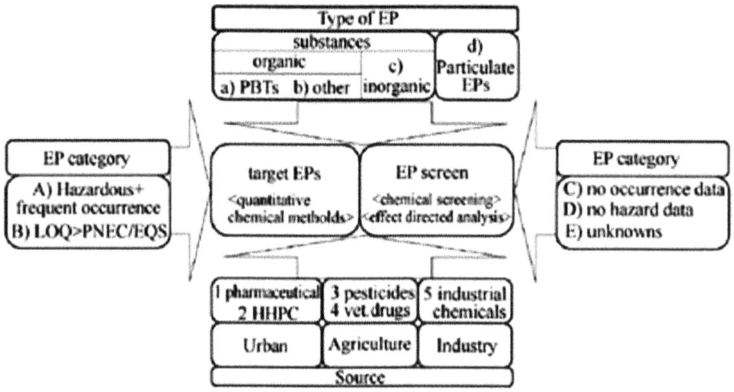

Figure 2. Concept of management of emerging pollutants

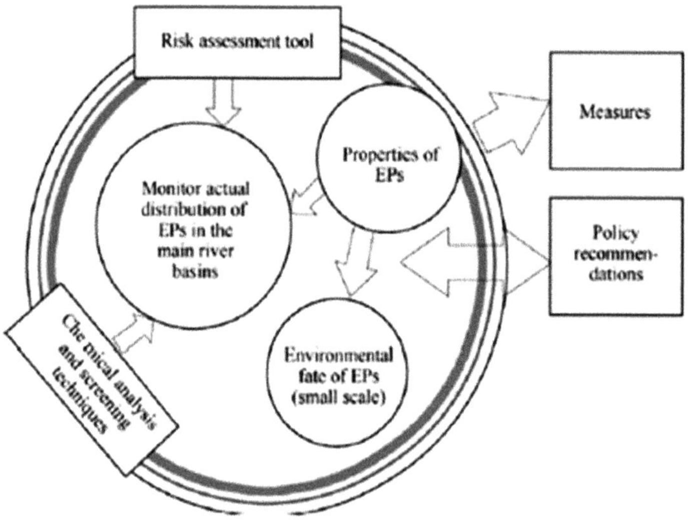

Figure 3. Multi- scale assessment of the ecological and human health risks

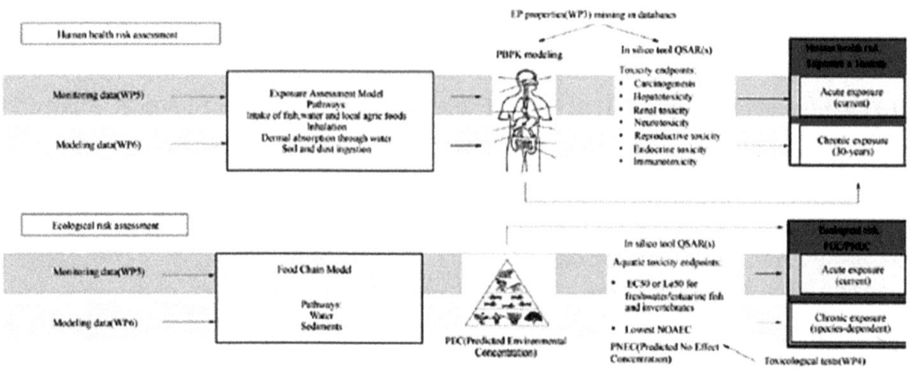

a) Municipal Wastewater (activated sludge, fixed film biological treatment, chemical phosphorus removal, biological phosphorus removal, denitrification, nitrification, chlorine disinfection, granular activated carbon, ozonation, reverse osmosis, and ultraviolet disinfection);
b) Drinking Water (aeration, coagulation, flocculation, sedimentation, filtration, chlorine disinfection, granular activated carbon, ozonation, and ultraviolet disinfection);
c) Treated effluent (secondary or tertiary treated): activated sludge, fixed film biological treatment, chlorine disinfection, granular activated carbon, ozonation, reverse osmosis, ultrafiltration, and ultraviolet disinfection).

Other water and wastewaters treatment techniques are as follows:

a) Hybrid Membrane Processes. The most popular is the use of hybrid processes, which integrate (a) two or more membrane processes and (b) membrane processes with other water treatment processes;
b) Coagulation-Membrane Process. The purpose of using coagulants is to increase the size of particles, in order to enable settling or removal by sand or dual layer filtration or membrane filtration
c) Adsorption-Membrane Process. Similar to the hybrid process of coagulation membrane filtration, adsorption can be combined with membrane filtration with or without clarification (Stylianos et al., 2015)
d) Electrolysis, heterogeneous photocatalysis, or photo-assisted electrolysis may be regarded as advanced oxidation processes (AOPs) and used in the supplementary treatment of wastewaters. The efficiency of the electrochemical oxidation depends on the anode material and the operating conditions, such as current density or potential (Adina et al.,2013).
e) Electrochemical oxidation is one of the popular advanced oxidation processes (AOPs) in recent research. AOPs offers a lot of advantages and benefits, which include: the environmental compatibility, versatility, energy efficiency and amenability to automation (Hongna et al., 2014).
f) Ahmed et al., (2015) stated that The electrocoagulation process was found to be more efficient than the traditional chemical coagulation process

Over the past years, the presence of antibiotics in the environment has been of particular interest to scientists since they can have harmful effects on land and sea ecosystems. Literature has provided several removal and analysis processes to solve this problem. In addition, many studies had been done on the presence of antibiotics in the environment. Although, there are extensive researches on presence and removal of emerging pollutants, there is still a considerable lack of integrated and classified information to assess these documents. Therefore, in the current chapter, the removal of these emerging pollutants from aqueous solutions was presented.

MATERIALS AND METHODS

For the development this chapter, survey of literature was conducted on the basis of accomplished information on emerging pollutants, presence and removal from environment using various water and wastewater treatment techniques. Information was collected from published in databases such as Google Scholar, PubMed, Elsevier, Scopus, Springer and Magiran using emerging pollutants, antibiotic, wastewater treatment methods and pollution control as the keywords.

Previous Treatment of Emerging Pollutants

Martin *et al.,* (1966) revealed factors influencing the occurrence of high number of iodine-resistant bacteria in iodinated swimming pool. It was reported that iodine appeared to be more effective than chlorine against both the standard faecal indicators (coliform bacteria and enterococci). It was documented that against *staphylococci* derived from the mouth, nose, and skin iodine appeared to be more effective than chlorine. Some microorganisms inherently developed resistance to disinfectant and even grow in disinfectant solutions. It has been documented that resistant cells of *Pseudomonas aeruginosa* and waterborne *Pseudomonas sp.*(strain Z-R) were able to multiply in nitrogen-free minimal salts solution containing various concentrations of commercially prepared, ammonium acetate buffered benzalkonium chloride (CBC), which is a potent antimicrobial agent (Frank *et al.,* 1969).

In Philadelphia, examination on the transfer of plasmids pBR322 and pBR325 in wastewater from laboratory strains of *Escherichia coli* to bacteria indigenous to the waste disposal system revealed that bacterial strains isolated from raw wastewater or a plasmid free *E. coli* laboratory strain served as recipients. The transfer of the pBR plasmid into the recipient strain occurred within 25 hours co-incubation in either L broth or sterilized wastewater (Michael *et al.,* 1985).

In France, comparison between vitro activity of Cefoxitin with Metronidazole and Clindamycin against 322 strains of anaerobic bacteria collected from several hospitals. The antibiotic-resistant were tested using an agar dilution method. The result showed that Metronidazole and Cefoxitin inhibited at least 89% of antibiotic-resistant strains tested. Clindamycin was found to be less active in inhibited antibiotic-resistant (Dubreuil *et al.,* 1984).

Research in Nigeria revealed that organisms belonging to seven genera of public health importance such as Pseudomonas, Streptococcus, Serratia, Staphylococcus, Klebsiella, Proteus and Bacillus showed different degrees of resistance to the test of antimicrobial agents ranging from 0% to 77.8%. Addition susceptibility of strains to phenol and to one quaternary ammonium compound was variable. Out of twenty-one antibiotic-multiresistant strains (methicillin-resistant *Staphylococci, Enterococcus spp, Pseudomonas aeruginosa, Klebsiella pneumoniae, Proteus mirabilis, Enterobacter cloacae, Serratia marcescens and Escherichia coli)* evaluated, eleven (52%) and eight (38%) strains were resistant to the quaternary ammonium and phenol compounds respectively. Six out of twenty –one organism isolated demonstrated susceptibility to antibiotics (*Staphylococci, Enterococcus spp, P. mirabilis, E. cloacae and E. coli*) and two strains (33%) out of six showed resistance to these disinfectants. This result revealed that there was no correlation between antibiotic-susceptibility and susceptibility to disinfectants in hospital strains (Mesdaghinia *et al.,* 2009).

Roberto *et al.* (2005) reported that amoxicillin is a widely used antibiotic, which has been found in effluents from Sewage Treatment Plant. Evaluation of ozonation treatment of pharmaceuticals was proposed as a process for its abatement from these pharmaceuticals effluents. The results of ozonation treatments of amoxicillin containing wastewaters indicated that ozone attack was mainly directed towards the phenolic ring of the studied molecule leading to the formation of hydroxyl-derivative intermediates. In addition the study revealed that there was no strong evidences of attack of ozone on sulphur atom with sulphur-oxide formation. Pauwels and Verstraete (2006) reviewed different techniques in use for hospital generated wastewater treatment and evaluated actions of technologists and ecotoxicologists as well as public health specialists.

In Aboulhassan *et al.* (2006), Jar-test experiments were employed to establish the optimum dosage of coagulants for the removal of surfactants, Chemical Oxygen Demand (COD) and turbidity from selected wastewaters at pH control. It was reported that coagulation and flocculation with $FeCl_3$ proved to be effective in a pH range between 7.00 and 9.00. The process reduced surfactants and COD by 99.00% and 88.00% respectively, and increased BOD_5 to COD ratio from 0.17 to 0.41. It was concluded that precipitation coagulation process, adsorptive micellar flocculation mechanism contributed to the removal of surfactants and organic matters from the wastewaters.

Zhimin *et al.* (2006) revealed that aeration significantly improves lagoon functionality, by increasing both organic and ammonia removal. Ammonia present in the wastewaters played a critical role in antibiotics disintegration and bacterial reduction due to conversion of free chlorine to monochloramine and dichloroamine. Commonly, a chlorine dose close to breakpoint chlorination is required to achieve effective removal of antibiotics. This action usually leads to higher consumption of free chlorine than necessary in most of the wastewaters treatment.

Most of the antibiotics taken by the patients are partially metabolized and excreted through faeces and urine. Residual quantities of these antibiotics in faeces and urine reached the wastewater. The residual quantity of antibiotics exposes the bacteria in hospital wastewaters to a wide range of biocides that could act as a selective pressure for the development of resistance (Nuñez and Moretton, 2007).

In Brazil, the bacterial population resistant to disinfectants were *Enterobacteriaceae, Staphylococcus spp*, and *Bacillus spp*, which are highly associated to nosocomial infections (Nuñez and Moretton, 2007). In Nepal, Aina *et al.*(2002) revealed that healthcare liquid wastes were loaded with multiple drug resistance bacteria and seemed to pose a huge public health threat in the transfer of such resistance to the bacterial pathogens causing community acquired infections, thereby limiting our antibiotic pool (Sharma *et al.,* 2010). Aina *et al.*(2002) reported that in Sweden high prevalence of Vancomycin Resistant Enterococci (VRE) was demonstrated in Swedish sewage, which can be attributed to antimicrobial drugs or chemicals released into the sewage system that sustained VRE in the system.

It has been documented that in Brazil extended-spectrum beta-lactamase-producing Klebsiella pneumoniae were detected in effluents and sludge of a hospital sewage treatment plant. This result indicated the hospital wastewater treatment plant did not demonstrate a satisfactory significant efficacy in removing pathogenic micro-organisms by allowing for the dissemination of multi-resistant bacteria into the environment.

Further documentation revealed that in U.S (wastewater research division, municipal environmental research laboratory, U.S. Environmental Protection Agency) revealed effect of ultra-violet (UV) light disinfection on antibiotic resistant coliforms in wastewater effluents. The document showed that UV irradiation effectively disinfected the wastewater effluent (the percentage of the total surviving coliform population resistant to Tetracycline or Chloramphenicol was significantly higher than the percentage of the total coliform population resistant to those antibiotics before UV irradiation and the finding was attributed to the mechanism of R-factor mediated resistance to tetracycline, Mark, 1982).

In Belgium, comparisons between the antimicrobial tolerance of Oxytetracycline-resistant heterotrophic bacteria isolated from hospital sewage and freshwater fish farm water showed that Oxytetracycline-resistant hospital heterotrophs displayed a higher frequency (84%) of Ampicillin (Amp) tolerance compared to the Oxytetracycline-resistant heterotrophs from the freshwater fish-farm site (22%, Aina *et al.,* 2002)

In Austria, evaluations of antibiotic resistance of *E. coli* in sewage and sludge were conducted. Aina *et al.*(2002) revealed that the highest resistance rates were established in *E. coli* strains of a sewage treatment plant which treated both municipal sewage and sewage from a Hospital.

In France, a study was conducted to trace whether antibiotic-resistant *Pseudomonas aeruginosa* isolated from hospitalized patients recovered in the hospital effluents. The study of genotyped both clinical and wastewater isolated antibiotic-resistant was determined by using pulsed-field gel electrophoresis (PFGE). The study revealed that there was no common PFGE pattern in antibiotic-resistant *P. aeruginosa* from humans and wastewater. It was concluded that antibiotic resistance profile of wastewater isolated antibiotic-resistant was different from that of clinical isolates (Tumeo *et al.,* 2008).

Jara *et al.* (2007) presented electro-oxidation treatment of aqueous solutions containing the antibiotics Ofloxacin and Lincomycin with different anodes (Ti/Pt, DSA1 type, graphite and three-dimensional (3D) electrode made of a fixed bed of activated carbon pellets). It was reported that effectiveness of the treatment of wastewater containing pharmaceuticals was assessed based on antibiotics and bacteria removal, as well as the electro-oxidation mechanism. The electro-oxidation was found to follow first order kinetics at the anode surface when Na_2SO_4 was at a low concentration (0.02 Normality). At lower concentration of Na_2SO_4 Ofloxacin was efficiently oxidized at the rate of 50 $mgcm^{-2}A^{-1}$ h^{-1} for the bi-dimensional Ti/Pt electrode. It was observed that Lincomycin was oxidized with slow overall kinetics, which can be attributed to difficult deprotonation (a step that precedes the primary electron transfer stage of the oxidation process). It was concluded that the three-dimensional electrode would be the most appropriate for continuous industrial-scale process of aqueous solutions containing the antibiotics Ofloxacin and Lincomycin treatment.

In Watkinson *et al.* (2007), antibiotic resistance (AR) patterns of 462 *Escherichia coli* isolates from various wastewater, different surface water sources, and oysters were evaluated. The study revealed that the rates of AR and multiple-AR among isolates from surface water sources adjacent to wastewater treatment plant discharge sources were significantly higher ($p < 0.05$) than other *Escherichia coli* isolated from other sources. The rates of AR among *Escherichia coli* isolated from oysters exposed to wastewater treatment plant discharges were lower (less than 10%) than other *Escherichia coli* isolated from other sources.

Pedro *et al.* (2009) gathered records on the occurrence of anti-infective available in literature for 25 years (1984 – 2009) in environmental water matrices. The collected information was compared with available ecotoxicologic values to appraise potential environmental concerns for anti-infective organisms. Pedro *et al.* (2009) utilized Web of Science and Google Scholar to search for publication (articles published) in peer-reviewed journals written in the English language starting from 1984 to 2009. Information on compound concentrations in wastewaters and natural and potable (drinking) waters, the source of contamination, country of provenance of the samples, year of publication, limits of quantification, and method of analysis were extracted. In the study 126 different substances were analysed in environmental waters, 68 different parent compounds and 10 degradation products or metabolites were quantified. Environmental concentrations of the compounds varied from about 10^{-1} to 10^{9} mg/L, depending on the type of compound, the environmental matrix, and the source of contamination. It was concluded that detrimental effects of anti-infective on aquatic microbiota are possible with the constant exposure of sensitive species. Indirect impact on human health cannot be ruled out when considering the potential contribution of high anti-infective concentrations to the spreading of anti-infective resistance in bacteria.

Jury *et al.* (2010) examined and considered the presence of antibiotics in STPs facilitates the acquisition and proliferation of resistance characteristics amongst bacteria in that environment. The main objective of the study was to establish whether it is just simply imparts a selective pressure for the propagation of clonal resistance. The study evaluated available literature on the acquisition of antibiotic resistance by previously non-resistant bacteria and the role of antibiotics in the transfer of resistant genes in STPs. It was concluded that precise explanations would be required to confirm these phenomena and experimental approaches, including culture-independent methods are suggested.

Emad and Malay (2010) examined photo-Fenton treatment of an antibiotic wastewater containing amoxicillin and cloxacillin, and evaluated the effect of operational conditions (H_2O_2: COD molar ratio and H_2O_2: Fe^{2+}molar ratio) on biodegradability (BOD_5: COD ratio) improvement and mineralization. The study revealed that optimum operating conditions for treatment of the antibiotic wastewater were observed to be H_2O_2: COD molar ratio of 1: 2.5 and H_2O_2: Fe^{2+} molar ratio of 1: 20 (COD: H_2O_2: Fe^{2+} molar ratio 1: 2.5: 0.125) at pH 3 and reaction period 30 minutes. At optimum operating conditions, complete degradation of amoxicillin and cloxacillin occurred spontaneously (within one minute), biodegradability increased from 0.09 to 0.50 ± 0.01 in 30 minutes, COD and DOC removal were 67 ± 1% and 51 ± 2%, respectively in 30 minutes. The study concluded that the photo-Fenton process can be used as pre-treatment for antibiotic degradation and biodegradability enhancement of antibiotic wastewater containing amoxicillin and cloxacillin.

In Chelliapan *et al.* (2006; 2011), reported that anaerobic technology is a promising alternative for pharmaceutical wastewater treatment. In Chelliapan and Golar (2011), an Up-Flow Anaerobic Packed Bed reactor was employed to treat pharmaceutical wastewater containing antibiotics. Effect of organic loading rate (OLR) was assessed by modifying feed substrate concentration and hydraulic retention time (HRT). Performance of the system was based on COD removal, volatile fatty acid (VFA), gas production, methane yield and pH. The study showed that COD reduction was between 60 and 70% at an OLR between 1.5 and 4.6 kg $COD.m^{-3}.d^{-1}$, which indicates that the biomass had acclimated to the antibiotics. An increase in the OLR soluble COD removal efficiency decreased to 50% at an OLR of 5.6 kg $COD.m^{-3}.d^{-1}$, which indicates that OLR have effect on the methanogens.

Honarbakhsh (2011) examined the impact of antibiotic resistant bacteria from final sewages on water receiving bodies. On wastewater collected Phylogenetic analysis of the 16S rRNA gene sequences was performed to develop the composition and structure of the four microbial communities. The distribution of β-lactamase genes across the four sites was also assessed. Clone library information and TRFLP profiles revealed that the four communities were all controlled by Betaproteobacteria, which constituted approximately two thirds of the final effluent community. The diversity and abundance of other bacterial phylotypes varied across sites. Bacterial diversity included 9 phylotypes in the final effluent, 15 and 13 phylotypes respectively in the downstream. Bacteria were isolated from the same samples using culture-dependent techniques, which allowed to link antibiotic resistance to particular bacterial species. Clone libraries were dominated by Betaproteobacteria, culturable isolates mainly belonged to the class of Gammaproteobacteria. Analysis of the isolates from the final effluent exhibited the presence of *Bacillus, Enterobacter, Acinetobacter, and Staphylococcus* strains, while samples collected downstream from the plant were categorized by species belonging to the genera *Brevibacterium, Chryseobacterium, Aeromonas and Delftia.* All the isolates were resistant to amoxicillin, and most displayed resistance to multiple antibiotics. Phylogenetic analysis revealed that the distribution of the *bla* TEM4 gene is divided to two main clusters. The first cluster contains sequences that are exclusively detected at the plant and receiving water, but not at the upstream site, while sequences of the second cluster were present at all of the four sites.

In Shreeshivadasan and Paul (2011), showed efficient substrate removal at low OLR between 0.43 and 1.86 kg $COD.m^{-3}d^{-1}$ and promote efficient chemical oxygen demand (COD) reduction between 70 and 75%. At these operational situations, an average of 95% Tylosin reduction was achieved in the upward anaerobic sludge reactor (UASR). With an increase in the OLRs to 3.73 kg $COD.m^{-3}.d^{-1}$ and by reducing the hydraulic retention time (HRT) between 4 and 2 days, there was reduction in the COD removal efficiency to 45%. It was concluded that changes in the organic loading of pharmaceutical wastewater affected the treatment performance of the anaerobic reactor, and at high OLRs, anaerobic reactor was not able to withstand the short HRT, probably due to the intricacy of pharmaceutical wastewater.

Thompson *et al.* (2013) isolated 224 *Staph*aureus strains from UHWW-1, UHWW-2 and its receiving STP inlet (SI) and post-treatment outlet (SO). These strains were categorized using the PhP and RAPD-PCR methods. The strains were tested for their antibiotic resistance patterns. Resistance of the microbails to cefoxitin and the presence of mecA gene were used to identify MRSA isolates. In the study, 11 common (C) isolates and 156 single (S) PhP-RAPD isolates were identified, with two multi-drug resistant (MDR) C- isolates found in H2, SI and SO. These C- isolates strains showed resistance to cefoxitin and vancomycin. The mean number of antibiotics to which the strains from UHWW were resistant (5.14 ± 2) was significantly higher than the STP isolates (2 .9 ± 1 .9, $p < 0.0001$). Out of the total samples, 131 (68%) isolates were MRSA strains, 24 isolates were vancomycin resistant. MDR strains (including MRSA) were identified to be more prevalent in hospital wastewaters than in the STP. The study provided evidence of the survival of MRSA strains in UHWWs and their transit to the STP. It was reported that MRSA strains survived in the final treated effluent and chlorination stage. It was concluded that there is the needs to further investigate the load of MRSA in hospitals' wastewaters and possible their survival in STPs.

Katouli et al.(2012) revealed that in the E.coli strains collected from UHWW, seven common (C) PhP-RAPD species were found frequently and in most of weekly samples with multi-drug resistant (MDR) C-species found in STP-I. It was revealed that on three occasions multi-drug resistant (MDR) C-species found in STP-O. In S. aureus strains, seven C- species were frequently found in a majority of samples with two MDR C-species in STP-I. It was showed that on two occasions two MDR C-species in STP-O. The study concluded that some MDR bacterial strains found in UHWW may have the ability to survive, transmit to the STP and to the final treated effluent before being released into surface water bodies.

Abargues et al. (2013) revealed that the removal rates of these compounds differed with each post-treatment method and that aeration (oxygen) resulted in the highest removal ratios indicating the importance of an aerobic step in EDC degradation; a mass balance also showed that chemical oxidation was more important than sorption processes.

Yangming et al. (2006) reported that a semiconductor catalyst had been used in electrochemical processes as well as in a photo-degradation system (Qamar et al., 2005; Shen et al., 2002). The removal of xenobiotic compounds, such as chlorophenols and pesticides, from municipal and industrial wastewater was achieved by electrochemical treatment processes.

Choe et al. (2013) compared perchlorate-selective ion exchange, biological reduction and catalytic treatment for perchlorate removal based on the environmental impacts of consumable inputs normalized to the mass treated, in a lifecycle analysis. Because resource consumption during the operational phase comprises greater than 80% of the total impacts of these technologies, the analysis indicates that ion exchange is competitive, but emerging technologies also show great promise from an environmental sustainability perspective.

Kovalova et al. (2013) compared the removal efficiencies of micro-pollutants in hospital wastewater using O_3, O_3/H_2O_2, PAC, and UV light with or without TiO_2. They found that the majority of wastewater micropollutants can be removed by ozone and PAC, while some recalcitrant compounds, like diatrizoate, can only be removed by high UV doses. Combined treatments of O_3/H_2O_2 and UV/TiO_2 did not improve removal efficiency over single treatments of ozone and UV. The efficiency of MBRs and ozonation in removing endocrine disruptive potentials was studied for use in treatment of hospital wastewater (Maletz et al., 2013). Treatment by MBRs reduced estrogenicity of sewage while ozonation efficiently removed most measured compounds. Future investigation may require establishing whether either of these processes generates endocrine active metabolites. Yet another subset of studies in this category evaluated whole plants and either compared different treatment trains or compared processes within a single plant to each other in terms of effectiveness for emerging pollutant abatement. The concentrations and removal rates of ARGs were measured in three municipal WWTPs employing conventional treatment followed by different advanced treatment systems (biological aerated filter, constructed wetland, and UV disinfection, Chen and Zhang 2013).

Flores et al. (2013) evaluated a full-scale drinking water treatment plant and several pilot treatment systems with a number of conventional (pre-chlorination, coagulation, sand filtration and ozonation) and advanced (RO, reverse electrodialysis and GAC adsorption) water treatment processes for their ability to remove perfluorinated compounds. Authors found the most efficient process was RO, which resulted in 99% removal of the studied pollutants. Other processes that showed removal of both the pollutants were adsorption onto GAC and reverse electrodialysis. All of the conventional processes proved to be fully ineffective. Concentrations of antibacterial chemicals triclocarbon and triclosan, and a triclosan biodegradation product methyltriclosan (MeTCS) were monitored in the liquid phase throughout a wastewater plant by Lozano et al. (2013). Triclocarbon and triclosan were primarily removed by sorption and settling, though concentrations of both also decreased during nitrification denitrification and triclosan concentration decreased during the activated-sludge stage. Ozonation followed by sand filtration and PAC followed by either sand filtration or UF were tested in parallel for removal of 70 potentially problematic substances and reduction of ecotoxicological effects at large scale WWTPs over a year (Margot et al., 2013). Both technologies significantly reduced toxicity of wastewater, but PAC/UF technologies were both more costly and more effective.

Adina et al.,(2013) reported that the degradation of 4-chlorophenol by an electrochemical advanced oxidation process. It was stated that several authors have proposed the following possible pathways of oxidation of some emerging pollutants by electrochemical processes. The pathways are as follows:

i. 4-chlorophenol dechlorination to phenol;
ii. hydroxylation of phenol to hydroquinone;
iii. dehydrogenation of hydroquinone to benzoquinone;
iv. oxidation of benzoquinone (with aromatic ring cleavage) to aliphatic carboxylic acids such as maleic acid, fumaric acid, malonic acid;
v. oxidation of maleic and fumaric acids to oxalic acid,
vi. formic acid to carbon dioxide and water

Rahim *et al.* (2014) reported that human exposure to antibiotic resistant bacteria (ARB) is a public health and environmental concern which could occur in a number of ways. The aim of Rahim *et al.* (2014) was to evaluate the occurrence of three groups of ARB and resistance genes of the three groups of ARB in hospital and municipal wastewaters (MWs) as possible sources of environmental pollution. A total of 66 wastewater samples were collected from raw MWs and hospital wastewaters (HWs) and final effluents of related wastewater treatment plants (WWTPs). The wastewater samples were analyzed to identify the three groups of ARB (gentamicin (GM), chloramphenicol (CHL) and ceftazidime resistant bacteria) and their ARGs (aac $(3)^{-1}$, $cm^{l}A^{1}$ and ctx-m-32, respectively). Rahim *et al.* (2014) revealed that the mean concentrations of GM, CHL and ceftazidime resistant bacteria in raw wastewater samples were 1.24×10^7, $3.29 \times 10^7$and 5.54×10^7 colony forming unit per100 ml, respectively. There was a variation in prevalence of diverse groups of ARB in MWs and HWs. All WWTPs, there were decreased in the concentrations of ARB. It was established that high concentrations of ARB were found in the final effluent of WWTPs. In line with ARB, different groups of ARGs were found frequently in both MWs and HWs. It was concluded that discharge of final effluent from conventional WWTPs is a potential route for the spreading of ARB and ARGs into the natural environment (surface water bodies and other sources) and poses a hazard to environmental and public health.

Sourav and Alok (2014) stated that the major area of concern is the presence of the incipient pollutants in wastewaters, surface waters and ground waters. The incipient contaminants denote the unregulated pollutants, which can be included for future regulation depending on their health effects on human being, plants and environment. These pollutants are surfactants, illicit drugs, endocrine disruptors, pharmaceuticals and personal care products (PPCPs), gasoline additives and many other groups of compounds. The presence of these endocrine disruptors and pharmaceutical materials (antibiotics and other materials) in the wastewater poses a great threat to the human beings and environment if not treated or removed. The methods of removing pharmaceutical wastes from the environment are activated sludge treatment, activated carbon columns, ion exchange columns, membrane processes.

Sourav and Alok (2014) reviewed few treatment processes that are effective in remediating the pollution load in the pharmaceutical wastewater in an environmental sustainable manner (Table 2).

Yamina *et al.* (2014) revealed that Minimum Inhibitory Concentration of isolated bacteria was between 250 and 950 micrograms per millilitres. All isolated bacteria showed co-resistance to zinc and mercury in aqueous solutions. Isolated bacteria were antibiotic resistant and tolerated NaCl to a significant concentration of NaCl. The cadmium-resistant species (*Klebsiella pneumoniae* and *Pseudomonas aeruginosa*) were the most resistant to cadmium. The species *Klebsiella pneumonia* resisted to the highest concentrations of zinc and mercury. The results of toxicity tests on *Vibrio fischeri* revealed that the DI_{50} (15 minutes) as low as 0.07 carried away 50% luminescence inhibition.

Amenu (2014) reviewed articles on antibiotics and disinfectants. It was reported that large quantities of antibiotics and disinfectants are used in hospitals for patient care and disinfection process respectively. It was documented that the reviewed article contain articles on hospital effluents with contain antibiotic resistant bacteria which are released to receiving water bodies resulting in huge public health threat.

Rongjun *et al.* (2015) revealed that at a low ambient temperature, 45% of initial COD, 40% initial NH_4^+-N concentrations were removed and BOD_5 removal efficiency was 85% of initial BOD_5 concentration. Microscopic observations of the wastewaters revealed that there were various active microorganisms showed in the biofilms and their distributions changed from stage to stage. Antimicrobial resistance in microorganisms are responsible for substantial cases of therapeutic failures. Resistance of species of *Enterococcus* to antibiotics has been linked to their ability to acquire and disseminate antimicrobial resistance determinants in nature (environment). Wastewater treatment plants (WWTPs) are known to be one of the main sources of antibiotic resistant bacteria such as species of *Enterococcus*.

Benson et al. (2015) revealed that there was a very high phenotypic multiple resistance within the isolates, which were confirmed by genetic analyses.

Birošová et al. (2015) revealed that water from alms- house, had no *E. coli* detected. Greater than 70% of *E. coli* isolated antibiotic resistant bacteria from hospital wastewater were resistant to all applied antibiotics (ampicillin, ciprofloxacin, gentamicin, tetracycline, and chloramphenicol). In treated (stabilized) sludge, ampicillin resistant E. coli was the dominated antibiotic resistant bacteria. Number of total *S. aureus* was the same in all studied wastewater and treated sludge samples. The total number *S. aureus* ranged from 1.4 to 2.0 logarithmic orders per ml or gram for treated sludge. In influent wastewater predominated antibiotic resistant bacteria were methicillin (MRSA), penicillin and gentamicin resistant S. aureus. In hospital wastewater the dominated antibiotic resistant bacteria were ciprofloxacin, penicillin, methicillin, ampicillin and vancomycin resistant. The study concluded that stabilized sludge represent sources of MRSA, penicillin, ampicillin and ciprofloxacin resistant *S. aureus*.

Derakhshan *et al.* (2016) reported that biological and chemical processes alone are not effective for antibiotics' removal from aquatic environments while combination of advanced oxidation and biological treatment processes can effectively reduce the amount of antibiotic. It was concluded that introduction of advanced treatment processes such as ozonation, advanced oxidation, activated carbon, Nano filtration and reverse osmosis can remove higher levels of antibiotics.

Huan et al. (2018) showed that combining photodegradation with biodegradation rates of 17α-ethinylestradiol (EE2), mediated by dissolved organic matter (DOM), and electrochemically modified DOM (EDOM) were promoted obviously. The efficiency of EE2 biodegradation was shown to be strongly correlated with electron accepting capacity (EAC) of DOM. Electrochemical modification can increase the EAC of DOM leading to EE2 biodegradation accelerated, and it also can form more triplet-state DOM moieties to promote the EE2 photodegradation in irradiation conditions, due to the increasing of quinone-type structures in DOM. Moreover, cell polymeric secretion (CPS) secreted from the microorganism could be stimulated to an excited state by irradiation, and that also accelerated EE2 degradation. Photolysis combined with biochemical degradation yielded less toxic degradation products. This study shows that the emission of DOM in wastewater after electrochemical treatment could accelerate estrogen degradation and play a positive role on the pollutant transformation in the environment.

CONCLUSION

It can be concluded that

a) biological and chemical processes alone are not effective for antibiotics' removal from aquatic environments while combination of advanced oxidation and biological treatment processes can effectively reduce the amount of antibiotic.
b) introduction of combination of electrochemical and advanced treatment processes such as ozonation, advanced oxidation, activated carbon, Nano filtration and reverse osmosis can remove higher levels of antibiotics

FUTURE RESEARCH DIRECTIONS

Conduct pilot studies to assess the extent of environmental contamination by antimicrobial residues and resistant organisms that enter the soil or water from human and animal waste. Evaluate the nature and magnitude of the impact of antimicrobial growth promoters and use the information to assist in risk-benefit assessments of each use. Evaluate the effect of current food processing and distribution methods on the emergence and spread of resistant organisms.

REFERENCES

Abargues, M. R., Ferrer, J., Bouzas, A., & Seco, A. (2013). Removal and Fate of Endocrine Disruptors Chemicals under Lab- Scale Postreatment Stage. Removal Assessment Using Light, Oxygen and Microalgae. *Bioresource Technology*, *149*, 142–148. doi:10.1016/j.biortech.2013.09.051 PMID:24096281

Aboulhassan, M. A., Souabi, S., Yaacoubi, A., & Baudu, M. (2006). Removal of surfactant from industrial wastewaters by coagulation flocculation process. *International Journal of Environmental Science and Technology*, *3*(4), 327–332. doi:10.1007/BF03325941

Adams, C., Asce, M., Wang, Y., Loftin, K., & Meyer, M. (2002). Removal of antibiotics from surface and distilled water in conventional water treatment processes. *Journal of Environmental Engineering*, *128*(3), 253–260. doi:10.1061/(ASCE)0733-9372(2002)128:3(253)

Adina, E. S., Cristina, O., Carmen, L, Paula, S., Paulina, V., Cornelia, B., & Ioan, G. (2013). *Waste Water Treatment Methods*. doi:10.5772/53755

Ahmed, S. N., Shreeshivadasan, C., Zuriati, Z., & Saad, A. A. (2015). Treatment Performance of Textile Wastewater Using Electrocoagulation (EC) Process under Combined Electrical Connection of Electrodes Int. *Journal of the Electrochemical Society*, *10*, 5924–5941.

Aina, I., Inger, K., Anders, F., & Roland, M. (2002). High Prevalence of Vancomycin-Resistant Enterococci in Swedish Sewage. ASM. *Applied and Environmental Microbiology*, 2838–2842.

Alder, A. C., McArdell, C. S., Golet, E. M., Molnar, E. M., Kuch, H. M., & Giger, W. (2003). Environmental exposure of antibacterial agents in hospital and municipal wastewater and river water in Switzerland (Abs.). *Proceedings of 3rd International Conference on Pharmaceuticals and Endocrine Disrupting Chemicals in Water*, 236–237.

Amenu, D. (2014). Assessment of Disinfectant and Antibiotic Resistant Bacteria in Hospital Wastewater. *World Journal of Life Sciences Research.*, *1*(1), 1–7.

American Public Health Association (APHA). (1998). Standard methods for examination of water and wastewater. American Public Health Association, American Water Works Association and Water Pollution Control Federation.

Andreozzi, R., Canterino, M., Marotta, R., & Paxeus, N. (2005). Antibiotic removal from wastewaters: The ozonation of amoxicillin. *Journal of Hazardous Materials*, *122*(3), 243–250. doi:10.1016/j.jhazmat.2005.03.004 PMID:15967280

Benson, C. I., Sisipho, G., Larry, C. O., Uchechukwu, U. N., & Anthony, I. O. (2015). Antibiotic Susceptibilities of Enterococcus Species Isolated from Hospital and Domestic Wastewater Effluents in Alice, Eastern Cape Province of South Africa. *International Journal of Environmental Research and Public Health*, *12*(4), 4231–4246. doi:10.3390/ijerph120404231 PMID:25893999

Berge, A. C. B., Dueger, E. L., & Sischo, W. M. (2006). Comparison of Salmonella enterica serovar distribution and antibiotic resistance patterns in wastewater at municipal water treatment plants in two California cities. *Journal of Applied Microbiology*.

Birošová, L., Nagyová, K., Mackuľak, T., & Medveďová, A. (2015). The Occurrence of Antibiotic Resistant Escherichia Coli and Staphylococcus Aureus In Wastewater and Sludge From Treatment Plant In Czech Republic. *Proceedings of the 14th International Conference on Environmental Science and Technology*.

Cabezaa, Y., Candelaa, L., & Ronenb, D. (2012). Monitoring the occurrence of emerging contaminants in treated wastewater and groundwater between 2008 and 2010. The Baix Llobregat. *Journal of Hazardous Materials*, *239*, 32–39. doi:10.1016/j.jhazmat.2012.07.032 PMID:22877748

Chang, X., Meyer, M. T., Liu, X., Zhao, Q., Chen, H., Chen, J., ... Shu, W. (2010). Determination of antibiotics in sewage from hospitals, nursery and slaughter house, wastewater treatment plant and source water in Chongqing region of Three Gorge Reservoir in China. *Environ.Pollut.*, *158*(5), 1444–1450. doi:10.1016/j.envpol.2009.12.034 PMID:20096493

Chelliapan, S., & Golar, S. (2011). Pre-treatment of Antibiotic Wastewater Using an Anaerobic Reactor. *Research Journal of Chemical Sciences*, *1*(6), 59–63.

Chelliapan, S., Wilby, T., & Sallis, P. J. (2006). Performance of an up-flow anaerobic stage reactor (UASR) in the treatment of pharmaceutical wastewater containing macrolide antibiotics. *Water Research*, *40*(3), 507–516. doi:10.1016/j.watres.2005.11.020 PMID:16387347

Chen, G. (2004). Electrochemical technologies in wastewater treatment. *Separation and Purification Technology*, *38*(1), 11–41. doi:10.1016/j.seppur.2003.10.006

Chen, H., & Zhang, M. (2013). Effects of Advanced Treatment Systems on the Removal of Antibiotic Resistance Genes in Wastewater Treatment Plants from Hangzhou, China. *Environmental Science & Technology*, *47*(15), 8157–8163. PMID:23802698

Choi, K.-J., Son, H.-J., & Kim, S.-H. (2007). Ionic treatment for removal of sulphonamide and tetracycline classes of antibiotic. *The Science of the Total Environment*, *387*(1-3), 247–256. doi:10.1016/j.scitotenv.2007.07.024 PMID:17764725

Collado, N., Buttiglieri, G., Marti, E., Ferrando-Climent, L., Rodriguez-Mozaz, S., Barcelo, D., ... Rodriguez-Roda, I. (2013). Effects on Activated Sludge Bacterial Community Exposed to Sulfamethoxazole. *Chemosphere*, *93*(1), 99–106. doi:10.1016/j.chemosphere.2013.04.094 PMID:23726012

Deniz, N., Angela, R., Dimitrios, B., & Viviane, Y. (2012). Removal of the antibiotic levofloxacin (LEVO) in water by ozonation and TiO_2 photocatalysis. *Chemical Engineering Journal, 189– 190*, 41– 48.

Derakhshan, Z., Mokhtari, M., Babaei, F., Roya, M. A., Mohammad, H. E., & Mohammad, F. (2016). Removal Methods of Antibiotic Compounds from Aqueous Environments–A Review. *J Environ Health Sustain Dev.*, *1*(1), 43–62.

Doolette, C., McLaughlin, M., Kirby, J., Batstone, D., Harris, H., Ge, H., & Cornelis, G. (2013). Transformation of PVP Coated Silver Nanoparticles in a Simulated Wastewater Treatment Process and the Effect on Microbial Communities. *Chemistry Central Journal*, *7*(1), 1–18. doi:10.1186/1752-153X-7-46 PMID:23497481

Dubreuil, L., Devos, J., Neut, C., & Romond, C. (1984). Susceptibility of Anaerobic Bacteria fromSeveral French Hospitals to Three Major Antibiotics.ASM. *Antimicrobial Agents and Chemotherapy*, *25*(6), 764–766. doi:10.1128/AAC.25.6.764 PMID:6742819

Ebele, A. J., Mohamed, A. A., & Stuart, H. (2017). Pharmaceuticals and personal care products (PPCPs) in the freshwater aquatic environment. *Emerging Contaminants*, *3*(1), 1–16. doi:10.1016/j.emcon.2016.12.004

Elmolla, E., & Chaudhuri, M. (2009a). Improvement of biodegradability of synthetic amoxicillin wastewater by photo-Fenton process. *World Applied Sciences Journal*, *5*, 53–58.

Elmolla, E., & Chaudhuri, M. (2009b). Optimization of Fenton process for treatment of amoxicillin, ampicillin and cloxacillin antibiotics in aqueous solution. *Journal of Hazardous Materials*, *170*(2-3), 666–672. doi:10.1016/j.jhazmat.2009.05.013 PMID:19515483

Emad, S. E., & Malay, C. (2010). Photo-Fenton Treatment of Antibiotic Wastewater. Nature Environment and Pollution. *An International Quarterly Scientific Journal Technology.*, *9*(2), 365–370.

EPA. (2010). Treating Contaminants of Emerging Concern A Literature Review Database. EPA.

Flores, C., Ventura, F., Martin-Alonso, J., & Caixach, J. (2013). Occurrence of Perfluorooctane Sulfonate (PFOS) and Perfluorooctanoate (PFOA) in N.E. Spanish Surface Waters and Their Removal in a Drinking Water Treatment Plant That Combines Conventional and Advanced Treatments in Parallel Lines. *The Science of the Total Environment*, *461–462*, 618–626. doi:10.1016/j.scitotenv.2013.05.026 PMID:23764674

Frank, W., Adair, S., Geftic, G., & Justus, G. (1969). Resistance of Pseudomonas to Quaternary Ammonium Compounds. ASM. *Applied Microbiology*, 299–302.

Geissena, V., Hans, M., Erwin, K., Günter, U., & Marti, N. (2015). Emerging pollutants in the environment: A challenge for water resource management. *International Soil and Water Conservation Research*, *3*(1), 57–65. doi:10.1016/j.iswcr.2015.03.002

Gerald, M., & Russell, A. D. (1999). Antiseptics and Disinfectants: Activity, Action, and Resistance. ASM. *Clinical Microbiology Reviews*, *12*(1), 147–179. doi:10.1128/CMR.12.1.147 PMID:9880479

Ghauch, A., Tuqan, A., & Assi, H. A. (2009). Antibiotic removal from water: Elimination of amoxicillin and ampicillin by microscale and nanoscale iron particles. *Environ. Pollution.*, *157*(5), 1626–1635. doi:10.1016/j.envpol.2008.12.024 PMID:19168269

Golet, E. M., Alder, A. C., & Giger, W. (2002). Environmental exposure and risk assessment of fluoroquinolone antibacterial agents in wastewater and river water of the Glatt Valley Watershed. Switzerland. *Environmental Science & Technology*, *36*(17), 3645–3651. doi:10.1021/es0256212 PMID:12322733

Golet, E. M., Xifra, I., Siegrist, H., Alder, A. C., & Giger, W. (2003). Environmental exposure assessment of fluoroquinolone antibacterial agents from sewage to soil. *Environmental Science & Technology*, *37*(15), 3243–3249. doi:10.1021/es0264448 PMID:12966965

Hirsch, R., Ternes, T., Haberer, K., & Kratz, K. L. (1999). Occurrence of antibiotics in the aquatic environment. *The Science of the Total Environment*, *225*(1-2), 109–118. doi:10.1016/S0048-9697(98)00337-4 PMID:10028708

Holm, J. V., Rügge, K., & Bjerg, P. L. (1995). Occurrence and distribution of pharmaceutical organic compounds in the groundwater downgradient of a landfill (Grindsted, Denmark). *Environmental Science & Technology*, *28*(5), 1415–1420. doi:10.1021/es00005a039 PMID:22192041

Honarbakhsh, M. (2011). *Impact of Wastewater Effluent on the Structure and Antibiotic Resistance of Native Microbial Communities*. A Master of Science Graduate Program in Microbiology and Molecular Genetics thesis submitted to the Graduate School-New Brunswick Rutgers, The State University of New Jersey and The Graduate School of Biomedical Sciences University of Medicine and Dentistry.

Hong, A. D., Ngoc, H. P., Hoang, T. N., Thi, T. H., Hung, V. P., Van, C. P., ... Alfredo, C. A. (2008). Occurrence, fate and antibiotic resistance of fluoroquinolone antibacterials in hospital wastewaters in Hanoi, Vietnam. *Chemosphere*, *72*(6), 968–973. doi:10.1016/j.chemosphere.2008.03.009 PMID:18485444

Hongna, L., Zhiwei, S., & Changxiong, Z. (2014). Trends in Research on Electrochemical Oxidation. *Croatica Chemica Acta*, *87*(2), 185–194. doi:10.5562/cca2446

Huan, H., Bin, H., Gen, F., Dan, X., Zhixiang, X., Xinhao, W., & Xuejun, P. (2018). Electrochemically modified dissolved organic matter accelerates the combining photodegradation and biodegradation of 17α- ethinylestradiol in natural aquatic environment. *Water Research*, *137*, 251–261. doi:10.1016/j.watres.2018.03.019 PMID:29550728

Ivan, S., Marin, M., Helena, K., Jakopovic, S. T., Josip, C., Ivan, M., & Marijan, A. (2011). Removal of antimicrobials using advanced wastewater treatment. *Journal of Hazardous Materials*, *192*, 319–328. PMID:21641718

Jara, C. C., Fino, D., Specchia, V., Saracco, G., & Spinelli, P. (2007). Electrochemical removal of antibiotics from wastewaters. *Applied Catalysis B: Environmental*, *70*(1-4), 479–487. doi:10.1016/j.apcatb.2005.11.035

Jurado, A., Vàzquez-Suñé, E., Carrera, J., López de Alda, M., Pujades, E., & Barceló, D. (2012). Emerging organic contaminants in groundwater in Spain: A review of sources, recent occurrence and fate in a European context. *The Science of the Total Environment*, *440*, 82–94. doi:10.1016/j.scitotenv.2012.08.029 PMID:22985674

Jury, K.L. Vancov, T., Stuetz, R.M., & Khan, S.J. (2010). Antibiotic resistance dissemination and sewage treatment plants. *Current Research, Technology and Education Topics in Applied Microbiology and Microbial Biotechnology*, 509 – 519.

Katouli, M., Thompson, J.M., Gündoğdu, A., & Stratton, H.M. (2012). Antibiotic Resistant Bacteria in Hospital Wastewaters and Sewage Treatment Plants. *Science Forum and Stakeholder Engagement: Building Linkages, Collaboration and Science Quality*, 225-229.

Kosma, C. I., Lambropoulou, D. A., & Albanis, T. A. (2010). Occurrence and removal of PPCPs in municipal and hospital wastewaters in Greece. *Journal of Hazardous Materials*, *179*(1-3), 804–817. doi:10.1016/j.jhazmat.2010.03.075 PMID:20395039

Kosutic, K., Dolar, D., Asperger, D., & Kunst, B. (2007). Removal of antibiotic from model wastewater by RO/NF membrane. *Separation and Purification Technology*, *53*(3), 244–249. doi:10.1016/j.seppur.2006.07.015

Kovalova, L., Siegrist, H., Von Gunten, U., Eugster, J., Hagenbuch, M., Wittmer, A., ... McArdell, C. S. (2013). Elimination of Micropollutants During Post- Treatment of Hospital Wastewater with Powdered Activated Carbon, Ozone, and UV. *Environmental Science & Technology*, *47*(14), 7899–7908. doi:10.1021/es400708w PMID:23758546

Lapworth, D. J., Baran, N., Stuart, M. E., & Ward, R. S. (2012). Emerging organic contaminants in groundwater: A review of sources, fate and occurrence. *Environ. Pollut.*, *163*, 287–303. doi:10.1016/j.envpol.2011.12.034 PMID:22306910

Li, H., Zhiwei, S., & Changxiong, Z. (2014). Trends in research on electrochemical oxidation. *Croatica Chemica Acta*, *87*(2), 185–194. doi:10.5562/cca2446

Liu, M., Zhang, Y., Ding, R., Gao, Y., & Yang, M. (2013). Response of Activated Sludge to the Treatment of Oxytetracycline Production Waste Stream. *Applied Microbiology and Biotechnology*, *97*(19), 8805–8812. doi:10.100700253-012-4589-8 PMID:23188460

Lopez-Serna, R., Jurado, A., Vazquez-Sune, E., Carrera, J., Petrović, M., & Barceló, D. (2013). Occurrence of 95 pharmaceuticals and transformation products in urban groundwater underlying the metropolis of Barcelona, Spain. *Environ. Pollut.*, *174*, 305–315. doi:10.1016/j.envpol.2012.11.022 PMID:23302545

Lozano, N., Rice, C. P., Ramirez, M., & Torrents, A. (2013). Fate of Triclocarban, Triclosan and Methyltriclosan During Wastewater and Biosolids Treatment Processes. *Water Research*, *47*(13), 4519–4527. doi:10.1016/j.watres.2013.05.015 PMID:23764601

Maletz, S., Floehr, T., Beier, S., Klumper, C., Brouwer, A., Behnisch, P., ... Hollert, H. (2013). In Vitro Characterization of the Effectiveness of Enhanced Sewage Treatment Processes to Eliminate Endocrine Activity of Hospital Effluents. *Water Research*, *47*(4), 1545–1557. doi:10.1016/j.watres.2012.12.008 PMID:23305681

Manisankar, P., Rani, C., & Viswanathan, S. (2004). Effect of halides in the electrochemical treatment of distillery effluent. *Chemosphere*, *57*(8), 961–966. doi:10.1016/j.chemosphere.2004.07.026 PMID:15488586

Margot, J., Kienle, C., Magnet, A., Weil, M., Rossi, L., de Alencastro, L. F., ... Barry, D. A. (2013). Treatment of Micropollutants in Municipal Wastewater: Ozone or Powdered Activated Carbon? *The Science of the Total Environment*, *461–462*, 480–498. doi:10.1016/j.scitotenv.2013.05.034 PMID:23751332

Mark, C. M. (1982). Effect of UV Light Disinfection on Antibiotic-Resistant Coliforms in Wastewater Effluents.ASM. *Applied and Environmental Microbiology*, *43*(2), 371–377. doi:10.1128/AEM.43.2.371-377.1982 PMID:7059170

Martin, S. F., & Charles, H. D. (1966). Factors Influencing the Occurrence of High Numbers of Iodine-Resistant Bacteria in iodinated Swimming Pools.ASM. *Applied Microbiology*, *14*(4), 627–635. doi:10.1128/AEM.14.4.627-635.1966 PMID:5927040

Meffe, R., & de Bustamante, I. (2014). Emerging organic contaminants in surface water and groundwater: A first overview of the situation in Italy. *The Science of the Total Environment*, *481*, 280–295. doi:10.1016/j.scitotenv.2014.02.053 PMID:24602913

Melanie, V., Indra, B., Anna, N., & Martin, J. (2017). Comparison of UVA and UVC irradiation of FQs and their MICs for sustainable water treatment. *Conference: 2nd Green and Sustainable Chemistry Conference at Berlin*. Retrieved from https://www.researchgate.net/publication/316977230_

Mesdaghinia, A. R., Naddafi, K., Nabizadeh, R., Saeedi, R., & Zamanzadeh, M. (2009). Wastewater Characteristics and Appropriate Method for Wastewater Management in the Hospitals. *Iranian J Pub. Health*, *38*(1), 34–40.

Michael, A. G., Mark, D. C., Kevin, B. A., & Jayne, C. B. (1985). Transfer of Plasmids pBR322 and pBR325 in Wastewater from Laboratory Strains of Escherichia coli to Bacteria Indigenous to the Waste Disposal System. *ASM. Applied and Environmental Microbiology*, 836–841.

Micheal, V. W., & John, W. V. (1985). Occurrence of Multiple-Antibiotic-Resistant Enteric Bacteria in Domestic Sewage and Oxidation Lagoons. *Applied and Environmental Microbiology*, 930–933. PMID:4083887

Miller, J. H., Novak, J. T., Knocke, W. R., Young, K., Hong, Y., Vikesland, P. J., ... Pruden, A. (2013). Effect of Silver Nanoparticles and Antibiotics on Antibiotic Resistance Genes in Anaerobic Digestion. *Water Environment Research*, *85*(5), 411–421. doi:10.2175/106143012X13373575831394 PMID:23789571

Morasch, B. (2013). Occurrence and dynamics of micropollutants in a karst aquifer. *Environ. Pollut.*, *173*, 133–137. doi:10.1016/j.envpol.2012.10.014 PMID:23202643

Naumczyk, J., Szpyrkowicz, L., & Zilio-Grandi, F. (1996). Electrochemical treatment of textile wastewater. *Water Science and Technology*, *34*(11), 17–24. doi:10.2166/wst.1996.0258

NORMAN. (2019). *Network of reference laboraties, research centres and related organisations for Monitoring of emerging environmental substances*. Retrieved from https://www.norman-network.net/?q=Home

Nuñez, L., & Moretton, J. (2007). Disinfectant-resistant bacteria in Buenos Aires city hospital wastewater. *Brazilian Journal of Microbiology*, *38*(4), 123–132. doi:10.1590/S1517-83822007000400012

Obst, U., Schwartz, T., & Volkmann, H. (2006). Antibiotic resistant pathogenic bacteria and their resistance genes in bacterial biofilms. *The International Journal of Artificial Organs*, *29*(4), 387–394. doi:10.1177/039139880602900408 PMID:16705607

Olowe, O. A., Olayemi, A. B., Eniola, K. I. T., & Adeyeba, O. A. (2004). Antibacterial activity of some selected disinfectants regularly used in hospitals. *African Journal of Clinical and Experimental Microbiology*, *5*(1), 126–130. doi:10.4314/ajcem.v5i1.7365

Olya, S. K., Katherine, Y. B., Carla, C., Benjamin, J. F., Meagan, S. M., Austa, M. P., ... Holly, A. S. (2014). Emerging Pollutants – Part II: Treatment. *Water Environment Research*, *86*(10), 2036–2096. doi:10.1002/j.1554-7531.2014.tb00240.x

Ötker, H. M., & Akmehmet-Balcioglu, I. (2005). Adsorption and degradation of enrofloxacin, a veterinary antibiotic on natural zeolite. *Journal of Hazardous Materials*, *122*(3), 251–258. doi:10.1016/j.jhazmat.2005.03.005 PMID:15967281

Pablo, C. F., Mart'ınez, C. J., Cristina, S., & Manuel, A. R. (2009). Technical and economic comparison of conventional and electrochemical coagulation processes. *Journal of Chemical Technology and Biotechnology (Oxford, Oxfordshire)*, *84*(5), 702–710. doi:10.1002/jctb.2102

Panizza, M., & Cerisola, G. (2003). A comparative study on direct and indirect electrochemical oxidation of polyaromatic compounds. *Ann Chim.*, *93*(12), 977–984. PMID:14756253

Park, S., & Choi, K. (2008). Hazard assessment of commonly used agricultural antibiotics on aquatic ecosystems. *Ecotoxicology (London, England)*, *17*(6), 526–538. doi:10.100710646-008-0209-x PMID:18449638

Pasquini, L., Merlin, C., Hassenboehler, L., Munoz, J.-F., Pons, M.-N., & Görner, T. (2013). Impact of Certain Household Micropollutants on Bacterial Behavior. Toxicity Tests/Study of Extracellular Polymeric Substances in Sludge. *The Science of the Total Environment*, *463–464*, 355–365. doi:10.1016/j.scitotenv.2013.06.018 PMID:23827359

Pauwels, B., & Verstraete, W. (2006). The treatment of hospital wastewater: an appraisal The treatment of hospital wastewater: an appraisal. *Journal of Water and Health*, *04*(4), 406–416. doi:10.2166/wh.2006.0024 PMID:17176811

Pedro, A. S., Matthieu, F., Christian, G., & Sébastien, S. (2009). Review of the Occurrence of Anti-infectives in Contaminated Wastewaters and Natural and Drinking Waters. *Environmental Health Perspectives*, *117*(5), 675–684. doi:10.1289/ehp.11776 PMID:19479007

Peng, X., Ou, W., Wang, C., Wang, Z., Huang, Q., Jin, J., & Tan, J. (2014). Occurrence and ecological potential of pharmaceuticals and personal care products in groundwater and reservoirs in the vicinity of municipal landfills in China. *The Science of the Total Environment*, *490*, 889–898. doi:10.1016/j.scitotenv.2014.05.068 PMID:24908648

Petrie, B., Barden, R., & Kasprzyk-Hordern, B. (2015). A review on emerging contaminants in wastewaters and the environment: Current knowledge, understudied areas and recommendations for future monitoring. *Water Research*, *72*, 3–27. doi:10.1016/j.watres.2014.08.053 PMID:25267363

Pomati, F., Castiglioni, S., Zuccato, E., Fanelli, R., Vigetti, D., Rossetti, C., & Davide, C. (2006). Effects of a complex mixture of therapeutic drugs at environmental levels on human embryonic cells. *Environmental Science & Technology*, *40*(7), 2442–2447. doi:10.1021/es051715a PMID:16646487

Putra, E. K., Pranowo, R., Sunarso, J., Indraswati, N., & Ismadji, S. (2009). Performance of activated carbon and bentonite for adsorption of amoxicillin from wastewater: Mechanism, isotherms and kinetics. *Water Research*, *43*(9), 2419–2430. doi:10.1016/j.watres.2009.02.039 PMID:19327813

Qamar, M., Saquib, M., & Muneer, M. (2005). Photocatalytic degradation of two selected dye derivatives, chromotrope 2B and amido black 10B, in aqueous suspensions of titanium dioxide. *Dyes and Pigments*, *65*(1), 1–9. doi:10.1016/j.dyepig.2004.06.006

Qian, S., Xuqi, C., Shuguang, L., Wentao, Z., Zhaofu, Q., & Gang, Y. (2015). Occurrence, sources and fate of pharmaceuticals and personal care products in the groundwater: A review. *Emerging Contaminants*, *1*(1), 14–24. doi:10.1016/j.emcon.2015.07.001

Rahim, A., Mahnaz, N., Hossein, K., & Akbar, H. (2014). Monitoring and Comparison of Antibiotic Resistant Bacteria and Their Resistance Genes in Municipal and Hospital Wastewaters. *International Journal of Preventive Medicine*, *5*(7), 887–894. PMID:25105001

Rana, P., Mohan, N., & Rajagopal, C. (2004). Electrochemical removal of chromium from wastewater by using carbon aerogel electrodes. *Water Research*, *38*(12), 2811–2820. doi:10.1016/j.watres.2004.02.029 PMID:15223274

Ravera, M., Ciccarelli, C., Gianotti, V., Scorza, S., & Osella, D. (2004). Electroassisted methods for waste destruction: Silver (II) and peroxydisulfate reagents in the electrochemically mediated oxidation of polyaromatic sulfonates. *Chemosphere*, *57*(7), 587–594. doi:10.1016/j.chemosphere.2004.08.035 PMID:15488920

Roberto, A., Marisa, C., Raffaele, M., & Nicklas, P. (2005). Antibiotic removal from wastewaters: The ozonation of amoxicillin. *Journal of Hazardous Materials*, *122*(3), 243–250. doi:10.1016/j.jhazmat.2005.03.004 PMID:15967280

Rodney, M. D., & William, J. C. B. (2002). Survival Mechanisms of Clinically Relevant Microorganisms.ASM. *Clinical Microbiology Reviews*, *15*(2), 167–193. doi:10.1128/CMR.15.2.167-193.2002 PMID:11932229

Rongjun, S., Guangshan, Z., Peng, W., Shixiong, L., Ryan, M. R., & Crittenden, J. C. (2015). Treatment of Antibiotic Pharmaceutical Wastewater Using a Rotating Biological Contactor. *Journal of Chemistry*, *2015*, 705275. doi:10.1155/2015/705275

Saichang, Z., Hong, C., & Jianan, L. (2013). Sources, distribution and potential risks of pharmaceuticals and personal care products in Qingshan Lake basin, Eastern China. *Ecotoxicology and Environmental Safety*, *96*, 154–159. doi:10.1016/j.ecoenv.2013.06.033 PMID:23871206

SalmaJ.RafikK. (2015). *Antibiotics*. Retrieved from https://www.researchgate.net/publication/272820065

Sánchez-Polo, M., Rivera-Utrilla, J., Prados-Joya, G., Ferro-García, M. A., & Bautista-Toledo, I. (2008). Removal of pharmaceutical compounds, nitroimidazoles, from waters by using the ozone/carbon systems. *Water Research*, *42*(15), 4163–4171. doi:10.1016/j.watres.2008.05.034 PMID:18762314

Schaider, L.A. Rudel, R.A., Ackerman, J.M. (2014). Pharmaceuticals, perfluorosurfactants, and other organic wastewater compounds in public drinking water wells in a shallow sand and gravel aquifer. *Sci. Total Environ., 468 – 469*, 384 – 393.

Sendzik, J., Shakibaei, M., Schafer-Korting, M., & Stahlmann, R. (2005). Fluoroquinolones cause changes in extracellular matrix, signalling proteins, metalloproteinases and caspase-3 in cultured human tendon cells. *Toxicology*, *212*(1), 24–36. doi:10.1016/j.tox.2005.04.002 PMID:15890441

Sharma, D. R., Pradhan, B., & Mishra, S. K. (2010). Multiple drug resistance in bacterial isolates from liquid wastes generated in central hospitals of Nepal. *Kathmandu University Medical Journal, 8*(1), 29, 40-44.

Shen, Z., Wang, W., Jia, J., Feng, X., Hu, W., & Peng, A. (2002). Catalytically assisted electrochemical oxidation of dye Acid Red B. *Water Environment Research*, *74*(2), 117–121. doi:10.2175/106143002X139811 PMID:12043967

Shreeshivadasan, C., & Paul, J. S. (2011). Application of Anaerobic Biotechnology for Pharmaceutical Wastewater Treatment. *Special Issue on Environmental Management For Sustainable Development.*, *2*(1), 13–21.

Sim, W. J., Lee, J. W., Lee, E. S., Shin, S. K., Hwang, S. R., & Oh, J. E. (2011). Occurrence and distribution of pharmaceuticals in wastewater from households, livestock farms, hospitals and pharmaceutical manufactures. *Chemosphere*, *82*(2), 179–186. doi:10.1016/j.chemosphere.2010.10.026 PMID:21040946

Sourav, M., & Alok, S. (2014). Treatment of Pharmaceutical Waste with Special Emphasis to Treatment Processes - A Review. *International Journal of Environmental Research and Development*, *4*(2), 171–176.

Stackelberg, P. E., Gibs, J., Furlong, E. T., Meyer, M. T., Zaugg, S. D., & Lippincott, R. L. (2007). Efficiency of conventional drinking-water-treatment processes in removal of pharmaceuticals and other organic compounds. *The Science of the Total Environment*, *377*(2-3), 255–272. doi:10.1016/j.scitotenv.2007.01.095 PMID:17363035

Stalder, T., Alrhmoun, M., Louvet, J.-N., Casellas, M., Maftah, C., Carrion, C., ... Dagot, C. (2013). Dynamic Assessment of the Floc Morphology, Bacterial Diversity, and Integron Content of an Activated Sludge Reactor Processing Hospital Effluent. *Environmental Science & Technology*, *47*(14), 7909–7917. doi:10.1021/es4008646 PMID:23789899

Stuart, M., Lapworth, D., Crane, E., & Hart, A. (2012). Review of risk from potential emerging contaminants in UK groundwater. *The Science of the Total Environment*, *416*, 1–21. doi:10.1016/j.scitotenv.2011.11.072 PMID:22209399

Stylianos, K. S., Katarzyna, S., Ioannis, A. K., & Anastasios, I. Z. (2015). Novel Water Treatment Processes Based on Hybrid Membrane-Ozonation Systems: A Novel Ceramic Membrane Contactor for Bubbleless Ozonation of Emerging Micropollutants. *Journal of Chemistry*, 1–12.

Thompson, J. M., Gundog˘du, A., Stratton, H. M., & Katouli, M. (2013). Antibiotic resistant Staphylococcus aureus in hospital wastewaters and sewage treatment plants with special reference to methicillin-resistant Staphylococcus aureus (MRSA). *Journal of Applied Microbiology*, *114*(1), 44–54. doi:10.1111/jam.12037 PMID:23057695

Trovó, A. G., Melo, S. A. S., & Nogueira, R. F. P. (2008). Photodegradation of the pharmaceuticals amoxicillin, bezafibrate and paracetamol by the photo-Fenton process - application to sewage treatment plant effluent. *Journal of Photochemistry and Photobiology A Chemistry*, *198*(2-3), 215–220. doi:10.1016/j.jphotochem.2008.03.011

Tumeo, E., Gbaguidi-Haore, H., Party, I., Bertand, X., Thouverez, M., & Talon, D. (2008). Are antibiotic-resistant Pseudomonas aeruginosa isolated from hospitalised patients recovered in the hospital effluents? *International Journal of Hygiene and Environmental Health*, *211*(1-2), 200–204. doi:10.1016/j.ijheh.2007.02.010 PMID:17459769

Van Loosdrecht, M. C. M., Nielsen, P. H., Lopez-Vazquez, C. M., & Brdjanovic, D. (2016). *Experimental Methods in Wastewater Treatment* (1st ed.). London: International Water Publishing Alliance House. doi:10.2166/9781780404752

Vieno, N. M., Hrkki, H., Tuhkanen, T., & Kronberg, L. (2007). Occurrence of pharmaceuticals in river water and their elimination in a pilot-scale drinking water treatment plant. *Environmental Science & Technology*, *41*(14), 5077–5084. doi:10.1021/es062720x PMID:17711226

Watkinson, A. J., Micalizzi, G. B., Graham, G. M., Bates, J. B., & Costanzo, S. D. (2007). Antibiotic-Resistant Escherichia coli in Wastewaters, Surface Waters, and Oysters from an Urban Riverine System. *Applied and Environmental Microbiology*, *73*(17), 5667–5670. doi:10.1128/AEM.00763-07 PMID:17616617

Xiaosong, C., Michael, T., Meyer, X., Liu, Q., Zhao, H., Chen, J., ... Weiqun, S. (2010). Determination of antibiotics in sewage from hospitals, nursery and slaughter house, wastewater treatment plant and source water in Chongqing region of Three Gorge Reservoir in China. *Environmental Pollution*, *158*(5), 1444–1450. doi:10.1016/j.envpol.2009.12.034 PMID:20096493

Yamina, B., Tahar, B., Lila, M., Hocine, H., & Laure, F. M. (2014). Study on Cadmium Resistant-BacteriaIsolated from Hospital Wastewaters. *Advances in Bioscience and Biotechnology*, *5*(08), 718–726. doi:10.4236/abb.2014.58085

Yang, S., Hai, F. I., Nghiem, L. D., Nguyen, L. N., Roddick, F., & Price, W. E. (2013). Removal of Bisphenol A and Diclofenac by a Novel Fungal Membrane Bioreactor Operated under Non-Sterile Conditions. *International Biodeterioration & Biodegradation*, *85*, 483–490. doi:10.1016/j.ibiod.2013.03.012

Yangming, L., Zhemin, S., Xuejun, C., Jinpin, J., & Wenhua, W. (2006). Preparation and application of nano-TiO2 catalyst in dye electrochemical treatment. *Water S.A.*, *32*(2), 205–210.

Zhang, G., Ji, S., & Xi, B. (2006). Feasibility study of treatment of amoxicillin wastewater with a combination of extraction, Fenton oxidation and reverse osmosis. *Desalination*, *196*(1-3), 32–42. doi:10.1016/j.desal.2005.11.018

Zhen, T., Wang, J., Wang, Q., Meng, H., & Wang, L. (2015). Research trends in electrochemical technology for water and wastewater treatment. *Applied Water Science*. doi:10.100713201-015-0280-4

Zheng, T., Wang, J., Wang, Q., Meng, H., & Wang, L. (2015). Research trends in electrochemical technology for water and waste water treatment. *Applied Water Science*, 1–18.

Zhimin, Q., John, J. M., Melanie, R. M., Rao, S., & Craig, D. A. (2006). Treatment of Antibiotics and Antibiotic Resistant Bacteria in Swine Wastewater with Free Chlorine. *Journal of Agricultural and Food Chemistry*, 8144–8154. PMID:17032022

ADDITIONAL READING

Aali, R., Baragh, S., Asgari, E., Fouladi Fard, R., Izanloo, H., & Hoseinpoor, S. (2019). Tracking of chloramphenicol, erythromycin, and sulfamethoxazole antibiotic-resistant bacteria from untreated wastewater effluents to receiving river. *Environmental Health Engineering and Management Journal*, *6*(2), 89–96. doi:10.15171/EHEM.2019.10

Ahmad, S. B. (2015). Electrochemical studies of 3-aminophenol and 3-aminobenzyl Alcohol in Aqueous Solutions at Different pH Values. *International Journal of Electrochemical Science*, *10*, 4742–4751.

Zhang, T. (2016). Antibiotics And Resistance Genes In Wastewater Treatment Plants. Amr In Food, Water And The Environment. *AMR CONTROL*, *2016*, 118–126.

KEY TERMS AND DEFINITIONS

ARB: Antibiotic resistant bacteria.

Chemical Oxygen Demand: Amount of oxygen required to oxidize both organic and inorganic substances in wastewaters.

CHL: Chloramphenicol.

DOM: Dissolved organic matter.

EAC: Electron accepting capacity.

EE2: 17α-ethinylestradiol.

GM: Gentamicin.

HWs: Hospital wastewaters.

MWs: Municipal wastewaters.

Ozonation Treatment: Oxidation process that requires the use of ozone.

PPCPs: Personal care products.

WWTPs: Wastewater treatment plants.

Chapter 9
Use of Low Cost Materials to Remove Chemicals of Emerging Concern From Wastewater Effluents

Sinmi Abosede
Pan Atlantic University, Nigeria

ABSTRACT

The use and production of synthetic organic chemicals for use in households and various industries has led to an increase in the occurrence and concentration of chemicals of emerging concern (CEC) in wastewater effluents and consequently the environment. Due to the refractory nature of some CECs, conventional physicochemical and biological treatments are not able to provide adequate elimination of these compounds. It is beneficial that these pollutants are removed by other methods before they are discharged to the environment or before the effluent can be recycled for reuse. The use of low-cost materials for the removal of CECs is an unexploited opportunity in developing countries, as these compounds possess the necessary functionalities to make them an ideal choice for water and water treatment in countries with limited economic and technical resources. This chapter reviews the various treatment processes that has been used for the removal of CECs in literature with a particular emphasis on low-cost materials.

INTRODUCTION

A recent study has revealed that at least 4 billion people (66% of global population) around the world, experience water scarcity at some point during the year (Mekonnen and Hoekstra 2016). In its 2018 report, the World Economic Forum ranked water crises (a significant decline in the available quality and quantity of fresh water, resulting in harmful effects on human health and/or economic activity) as one of the top 5 global risks in terms of societal impact (World Economic Forum 2018). It has also been predicted that there will be a 40% gap in the demand and availability of freshwater by 2030 (Addams et al. 2009). The reuse of wastewater effluent holds the key to the efficient utilization of the limited

DOI: 10.4018/978-1-7998-1871-7.ch009

global freshwater resource, by making available an alternative source of water to supplement existing supplies. It has been shown in several research studies that treated wastewater, can be a major component of the water resources strategy to meet the needs of a growing population and wastewater effluent has been used for purposes such as irrigation and groundwater recharge (Miller, G 2006; Levine and Asano 2004). The presence of toxic organic contaminants in the final effluent of wastewater plants is detrimental to organic life in the receiving waters and is a major obstacle to the recycle of the effluent for reuse (Pedersen et al. 2003).

Contaminants of Emerging Concern (CEC) are newly discovered naturally occurring, manufactured or manmade chemicals or materials, suspected to be present in various environmental compartments and whose toxicity or persistence are likely to significantly alter the metabolism of a living being (Sauvé and Desrosiers 2014). They include Pharmaceuticals and Personal care products (PPCPs), Endocrine Disrupting Chemicals (EDCs), steroids, hormones, surfactants and surfactant metabolites, flame retardants, pesticides, industrial additives, nanomaterials and gasoline additives (Stefanakis & Becker, 2016). The use and production of synthetic organic chemicals for use in households and various industries, has led to an increase in the occurrence and concentration of these compounds, in wastewater effluents and consequently the environment. Many studies in recent years have shown that treated wastewater effluent, is the main contributor to water pollution because of the presence of toxic organic pollutants (Ashton et al. 2004; Ternes et al. 2004; Gulkowska et al. 2008; Zhang et al. 2008; Launay et al. 2013).

In recent years, the presence of Endocrine-Disrupting Chemicals (EDC) in the environment has been of particular concern, because of the severe threat they pose to animals and humans. These compounds disturb the endocrine system by mimicking, blocking or disrupting the functions of hormones, affecting the health of humans and animals species (Bolong et al. 2009). For example, hermaphrodite fish has been detected in the lagoons of sewage treatment works, where male fish were found to be producing the female yolk precursor protein vitellogenin (Purdom et al. 1994) and it has been suggested that male and female reproductive system development and function may be susceptible to the effects of EDCs (Fowler et al. 2012; Knez, 2013). Pharmaceuticals and Personal Care Products (PPCPs) are also of particular concern. These substances are natural or man-made chemicals used in prescription medicines and veterinary drugs. The presence of antibiotics in wastewater effluent and aquatic environments, may increase bacterial antibiotic resistance, which can represent a potential threat for human and animal health and create challenges in the control of infections (Stefanakis & Becker, 2016).

Due to the refractory nature of some CECs, conventional physicochemical and biological treatments are not able to provide adequate elimination of these compounds (Luo et al., 2014). The use of low cost materials in the removal of toxic organic compounds from wastewater has gained widespread in recent years, as they offer potentially simple and economic solutions to the challenges imposed by the presence of toxic compounds in the environment. A material can be considered low cost, if it requires minimal pre-treatment and can be found in large quantities in nature (Bailey et al., 1999). The use of low cost materials for the removal of CECs is an unexploited opportunity in developing countries, as these materials possess the necessary functionalities to make them an ideal choice for water and water treatment in countries with limited economic and technical resources. This chapter reviews the various treatment processes that have been used for the removal of CECs in literature, with a particular emphasis on low cost materials.

BACKGROUND

Conventional Wastewater Treatment for the Removal of CEC

Conventional wastewater treatment involves the use of various physical, biological and chemical processes to remove organic pollutants from wastewater before discharging the treated effluent to the environment. Majority of wastewater treatment plants all over the world, treat wastewater via two main stages – Primary Treatment and Secondary Treatment. Primary Treatment is used to remove gross and suspended solids from the wastewater using mainly physical processes such as Screening, Grit Chamber, Maceration, Settlement Tanks and Coagulation. Screens, grit chamber and macerators are used to remove solids such as rags, sticks, sand and stones. This is essential to protect mechanical equipment such as pumps from abrasive wear and to avoid clogging of pipes. Suspended solids are removed by sedimentation in rectangular or circular tanks, this settling process can be improved by the addition of chemical coagulants such as alum and ferric chloride. Primary treatment typically removes about 70% of total suspended solids from wastewater (Kiely, 1997).

Secondary Treatment follows Primary Treatment and is used to remove residual dissolved and colloidal organic matter not removed by primary treatment. The primary overflow is contacted with air in the presence of aerobic bacteria and other micro-organisms, which converts the biodegradable organic matter present to carbon dioxide and water. The removal mechanisms include biodegradation, biosorption and biotransformation. The organic pollutants can be biodegraded to different degrees, resulting in mineralisation or partial degradation and the formation of by-products (Luo et al., 2014). The two commonly used plants for this process are; the trickling-bed filter and the activated sludge process. Trickling-bed filters contain beds of filter media usually stones, through which the wastewater passes. Bacteria gather and multiply on the filter media until they consume most of the organic matter in the wastewater. The resulting partially treated water trickles out of the bed through pipes to a sedimentation tank, where the excess bacteria is removed through sedimentation. In the activated sludge-process, air and sludge heavily laden with bacteria are brought into close contact with the wastewater in an aerated tank and allowed to remain for a few hours. During this time, the organic matter in the wastewater is biodegraded and is converted into harmless by-products. In combination with Primary Treatment, Secondary Treatment processes can typically remove up to 85% of the BOD_5 (5 day biological oxygen demand) and suspended solids present in the wastewater (Pescod, 1992).

Most biodegradable contaminants will be easily removed by Primary and Secondary Treatment, however due to their recalcitrant nature, most CECs are not efficiently removed by these processes and persist in the receiving environments (Gupta & Thakur, 2017). Primary treatment was found to be ineffective in the removal of hydrophilic PPCPs, however PPCPs which are highly lipophilic were removed through biosorption onto the lipophilic layer of the primary sludge (Carballa et al., 2005). Removal rates of up to 70% were achieved for lipophilic compounds, while moderate removal rates of not more than 25% were achieved for the hydrophilic compounds. Different removal rates ranging from 25% - 75%, were reported for the removal of different PPCPs from wastewater using the activated sludge process (Salgado et al., 2012). Low removal rates were observed for fluoxetine, clorazepate, enalapril, captopril and ampicillin, while high removal rates were found for ibrupofen, ketoprofen, atenolol and clofibric acid. The researchers found that the removal of the PPCPs was related to the hydrophobicity of the compounds. The removal of estrogens, endocrine disruptors, and pharmaceutical substances by activated sludge was studied by Urase & Kikuta (2005). Removal rates ranged from 5% - 60%, removal was predominantly under acidic conditions by adsorption unto the sludge and not by biodegradation. A disadvantage of using biological treatment for the removal of CECs is that the plants need to be designed with long sludge and hydraulic retention time, which results in plants with large footprint (Clara et al., 2005; Servos et al., 2005).

Tertiary Treatment is an additional treatment step used to remove any residual pollutants present in the wastewater effluent after conventional treatment, thus achieving a higher effluent quality. It is a very important step in the reclamation of water intended for reuse. Typical tertiary processes include: Membrane Filtration, Chemical Oxidation, Ion Exchange, Nanotechnology and Adsorption. Oxidation with chlorine and ozone has been successfully used to remove various EDCs and pharmaceuticals from wastewater (Huber et al., 2005). Hernández-Leal et al., (2011) investigated the potential of ozone to remove a wide range of various personal care products from grey water. Removal rates of more than 79% was observed for most of the compounds. In another study, removal rates of more than 95% were achieved for the removal of PPCP products using ozonation (Sui et al., 2010). Kim, Yamashita, & Tanaka (2009) tested the effectiveness of UV treatment for the removal of 41 pharmaceutical products. Removal rates of up to 90% were achieved for majority of the compounds. Adams et al., (2002) found that Ultraviolet (UV) photolysis was inefficient for the removal of a group of antibiotics from drinking water. Removal rates of 50 - 80% were only achieved when dose rates of the order of 100 times greater than the typical disinfection dosages were used. The same study showed that using reverse osmosis, removal rates of up to 90% could be achieved for the removal of the antibiotics. Though effective in the removal of CECs from wastewater, Ozonation and UV processes are energy intensive and expensive to operate.

Membrane Filtration utilises hydrostatic pressure to remove suspended solids and high molecular weight solutes and allow water and low molecular weight solutes to pass through Rodriguez-Narvaez et al, (2017). The main membrane processes used for water and wastewater treatment include - ultra-filtration, nanofiltration, microfiltration, and reverse osmosis. Micro filtration is not effective in the removal of CECs, as it cannot remove contaminants with size <1μm (Zhou et al., 2010). Nanofiltration and ultrafiltration have been successfully used in the removal of CECs. Acero et al., (2010) investigated the removal of 11 CECs (pharmaceuticals and pesticides) from aqueous solution by ultrafiltration and nanofiltration membranes. They found that most of the CECs achieved removal rates of above 70% with the nanofiltration membranes, while lower removal rates (<50%) were obtained for the ultrafiltration membranes. Yoon, et al., (2007) investigated the removal of 27 pharmaceuticals, personal care products and endocrine disruptors using ultrafiltration and nanofiltration. The removal rates for most of the compounds with nanofiltration ranged from 44 – 93%, while lower removal rates (< 40%) were achieved using ultrafiltration. It was suggested that the dominant mechanism for removal of the compounds was hydrophobic adsorption. Garcia et al, (2013) used a two stage process consisting of microfiltration and reverse osmosis to remove 22 different CECs from wastewater effluent. The combined effect of both processes resulted in removal rates ranging from 65%-90%. The removal of fifteen PPCPs using a combined microfiltration and reverse osmosis process was investigated (Sui et al., 2010), with removal rates of more 95% for most of the compounds. Membrane processes are very attractive for the removal of CECs, however, they are very energy intensive and subject to membrane fouling.

Adsorption is considered to be the best wastewater treatment process due to its versatility, high adsorptive capacity and ease of operation (Ali et al., 2012). Many adsorbents are used for the removal of pollutants, however, activated carbon adsorption is the most popular. Kovalova et al., (2013) investigated the removal of various CECs (pharmaceuticals, metabolites and industrial chemicals) using Powdered Activated Carbon (PAC), from a hospital's wastewater effluent. Removal rates of up to 86%

were achieved for most of the compounds. PAC also successfully eliminated various CECs such as Bisphenol-A, triclosan, tonalide, galaxolide from wastewater effluent, achieving removal rates of more than 94% for most of the compounds (Hernández-Leal et al., 2011). The removal of steroidal estrogens and pharmaceuticals from the wastewater effluent of a sewage treatment works was assessed using a full scale Granular Activated Plant (GAC) (Grover et al, 2011). High removal rates were observed for many of the pharmaceuticals such as mebeverine, indomethacine, and diclofenac; moderate removal rates (43-64%) were observed for the steroidal estrogens and low removal rates (17-23%) for some pharmaceuticals such as carbamazepine and propranolol.

The conversion of biomass to activated carbon involves the use of physical and/or chemical processes to change the surface properties and pore structure of the biomass, to make it suitable for the adsorption of pollutants, this comes with associated energy and chemical costs. This makes the use of commercial activated carbon for the removal of organic pollutants prohibitive on a large scale, because of the associated high absorbent and regeneration costs associated with the process (Carmalin & Eder, 2018; Gupta et al., 2011). The most common feedstocks for the production of commercial activated carbon are wood, coal, lignite and peat (Ali et al., 2012), which have a high carbon content. Research on cheaper alternative carbon sources for the production of activated carbon has intensified in recent years. Agricultural, industrial and municipal waste have been used as alternative sources of carbon (Ozdemir et al, 2014). The activated carbon is manufactured by the pyrolysis of biomass, followed by chemical or physical activation (Carmalin & Eder, 2018). An ammonium chloride activated carbon, derived from a waste biomass, was used to remove diazinon, a pesticide from contaminated water (Moussavi et al, 2013), with a removal rate of about 97.5%. The adsorption isotherm fitted the Langmuir model and the reaction kinetics could be described by the pseudo-second order model. A potassium activated carbon produced from rubber tire, was successful in the removal of three pesticides (methoxychlor, atrazine and methyl parathion) from wastewater (Gupta et al., 2011), with removals rates ranging from 74.1 – 93.3%. The adsorption isotherm fitted the Langmuir model and the kinetic reactions, fitted the pseudo-first order model.

An activated carbon derived from peach stones, activated with phosphoric acid was used to remove three pharmaceuticals caffeine, diclofenac and carbamazepine from aqueous solution (Torrellas et al., 2015). Removal rates ranging from 48 – 96% were achieved and the adsorption isotherm was found to fit the Sipps model well. Rovani et al., (2014) prepared an activated carbon prepared from a mixture of coffee grounds, eucalyptus sawdust and soybean oil, activated with lime. This material was used for the removal of EDCs compounds, 17β-estradiol and 17α-ethinylestradiol from aqueous solutions. Removal rates of up to 91% were achieved for both EDCs and it was found that the adsorption isotherm fitted the Sipps model and the reaction kinetics was well described by the General Order kinetic model. He et al., (2016) prepared a highly porous carbon by the potassium hydroxide (KOH) activation of α-cellulose and used it to study the adsorption of two antibiotics tetracycline (TC) and sulfamethazine (SMZ) from aqueous solutions. Adsorption kinetics for the two compounds, showed that the kinetic results could be described by the pseudo-second model, while the adsorption isotherm results were found to fit the Langmuir model.

Each tertiary treatment method has its merits and limitations in application, however most tertiary treatment processes are expensive and their use is often restricted due to economic considerations. They involve complex maintenance schedules, high capital costs and generate effluent with high concentrations of toxic contaminants (Carmalin & Eder, 2018).

USE OF LOW COST MATERIALS FOR THE REMOVAL OF CEC

The use of low cost materials to remove heavy metals (Anastopoulos & Kyzas, 2015b; Dhankhar & Hooda, 2011) and dyes (Anastopoulos & Kyzas, 2015a; Saba et al., 2016) from wastewater is well documented, however, there are relatively few studies on the use of low cost materials to remove CECs from wastewater effluents. Various agricultural, industrial and municipal waste have been successfully used in their natural state, for the adsorption of pollutants from wastewater effluent. These low cost adsorbents require little processing and are usually abundant in nature and their use for the removal of CECs is an unexploited opportunity in developing countries, as these materials possess the necessary functionalities to make them an ideal choice for wastewater treatment in countries with limited economic and technical resources to provide advanced wastewater treatment systems for the removal of pollutants (Chaukura et al., 2016).

The adsorption of 22 different pesticides on rice bran was studied by Adachi et al., (2001). The removal rates ranged from 22.2-98.8% and the adsorption isotherm fitted the Freundlich model. The mechanism of adsorption was attributed to the uptake of the pesticides into the intracellular particles of the rice bran. Several organochlorine compounds were also successfully removed from aqueous solution by rice bran (Adachi et al., 2001). Removal rates ranged from 65-95% and the adsorption isotherm fitted the Freundlich isotherm. Lazim et al., (2015) investigated the potential of acid treated coir pith, coconut shell and durian peel on the removal of bisphenol A (BPA), an Endocrine Disrupting Compound, from aqueous solution. Removal rates of up to 72% were achieved. Nanseu-Njiki & Ngameni (2010) studied the removal of paraquat from aqueous solution using sawdust. The adsorption isotherm fitted the Langmuir model and the reaction kinetics followed the pseudo-second order model. Cork, a lignocellulosic natural material, has been used to remove CECs such as triclosan, naproxen, ketoprofen, carbamazepine and methyl paraben (Mallek et al., 2018). Adsorption isotherms were found to fit the Freundlich model and removal rates ranging from 50 – 100% were achieved.

Biochar (BC) is a stable source of carbon, which is obtained from biomass by thermal or hydrothermal processes at high temperature with low or zero oxygen atmosphere (Klinar, 2016). It has a high specific area and possesses various functional groups such as amino, carboxyl, hydroxyl, etc. This makes them an attractive remediation agent for contaminated soils and water. To improve their adsorption capacity, biochar can be modified using various processes such as impregnation of metal oxides and nanomaterials (Agrafioti et al., 2014; Zhou et al., 2013), steam activation (Rajapaksha et al., 2015), chemical modification (Jin et al., 2014; Kim et al., 2016) and heat treatment (Shen & Liu, 2008). The modified biochar usually has enhanced chemical and physical properties such as surface area, porosity, functionality and chemical interactions in comparison to the unmodified biochar (Ahmed et al., 2016). The raw materials (agricultural biomass and animal solid waste) used for biochar production are also abundant and low-cost (Li et al., 2019) and have used to successfully used to treat wastewater contaminated with emerging contaminants.

Yi et al., (2016) investigated the use of biochar from rice husk and wood chip to remove levofloxacin, an antibiotic from solution. These biochars were successful in removing the antibiotic from solution and sorption capacities ranged from 1.49 to 7.72 mg/g. Kim et al., (2016) studied the adsorption of some sunscreen compounds (benzophene and benzotriaxole) and some endocrine-disrupting compounds (bisphenol and 17 β – estradiol) on commercial powdered activated carbon (PAC) and activated biochar. The found that the biochar was more effective than the PAC in removing the CECs from solution, by about 5-30%. This was attributed to the higher surface area and pore volume of the biochar in comparison to the PAC. Biochar derived from wood biomass was used successfully to remove 4 antibiotics namely sulfamethazine, sulfamethoxazole, sulfathiazole and chloramphenicol from lake water and synthetic wastewater samples (Ahmed et al., 2017). Removal rates ranged from 58% - 100%. Weidemann et al, (2018) studied the removal of 10 various CECs (octhilinone, triclosan, trimethoprim, sulfamethoxasole, ciprofloxacin, diclofenac, paracetamol, diphenhydramine, fluconazole, and bisphenol A) from solution using modified biochar obtained from agricultural waste such as tomato and olive press wastes, rice husks, and horse manure. The different biochars exhibited differing removal rates for the CECs. All four biochars achieved approximately 100% removal of Diclofenac and bisphenol A, and rice husk was very successful in the removal of ciprofloxacin, diphenhydramine, octhilinone and triclosan, with removal rates of more than 83%.

Microorganisms have the ability to breakdown or transform pollutants through biodegradation and bioremediation (Asgher et al., 2014). Chang et al., (2018) investigated the use of spent mushroom compost to remove two pharmaceutical products, acetaminophen and sulfonamides from wastewater. The removal of the compounds were achieved by degradation with the extracellular enzymes of white rot fungi present in the compost and adsorption on the compost, with removal rates ranging from 27.1% - 64%. Kebede et al., (2019) found that the water-soluble proteins from the seeds of Moringa stenopetala, was successful in the removal of antibiotics from wastewater. The dominant functional groups of the protein were the amide and amines, and the study discovered that these were the main groups responsible for the removal of the antibiotics through hydrogen bonding or electrostatic force of attraction between the proteins and antibiotics. Removal rates in the range of 70.4- 96.3% were achieved for some of the antibiotics. Micro-algae based systems have demonstrated a potential to remove contaminants from aqueous solution (Craggs et al., 2012). Microalgae are unicellular photosynthetic micro-organisms, living in saline or freshwater environments, that convert sunlight, water and carbon dioxide to algal biomass (Ruane et al., 2010). The removal of two EDCs 17α-ethynylestradiol (EE2) and 17β- estradiol (E2), from aqueous solutions, was investigated using three different algae species (Ge et al, 2009). The reaction mechanism was photodegradation of the EDCs, which led to a reduction in the toxicity of the solution. Freshwater Microalgae demonstrated the potential to remove Bisphenol A, an EDC from aqueous solution through bioaccumulation and biodegradation (Ji et al., 2014). Various pharmaceuticals and estrogens (diclofenac, ibuprofen, paracetamol, metoprolol, carbamazepine and trimethoprim, estrone, 17β-estradiol and ethinylestradiol) were removed from wastewater using microalgae (de Wilt et al., 2016). The reaction mechanisms included biodegradation, photolysis and biosorption. Coimbra et al., (2018) studied the removal of diclofenac from solution using microalgae strains. The adsorption kinetics fitted the pseudo-second order model and the adsorption isotherm fitted the Lagmuir isotherm model.

Phytoremediation is the use of living plants for the treatment of contaminated soils and waters (Vasavi et al., 2010). The biological, chemical and physical properties of the plants give them the ability to remove the contaminants from soil or water, through the immobilization, degradation, extraction and or/volatilization of the contaminants. It is an especially attractive method of treating contaminated soils and waters because the plants are low cost, widely available and there is the possibility to recover and reuse the compounds recovered (Eskander & Saleh, 2017). Phytoremediation has been used predominantly for the removal of heavy metals from contaminated soil and waters (Ashraf, Ali, Zahir, Ashraf, & Asghar, 2019; Tang et al., 2019). However, a few researchers have used phytoremediation for the removal of some CECs. Supalkova et al., (2008) reported that duckweed (*L. minor*) was effective in the removal of cisplatin, a chemotherapy drug from wastewater. Iatrou et al., (2017) conducted a study on the phytoremediation of four antimicrobials (cefadroxil, CFD; metronidazole, METRO; trimethoprim, TRI;

sulfamethoxazole, SMX) by *L. minor* in wastewater effluent. They observed that the duckweed was able to achieve removal rates ranging from 59% to 100% and the dominant mechanism was plant uptake and biodegradation. Dolphen & Thiravetyan (2015) found that various plants such as *Cyperus alternifolius* (Umbrella papyrus), *Echinodorus cordifolius* (Creeping Burrhead), *Thalia geniculata* (Alligator Flag), *Acorus calamus* (Sweet Flag), and *Dracaena sanderiana* (Lucky Bamboo) were effective in the removal of ethanolamines from wastewater. Removal rates of up to 91% was achieved and the main mechanism of removal of the ethanoalmines was through plant uptake and biodegradation. Horseradish hairy root culture (Armoracia rusticana) was shown to have the potential to degrade benzophenone, a CEC found in skin care products (Chena et al. 2016)`. More than 20% of Benzophenone was degraded after 3 hours.

Limitations of the Use of Low Cost Materials for the Removal of CEC

The use of low cost materials in the removal of CECs is very attractive because of the lower associated costs, wide availability and ease of operation compared with more convention treatment methods. However, there are some issues related to their use in wastewater treatment. The management of the spent material and the removed pollutant after treatment is a challenge that must be addressed, in the interest of the environment. Ali et al., (2012) suggested that the removed pollutants should be recycled or dumped deep into the soil or filled in steel containers. The materials can be regenerated, buried in a landfill or further degraded by plants.

There is limited scope to use most low cost low cost materials in fixed beds, because of their poor mechanical properties. For example, micro-organisms used for the removal of CECs from wastewater need a continuous supply of nutrient (Vijayaraghavan & Yun, 2008). The continuous supply of nutrients might be just as expensive as the conventional tertiary treatment processes. Batch adsorption processes using most low cost materials are not effective for long term applications, due to the difficulty of separating the spent material from solution. Physical and chemical processes might be necessary to achieve separation of the low cost material from the effluent. If the low cost materials are used in their natural state, there is a likelihood that they may exert a biological oxygen demand during decomposition. This will have an adverse effect on the aquatic organisms in the natural waters, the effluent will be discharged to. To prevent this, it might be necessary to improve the performance of the low cost materials through physical and chemical modifications.

Cost Benefit Analysis

The economic feasibility of the use of low cost materials in wastewater treatment is not well documented and is an area for further research (Zaman et al., 2020). Cost benefit analysis is a vital tool used in water and wastewater management to aid decision making in the determination of the appropriate treatment process to remove particular contaminants. Cost benefit analysis comparing the performance of low cost materials with standard and commercial processes are limited (Huggins at al.,2016). Most of the analysis in literature as focused on the analysis for the use of low cost materials to remove metals from wastewater (Gupta & Babu, 2008; Kurniawan et al., 2006).

However, due to wide availability and low cost, the use of these materials for the removal of CECs has the potential to reduce the overall cost of the treatment of wastewater, in addition to the added benefit of using a potential waste product in a sustainable manner for the removal of organic contaminants from wastewater.

CONCLUSION

Due to the refractory nature of some CECs, conventional physicochemical and biological treatments are not able to provide adequate elimination of these compounds. Tertiary treatment is an additional treatment step used to remove any residual pollutants present in the wastewater effluent after conventional treatment, thus achieving a higher effluent quality. It is beneficial that a chosen Tertiary Treatment process is cost effective, simple and compatible with existing operations. The greatest challenge in implementing the strategy to reuse treated wastewater to augment existing water supplies, is the adoption of low cost wastewater treatment technologies, which will maximize the efficiency of limited water resources and ensure compliance with all health and safety standards regarding reuse of treated wastewater effluents (Grassi et al. 2012). The use of low cost materials in the removal of toxic organic pollutants from wastewater effluent has gained widespread acceptance in recent years, as they offer potentially simple and economic solutions to the challenges imposed by the presence of CECs in the environment or in effluent intended for reuse. In this review, an attempt has been made to draw attention to the different treatment processes used in the treatment of effluent containing CECs, with a particular focus on the use of low-cost materials.

REFERENCES

Acero, J. L., Benitez, F. J., Teva, F., & Leal, A. I. (2010). Retention of emerging micropollutants from water and a municipal secondary effluent by ultrafiltration and nanofiltration. *Chemical Engineering Journal*, *163*(3), 264–272. doi:10.1016/j.cej.2010.07.060

Adachi, A., Ikeda, C., Takagi, S., Fukao, N., Yoshie, E., & Okano, T. (2001). Efficiency of Rice Bran for Removal of Organochlorine Compounds and Benzene from Industrial Wastewater. *Journal of Agricultural and Food Chemistry*, *49*(3), 1309–1314. doi:10.1021/jf001147c PMID:11312856

Adachi, A., Takagi, S., & Okano, T. (2001). Studies on Removal Efficiency of Rice Bran for Pesticides. *Journal of Health Science*, *47*(2), 94–98. doi:10.1248/jhs.47.94

Adams, C., Wang, Y., Loftin, K., & Meyer, M. (2002). Removal of Antibiotics from Surface and Distilled Water in Conventional Water Treatment Processes. *Journal of Environmental Engineering*, *128*(3), 253–260. doi:10.1061/(ASCE)0733-9372(2002)128:3(253)

Agrafioti, E., Kalderis, D., & Diamadopoulos, E. (2014). Arsenic and chromium removal from water using biochars derived from rice husk, organic solid wastes and sewage sludge. *Journal of Environmental Management*, *133*, 309–314. doi:10.1016/j.jenvman.2013.12.007 PMID:24412594

Ahmed, M. B., Zhou, J. L., Ngo, H. H., Guo, W., & Chen, M. (2016). Progress in the preparation and application of modified biochar for improved contaminant removal from water and wastewater. *Bioresource Technology*, *214*, 836–851. doi:10.1016/j.biortech.2016.05.057 PMID:27241534

Ali, I., Asim, M., & Khan, T. A. (2012). Low cost adsorbents for the removal of organic pollutants from wastewater. *Journal of Environmental Management*, *113*, 170–183. doi:10.1016/j.jenvman.2012.08.028 PMID:23023039

Anastopoulos, I., & Kyzas, G. Z. (2015a). Composts as Biosorbents for Decontamination of Various Pollutants: A Review. *Water, Air, and Soil Pollution*, *226*(3), 61. doi:10.100711270-015-2345-2

Anastopoulos, I., & Kyzas, G. Z. (2015b). Progress in batch biosorption of heavy metals onto algae. *Journal of Molecular Liquids*, *209*, 77–86. doi:10.1016/j.molliq.2015.05.023

Asgher, M., Shahid, M., Kamal, S., & Iqbal, H. M. N. (2014). Recent trends and valorization of immobilization strategies and ligninolytic enzymes by industrial biotechnology. *Journal of Molecular Catalysis. B, Enzymatic*, *101*, 56–66. doi:10.1016/j.molcatb.2013.12.016

Ashraf, S., Ali, Q., Zahir, Z. A., Ashraf, S., & Asghar, H. N. (2019). Phytoremediation: Environmentally sustainable way for reclamation of heavy metal polluted soils. *Ecotoxicology and Environmental Safety*, *174*, 714–727. doi:10.1016/j.ecoenv.2019.02.068 PMID:30878808

Bailey, S. E., Olin, T. J., Bricka, R. M., & Adrian, D. D. (1999). A review of potentially low-cost sorbents for heavy metals. *Water Research*, *33*(11), 2469–2479. doi:10.1016/S0043-1354(98)00475-8

Carballa, M., Omil, F., & Lema, J. M. (2005). Removal of cosmetic ingredients and pharmaceuticals in sewage primary treatment. *Water Research*, *39*(19), 4790–4796. doi:10.1016/j.watres.2005.09.018 PMID:16246396

Carmalin, S. A., & Eder, C. L. (2018). Removal of emerging contaminants from the environment by adsorption. *Ecotoxicology and Environmental Safety*. PMID:29253687

Chang, B.-V., Fan, S.-N., Tsai, Y.-C., Chung, Y.-L., Tu, P.-X., & Yang, C.-W. (2018). Removal of emerging contaminants using spent mushroom compost. *The Science of the Total Environment*, *634*, 922–933. doi:10.1016/j.scitotenv.2018.03.366 PMID:29660886

Chaukura, N., Gwenzi, W., Tavengwa, N., & Manyuchi, M. M. (2016). Biosorbents for the removal of synthetic organics and emerging pollutants: Opportunities and challenges for developing countries. *Environmental Development*, *19*, 84–89. doi:10.1016/j.envdev.2016.05.002

Chena, F., Hubera, C., May, R., & Schröder, P. (2016). Metabolism of oxybenzone in a hairy root culture: Perspectives for phytoremediation of a widely used sunscreen agent. *Journal of Hazardous Materials*, *306*, 230–236. doi:10.1016/j.jhazmat.2015.12.022 PMID:26736174

Clara, M., Strenn, B., Gans, O., Martinez, E., Kreuzinger, N., & Kroiss, H. (2005). Removal of selected pharmaceuticals, fragrances and endocrine disrupting compounds in a membrane bioreactor and conventional wastewater treatment plants. *Water Research*, *39*(19), 4797–4807. doi:10.1016/j.watres.2005.09.015 PMID:16242170

Coimbra, R., Escapa, C., Vázquez, N., Noriega-Hevia, G., Otero, M., Coimbra, R. N., & Otero, M. (2018). Utilization of Non-Living Microalgae Biomass from Two Different Strains for the Adsorptive Removal of Diclofenac from Water. *Water (Basel)*, *10*(10), 1401. doi:10.3390/w10101401

Craggs, R., Sutherland, D., & Campbell, H. (2012). Hectare-scale demonstration of high rate algal ponds for enhanced wastewater treatment and biofuel production. *Journal of Applied Phycology*, *24*(3), 329–337. doi:10.100710811-012-9810-8 de Wilt, A., Butkovskyi, A., Tuantet, K., Leal, L. H., Fernandes, T. V., Langenhoff, A., & Zeeman, G. (2016). Micropollutant removal in an algal treatment system fed with source separated wastewater streams. *Journal of Hazardous Materials*, *304*, 84–92. doi:10.1016/j.jhazmat.2015.10.033 PMID:26546707

Dhankhar, R., & Hooda, A. (2011). Fungal biosorption – an alternative to meet the challenges of heavy metal pollution in aqueous solutions. *Environmental Technology*, *32*(5), 467–491. doi:10.1080/09593330.2011.572922 PMID:21877528

Dolphen, R., & Thiravetyan, P. (2015). Phytodegradation of Ethanolamines by *Cyperus alternifolius* : Effect of Molecular Size. *International Journal of Phytoremediation*, *17*(7), 686–692. doi:10.1080/15226514.2014.964839 PMID:25976882

Eskander, S., & Saleh, H. E.-D. (2017). Phytoremediation: An Overview. Environmental Science and Engineering Vol. 11: Soil Pollution and Phytoremediation. Studium Press LLC.

Garcia, N., Moreno, J., Cartmell, E., Rodriguez-Roda, I., & Judd, S. (2013). The application of microfiltration-reverse osmosis/nanofiltration to trace organics removal for municipal wastewater reuse. *Environmental Technology*, *34*(24), 3183–3189. doi:10.1080/09593330.2013.808244 PMID:24617078

Ge, L., Deng, H., Wu, F., & Deng, N. (2009). Microalgae-promoted photodegradation of two endocrine disrupters in aqueous solutions. *Journal of Chemical Technology and Biotechnology (Oxford, Oxfordshire)*, *84*(3), 331–336. doi:10.1002/jctb.2043

Grover, D. P., Zhou, J. L., Frickers, P. E., & Readman, J. W. (2011). Improved removal of estrogenic and pharmaceutical compounds in sewage effluent by full scale granular activated carbon: Impact on receiving river water. *Journal of Hazardous Materials*, *185*(2), 1005–1011. doi:10.1016/j.jhazmat.2010.10.005 PMID:21035257

Gupta, A., & Thakur, I. S. (2017). Treatment of Organic Recalcitrant Contaminants in Wastewater. In *Biological Wastewater Treatment and Resource Recovery*. InTech. doi:10.5772/66346

Gupta, S., & Babu, B. V. (2008). Economic feasibility analysis of low cost adsorbents for the removal of Cr (VI) from waste water. In *Proceedings of International Convention on Water Resources Development and Management (ICWRDM), BITS Pilani*. Citeseer.

Gupta, V. K., Gupta, B., Rastogi, A., Agarwal, S., & Nayak, A. (2011). Pesticides removal from waste water by activated carbon prepared from waste rubber tire. *Water Research*, *45*(13), 4047–4055. doi:10.1016/j.watres.2011.05.016 PMID:21664639

He, J., Dai, J., Zhang, T., Sun, J., Xie, A., Tian, S., & Huo, P. (2016). *Preparation of highly porous carbon from sustainable a-cellulose for superior removal performance of tetracycline and sulfamethazine from water*. Academic Press.

Hernández-Leal, L., Temmink, H., Zeeman, G., & Buisman, C. J. N. (2011). Removal of micropollutants from aerobically treated grey water via ozone and activated carbon. *Water Research*, *45*(9), 2887–2896. doi:10.1016/j.watres.2011.03.009 PMID:21453950

Huber, M. M., Göbel, A., Joss, A., Hermann, N., Löffler, D., McArdell, C. S., & Von Gunten, U. (2005). Oxidation of pharmaceuticals during ozonation of municipal wastewater effluents: A pilot study. *Environmental Science & Technology*, *39*(11), 4290–4299. doi:10.1021/es048396s PMID:15984812

Huggins, T. M., Haeger, A., Biffinger, J. C., & Ren, Z. J. (2016). Granular biochar compared with activated carbon for wastewater treatment and resource recovery. *Water Research*, *94*, 225–232. doi:10.1016/j.watres.2016.02.059 PMID:26954576

Iatrou, E. I., Gatidou, G., Damalas, D., Thomaidis, N. S., & Stasinakis, A. S. (2017). Fate of antimicrobials in duckweed Lemna minor wastewater treatment systems. *Journal of Hazardous Materials*, *330*, 116–126. doi:10.1016/j.jhazmat.2017.02.005 PMID:28214647

Ji, M.-K., Kabra, A. N., Choi, J., Hwang, J.-H., Kim, J. R., Abou-Shanab, R. A. I., & Jeon, B.-H. (2014). Biodegradation of bisphenol A by the freshwater microalgae Chlamydomonas mexicana and Chlorella vulgaris. *Ecological Engineering*, *73*, 260–269. doi:10.1016/j.ecoleng.2014.09.070

Jin, H., Capareda, S., Chang, Z., Gao, J., Xu, Y., & Zhang, J. (2014). Biochar pyrolytically produced from municipal solid wastes for aqueous As(V) removal: Adsorption property and its improvement with KOH activation. *Bioresource Technology*, *169*, 622–629. doi:10.1016/j.biortech.2014.06.103 PMID:25103038

Kebede, T., Dube, S., Nindi, M., Kebede, T. G., Dube, S., & Nindi, M. M. (2019). Removal of Multi-Class Antibiotic Drugs from Wastewater Using Water-Soluble Protein of Moringa stenopetala Seeds. *Water (Basel)*, *11*(3), 595. doi:10.3390/w11030595

Kiely, G. (1997). *Environmental engineering*. London: McGraw-Hill.

Kim, E., Jung, C., Han, J., Her, N., Park, C. M., Jang, M., & Yoon, Y. (2016). Sorptive removal of selected emerging contaminants using biochar in aqueous solution. *Journal of Industrial and Engineering Chemistry*, *36*, 364–371. doi:10.1016/j.jiec.2016.03.004

Kim, I., Yamashita, N., & Tanaka, H. (2009). Performance of UV and UV/H_2O_2 processes for the removal of pharmaceuticals detected in secondary effluent of a sewage treatment plant in Japan. *Journal of Hazardous Materials*, *166*(2), 1134–1140. doi:10.1016/j.jhazmat.2008.12.020 PMID:19147286

Kovalova, L., Siegrist, H., von Gunten, U., Eugster, J., Hagenbuch, M., Wittmer, A., & McArdell, C. S. (2013). Elimination of Micropollutants during Post-Treatment of Hospital Wastewater with Powdered Activated Carbon, Ozone, and UV. *Environmental Science & Technology*, *47*(14), 7899–7908. doi:10.1021/es400708w PMID:23758546

Kurniawan, T. A., Chan, G. Y. S., Lo, W., & Babel, S. (2006). Comparisons of low-cost adsorbents for treating wastewaters laden with heavy metals. *The Science of the Total Environment*, *366*(2–3), 409–426. doi:10.1016/j.scitotenv.2005.10.001 PMID:16300818

Li, L., Zou, D., Xiao, Z., Zeng, X., Zhang, L., Jiang, L., & Liu, F. (2019). Biochar as a sorbent for emerging contaminants enables improvements in waste management and sustainable resource use. *Journal of Cleaner Production*, *210*, 1324–1342. doi:10.1016/j.jclepro.2018.11.087

Luo, Y., Guo, W., Ngo, H. H., Nghiem, L. D., Hai, F. I., Zhang, J., & Wang, X. C. (2014). A review on the occurrence of micropollutants in the aquatic environment and their fate and removal during wastewater treatment. *The Science of the Total Environment*, *473–474*, 619–641. doi:10.1016/j.scitotenv.2013.12.065 PMID:24394371

Mallek, M., Chtourou, M., Portillo, M., Monclús, H., Walha, K., Salah, A., & Salvadó, V. (2018). Granulated cork as biosorbent for the removal of phenol derivatives and emerging contaminants. *Journal of Environmental Management*, *223*, 576–585. doi:10.1016/j.jenvman.2018.06.069 PMID:29975884

Mat Lazim, Z., Hadibarata, T., Hafiz Puteh, M., & Yusop, Z. (2015). Adsorption Characteristics of Bisphenol A onto Low-Cost Modified Phyto-Waste Material in Aqueous Solution. *Water, Air, and Soil Pollution*, *226*(34).

Moussavi, G., Hosseini, H., & Alahabadi, A. (2013). The investigation of diazinon pesticide removal from contaminated water by adsorption onto NH4Cl-induced activated carbon. *Chemical Engineering Journal*, *214*, 172–179. doi:10.1016/j.cej.2012.10.034

Nanseu-Njiki, C. P., Dedzo, G. K., & Ngameni, E. (2010). Study of the removal of paraquat from aqueous solution by biosorption onto Ayous (Triplochiton schleroxylon) sawdust. *Journal of Hazardous Materials*, *179*(1), 63–71. doi:10.1016/j.jhazmat.2010.02.058 PMID:20334968

Ozdemir, I., Şahin, M., Orhan, R., & Erdem, M. (2014). Preparation and characterization of activated carbon from grape stalk by zinc chloride activation. *Fuel Processing Technology*, *125*, 200–206. doi:10.1016/j.fuproc.2014.04.002

Pescod, M. B. (1992). *Wastewater treatment and use in agriculture*. Food and Agriculture Organization of the United Nations.

Rajapaksha, A. U., Vithanage, M., Ahmad, M., Seo, D.-C., Cho, J.-S., Lee, S.-E., ... Ok, Y. S. (2015). Enhanced sulfamethazine removal by steam-activated invasive plant-derived biochar. *Journal of Hazardous Materials*, *290*, 43–50. doi:10.1016/j.jhazmat.2015.02.046 PMID:25734533

Rodriguez-Narvaez, O. M., Peralta-Hernandez, J. M., Goonetilleke, A., & Bandala, E. R. (2017). Treatment technologies for emerging contaminants in water: A review. *Chemical Engineering Journal*, *323*, 361–380. doi:10.1016/j.cej.2017.04.106

Rovani, S., Censi, M. T., Pedrotti, S. L., Lima, É. C., Cataluña, R., & Fernandes, A. N. (2014). Development of a new adsorbent from agro-industrial waste and its potential use in endocrine disruptor compound removal. *Journal of Hazardous Materials*, *271*, 311–320. doi:10.1016/j.jhazmat.2014.02.004 PMID:24647264

Ruane, J., Sonnino, A., & Agostini, A. (2010). Bioenergy and the potential contribution of agricultural biotechnologies in developing countries. *Biomass and Bioenergy*, *34*(10), 1427–1439. doi:10.1016/j.biombioe.2010.04.011

Saba, B., Christy, A. D., & Jabeen, M. (2016). Kinetic and Enzymatic Decolorization of Industrial Dyes Utilizing Plant-Based Biosorbents: A Review. *Environmental Engineering Science*, *33*(9), 601–614. doi:10.1089/ees.2016.0038

Salgado, R., Marques, R., Noronha, J. P., Carvalho, G., Oehmen, A., & Reis, M. A. M. (2012). Assessing the removal of pharmaceuticals and personal care products in a full-scale activated sludge plant. *Environmental Science and Pollution Research International*, *19*(5), 1818–1827. doi:10.100711356-011-0693-z PMID:22207237

Servos, M. R., Bennie, D. T., Burnison, B. K., Jurkovic, A., McInnis, R., Neheli, T., & Ternes, T. A. (2005). Distribution of estrogens, 17β-estradiol and estrone, in Canadian municipal wastewater treatment plants. *The Science of the Total Environment*, *336*(1–3), 155–170. doi:10.1016/j.scitotenv.2004.05.025 PMID:15589256

Shen, W., Li, Z., & Liu, Y. (2008). Surface Chemical Functional Groups Modification of Porous Carbon. *Recent Patents on Chemical Engineeringe*, *1*(1), 27–40. doi:10.2174/2211334710801010027

Stefanakis, A., & Becker, J. (2016). A Review of Emerging Contaminants in Water. In *Impact if Water Pollution on Human Health and Environmental Sustainability* (pp. 55–80). IGI Global. doi:10.4018/978-1-4666-9559-7.ch003

Stefanakis, A. I., & Becker, J. A. (2016). A review of emerging contaminants in water: classification, sources, and potential risks. In *Impact of Water Pollution on Human Health and Environmental Sustainability* (pp. 55–80). IGI Global. doi:10.4018/978-1-4666-9559-7.ch003

Sui, Q., Huang, J., Deng, S., Yu, G., & Fan, Q. (2010). Occurrence and removal of pharmaceuticals, caffeine and DEET in wastewater treatment plants of Beijing, China. *Water Research*, *44*(2), 417–426. doi:10.1016/j.watres.2009.07.010 PMID:19674764

Supalkova, V., Beklova, M., Baloun, J., Singer, C., Sures, B., Adam, V., ... Kizek, R. (2008). Affecting of aquatic vascular plant Lemna minor by cisplatin revealed by voltammetry. *Bioelectrochemistry (Amsterdam, Netherlands)*, *72*(1), 59–65. doi:10.1016/j.bioelechem.2007.11.012 PMID:18201947

Tang, L., Hamid, Y., Zehra, A., Sahito, Z. A., He, Z., Hussain, B., ... Yang, X. (2019). Characterization of fava bean (Vicia faba L.) genotypes for phytoremediation of cadmium and lead co-contaminated soils coupled with agro-production. *Ecotoxicology and Environmental Safety*, *171*, 190–198. doi:10.1016/j.ecoenv.2018.12.083 PMID:30605848

Torrellas, S. Á., García Lovera, R., Escalona, N., Sepúlveda, C., Sotelo, J. L., & García, J. (2015). Chemical-activated carbons from peach stones for the adsorption of emerging contaminants in aqueous solutions. *Chemical Engineering Journal*, *279*, 788–798. doi:10.1016/j.cej.2015.05.104

Urase, T., & Kikuta, T. (2005). Separate estimation of adsorption and degradation of pharmaceutical substances and estrogens in the activated sludge process. *Water Research*, *39*(7), 1289–1300. doi:10.1016/j.watres.2005.01.015 PMID:15862328

Vasavi, A., Usha, R., & Swamy, P. M. (2010). Phytoremediation- An overview review. *Journal of Industrial Pollution Control*, *26*.

Vijayaraghavan, K., & Yun, Y.-S. (2008). Bacterial biosorbents and biosorption. *Biotechnology Advances*, *26*(3), 266–291. doi:10.1016/j.biotechadv.2008.02.002 PMID:18353595

Weidemann, E., Niinipuu, M., Fick, J., & Jansson, S. (2018). Using carbonized low-cost materials for removal of chemicals of environmental concern from water. *Environmental Science and Pollution Research International*, *25*(16), 15793–15801. doi:10.100711356-018-1781-0 PMID:29582326

Yi, S., Gao, B., Sun, Y., Wu, J., Shi, X., Wu, B., & Hu, X. (2016). Removal of levofloxacin from aqueous solution using rice-husk and wood-chip biochars. *Chemosphere*, *150*, 694–701. doi:10.1016/j.chemosphere.2015.12.112 PMID:26796588

Yoon, Y., Westerhoff, P., Snyder, S. A., Wert, E. C., & Yoon, J. (2007). Removal of endocrine disrupting compounds and pharmaceuticals by nanofiltration and ultrafiltration membranes. *Desalination*, *202*(1-3), 16–23. doi:10.1016/j.desal.2005.12.033

Zaman, D., Tiwari, M. K., & Mishra, S. (2020). Low-Cost Adsorptive Removal Techniques for Pharmaceuticals and Personal Care Products. In T. Gupta, S. Singh, P. Rajput, & A. Agarwal (Eds.), *Measurement, Analysis and Remediation of Environmental Pollutants. Energy, Environment and Sustainability* (pp. 397–421). Singapore: Springer. doi:10.1007/978-981-15-0540-9_19

Zhou, J., Chang, Q., Wang, Y., Wang, J., & Meng, G. (2010). Separation of stable oil–water emulsion by the hydrophilic nano-sized ZrO2 modified Al2O3 microfiltration membrane. *Separation and Purification Technology*, *75*(3), 243–248. doi:10.1016/j.seppur.2010.08.008

Zhou, Y., Gao, B., Zimmerman, A. R., Fang, J., Sun, Y., & Cao, X. (2013). Sorption of heavy metals on chitosan-modified biochars and its biological effects. *Chemical Engineering Journal*, *231*, 512–518. doi:10.1016/j.cej.2013.07.036

Chapter 10
Emerging Contaminants:
Pollution Control and Abatement

Emily Ng'eno
Masinde Muliro University of Science and Technology, Kenya

Victor Shikuku
Kaimosi Friends University College, Kenya

ABSTRACT

This chapter discusses water pollution control and abatement strategies in relation to emerging contaminants. The pollution prevention measures described in the chapter can broadly be categorized into four main areas that are highly interdependent: behavioral changes, new technical solutions to aid remediation of the environment, further research and data availability, and legislation or policy reforms. These main areas have been expanded in detail under 13 subtitles that are not only interdependent but also practical and achievable. The chapter demonstrates that if the proposed measures are collectively taken into consideration, then most of the United Nations Sustainable Development goals, especially the goals relating to water quality, would become a reality. The benefits of pollution control and abatement are widespread and far-reaching and can better the quality of life on the planet.

INTRODUCTION

Pollution is when hazardous substances are present in the environment. It is the world's major cause of illnesses, disorders, disease and early death (Landrigan et al., 2018). Environmental scientists have, for a long time, majored in environmental cleanup approaches and ignored pollution control and abatement (Hans, 2012) which is essential for a healthy environment necessary for the sustenance of human health and livelihoods. Pollution control and abatement can be described as any practice or measure that minimizes, eradicates, prevents pollution at the point source or control waste after it has been generated. In this particular chapter, the distinction between preventing wastes from being generated and controlling waste after it is generated is blurred as both imply waste reduction and therefore prevention and control may be used interchangeably. As the adage says *prevention is better than cure*, reduction of pollution at

DOI: 10.4018/978-1-7998-1871-7.ch010

the point source implies less waste generation hence less investment in waste treatment or disposal and fewer hazards posed to the ecosystem (United States Environmental Protection Agency (USEPA), 2016).

Water, air and soil pollution poses significant human health risks and threats to the ecosystem's sustainability. The contaminants of interest to be discussed are water and soil pollutants. Besides, soil contaminants, in most instances, find their way into water sources through runoffs and leaching. Most of the approaches discussed therefore major in alleviating water pollution. With the global challenges of access to clean water, especially in emerging economies, the reversal of the declining water quality worldwide and enhancement of integrated water and wastewater management systems together with safe reuse are subjects of significant concern (Sarantuyaa & Frank-Kamenetsky, n.d.). This important aspect was captured by the United Nations' Sustainable Development Goals (UN-SDGs) Target 6.3 which demands countries *to improve water quality by reducing pollution, eliminating dumping and minimizing release of hazardous chemicals and materials, halving the proportion of raw wastewater and substantially increasing recycling and safe reuse globally (*United Nations Sustainable Development Goals (UN-SDGs), (2015). The direct and indirect outputs of improved water quality and wastewater management and reuse include poverty alleviation, health improvement, food security, ecosystems unity and biodiversity conservation (Kanangire et al., 2016).

This chapter's main objective is to have a look at different approaches which has been applied or can be applied in controlling and abating emerging contaminants (ECs) in the environment.

BACKGROUND INFORMATION

ECs, mainly found in wastewater, aquatic environment and soil bring in the latest potential ecological and human health risks. ECs comprise synthetic or naturally occurring mass-produced hazardous chemicals present in our everyday consumer products and microorganisms that have been detected in drinking water at trace concentrations and the associated threat to biota is unknown and in dispute. These uncommonly monitored contaminants have the potential to enter the environment and cause adverse ecological and human health effects (Raghav et al., 2013, Thomaidis et al., 2012). They include pharmaceuticals, personal care products, biocides (pesticides and herbicides), fragrances, plasticizers, hormones, flame retardants, nanoparticles, perfluoroalkyl compounds, chlorinated paraffins, siloxanes, algal toxins, various trace elements including rare earths, radionuclides, endocrine disrupting compounds as well as microorganisms that are less monitored or regulated in the environment among others (Gavrilescu et al., 2015; Dulio et al., 2018; Petri et al.,2015; Sarantuyaa & Frank-Kamenetsky, n.d.).

Relative to the conventional or traditional contaminants, ECs have five distinguishing characteristics:

- The existing ECs are numerous and of high diversity with new ones emerging
- Potential health threat to biota
- Trace concentrations (up to nanograms per litre)
- Varying and fluctuating concentrations in different environmental matrices, attributable to varied quantities applied in different regions and inefficient wastewater treatment plants (Gavrilescu et al., 2015).
- Absence of regulation due to scarce information and understanding of ECs on the occurrence, fate, transport and toxicity individually and as complex mixtures (Sarantuyaa & Frank-Kamenetsky, n.d.).

ECs are widely dispersed throughout the various environmental compartments, far from the point of application or discharge, due to their mobility and persistence during long-distance transport (Gavrilescu et al., 2015). Their presence in the environment has been brought about by increased industrial activities, intensified chemical driven agricultural production and higher consumer demand which has soared their production over the last decades with concomitant negative environmental impact. Additionally, biological micropollutants such as viruses, bacteria, mycoplasmas and protozoa of divergent strains have been brought about by anthropogenic activities which have resulted in contamination of water sources. These emerging or reemerging pathogens are the source of many waterborne diseases causing death worldwide (Theron & Cloete, 2002).

The point sources of ECs include household septic systems, agricultural runoffs, treated and raw municipal wastewater treatment plants (WWTPs), hospital discharges and industrial effluents. They can also be introduced into the water supply by chance during manufacturing, handling, storage or use or be deliberately discarded, either unutilized or as a residual in the packaging (Shala and Foster, 2010). Various classes of ECs have been detected in drinking water, a testament that conventional water purification and wastewater treatment processes are incapable of completely removing these complex chemical compounds despite the fact that they result from similar domestic, commercial and industrial activities, as conventional contaminants (Kolpin et al., 2002; Petrie et al., 2015). This is due to their physico-chemical features such as solubility, polarity and low bioavailability which hinders biodegradation (Stamatelatou et al., 2011; Madadi et al., 2005). Advanced technologies, such as membrane filtration, ultrafiltration, nanofiltration, and reverse osmosis can partially remove some endocrine-disrupting chemicals and pharmaceuticals (Raghav et al., 2013). However, the high capital investment associated with these technologies prohibits their wide application on a large scale (Sarantuyaa & Frank-Kamenetsky, n.d.). Consequently, humans and ecosystems are continuously vulnerable to these ECs via ingestion of contaminated food and drinking water, exposure to household chemicals and pollutant-laden wastewater used to irrigate crops in water-deficient areas (Kanangire et al., 2016).

Currently, there is poor scientific understanding of the fate, transport, and risk of ECs, individually and as complex mixtures, and their cumulative and long term ecotoxicological effects in the environment (Kanangire et al., 2016; Petrie et al., 2015). There is however scientific corroboration that these compounds could cause acute and chronic human health dangers through ingestion and exposure *via* drinking water. They are associated with chronic toxicity, cancerous tumors, birth defects, and developmental disorders, development of pathogen resistance, fertility, and reproductive health disorders. Some ECs even in very low quantities are capable of potentially causing endocrine disruption in humans and aquatic life while others are associated with the development of bacterial pathogen resistance (Fent et al., 2006; Pruden et al., 2006, Rosal et al., 2010). It is also worrying since they not only impact the living organisms but also their progeny.

ECs' water pollution affects both developing countries and developed countries. Although there is increasing research focus on ECs in developed countries, effective tools and policies to monitor, regulate and control these pollutants in water resources have not been implemented (Sarantuyaa & Frank-Kamenetsky, n.d.). The situation is dire in developing countries, where large quantities of insufficiently treated or raw municipal wastewater and industrial effluents are discharged into surface waters, lakes and coastal zones daily, compounded by lack of analytical infrastructure for detection and periodic monitoring. The lack of state of the art analytical tools in developing countries explains the paucity of data on the occurrence and fate of ECs in these countries. Nevertheless, the situation is slowly changing in an upward trajectory.

Environmental chemists in most instances put a lot of effort in occurrence, fate, transport and remediation of pollutants and forget that if the origin of pollutants is curbed, then most environmental problems would be avoided. This is a gap that this chapter addresses by reviewing ways of strengthening scientific knowledge on human health and environmental dangers of ECs by nudging behavioral changes. It also encourages researchers, governments and the general public to embrace appropriate technological, regulatory and policy approaches to monitor, prevent and control ECs in water resources and wastewater. A successful approach in tackling pollution by ECs necessitates a highly interdisciplinary approach between governmental, non-governmental institutions and the general population. International cooperation is also key in mobilizing and building capacities at technical and policy levels to support countries in tackling global water pollution by ECs. This cooperation will also promote safe wastewater recycling, control transboundary water pollution and ultimately enhance water quality and food security (Sarantuyaa & Frank-Kamenetsky, n.d.).

Pollution prevention has monetary benefits as well in that waste management and cleanup costs will subside. Additionally, environmental and health costs will also subsidize due to lesser health problems and environmental damage. Consequently, there will be less need for households, businesses, and communities to handle waste and attend to those affected by environmental pollution leading to more manpower and hence economic growth. Furthermore, natural resources will be conserved and protected thereby strengthening economic growth through more efficient production in the industry.

Generally, pollution control and abatement can broadly be summarized into four main areas: behavioral changes; new technologies tailored for sequestration of contaminants during wastewater treatment; new research and policy reforms. The different ways of pollution control and abatement are interdependent for instance, preventing discharges will require legislation. The next section discusses in detail the different ways of pollution prevention and abatement.

VARIOUS POLLUTION CONTROL AND ABATEMENT MEASURES

Behavioral Changes

These are measures that are achievable by changing our behavior towards handling waste as consumers, industries and/or government. These approaches are discussed in detail below:

Controlling Discharges of Emerging Contaminants into Water Bodies

The most effective strategy in controlling discharges is to ensure that all dischargers enact measures aimed at reducing pollutants at their source (San Francisco Estuary Project (SFEP), 2007). That is, the point sources of these micropollutants especially industries should be tasked with the duty of removing these pollutants before releasing their effluent to streams, lakes or oceans. Figure 1 shows the effluent release from a water treatment company. The release should be subjected to periodic chemical analysis to ensure it is, if possible, free from ECs. The plant operations should also be modified to promote recycling and production of less waste. Industries that meet the environmental and safety requirements should be recognized periodically (e.g annually) by providing them with incentives so that compliance is stimulated.

It is a known fact that chemical and biological ECs are abundant in WWTPs (De Gusseme et al., 2010). This should make these micropollutants easier to eliminate and thus control discharges but due

Figure 1. Effluent release into a lake

to their movement or retention by sorption, volatilization or dispersion in their natural receptors, there is movement from one environmental compartment to another (Gavrilescu et al., 2015). To curb this challenge, the effluents should be freed from these pollutants at the point source. Regulations for ECs, which are predominantly lacking, that ensure their elimination from effluents should thus be formulated and implemented by the relevant authorities.

To further address the reduction of these pollutant discharges into the ecosystem from the point and nonpoint sources, the institutions mandated with ensuring water quality should ensure that companies which carry out pretreatment programs should have their technologies ameliorated and expanded to control persistent, accumulative toxics and to be able to handle mass limits. Furthermore, numerical effluent limitations must be described to ensure that the capacity of the ecosystem to accept pollutants is not exceeded. Toxicity control requirements for point sources should also be enforced. On the other hand, best management practices for nonpoint sources should be implemented and other regulatory and enforcement mechanisms should be adopted to assure compliance with the required standards (SFEP, 2007).

The control of discharges into water sources cannot be complete without tackling runoffs. This is because ECs, like most other chemical pollutants, are introduced into water sources through runoffs. Improvement of the management and control of runoffs, public highways, and drainage infrastructure cannot be overemphasized. This can be done by considering the following action points:

i. Routing off runoffs from commercial, residential, and industrial areas into the sanitary sewers with treatment systems to enable a net reduction in emerging pollutants' loading into the water reservoirs.
ii. Improper connections and illegal dumping into storm drainage systems should be minimized if not wholly eliminated. Here, collaboration between municipal wastewater treatment plants and urban runoff management agencies through the provision of incentives is indispensable.
iii. Farmers using agrochemicals should be educated on the need to reduce the use of toxic pesticides and herbicides that threaten water quality. Less toxic alternatives should also be introduced to them.
iv. Agricultural extension officers should sensitize farmers on best application methods, compliance with label instructions and other use restrictions and new control measures.
v. Once awareness has been created, these farmers should be monitored and regulated to ensure that runoffs from their farmlands are managed effectively such that they do not end up in water sources. Failure to comply should result in consequences and those who comply should be rewarded.
vi. Contamination of surface water as a result of drift from aerial spraying of agricultural chemicals should be quantified and mitigated (SFEP, 2007).
vii. Fill data paucity and measure progress relating to the management of runoffs through continuous research and monitoring. The research should focus on environmental concentrations of the target compounds in runoffs and their chemical toxicity. Further action points regarding research and monitoring to control and abate discharges are captured in the following bullets:
    - Biotoxicity monitoring should be used to identify waters where agrochemicals and other toxic materials are negatively affecting aquatic life. These evaluations can then be used to find the chemicals that are causing adverse impacts, and control strategies should be developed to address those effects.
    - Governments through Environmental Management Authorities should ensure that there is an approved laboratory analysis method for every emerging pollutant and significant metabolic products. The detection level should be below concentrations that may impact beneficial uses.
    - When an EC is detected in the waters, the appropriate parties should work to determine whether water quality objectives are violated and to develop control measures, if necessary, that will result in compliance with the prescribed regulations.
    - In the case of the detection of questionable contaminants in the water systems, technical and monetary support should be provided for the development of any necessary control and mitigation measures.

Noteworthy, measures to control the inputs of ECs to water bodies and the aquatic environment should address all stages of the product lifecycle from manufacturing to consumption to waste management (Sarantuyaa & Frank-Kamenetsky, n.d.). Such measures should include both technical solutions and policy tools for the prevention of the discharge of ECs into wastewater. This may require control programs to prevent consumer product-related discharges from exceeding the environmental standards. In such a situation, the costs for the development of the control program and the monitoring needed to verify that the control program is effective should generally be borne directly or indirectly by the users and manufacturer of the product. This may need legislation to be enacted to provide adequate funds to address this kind of environmental pollution, especially in the case of bioaccumulative pollutants (SFEP, 2007).

Product stewardship programs for new commercial products or new uses of existing chemicals should be developed to minimize future pollutant releases. The program should be the collaboration between environmental non-governmental organizations, government, manufacturers, and retailers. This may require the manufacturers to redesign products to be less hazardous or provide for take-back at the end of the product life, or a combination of the two (Sarantuyaa & Frank-Kamenetsky, n.d.). The Legislature may be roped in for the implementation of the program to be successful. It will also be necessary to conduct public education and outreach to make the program effective (SFEP, 2007).

Reduction and/or Substitution of Emerging Contaminants

If the total control of discharges into water bodies is not feasible, then reduction and/or substitution of substances containing ECs can be envisaged. This can be achieved by carrying out the following in the household set up:

i. Decreasing the total consumption of pharmaceuticals by looking into alternative methods of treatment.
ii. Reducing chemical use in homes or using non-toxic or less toxic chemicals like detergents, degreasers and other maintenance chemicals.
iii. Avoiding products with microplastics, for example, personal care products such as hand washing detergents, toothpastes, and facial cleansers which have tiny little plastic beads in them which are sources of ECs.
iv. Avoiding the use of antibacterial soaps which increases pathogen resistance (USEPA, 2016).

In the agricultural sector, reducing ECs' use involves:

i. Lowering the use of contaminant laden wastewater and agrochemical inputs.
ii. Adoption of less environmentally harmful agrochemicals or cultivation of crop strains with natural resistance to pests.
iii. Surface runoffs that are laden with silt, residual fertilizers, agrochemicals and other indeterminate range of pollutants that pollute water sources should be avoided by improving agricultural practices that encourage efficient water use (kanangire et al., 2016).
iv. Enacting control measures to abate pollutant loadings from certain agricultural practices for example confined animal feeding operations, irrigated agriculture and open pasture livestock grazing which all have the potential to discharge pollutants to water systems. This can be achievable by limiting effluent, regulating discharges from irrigated lands, determining the water quality impacts of discharges from agricultural drains and irrigation runoffs, determining compliance with water quality standards and identifying the need for additional control measures (SFEP, 2007).

On the industrial front, manufacturers should consider a balance between the use of commercial products and their pollution impacts. This will inform whether the product should be redesigned, reformulated, substituted or banned entirely. When a potentially harmful chemical associated with consumer products is detected in the ecosystem, appropriate parties, users or manufacturers should develop control strategies, including possible phase-out that will result in compliance with the environmental requirements (SFEP, 2007). This may involve making commercially viable environmentally benign replacements reflecting the green chemistry principles of *less hazardous chemical synthesis* and *designing safer chemicals*. This will require collaboration where the industry communicates what they need their chemicals to have and the environmental scientists identifying which of the potentially commercially viable alternatives are benign. In this role, the environmental scientists could be consultants, industry employees or part of an interdisciplinary research project with synthetic chemists and material scientists and the regulator providing incentives (Hans, 2012).

Proper Disposal

In some situations, controlling discharge into the water bodies or reducing their use may prove to be difficult, impossible or impractical. This, therefore, calls for ways of proper disposal and elimination of dumping. Figure 2 shows an example of improper waste disposal which should be avoided at all costs. Such cases can be alleviated by the introduction of pollution prevention measures such as banning the use of plastics which may contain bisphenol A, an EC. Other measures include take-back of unused medicines and pesticides by pharmaceutical and agrochemical companies and other distributors. This should be implemented in countries where such regulations are not laid down in order to reduce the disposal of unused pharmaceutically active products and agrochemicals in landfills or surface water (Sarantuyaa & Frank-Kamenetsky, n.d.). If no take-backs are provided, then proper disposal instructions should be provided. Figure 3 is a picture of unused/expired pharmaceuticals in a household. In most instances, such will be disposed of by being flushed down the drain or being disposed of with the household trash which will eventually end up in water sources, dumpsites or landfills.

Figure 2. Improper dumping of solid waste

Furthermore, educational campaigns should be carried out to increase public awareness of the correct disposal of ECs' containing products. Patients should also be encouraged to purchase just enough drugs or ask for drugs with low environmental impact in addition to educating them on proper use. The general public should also commit to health and wellness strategies to reduce the need for drugs (USEPA, 2016). Eco-labeling of pharmaceutical products, industrial products, agrochemicals, and other consumables can help the public to consider environmental perspectives when choosing products (Sarantuyaa & Frank-Kamenetsky, n.d.).

Products containing ECs should be sorted, defragmented and then sent either for recycling, incineration or landfilling. Recycling should be done at the point source; incineration will turn them into gases, fly ash or bottom ash which can be used for the manufacture of other products (Tome et al., 2018) while landfilling will isolate them. Caution should be practiced while doing landfilling because it can be a source of groundwater contamination. In arid and semi-arid areas, groundwater is the main source of drinking and household water (Ebele et al., 2017).

New Technological Inventions

Technical Solutions

Besides controlling discharges, reducing their use and enhancing proper disposal, then technical solutions should be provided for the removal of these chemicals and pathogens from the general environment and specifically the water sources.

To understand the full range of potential emerging contaminant effects, it is important to quantify and monitor pollutant levels at the emission source, within the environmental compartments as well as in biota (Petrie et al., 2015). Although great strides have been made in the detection and analysis of trace pollutants during recent decades (Daughton, 2005) due to advancements in analytical techniques, a wide array of undetected ECs need to be identified and quantified in various environmental matrices and biological tissues. In this regard, more effort should be put into finding new technical and cost-effective solutions to be applied mainly as tertiary advanced wastewater treatment methods (Sarantuyaa & Frank-Kamenetsky, n.d.). Here, emerging economies need to invest in technological advancement in analytical tools to bridge the paucity of information on the occurrence and fate of ECs in these countries.

Figure 3. Unused or expired pharmaceuticals in a household

Although wastewater discharges present a significant point source of ECs into aquatic ecosystems, their determination at these point sources, especially in emerging economies remains limited. In established economies, however, some advanced physico-chemical treatment methods have been applied, coupled with toxicity tests after each stage of treatment, and this has resulted in the removal of some toxic compounds, despite the time and financial implications (Gavrilescu et al., 2015). Here, analytical methods that can determine ECs at the enantiomeric level are needed (Petrie et al., 2015). This, therefore, requires more advanced technical skills that should be considered so that more ECs are detected even at trace levels.

With technology taking leaps every other moment, online monitoring is another viable technical solution as it can improve the reliability of monitoring data. It however involves expensive equipment with relatively high maintenance costs (Gavrilescu et al., 2015). An efficient option now being applied is referred to as passive sampling methods such as Polar Organic Chemical Integrative Samplers (POCIS), which can sample water over an extensive period, providing time-weighted average (TWA) concentrations (Yoo et al., 2007; Macleod et al., 2007).

Other advanced procedures include chemical degradation coupled with specialized microorganisms or UV light action. A study done by De Gusseme et al., (2010) showed that biologically produced zerovalent silver nanoparticles (bio-Ag^0) can be a very effective disinfectant that may be used at WWTPs.

Oxidation, adsorption, reverse osmosis, ozonation, membrane, and nano-filtration methods could also be used for the pre-treatment of hospital wastewater and industrial effluents before discharging to the sewer system although some techniques are cost-prohibitive (Sarantuyaa & Frank-Kamenetsky, n.d.).

Other than chemical pollutants, the development of biomonitoring tools and biosensors for environmental monitoring in real-time mode should be considered. Biomonitoring tools (e.g. bioassays, biomarkers, microbial community analyses) have great potential in the risk assessment of both regulated pollutants and ECs. Biosensors are capable of detecting the presence of a substrate by using biological components, which then provides a quantifiable signal (Gu et al., 2004). With more research and financial input, the development of large-scale biosensor clusters which determine several multiple species in real-time, along with more effective exploitation of specialized microbes or enzymes capable of degrading endocrine disruptors and other micropollutants is feasible The development of molecular biomarkers as detection tools for nanomaterials has been found to be effective in quantifying their concentration levels (Gavrilescu et al., 2015).

Another technical option is the use of molecular self-assembly. The self-assembly technique makes possible organized, patterned nanostructures which can involve biomaterials (proteins, lipids, nucleic acids), without external control and directions, which can then be applied to the development of amperometric immunosensors (Luz et al., 2013). This was inspired by nature and has been proposed to synthesize nanostructures with distinctive functions, because current detection methods for pathogenic bacteria, protozoa or viruses has proved to be sometimes inaccurate, but also costly in terms of resources and time (Gavrilescu et al., 2015).

Genomics is another new tool for recognizing and understanding the molecular pathways disturbed by ECs and can relate them to both the whole organism and population-level effects. An example is the use of DNA microarrays to understand the effects of pollutants at the molecular level (Poynton & Vulpe, 2009).

There is also an imperative need to boost research and innovation for new and cost-effective treatment technologies especially bioremediation, in line with the uptake, mode of action and consequences of each class of ECs. Bioremediation studies should take into account the following considerations: contaminant concentration; contaminant/contamination characteristics and category; scale and level of contamination; the risk intensity generated for health or the environment; the opportunity to be applied in situ or ex situ; the later use of the site; and available resources (Gavrilescu et al., 2015). Ngeno et al. (2016) among other numerous studies have demonstrated the use of various biosorbents in the removal of pollutants from wastewaters.

Activated sludge wastewater treatment, as carried out by Arboleda et al., (2013) demonstrated the use of fungi to decontaminate water polluted by bisphenol A. The isolation of activated sludge bacteria is capable of degrading endocrine disruptors which are sometimes difficult to degrade even by advanced treatment processes such as hydrolysis or photocatalysis. For recalcitrant compounds that cannot be biodegraded in activated sludge systems, the use of membrane bioreactors have proved to be efficient with some studies reporting a removal efficiency of 96% in municipal wastewater (Spring et al., 2007). Waterborne diseases can also be tackled by biotreatment of water and wastewaters using aerobic nitrification which has also been shown to remove pharmaceuticals and personal care products (Forrez et al., 2011).

The scientific world is constantly developing technical solutions to combat environmental pollution not only in water sources but also in soils and sediments where organic contaminants partition into (Petrie et al., 2015). New tools to improve future monitoring and regulation of ECs through the target and non-target screening should also be developed. Harmonization of liquid chromatography coupled to high-resolution tandem mass spectrometry [LC–HR-MS(MS)] and gas chromatography coupled to mass spectrometry (GC–MS) should be enhanced to assist in structure elucidation and pollution pattern recognition tools (Dulio et al., 2018; Petrie et al., 2015). Besides, mass spectra for the environmental contaminants should be in an open-access database to support the identification of compounds with an unidentified chemical structure (Dulio et al., 2018).

Generally, a holistic approach to the regulation of chemicals in the aquatic environment should be adopted. This is an effect-based method whereby instead of measuring a limited list of target individual substances known to be responsible for a given effect, it makes more sense to measure all substances (target substances plus other unknowns) that may contribute to that effect (Dulio et al., 2018). Additionally, more effort should be put into developing innovative tools for the evaluation of the effects of toxicity on human health.

It is an acknowledged fact that the removal of pollutants from any given environment would be made more successful by applying multidisciplinary techniques. This implies that the various bio-technologies discussed could be merged to eliminate accumulated ECs. Merging involves the action of microbes, plants, and animals under specific conditions that address both abiotic and biotic factors, to achieve contaminant mineralization, transformation or immobilization (Uhlı́k et al., 2010). For example, the combination of biological processes with adsorption on solids in the treatment of wastewaters can provide 45–99% removal efficiency of endocrine-disrupting chemicals from influents (Gavrilescu et al., 2015).

More scientifically validated, innovative and sustainable processes are necessary to tackle these challenges to eliminate these chemicals and pathogens from the environment.

Increase Funding for Technical Support and Infrastructure

The development of efficient, sustainable and state of the art technical solutions cannot be possible without funds that should be mobilized and increased. The development of most analytical methods and remediation techniques requires a sufficient amount of funding. Furthermore, continuous generation of up to date scientific data and monitoring requires more resources. These resources should be pooled from multidisciplinary sources.

There should also be funding of large-scale infrastructure improvements that aid in pollution control. Resources are needed to fund urban runoff control, pretreatment and waste minimization programs that are currently being started by international regulations, state requirements and local government initiatives (SFEP, 2007). Global organizations should mobilize, increase and focus the funding and international technical support to pollution control. Climate change and non-communicable disease control programs could be expanded to include pollution control, and new funding mechanisms developed (Landrigan, 2018). These financing and technical assistance programs should be tracked and measured to assess cost-effectiveness and enhance accountability.

Municipalities should invest in projects that reduce pollution. This can be achieved to a certain extent by establishing an equipped agency involving several government agencies and between governments and the private sector. This agency should be well-staffed and given enough resources to be able to regulate projects that can lead to environmental conservation.

Recycling of Substances Containing Emerging Contaminants

Substances containing ECs should be recycled at the site of their generation (closed-loop recycling) to prevent further pollution (SFEP, 2007). This requires the development of advanced systems to sort and recycle waste materials.

As much as reuse is being encouraged, wastewater reuse is not entirely recommended. This is because they are associated with the presence of non-regulated contaminants whose environmental fate and long term effects are not yet fully understood. Currently, there is agreement that reclaimed wastewater releases antibiotic-resistant bacteria and their genes (Dulio et al., 2018). There is, therefore, an urgent need for a better understanding of the presence and fate of micro-contaminants promoting the widespread antibiotic-resistant bacteria and genes in WWTPs effluents before their disposal or further reuse.

RESEARCH AND DATA AVAILABILITY

Some of the reasons that curtail the regulation of ECs include lack of information regarding their occurrence, fate and toxicity (Sarantuyaa & Frank-Kamenetsky, n.d.) catalyzed by lack of appropriate analytical methods for their determination due to problems generated by the physico-chemical properties of target compounds and the complexity of environmental characteristics which may determine an unexpected behaviour of the emerging pollutants in air, water, or soil. This creates paucity in knowledge necessary to guide policy (Gavrilescu et al., 2015). Additionally, the sheer number of existing and potential contaminants dwarfs the pace of research (Hans, 2012).

Landrigan, (2018) in a statement that emphasizes the need for more research says additional research is required that explores emerging causal links between pollution and disease; quantifies the global burden of disease associated with chemical pollutants of known toxicity; identifies and characterizes the adverse health outcomes caused by new and emerging chemical pollutants; identifies and maps pollution exposures, particularly in low- and middle-income countries; improves estimates of the economic costs of pollution and pollution-related disease; and quantifies the health and economic benefits and cost-effectiveness of interventions against pollution.

Moreover, technical solutions like monitoring and managing bioremediation requires the characterization of the fate and behavior of the compounds of interest in the environment to update the choice of bioremediation strategy (Gavrilescu et al., 2015). More research is therefore required for the assessment and treatment of emerging pollutants together with tools, equipment, and knowhow. Such an integrated approach should take into consideration the entire lifecycle of the pollutants, from the source of emission to their removal through treatment and remediation techniques, without neglecting the impacts and risks they may pose to the environment and human health (Sarantuyaa & Frank-Kamenetsky, n.d.).

Regulators and research funding agencies should, therefore, support ongoing and new research documenting the occurrence of ECs in drinking water, groundwater, soil and other environmental matrices in particular for substances with a high level of consumption. The scanty data and information on their nature, variability, transport and fate of these chemicals in wastewater and treatment facilities must necessarily be increased. Some ECs with high consumption that has not been included in screening campaigns (Dulio et al., 2018) and the occurrence and fate of metabolites are subjects of inquiry. Even limited data collected at national and local levels can help document pollution and track progress towards short and long term targets. Monitoring data are needed from every country and region globally.

While studying the health impacts of ECs, it is important to study and document the uptake, mode of action and biological endpoints of each EC to establish a correlation between contaminant and consequence.

Besides, once data has been obtained on their concentrations, more studies should be devoted to monitoring and understanding the processes involved in conventional wastewater treatment in eliminating or reducing the concentrations of a large diversity of ECs. This calls for analytical methods with lower detection limits for measuring concentrations of ECs.

More robust data is also required to evaluate their fate and dissipation in the environment, their long-term effects on biota and their appearance in the food chain. Studies on potential impacts of ECs on ecosystems are needed to identify and assess potential adverse effects of these substances and, consequently, to set the boundaries and thresholds for good environmental status. Moreover, the ecotoxicological significance predominantly concerning long-term, low-level exposure of some emerging micropollutants remains largely unknown, because satisfactory data to determine their risk often do not exist.

Further studies should be put into identifying and preparing a comprehensive list of ECs; characterizing the physical and chemical interactions of these ECs in soil, sediment and water and assessing the functioning of the water/soil system in the presence of these pollutants (Gavrilescu et al., 2015).

Establishing action plans for emerging and reemerging pathogens that are likely to alter the future biodiversity of the environment is an additional requirement worldwide. Current environmental challenges can restructure ecosystems in future and often lead to a decrease in biodiversity and a loss of important functions and services. In this context, further research into future hazardous or potentially dangerous chemicals and biological agents that will likely restructure the ecosystem is a necessity.

An inventory of the available information in terms of persistence, fate, fluxes, toxicity, and endocrine-disrupting potential of both individual compounds and complex mixtures at trace concentrations is lacking hence the development of new technologies for remediation and disinfection of water resources to drinking water standards is an ongoing challenge.

With increasing modernization, comes an increased demand for consumer products consequently leading to more new pollutants in the environment. The majority of these compounds have not yet been studied in detail attributable to the lack of appropriate standards for instrumental analysis techniques in the case of low concentrations of these micropollutants in environmental compartments (Gavrilescu et al., 2015). It is also important to note that existing preventative and remediation technologies may not be able to handle new chemicals that are being produced implying that continuous research is paramount.

DEVELOPMENT OF POLICIES AND LEGAL FRAMEWORK

Legislation

Major options for pollution control at one point or another require legislation for their success. When there is a provision in law or policy guiding a process, then implementation can be achieved and level of compliance monitored. National and International policies and laws should be put in place to ban several products containing toxic ECs and also regulate and restrict the use of others. This will prevent reliance on the industries, replete with conflict of interest, to regulating themselves as the public becomes their watchdog.

In most developing countries, the emphasis is laid on control of conventional pollutants and the protection of surface waters which entirely relies on municipal wastewater treatment and standards largely imposed on industrial point sources while non-point sources have been largely ignored and progress is not stimulated. More recently, the emphasis has slowly shifted to control of non-traditional contaminants and the protection of surface waters and the conversation between academicians, environmentalists, health experts and policy legislating agencies are bound to increase. Furthermore, there is indiscriminate, ad-hoc, unscientific and arguably careless disposal of contaminants and discharge of raw wastewater and sewage into aquatic ecosystems by some municipalities contrary to existing legislation and laws. This is compounded by the fact that most countries rely entirely on direct regulation rather than using permits or taxation (SFEP, 2007).

There also seems to be a disjoint between the existing legal tools and actual implementation due to how fragmented they are. Besides, most governments in developing countries have created a vacuum by taking a back seat and leaving pollution check in the hands of the public, environmental nongovernmental organizations and human rights groups who carry out inhomogeneous enforcement. The majority of industries and individuals, especially in emerging economies, have continued to take advantage of the aforementioned fragmentation and vacuum with the initial trend of self-checking on their terms and violating the existing pollution legal framework with impunity (Stadlinger et al., 2013). Reliance on self-regulation by the industries and the public to act as a watchdog of illegal acts is an impotent approach and demonstrably cannot deliver the desired results (SFEP, 2007).

In the absence of legislation and proper implementation, the prospects of curbing ECs into the environment will remain unrealized (Kanangire et al., 2016). For example, absence or limited surveillance programs for pesticide residue levels in the agricultural and fisheries products from water sources have a wider policy and market implications. It may detrimentally impact exports from countries with lesser controls following stringent regulation in the importing countries on residue limits in imported fisheries and other agricultural/horticultural products (Stadlinger et al., 2013).

Emphasis needs to be placed on policy approaches based on the precautionary principle rather than remediation, as the latter is less effective owing to the high costs involved in removing pollutants once they are discharged into the water systems. This entails policies and regulations that transcend the water sector to tackle the source of pollution pathways, as well as measures to ensure sustainable production, use, and disposal of ECs (Sarantuyaa & Frank-Kamenetsky, n.d.).

The current legislation requiring that removal of a chemical from the market must be accompanied by data showing a negative impact on human health and the environment underscores the need for a more effective and realistic shift in policy on risk assessment programs. Changing some of the policies governing the production and use of new chemicals and existing ECs in the marketplace may as well have the biggest impact. The need for legislation and compliance monitoring to legally mandate pollution prevention audits cannot be negotiated.

Establishing a Pollution Prevention Program

Through legislation, pollutants entering the water systems should be reduced by establishing pollution prevention programs spearheaded by a global body and run by both governmental and non-governmental institutions. These programs describe definite strategies for curbing the continuous discharge of toxic pollutants and dissuade dependence on toxic substances. A comprehensive pollution prevention program may include, but not limited to, the following strategies, where practicable:

- Establish goals for reducing EC's discharges into water sources
- Develop strategies to protect aquatic species and human health from ECs
- Implore industries to redesign or reformulate products to obtain compounds with lesser health risks
- Encourage industries and farmers to substitute potential pollutants with natural materials or alternative chemicals that introduce smaller quantities of hazardous substances into agricultural and industrial production processes
- Reward treatment and proper management of released effluents
- Develop environmental audit procedures for ECs
- Encourage industries to improve their process technology and equipment together with efficient plant operations to alter the primary source of waste generation
- Implore farmers to improve agricultural practices to reduce pollutants from farm runoffs and implement control measures to reduce agricultural pollution
- Ensure that industries recycle their waste at the site of their generation (closed-loop recycling).
- Develop commercial product stewardship program
- Ensure that mines reduce their toxic loadings
- Finally, make sure that construction waste material and pollutants are regulated.

Under its wings, there should be an exhaustive toxic reduction program with defined goals for curbing the continuous loading of toxic pollutants, identifying areas where pollution prevention techniques need implementation, and monitoring and reporting of success in meeting these objectives (SFEP, 2007).

Establish Systems to Audit Pollution and its Effects on Human Health and The Ecosystem

The health risks posed by chronic exposure to ECs on biota cannot be underestimated. It is an accepted fact that chemical data alone are not sufficient in assessing the potential threats of all ECs to the living organisms. The impacts of chemical pollutants on human health through analyses of pollution-induced biological and biochemical effects are desirable to evaluate these risks. In this regard, therefore, systematic processes and protocols requiring innovative approaches are needed for monitoring pollution and its effects, specifically on human health. These systems should be able to collect data both nationally and locally which can help document acute and chronic pollution effects. This should be a global initiative with a global body providing a bridge between governmental and non-governmental institutions. This initiative should aggregate, analyze and archive data on pollution and pollution-related disease worldwide and disseminate its findings widely to leaders of governments, the media and the global public (SFEP, 2007).

This global initiative should also carry out audits on ECs or mandate all industries that discharge significant toxic pollutants into the water sources to perform the audits. These findings should be made public to deter further pollution. This will encourage corporate management accountability and provide regulatory agencies with data on concentration levels of ECs in water sources. This initiative should also include mandatory annual reporting of pollution prevention and control activities.

The initiative should also look into the capacity of the receiving water together with the interactions of the aquatic species present and the incoming pollutants in addition to the bioaccumulation effects. Objectives should be developed to guard against potential adverse effects due to the accumulation of ECs through the food chain (SFEP, 2007). The initiative will thus be able to protect both human health and aquatic life against the adverse health effects of ECs.

Build Multi-Sectored Partnerships for Pollution Control

Broad-based partnerships across several government agencies and between governments and the private sector can advance pollution control and accelerate the development of bioremediation technologies. Cross-ministerial collaborations that bring together health, environment, finance, energy, agriculture, development, and transport ministries are also advisable (Landrigan et al., 2018).

There must also be collaboration between scientists, governments, non-governmental institutions and global organizations to ensure:

- Application of reasonable regulatory requirements and improvement of collaboration to facilitate pollution reduction projects, emphasizing aquatic habitat improvement and human community benefits;
- Strengthening of scientific knowledge, understanding and policy on emerging pollutants in water and wastewater through a series technical and policy case-studies on ECs in water and wastewater in different regions and the development of technical and policy guidelines (Sarantuyaa & Frank-Kamenetsky, n.d.);
- Promoting and facilitating knowledge sharing, scientific collaboration and joint research on issues related to ECs and the implications for wastewater reuse (Dulio et al., 2018);
- Capacity building and awareness-raising on ECs in water and wastewater.

These partnerships are achievable if policies are provided to allow for such. Pollution control and abatement should be made a priority nationally and internationally and integrated it into country and city planning processes. Governments at all levels should establish short-term and long-term targets and timetables. Useful policy tools including legally mandated regulation, the polluter-pays principle (when the party responsible for producing pollution is responsible for paying for the damage it causes) and an end to subsidies and tax breaks for polluting industries should be adopted by governments at all levels through partnerships (SFEP, 2007). These collaborations should establish forums to address and remedy, where appropriate, administrative and regulatory barriers that inhibit implementation of urban runoff control measures, including construction, operation, and maintenance of detention devices, wetlands, and paved surfaces.

Providing Incentives

The Legislature, regulatory agencies, and local agencies should recommend institutional and financial changes needed to place more focus on pollution control and abatement by providing economic incentives to discourage reliance on toxic materials and reduce the discharge of toxic pollutants over time (SFEP, 2007). More incentives should also be provided for industries with reduced emission of ECs. In the agricultural setup, farmers should also be given incentives for reducing ECs in agricultural drainage. These incentives could include tax exemption and other practical ways. Revenue enhancement measures, in the form of additional fees and direct cost measures could also provide local agencies with needed resources to adequately implement programs that reduce pollution. Economic measures for agricultural discharges should incorporate incentives in water pricing to reduce sediment loading and improve water quality (SFEP, 2007).

Voluntary retirement of marginal agricultural lands that currently yield a high discharge of toxic elements should be encouraged through public/private joint ventures and incentives given to landowners.

Other economic incentives to encourage pollution control are as follows:

- Encourage recycling of hazardous consumer products that might otherwise be released to the environment by coming up with a reward system for those who recycle;
- Implement effluent taxes based on mass loading to stimulate waste minimization by dischargers;
- Implement effluent credits based on mass loading to encourage municipal wastewater treatment plants to accept strategically routed urban runoff into the sanitary sewer to enable a net reduction in pollutants to the water sources;
- Increase regulatory incentives for municipalities, to invest in projects that manage runoffs, treatment of wastewaters and general improvement of water quality.

Public Education

Finally, pollution control cannot be successful without involving the public. The public should be educated on the health dangers of ECs and ways of reducing opportunities for human exposure through education on the proper handling and disposal of many ECs. These can be done through education, outreach, and forums to address the public on the human health effects of ECs. They can also be introduced into school curricula. Collaborative programs should be developed to inform the public, commercial entities, and industries about the proper use and disposal of materials and waste, and correct practices of urban runoff control. This is entirely possible by laying the legal framework for such educational campaigns.

CONCLUSION

Reducing pollution by ECs brings multiple benefits for poverty reduction, health improvement, food security, ecosystems' unity, and biodiversity conservation. Pollution control should be made a high priority nationally and internationally and should be integrated into the country and city planning processes. A more rigorous localized, multi-agency research program on ECs should be established to coordinate studies on occurrence, fate, and exposures to these contaminants and their impacts on people and the ecosystem. This implies the acquisition of new or improvement of existing technological infrastructure, especially in emerging economies.

Importantly, reducing pollution is both feasible and cost-effective. Many developed countries have successfully reduced pollution while still increasing their gross domestic product, which demonstrates that it is possible to control pollution without stifling economic growth. Moreover, pollution mitigation and prevention can also yield significant economic benefits, for example, in the USA, for every dollar invested in air pollution control since 1970, an estimated $30 has been returned to the economy in benefits (Landrigan, 2018).

In addition to benefits to human health and the economy, pollution control will advance attainment of many of the United Nations SDGs, including: improving health (SDG 3), improving access to clean water and improving sanitation (SDG 6), promoting social justice (SDG 10), building sustainable cities and communities (SDG 11), slowing the pace of climate change (SDG 13) and protecting water and land (SDGs 14 and 15) (UN-SDGs, 2015).

REFERENCES

Arboleda, C., Cabana, H., De Pril, E., Jones, J. P., Jimenez, G. A., & Mejia, A. I. (2013). *Elimination of Bisphenol A and Triclosan using the enzymatic system of autochthonous Colombian forest fungi.* Biotechnology. doi:10.5402/2013/968241

Daughton, C. G. (2005). "Emerging" Chemicals as Pollutants in the Environment: A 21st Century Perspective. *Renewable Resources Journal.*

De Gusseme, B., Sintubin, L., Baert, L., Thibo, E., Hennebel, T., Vermeulen, G., ... Boon, N. (2010). Biogenic silver for disinfection of water contaminated with viruses. *Applied and Environmental Microbiology*, *76*(4), 1082–1087. doi:10.1128/AEM.02433-09 PMID:20038697

Dulio, V., Bavel, B. V., Brorstrom-Lunden, E., Harmsen, J., Hollender, J., Schlabach, M., ... Koschorreck, J. (2018). Emerging pollutants in the EU: 10 years of NORMAN in support of environmental policies and regulations. *Environmental Sciences Europe*, *30*(1), 5. doi:10.118612302-018-0135-3 PMID:29568720

Ebele, A. J., Abdallah, M. A.-E., & Harrad, S. (2017). Pharmaceuticals and personal care products (PPCPs) in the freshwater aquatic environment. *Emerging Contaminants*, *3*(1), 1–16. doi:10.1016/j.emcon.2016.12.004

Fent, K., Weston, A. A., & Carminada, D. (2006). Ecotoxicology of human pharmaceuticals. *Aquatic Toxicology (Amsterdam, Netherlands)*, *76*(2), 122–159. doi:10.1016/j.aquatox.2005.09.009 PMID:16257063

Forrez, I., Boon, N., & Verstraete, W. (2011). *Biodegradation of Micropollutants and Prospects for Water and Wastewater Biotreatment. In Comprehensive Biotechnology* (2nd ed.; Vol. 6, pp. 485–494). Elsevier.

Gavrilescu, M., Demnerová, K., Aamand, J., Agathos, S., & Fava, F. (2015). Emerging pollutants in the environment: Present and future challenges in biomonitoring, ecological risks, and bioremediation. *New Biotechnology*, *32*(1), 147–156. doi:10.1016/j.nbt.2014.01.001 PMID:24462777

Gu, M. B., Mitchell, R. J., & Kim, B. C. (2004). Whole-cell-based biosensors for environmental biomonitoring and application. *Advances in Biochemical Engineering/Biotechnology*, *87*, 269–305. doi:10.1007/b13533 PMID:15217109

Hans, P. H. A. (2012). Emerging Decontaminants. *Environmental Science & Technology*, *46*(8), 4259–4260. doi:10.1021/es301074u PMID:22462768

Kanangire, C. K., Matano, A., Dida, G. O., & Anyona, D. (2016). *A Systematic Review of Effects of Emerging pollutants on Human Health and Livelihoods of Population living in the Lake Victoria Basin of Kenya. Case study report*. UNESCO.

Kolpin, D. W., Furlong, E. T., Meyer, M. T., Thurman, E. M., Zaugg, S. D., Barber, L. B., & Buxton, H. T. (2002). Pharmaceuticals, hormones, and other organic wastewater contaminants in U.S. streams, 1999–2000: A national reconnaissance. *Environmental Science & Technology*, *36*(6), 1202–1211. doi:10.1021/es011055j PMID:11944670

Landrigan, P. J., Fuller, R., Acosta, N. J. R., Adeyi, O., Arnold, R., Basu, N. N., ... Zhong, M. (2018). The Lancet Commission on pollution and health. *Lancet*, *391*(10119), 462–512. doi:10.1016/S0140-6736(17)32345-0 PMID:29056410

Luz, R. A. S., Lost, R. M., & Crespilho, F. N. (2013). *Nanomaterials for biosensors and implantable biodevices. In From implantable biosensors to green power generation* (pp. 27–48). Springer.

MacLeod, S. L., McClure, E. L., & Wong, C. S. (2007). Laboratory calibration and field deployment of the polar organic chemical integrative sampler for pharmaceuticals and personal care products in wastewater and surface water. *Environmental Toxicology and Chemistry*, *26*(12), 2517–2529. doi:10.1897/07-238.1 PMID:18020693

Madadi, V. O. (2005). *Chemodynamics studies and assessment of pesticides residue levels in Lake Victoria catchment area for Rivers Sio and Nzoia* (MSc. Thesis). Department of Chemistry, University of Nairobi.

Ngeno, E. C., Orata, F., Baraza, L. D., Shikuku, V. O., & Kimosop, S. J. (2016). Adsorption of Caffeine and Ciprofloxacin onto Pyrolitically Derivedwater Hyacinth Biochar: Isothermal, Kinetic, and Thermodynamic Studies. *Journal of Chemistry and Chemical Engineering*. doi:10.17265/1934-7375/2016.04.006

Petrie, B., Barden, R., & Kasprzyk-Hordern, B. (2015). A review on emerging contaminants in wastewaters and the environment: Current knowledge, understudied areas, and recommendations for future monitoring. *Water Research*, *72*, 3–27. doi:10.1016/j.watres.2014.08.053 PMID:25267363

Poynton, H. C., & Vulpe, C. D. (2009). Ecotoxicogenomics: Emerging technologies for emerging contaminants. *Journal of the American Water Resources Association*, *45*(1), 83–96. doi:10.1111/j.1752-1688.2008.00291.x

Pruden, A., Pei, R., Storteboom, H., & Carlson, K. H. (2006). Antibiotic resistance genes as emerging contaminants: Studies in Northern Colorado. *Environmental Science & Technology*, *40*(23), 7445–7450. doi:10.1021/es060413l PMID:17181002

Raghav, M., Eden, S., Mitchell, K., & Witte, B. (2013). *Contaminants of Emerging Concern in Water*. Tucson, AZ: Water Resources Research Center, College of Agriculture, University of Arizona. Retrieved from http://hdl.handle.net/10150/325905

Rosal, R., Rodrı'guez, A., Perdigo'n-Melo'n, J. A., Petre, A., Garcı'a-Calvo, E., Go'mez, M. J., ... Fernández-Alba, A. R. (2010). Occurrence of emerging pollutants in urban wastewater and their removal through biological treatment followed by ozonation. *Water Research*, *44*(2), 578–588. doi:10.1016/j.watres.2009.07.004 PMID:19628245

San Francisco Estuary Project. (2007). *Comprehensive Conservation And Management Plan: Pollution Prevention and Reduction*. San Francisco: Author.

Sarantuyaa, Z., & Frank-Kamenetsky, D. (n.d.). *Emerging Pollutants in Water and Wastewater: UNESCO-HELCOM Case Study on Pharmaceuticals in the Aquatic and Marine Environment in the Baltic Sea Region*. Academic Press.

Shala, L., & Foster, G. D. (2010). Surface Water Concentrations and Loading Budgets of Pharmaceuticals and Other Domestic-Use Chemicals in an Urban Watershed (Washington, DC, USA). *Archives of Environmental Contamination and Toxicology*, *58*(3), 551–561. doi:10.100700244-009-9463-z PMID:20098984

Spring, A. J., Bagley, D. M., Andrews, R. C., Lemanik, S., & Yang, P. (2007). Removal of endocrine disrupting compounds using a membrane bioreactor and disinfection. *Journal of Environmental Engineering and Science*, *6*(2), 131–137. doi:10.113906-049

Stamatelatou, K., Pakou, C., & Lyberatos, G. (2011). *Comprehensive biotechnology* (2nd ed.; Vol. 6). Elsevier. doi:10.1016/B978-0-08-088504-9.00496-7

Theron, J., & Cloete, T. E. (2002). Emerging waterborne infections: Contributing factors, agents and detection tools. *Critical Reviews in Microbiology*, *28*(1), 1–26. doi:10.1080/1040-840291046669 PMID:12003038

Thomaidis, N. S., Asimakopoulos, A. G., & Bletsou, A. A. (2012). Emerging contaminants: A tutorial mini-review. *Global NEST Journal*, *14*, 72–79.

Uhlı'k, O., Leewis, M. C., Kurzawova´, V., Lovecka´, P. Sˇtursa, P., & Demnerova´, K. (2010). Approaches to microbial diversity analysis in contaminated environments: Trends in bioremediation and phytoremediation. Research Signpost, 55–71.

United Nations Sustainable Development Summit. (2015). *United Nations Sustainable Development Goals*. Retrieved from https://www.un.org/sustainabledevelopment/sustainable-development-goals/

United States Environmental Protection Agency. (2016). *Pollution Prevention*. Retrieved from https://www.epa.gov/p2/learn-about-pollution-prevention

United States Environmental Protection Agency. (2007). *Pharmaceuticals in the Environment: Overview of Concerns and Stewardship Activities*. USDA Forest Service 2007 Sustainable Operations Summit.

Yoo, C. K., Villez, K., Van Hulle, S. W. H., & Vanrolleghem, P. A. (2007). Enhanced process monitoring for wastewater treatment systems. *Environmetrics*, *19*(6), 602–617. doi:10.1002/env.900

KEY TERMS AND DEFINITIONS

Ecosystem: Plants and animals and how they relate with their physical environment.

Effluent: The final release of a wastewater treatment plant.

Emerging Contaminants: Are synthetic or naturally-occurring hazardous chemicals present in most consumer products and microorganisms that have been detected in drinking water at trace concentrations and their risks to biota is unknown or mostly disputed.

Micropollutants: Are inorganic and organic environmental contaminants that are bioactive, nonbiodegradable and persistent even at very low concentrations, in the range of micro, nano or pico-grams per litre and cannot be fully eliminated with conventional wastewater treatment methods.

Non-Point Source: Is diffuse/discreet source of pollution which occur over a wide area and cannot be identified to a single source.

Point Source: Is a single identifiable source of pollution like a drainage pipe.

Pollution: It is the introduction of a harmful or hazardous substance into the environment.

Pollution Prevention: Is any practice that reduces, eliminates, or prevents pollution at its source.

Runoffs: Is the flow of water down a gradient that occurs when excess stormwater, meltwater, or other water sources flows over the Earth's surface.

Water Sources: Any water reservoir including surface waters like rivers, lakes, seas, oceans, or ground waters.

Chapter 11
Kinetics of Heavy Metals Adsorption on Gravels Derived From Subsurface Flow Constructed Wetland

Celestin Defo
University of Dschang, Cameroon

Ravinder Kaur
Indian Agricultural Research Institute, India

ABSTRACT

Adsorption kinetics of Ni, Cr, and Pb on gravels collected from constructed wetland was studied at varied metal concentrations and contact period for estimating the removal of heavy metals from wastewater. Batch experiments were conducted by shaking 120 ml of metal solutions having 5 concentration levels each of Ni (1.0, 2.0, 3.5, 5.0 and 6.0 mg l-1), Cr (1.0, 2.0, 3.0, 4.5 and 6.0 mg l-1), and Pb (1.0, 3.0, 6.0, 8.0 and 12.0 mg l-1) with 50 g of gravels for as function of time. Adsorption of Ni, Cr, and Pb on gravels ranged from 34.8 to 47.2, 42.7-54.9, and 47.5-56.9%, indicating their removal in the order: Pb > Cr > Ni. Freundlich model showed a good fit for Ni and Cr (R2>0.9) while Langmuir model fitted better for Pb (R2= 0.7). The pseudo-second-order model showed the best fit to simulate the adsorption rates of these metals on gravel.

INTRODUCTION

In water scarce regions such as India, arid and semi-arid Africa, reuse of wastewater has become an important component of water resources planning. A number of issues related to wastewater management need to be understood and researched before reuse. Scientists around the globe are working on new ways of conserving water. It is an opportune time, to refocus on one of the ways to recycle water through the reuse of urban treated wastewater, for irrigation and other purposes and other ecosystem services which

DOI: 10.4018/978-1-7998-1871-7.ch011

require lower water quality. It is a reliable and cheap source of nutrients but also contains heavy metals and other pollutants which contaminate soil-water-plant-animal continuum thus need removal. Conventional treatment plants for pollutants removal from wastewater are prohibitively expensive, generate huge quantities of toxic sludge and housed only in big cities (Vymazal, 2010). Decentralised constructed wetlands (CWs) are low cost and eco-friendly alternatives constructed wetland which removes more than 60% of the metals present in wastewater thus are suitable for rural areas and small towns (Marchand et *al.,* 2010; Vymazal, 2010). A wide range of centralized sewage treatment methods are used instead in developing countries, including stabilization pond systems, septic tanks, activated sludges, trickling filters, anaerobic systems and land application systems (Canter et *al.*, 1982). Kivaisi (2001) revealed that most developing countries have very few wastewater treatment facilities, mainly due to high costs of treatment processes and lack of effective environmental pollution control laws or law enforcement. Kivaisi (2001) showed constructed wetlands as the recently proven efficient technologies for wastewater treatment. Constructed wetlands are low cost (1% energy requirement, no chemical application, no skilled manpower requirement) and are easily operated and maintained, and have a strong potential for application in developing countries compared to conventional treatment systems. Alternative technologies to conventional wastewater management have been found to be environmentally sound and economically viable approach to wastewater management. Constructed wetlands have contributed to providing low cost sanitation in different countries. A limited database supports the capability of the subsurface flow wetlands process for effective removal of metals and other priority pollutants (Kovacic et *al.*, 2006; Reed et *al.*, 1987). It is thus important to establish through continued experimentation, the utility of constructed wetlands in the removal of heavy metals and other pollutants and the removal mechanisms involved. Keeping in view these observations, the objectives of the study are as follows:

i) to estimate the adsorption efficiency of a gravel materials and the removal time for different metals;
ii) to search the model which fits better the adsorption;
iii) to determine the intensity of adsorption and the types of bonds between metals and the substrate.

BACKGROUND

In constructed wetlands, interconnected physical, chemical and biological processes are involved in contaminants removal from wastewater (Marchand et *al.,* 2010). Sedimentation and filtration remove heavy metals present in particulate matter. Besides these, adsorption of heavy metals on the substrate surface is pointed as the main removal mechanism for these pollutants in their dissolved form (Mautner et *al.,* 2016). The other important removal mechanism is the plant uptake. CWs as a whole system were found to be very efficient in removing almost all of Cr, Cd, Ni Pb and other pollutants present in wastewater (Yadav et *al.,* 2010; Morari et *al.,* 2015, Mohamad et *al.*, 2018). In the adsorption process, heavy metals are adsorbed by surfaces either electrostatically by physical adsorption, which produces relatively weak complexes, or chemically by chemisorption, which produces strong complexes (Hussein et *al.*, 2003). Gravel and soil materials are well-known adsorbents generally used as substrate components for the removal of heavy metals in wastewater treatment in CWs.

The efficiency of the removal of heavy metals in constructed wetlands has been reported in previous investigations. In constructed wetland studies, many authors generally focussed their investigations of the efficiency of the whole system regardless of the contribution of the components of the system such as plant uptake and adsorption on the matrix (sand, soil, gravel). In the cases that the constructed wetland components (plants and substrates) contributions are investigated, these studies highlighted especially the contribution of different mycrophytes in the removal of heavy metals while the contribution of the substrates (soil, gravel) are less studied. The adsorption mechanism of heavy metals basically includes physical (relatively weak bonds between metal ions and adsorbent) and chemical adsorption (strong bonds between metal ions and the substrate) (Annadurai et *al.*, 2003; Gunawardana et *al.*, 2013). Gravel, sand and soil materials are generally used as substrate in CWs. Previous studies mainly focussed on the efficiency of the whole system regardless of the contribution of the components such as plant uptake, sedimentation, filtration or adsorption on the matrix (sand, soil and gravel) which are important to understand the mechanism and designing efficient CW (Morari et *al.*, 2015). Few studies conducted were confined to the role of different macrophytes in metal removal ignoring the contribution of the substrate (sand, soil and gravel). The available information concerning heavy metal adsorption on gravel materials used in constructed wetland does not satisfactorily explain the adsorption mechanism (metal specific adsorption, optimum residence time for heavy metal removal). Some of these studies have been focused on very limited numbers of metals and sorbents. Kabbashi et *al.* (2009) studied the capability of carbon nanotubes (CNTs) to adsorb lead (Pb) in aqueous solution through a batch mode adsorption experiment, conducted to control the effects of pH, speed, carbon dosage levels and contact time. The highest removal efficiency of Pb (II) was ranged between 85 -96.03% at pH 5 for 40 mg/L of CNTs, respectively. The model was better fitted Langmuir isotherm indicating the strong adsorptions of Pb (II) on the surface of carbon (adsorption capacity Xm = 102.04 mg/g). In addition, Tsibranska and Hristova (2001) investigated single and four-component adsorption kinetics for removal of Pb 2^+, Cd 2^+, Cu 2^+ and Zn 2^+ from aqueous solutions, using activated carbon from apricot stones as adsorbent. Their results showed that the intra-particle diffusion coefficients for the individual ions were in the range 1-3 × 10 -8cm^2/s, close to that obtained for single-component solutions with the same total initial ion concentration, while the lower values of the kinetic parameters were obtained in four-component solutions. Despite the fact that these previous studies released very comprehensive and pertinent analysis of the adsorption of metals in aqueous solutions, it does not cover the wide range of metals and possible adsorbents existing in the environment. Hence, this work attempted to assess the adsorption kinetics of Ni, Pb and Cr on gravel using pseudo-first and second order models as well as Langmuir and Freundlich adsorption isotherms.

MAIN FOCUS OF THE CHAPTER

Methodological Approach

It describes the experimental set up and operational conditions of vertical subsurface flow constructed wetlands, the field sampling processes and laboratory analysis procedures. Additionally, the procedures of data analysis and evaluation are presented, the determination of the kinetics of adsorption of heavy metals (Ni, Cr and Pb) on gravel using pseudo-first and second order models (These models depict better the adsorption coefficient and the absorption rate), Langmuir and Freundlich models and the determination of adsorption and plant uptake coefficients using linear models (simple relationship between regressor and dependent variable).

Experimental Set Up and Operational Conditions

Each research work on treatment wetlands required a description of the treatment wetland (type of wetland, wastewater applied loading schemes, plants used, etc.), even if the treatment performance is not the focus of the paper. In the present work, it is not only stated that gravel has been taken from a treatment wetland. A description of the treatment wetland is completely presented and thus the results of the adsorption of metals onto the matrix system (gravel materials) can be comprehend into the context.

The experimental setup consisted of vertical subsurface flow constructed wetlands (VSSF CW) in form of wetland microcosm (50 litres plastic container) at Indian Agricultural Research Institute, in New Delhi, India. Each wetland microcosms had a cylindrical shape, a diameter of 35 cm and a height equal to 50 cm. The wetland microcosms were filled with gravel (size ranging from 3 mm to 25 mm and a total weight of 60.1 kg per wetland) to a depth of 35 cm and planted with Typha, Phragmites, Vaccha, Arundo and Vetiver. These plant species had similar age and height at a spacing of 2-5 cm. The experiment was replicated thrice (to establish the relative effect of vegetation. The unvegetated wetland microcosm (containing only gravel) was also used for each treatment (this time without replication). These microcosms were irrigated with Ni, Cr and Pb solutions of three different strengths each. The solution concentrations of Ni (0.5, 1.0 and 5.0 ppm), Cr (1.5, 3.0 and 5.0 ppm) and Cr and (1.0, 5.0 and 10.0 ppm) were used. The levels of concentrations (low, medium and high) were designed to mimic their current concentration levels (low, medium and high) found in sewage water. Solutions containing different concentration levels of Ni and Pb were prepared using salts of Ni $(NO_3)_2{\cdot}6H_2O$ and Pb $(NO_3)_2.6H_2O$ while Cr was already available in standard solution (Atomic Absorption STD solution, 250 ml packed size, made by CDR®). The use of nitrates was justified by the fact that these ions do not form ion pair with metal ions. As a whole, there were 10 wetland microcosms for each metal concentration level, a total of 30 wetland microcosms/per metal specie and 90 microcosms for the whole experiment (Water Technology Centre, 2014).

Irrigation phase was scheduled at the beginning of each month and 20 l of synthetic wastewater containing Ni (0.5 ppm, 1.0 ppm and 5.0 ppm); Cr (1.5 ppm, 3.0 ppm and 5.0 ppm) and Pb (1.0 ppm, 5.0 ppm and 10.0 ppm) were applied to irrigate the sets of VSSF CWs available on the field. Each set of VSSF CWs (or wetland microcosms), taken metal-concentration level-plant wise contained 19 wetland microcosms. The water volume demand to supply each set of CWs (on the basis of 20l/wetland microcosm) was 380 l of synthetic wastewater. The synthetic wastewater was prepared on the field according to the following procedure (Water Technology Centre, 2014): (i) Measurement of appropriate volumes of metal solution from stock solutions described above, (ii) loading of the measured wastewater volume in 4 empty plastic containers of 100 l each. (ii) Volume make up of containers (three (03) containers make up to 100 l and 01 up to 80 l) using groundwater.

The depth of water was generally at 5 cm above the gravel bed after each irrigation. Samples of influent wastewater were collected (during irrigation phase) in 125 ml bottles previously washed and acidified, and 2-3 drops of nitric acid were mixed to the samples to avoid metal complexation. The hydraulic retention time (HRT) was 22 days. After this period, plant samples were collected (in order to assess the heavy metal uptake after 22 days), microcosm wetlands were emptied, and samples of effluent wastewater were collected. Then, the monthly irrigation of the microcosm wetlands was immediately planned. This experiment was conducted in the Vertical Subsurface Flow Constructed Wetlands (VSSF CWs) experimental plot at the Indian Agricultural Research Institute (IARI), New Delhi (Figure 1). The present work aimed at depicting the adsorption process of metals onto gravel materials therefore, it did not consider another processes involved in the wastewater treatment in these wetlands.

Gravel Materials Collection and Kinetics Experiment Preparation In The Laboratory

Gravel materials with size ranging from 1 mm to 25 mm (mixed fine and coarse gravel), all mixed together were collected from vertical subsurface flow constructed wetlands (VSSF CWs) of the Indian Agricultural Research Institute (New Delhi) which is in operation since two years. Gravel collected was air-dried at room temperature and used for the further steps of the present study.

Metal solutions of five different concentration levels of 1.0, 2.0, 3.5, 5.0 and 6.0 mg l^{-1} of Ni and 1.0, 3.0, 6.0, 8.0 and 12.0 mg l^{-1} of Pb were prepared from Ni $(NO_3)_2{\cdot}6H_2O$ and Pb $(NO_3)_2.6H_2O$. Solutions containing 1.0, 2.0, 3.0, 4.5 and 6.0 mg l^{-1} of Cr was prepared after dilution of 1000 mg l^{-1} Cr stock solution (250 ml). This was a mono-metal experiment were the metal species were present one by one in the containers. The metal species as well as their concentration ranges were designed to mimic the real situation dictated by the current environmental emergency in urban wastewater in India (Rawat et *al.*, 2003; Khajanchi et *al.*, 2016; Water Technology Centre, 2014; Shakunthala et *al.*, 2010).

Batch studies were performed in triplicates by shaking 120 ml of a solution of above-mentioned concentrations of Ni, Cr and with Pb with 50 g of gravel continuously for 48 hours on a shaker running 200 rounds per minutes (rpm) at room temperature. The pH was kept 7.0 to avoid formation of metals complexes. Samples were collected after 1, 5, 15, 30 minutes, 1, 2, 6, 12, 24 and 48 hours resulting 150 samples per metal and total 450 samples for Ni, Cr and Pb. Collected metal samples were digested in HNO_3 and determined their total contents (mg l^{-1}) using Atomic Absorption Spectrophotometer.

Figure 1. A view of the vertical subsurface flow constructed wetlands of IARI

Data Evaluation

Data obtained at varied metal solution concentration level and shaking times were used to estimate adsorption capacity and the equilibrium relationships between adsorbent and metal ions using Freundlich and Langmuir isotherms.

Estimation Of Equilibrium Concentration (Ceq) And The Amount Of Metal Adsorbed

The equilibrium concentration is determined by plotting variation in metal concentrations as a function of time. The amount of metals adsorbed on gravels was determined using the expression:

$$Q = \frac{\left[V * \left(Co - Ct\right)\right]}{W} \tag{1}$$

Where, Q is the amount of adsorbed ions (mg kg^{-1}), Co the initial concentration of the metal ions in solution (mg l^{-1}), C*t* is the concentration of the metal ions in solution (mg l^{-1}) at time *t*. At equilibrium, C*t* = *Ceq* (equilibrium concentration of the metal ions in solution (mg l^{-1}). V is the volume of the solution (ml) and W the weight of air-dried gravels/substrate (kg).

The adsorption percentage (AP) of a given metal on gravel was calculated as:

$$\mathrm{AP}(\%) = \frac{100x\left(Co - Ct\right)}{Co} \tag{2}$$

Fitting Of Data To Linear, Langmuir And Freundlich Isotherms

The Langmuir model is explained by the following expression (Fifi et *al.*, 2013):

$$Ceq / Qeq = 1 / \left(b * Qmax\right) + Ceq / Qmax \tag{3}$$

Where *b* is the constant (associated with the affinity pattern between the gravel and metal ions), *Qmax* is the maximum sorption (mg kg^{-1}). The Freundlich isotherm is given by the expression (4):

$$Qeq = K_F * Ceq^{1/n} \tag{4}$$

Where K_F is the Freundlich coefficient (l kg^{-1}) while *n* is the experimental constant (unitless).

In this equation, K_F and n are related to the maximum adsorption capacity and adsorption intensity of the metal ion respectively.

Fitting Data To Pseudo-First And Second Order Models

To evaluate the reaction rates of Pb, Cr and Ni on the gravel, pseudo-first-order and pseudo-second-order kinetics models were performed. The following equation expressed the linear form of pseudo-first-order (Lagergren, 1898):

$$\mathrm{Ln}\left(Qeq - Qt\right) = \log Qeq - k1 * t \tag{5}$$

Where Qeq (mg kg^{-1}) represents the adsorption capacity at equilibrium while Qt (mg kg^{-1}) is the quantity of the metal ions adsorbed at time t and k_1 (min^{-1}) represents the rate constant estimated from the slope of the linear plot.

The linearized form of pseudo-second-order equation is expressed as (Guo et *al.,* 2008):

$$\frac{t}{Q_t} = \frac{1}{\mathrm{k2Q_{eq}}^2} + \frac{\mathrm{t}}{\mathrm{Qeq}} \tag{6}$$

Where k_2 (g $mg^{-1}minute^{-1}$) represents the rate constant of pseudo-second-order model. The values of k_2 and Qeq are calculated from the slope and intercept of the plot obtained from the equation

i. Statistical analysis

Statistical analysis of the dataset was performed in Microsoft Excel and simple linear regression in SPSS software (version 8.02, 2008).

RESULTS AND DISCUSSION

In the adsorption of the tested metals on the gravel material, most of the metal ions were sequestered very fast from solutions within the first 12 hours and almost no increase in the level of bound metals have been occurred after this time interval (data not shown). The comparison of the sorption performance of the different metals was achieved under the same environmental conditions (pH, temperature, agitation speed, etc.). Adsorption equilibrium isotherms were plotted for the metal uptake Q against the residual metal concentrations in solution. The Q versus C_f sorption isotherm relationship was mathematically expressed by linearized Langmuir and Freundlich models. The higher the values of k and n and the lower the value of b, the higher the affinity of the biomass (Asku et *al.*, 1991; Jalaliet *al.,* 2002).

Adsorption Kinetics

Adsorption rate is important to understand the metal removal process, therefore; the influence of contact time on the adsorption percentage (%) of different metal ions on gravel materials was analysed. Irrespective of metal species, the adsorption was fast in the beginning (lesser duration), followed by a slower rate (longer duration) and finally reached the equilibrium state at which there was no change in adsorption with further increase in time. Similar observations are also corroborated by Rashid et *al.* (2015) and Fifi et *al.* (2013). The reduction in the adsorption with time is owed to the difficulty faced by metal ions to occupy the residual vacant active sites on the substrate (Mushtaq et *al.*, 2016). In addition to this, Ullah et *al.* (2013) reported that the reduced adsorption rate might also be associated with intra-particle dispersion during the second phase of the adsorption. Irrespective of solution Ni concentration, adsorption of Ni was the maximum in 12 hours of the contact period and achieved equilibrium as the adsorption was constant beyond 12 hours.

At 12 hours of contact time, the removal percentage of Ni varied from 34.5 to 47.5 mg l^{-1} with the change in Ni concentration in the solution.

Similarly to Ni, adsorption of Cr was not affected by the initial concentration and the maximum adsorption reached the equilibrium within 12 hours from the start of the experimentation (Figure 3). At 1.0 mg l^{-1} Cr level, the maximum adsorption percentage was 54.9%. With the further increase in Cr solution concentration from 2.0 to 6.0 mg l^{-1}, there was not a much variation in the maximum adsorption percentage and ranged from 42.7 to 45.4%.

The initial Pb concentration level of 1.0 mg l^{-1} was not used for plotting because of its concentrations in the supernatant was in traces. Compared to Ni and Cr, the adsorption maxima of Pb attained earlier (6 hours of contact).

Beyond 6 hours, the equilibrium state was reached and the adsorption rate became constant. Earlier studies also found short contact time appropriate to reach the equilibrium (Chong et *al.,* 2013; Rashid et *al.,* 2015). The adsorption maxima of Pb also varied with initial concentration adsorbing 47.0 to 57.0% of Pb on gravel. Allende et *al.* (2012) found that gravel had 43% pollutant removal efficiency. Similarly, Sahu (2014) also observed that constructed wetlands removed 51% Cr and 47% Ni from wastewater. The adsorption percentage of Ni, Cr and Pb at equilibrium varied with initial solution concentration but had no particular increasing or decreasing trend.

The factors such as metal solubility and substrate properties (permeability, mineralogy, porosity, surface area etc.) govern metal adsorption. Though gravels have low surface area but are porous and contain Ca, Si, Al, and Fe which can easily exchange with metal ions in the solution (Abdulfatah et *al.,* 2015; Sizirici and Yildiz, 2018). High metal (Ni, Cr and Pb) solubility and presence of Fe-Al oxides

Figure 2. Variation of the adsorption percentage of Ni as function of time

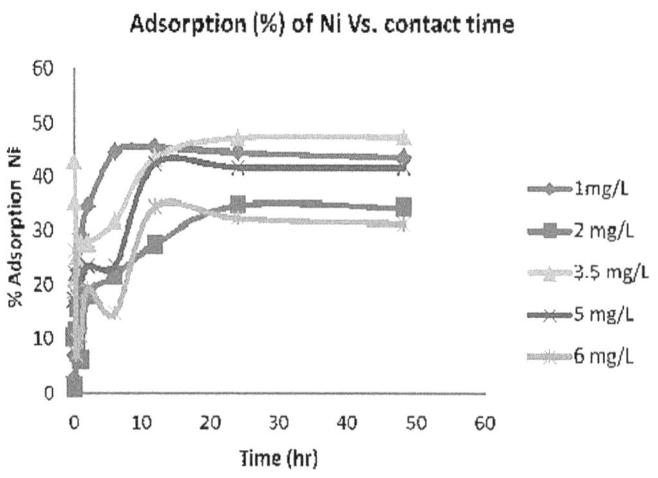

Figure 3. Variation of the adsorption percentage of Cr as function of time

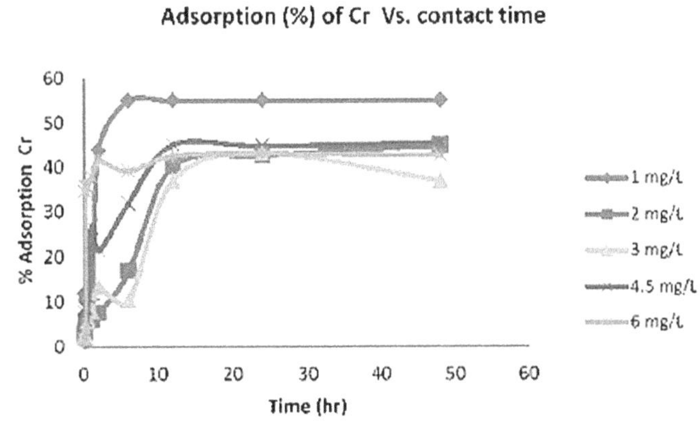

and organic matter coatings on gravel surfaces further add to the adsorption. The adsorption was predominantly physical that involved weak Van Der Waal forces (Abdulfatah et *al.,* 2015). In this type of adsorption, the adsorption sites are not specific; metals adsorbed freely cover the surface of adsorbent. Aziz et *al.* (2008) found coarse surfaces of gravels and pH (7.0 to 8.5) of solution beneficial for the adsorption of metal ions on gravels. Higher adsorption percentage and earlier attaining of equilibrium of Pb compared to Ni and Cr showed a better affinity between Pb ions and gravel.

Pseudo-First Order and Pseudo-Second Order Models

The data generated from adsorption of Ni, Cr and Pb were fitted to the pseudo-first-order [$\ln(Qep - Qt) = f(t)$] and pseudo-second order [$\frac{t}{Qt} = f(t)$] and presented in Figure 5 and Table 1). The values of the determination coefficients in pseudo-first-order ($R^2 = 0.05 - 0.10$) were lower than those of the pseudo-second-order ($R^2 > 0.99$) at pH 7.0 (Table 1).

The results indicated that pseudo-second order is suitable for predicting the adsorption rates of Ni, Cr and Pb on the gravels. Fifi et *al.* (2013) also found pseudo-second order applicable for the adsorption of Pb, Cu and Ni in soil columns. Similarly, Abdić et *al.* (2018) found that the pseudo-second-order

Figure 4. Variation of the adsorption percentage of Pb as function of time

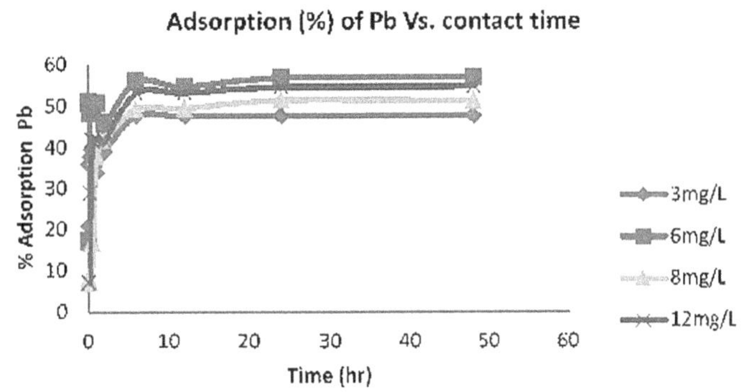

Table 1. Pseudo first and second order adsorption rate constants for Ni, Cr and Pb

	Pseudo-first order kinetics model			Pseudo-second order kinetics model		
Metal	**Qe (mg kg^{-1})**	**K_1 (min^{-1})**	**R^2**	**Qe (mg kg^{-1})**	**K_2 (mg kg^{-1} min^{-1})**	**R^2**
Ni	1.0	0.0001	0.15	11.63	0.82	0.99
Cr	5.5	0.0010	0.46	9.1	0.15	0.99
Pb	1.0	0.0100	0.54	28.6	4.10	1.0

kinetic model best fit the experimental data with high correlation coefficients ($R^2 > 0.9997$) while using tangerine peel aiming its use as a potential adsorbent of eight heavy metal ions (Cd, Co, Cr, Cu, Mn, Ni, Pb and Zn) from aqueous solution.

Equilibrium Concentration and the Adsorption of Ni, Cr and Pb (mg l^{-1}) on Gravel

Due to the adsorption on the gravel during the contact period ranging from 1 minute to 2 days, heavy metal concentration in the supernatant was found less than their respective initial concentration. The metal concentrations in supernatant varied with the increase in their initial levels. At equilibrium, amount of Ni adsorbed on gravel increased as a function of the initial solution concentration but not with increasing contact period (Figure 1). Corresponding to the initial concentration of 1.0, 2.0, 3.5, 5.0 and 6.0 mg l^{-1} of Ni, the equilibrium Ni concentrations was 0.6, 1.3, 1.8, 3.2 and 4.1 mg l^{-1} which led to the amount of 0.9, 1.7, 4.0, 4.3.0 and 4.6 mg kg^{-1} of Ni adsorbed on gravels (Table 2).

Similarly, 1.3, 3.0, 3.1, 4.8 and 62 mg kg^{-1} of Cr was adsorbed resulting equilibrium concentration values of Cr 0.4, 0.7, 1.7, 2.5 and 3.4 mg l^{-1} corresponding to 1.0, 2.0, 3.0, 4.5 and 6.0 mg l^{-1} initial solution concentration. The amount of Cr adsorbed was increasing (1.3-6.2 mg kg^{-1}) when Cr concentration was increased from 1.0 to 6.0 mg l^{-1}. Compared to Ni and Cr, equilibrium concentration of Pb varied with the change in solution Pb concentration and also followed an increasing pattern. It indicated high removal of Pb by the gravels. The amount of Pb adsorbed was 3.4, 7.7, 8.7 and 16.3 mg kg^{-1} corresponding to values of initial concentrations 3.0, 6.0, 8.0 and 12.0 mg l^{-1} respectively. Adsorption percentage calculated as a function of their initial and equilibrium concentration varied from 45.5 to 34.5% in case of Ni, 44.2-54.7% in Cr and 47.2–56.2% in Pb.

In agreement with the present study, Aziz et *al.* (2008) observed that adsorption percentage of Ni, Cr and Pb on gravel varied from 44 to73, 42-72 and 42-78% respectively. Allende et *al.* (2012) also recorded 43% removal of as in gravel columns.

Adsorption Isotherm Models

Langmuir and Freundlich isotherms were used to describe the adsorption intensity and the affinity between metal ions and the gravel. The values of different constants and parameters deduced from Freundlich and Langmuir adsorption isotherm for Ni, Cr and Pb are presented in Table 3 and Figure 5.

Langmuir isotherm fitted better for Pb adsorption (R^2=0.70) compared to Ni and Cr (R^2<0.5), while Freundlich isotherm suited more for Ni and Cr (R^2>0.90) than Pb (R^2 = 0.54). The adsorption

Figure 5. Pseudo-first-order and pseudo-second-order adsorption plots for Pb, Ni and Cr

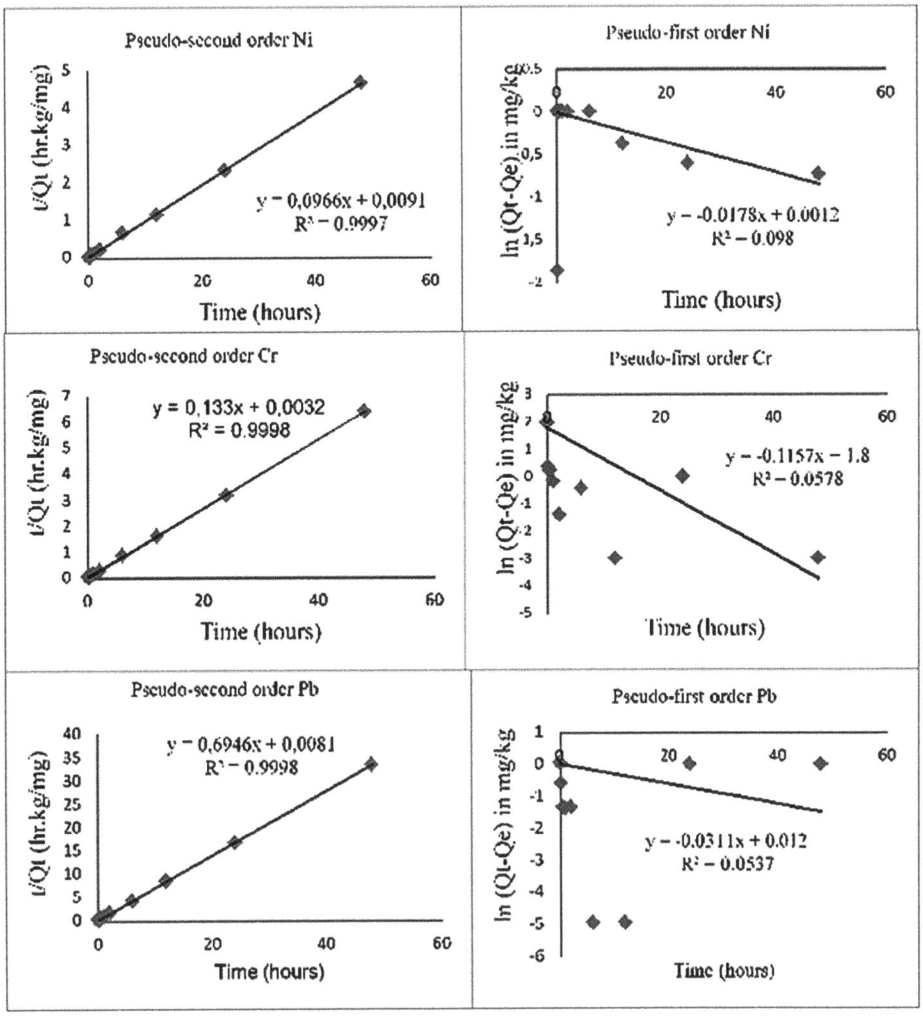

Table 2. Equilibrium concentrations (Ce) and amount adsorbed (Qe) of Ni, Cr and Pb

Ni			Cr			Pb		
Conc levels (mg l^{-1})	Equil. Conc (mg l^{-1})	Amount adsorbed (mg kg^{-1})	*Conc levels (mg l^{-1})	**Equil. Conc (mg l^{-1})	Amount adsorbed (mg kg^{-1})	Conc levels (mg l^{-1})	Equil. Conc (mg l^{-1})	Amount adsorbed (mg kg^{-1})
1.0	0.6	0.9	1.0	0.4	1.3	1.0	/[a]	/[a]
2.0	1.3	1.7	2.0	0.7	3.0	3.0	1.6	3.4
3.5	1.8	4.0	3.0	1.7	3.1	6.0	2.8	7.7
5.0	3.2	4.3	4.5	2.5	4.8	8.0	4.3	8.7
6.0	4.1	4.6	6.0	3.4	6.2	12.0	5.1	16.3

[a] Pb content in supernatant was in traces, *Concentration, **Equilibrium

Table 3. Langmuir and Freundlich adsorption isotherms values for Ni, Cr and Pb

	Langmuir model					Freundlich model				
Metal	**Equations**	**Qmax (mg kg⁻¹)**	**b**	**R_L**	**R^2**	**Equations**	**k_F (mg kg⁻¹)**	**1/*n***	***n***	**R^2**
Ni	Ce/Qe = -0.140Ce + 0.001	2.1	0.9	0.16	0.2	$Qe = 8.43Ce^{1.72}$	8.4	1.72	0.58	0.93
Cr	Ce/Qe = 0.187Ce + 0.001	7.0	128.7	0.007	0.2	$Qe = 4.610Ce^{1.69}$	3.6	1.69	0.59	0.96
Pb	Ce/Qe = 0.077Ce + 0.002	1.3	6443.3	0.00032	0.7	$Qe = 3.15Ce^{0.96}$	3.1	0.96	1.04	0.54

Freundlich coefficient (k_F) was 8.4 mg kg^{-1} for Ni and 3.6 mg kg^{-1} for Cr indicating that gravel had good adsorption capacity ($k_F > 2$). The results also showed preferential adsorption of Ni over Cr. This was in agreement with the adsorption percentages at equilibrium which ranged between (34.5 – 45.5%) for Ni and (43.0 – 52.6%) for Cr. The values of *n* values obtained were 0.59 for Cr and 0.58 for Ni. Values of $n<1$ indicated physical adsorption caused by van der Waals forces of attraction and weak bonds between the metal ions and the matrix (Tripathi and Ranjan, 2015). Langmuir models fitted Pb metal ions adsorption on gravel. Regarding the parameters of the Apart from R^2, high values of b again indicated

Figure 6. Langmuir and Freundlich isotherms for Ni, Pb and Cr

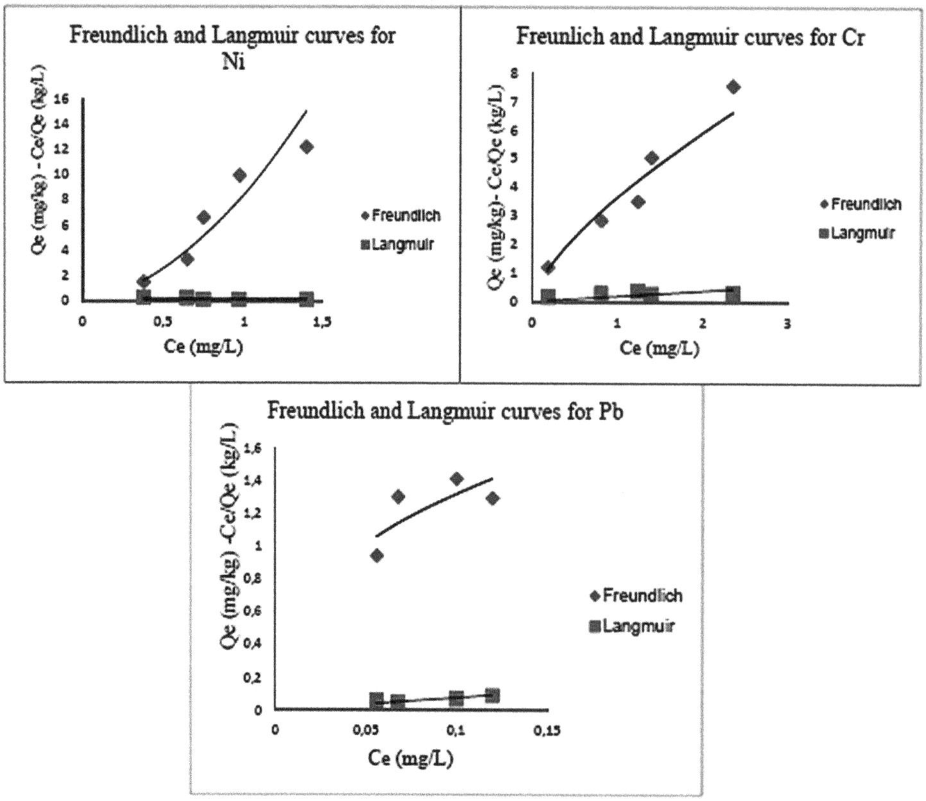

that Langmuir isotherm was better for Pb adsorption. However, other adsorbents like chitosan which had very high value of Langmuir coefficient b (60360) adsorbed Pb as high as 59000 mg kg^{-1} at equilibrium concentration (0.9 mg l^{-1}) were found to higher affinity between Pb and adsorbent (Chitana, 2002). The values of dimensionless constant R_L were <1 for all the three metals studied but it was the least in case of Pb (0.00032). It showed that similar to Ni and Cr the adsorption of Pb on gravel surface was physical which was due to weak bonds of low energy (Huang et *al.*, 2015). In another study conducted by Sizirici and Yildiz (2018), using Iron oxide-coated gravel in order to remove Cd(II), Cu(II), Pb(II), Fe(III) and Al(III) simultaneously in high-strength synthetic leachate samples, the experimental data obtained from batch study satisfactorily fitted to the Freundlich model indicating surface heterogeneity and multilayer adsorption process.

SOLUTIONS AND RECOMMENDATIONS

Water scarcity is still a universal concern in this century. Globally, about 80 countries and regions, representing 40% of the world's population, are experiencing water stress, and about 30 of these countries are suffering water scarcity during a large part of the year (Kivaisi, 2001). In India presently 15% of available water is consumed in domestic and industrial sectors which will be 30% in 2030 causing fresh water scarcity for agriculture. However, after use 70-80% of water supplied to municipalities is returned back as wastewater. Under water scarcity, reuse of wastewater in agriculture is inevitable and considered as a common component of water resource planning as the costs of wastewater disposal rise and opportunities for conventional water supply development dwindle. In water scarce conditions, though application of wastewater in agriculture improves crop productivity and livelihood but also contaminates natural resources and disproportionally affects health of farmers and consumers due to presence of pathogens, heavy metals and other pollutants (Kajanchi et *al.*, 2016). To overcome adverse effects, necessary technologies for wastewater treatment have been developed. One of the sustainable wastewater treatment alternatives is the constructed wetlands (CWs) which are efficient, low-cost, easy to use and eco-friendly (Marchand et *al.*, 2010).

Adsorption Kinetics

Adsorption rate is important to understand the process occurring in the experiments. Consequently, the influence of contact time on the adsorption percentage (%) of different metal ions on gravel materials was analysed. The time-based distribution of adsorption efficiencies for Ni, Cr and Pb and the results of kinetic models (pseudo-first order and pseudo-second order) are shown in this section.

Temporal Variation of Adsorption Efficiencies

Irrespective of metal species the adsorption of Ni, Cr and Pb on gravels was observed to be the fast in the beginning followed by a slower rate and finally reached the equilibrium state. Previous studies also indicated that the adsorption process occurs in two phases: initial fast phase (reduced duration) followed by the relaxed second phase (extended duration), which leads directly to the equilibrium state (Rashid et al., 2015; Fifi et al., 2013). Even though the second phase had a larger contact time but the adsorption capacity was not increased regardless of the metal ion. According to Rashid et al., (2015), the reduction in the adsorption rate owed to the difficulty faced by metal ions to occupy the residual vacant active sites on the substrate (Mushtaq et al., 2016). In addition, Ullah et *al.* (2013) revealed that the reduction in adsorption capacity with time might be associated with intra-particle dispersion process during the second phase of the adsorption. For Ni and Cr, equilibrium reached after 12 hours. However, in the case of Pb, equilibrium was attained in 6 hours which may be owed to good adsorption efficiency of gravel for it. Earlier studies also revealed that for some sorbents, a contact time ranging from 2 to 4 h is appropriate to reach the equilibrium (Chong et *al.*, 2013; Rashid et *al.*, 2015). Irrespective of its initial concentration, adsorption of Ni on gravels was the maximum at 12 hours of the contact period. Beyond 12 hours, the equilibrium state was reached and the adsorption was constant. At equilibrium stage, removal percentage of Ni ranged from 34.5 to 45.5% and had no particular increasing or decreasing trend with the increase in its initial concentration from 1.0 to 6.0 mg l^{-1}. Similar to Ni, adsorption of Cr was not affected by the initial concentration and the maximum adsorption reached the equilibrium within 12 hours from the start of the experiment. The maximum adsorption percentage of Cr was 54.9% at its 1.0 mg l^{-1} initial concentration and varied only from 42.7 to 45.4% at 2.0, 3.0, 4.5 and 6.0 mg l^{-1} of Cr. The adsorption of Ni and Cr on gravels is attributed to physical forces between the adsorbent and metal ions (Rashid et *al.*, 2015; Abdulfatah et *al.*, 2015)

The initial concentration level (1.0 mg l^{-1}) was not considered in Pb adsorption kinetics because of its very content in the supernatant traces. Compared to Ni and Cr, the adsorption maxima of Pb also varied with initial concentration. Within 6 hours of contact the adsorption of Pb reached to the maximum. Beyond 6 hours, the equilibrium state was reached and the adsorption rate became constant. With changing initial concentration of Pb from 3.0 to 12. 0 mg l^{-1}, the maximum adsorption of Pb on gravels varied between 47.0 to 57.0%. Higher removal percentage of Pb and earlier equilibrium attainment might be due to the better affinity between Pb ions and gravel in comparison to other metals.

Aziz et *al.* (2008) indicated that coarse surfaces of gravel and pH level of solution (7.0 to 8.5) assisted in the elimination (adsorption) of metal ions from wastewater. The factors responsible for metal adsorption could be (i) the solubility of the metal ions; (ii) the permeability, mineralogy, porosity and surface properties of gravel. On one hand, gravel contains Ca, Si, Al, and Fe, has a good porosity but a low surface area (Abdulfatah et *al.*, 2015). While on the other hand, the metal ions (Ni, Cr and Pb) have good solubility in aqueous solution. (iii) Another key component might be the association of heavy metals with Fe/oxides and organic matter coatings on solid surfaces. In this case, adsorption was favourable due to organic matter and Fe/oxydes coatings on solid surfaces rough surface of gravel, its porosity and high solubility of metals in solution. The gravels have low surface area and low cation exchange capacity. The forces responsible for adsorption of metals on gravels are predominantly physical (Van Der Waal forces) with weak bonds (Abdulfatah et *al.* (2014). In constructed wetlands, most of the macrophytes assist in maintaining oxidizing conditions by shoot-to-root oxygen transport (Marchand et *al.*, 2010). Such conditions promote formation of iron oxides, hydroxides and oxyhydroxides, such as the iron-plaques, and consequently result in metal removal by adsorption and precipitation. Adequate physico-chemical substrate conditions offer an efficient matrix for metal remov*al*. However, without any plants the substrate will become devoid of organic matter, thus decreasing the capacity of substrates to maintain sulphate reduction and metal immobilization (Marchand et *al.*, 2010).

Pseudo-First Order and Pseudo-Second Order Models

In this section, the aim of the adsorption kinetics modelling is to elucidate the speed of adsorption process and the factors influencing the reaction rate. The plots of the pseudo-first-order curves (ln (Qep - Qt) = f (t)) and pseudo-second order curves (t/Qt =f (t)) for all metals (Ni, Cr and Pb) generated the data necessary to analyse kinetics adsorption models of the present study.

The correlation coefficients of the pseudo-first-order ($R^2 = 0.05 - 0.10$) were less than those of the pseudo-second-order ($R^2 > 0.99$) at pH 7.0. According to Wang et *al.* (2009), first-order rate (k1) equation defines the adsorption rate on the basis of the adsorption capacity, while the pseudo-second-order rate expresses chemical sorption including valence forces and ion exchange. The pseudo-first order model was unsuitable to describe the adsorption rate processes due to the low values of coefficients of determination (Fifi et al., 2013). The pseudo-second order showed a good fitting ($R^2 > 0.99$) of the experimental data for Pb, Ni and Cr ions. These observations indicated that this model (pseudo-second order) is suitable for predicting the adsorption rates for each metal ion on the gravel materials. Similarly, Fifi et *al.* (2013) indicated the applicability of pseudo-second order for the kinetics adsorption of Pb, Cu and Ni in soil columns.

Equilibrium Isotherms

For assessing the adsorption of Ni, Cr and Pb on gravel materials, heavy metals concentrations in the supernatant and the amount adsorbed were plotted against the contact time to detect the equilibrium state. Due to the adsorption on the gravel during the contact period ranging from 1 minute to 2 days, heavy metal concentration in the supernatant was found less than their respective initial concentration. For all the three metals tested, concentrations of metals in supernatant varied with the increase in their initial levels. At equilibrium, amount of Ni adsorbed on gravel increased as function of the initial concentration in solution but not with increasing contact period. The values of equilibrium concentrations were: 0.6, 1.3, 1.8, 3.2 and 4.1 mg l^{-1} for 1.0, 2.0, 3.5, 5.0 and 6.0 mg l^{-1} respectively. In case of Cr the equilibrium concentration values were 0.4, 0.7, 1.7, 2.5 and 3.4 mg l^{-1} corresponding to 1.0, 2.0, 3.0, 4.5 and 6.0 mg l^{-1} initially introduced in solutions respectively. At 1.0 mg L^{-1} initial concentration level, the equilibrium concentration of Cr was lower (0.4 mg l^{-1}) than that of Ni (0.6 mg l^{-1}), and at 6.0 mg l^{-1} level, equilibrium concentration of Cr (3.4 mg l^{-1}) was also less than of Ni (4.1 mg l^{-1}).

The of equilibrium concentrations values of Pb were: 1.5, 2.8, 4.3 and 5.2 for 3.00, 6.00, 8.00 and 12.00 mg/l respectively. As observed for Ni and Cr, the equilibrium concentrations of Pb also varied as a function of the initial concentration in solution. Similar findings were also documented by Aziz et *al.* (2008) who used particle media with size ranging from 2.36 to 4.75 mm, to remove heavy metals from water or selected wastewater and Fifi et *al.* (2013) for the adsorption of Pb in the calcareous soil of Haiti.

At equilibrium, the amount of heavy metals adsorbed varied with ions species and the initial concentration. The amount of Ni adsorbed at equilibrium was: 0.9, 1.7, 4.0, 4.3 and 4.6 mg L^{-1} corresponding to the initial concentrations of 1.0, 2.0, 3.5, 5.0 and 6.0 mg l^{-1}. The maximum amount of Ni adsorbed was 4.3 mg kg^{-1} recorded for the highest con81 and 6.2 mg kg^{-1}, corresponding to the initial concentrations 1.0, 2.0, 3.0, 4.5 and 6.0 mg l^{-1}. Equally, the maximum amount was (6.2 mg kg^{-1}) was recorded for the highest concentration (6.0 mg l^{-1}). The range of Q_{eq} recorded for Pb was: 3.4, 7.7, 8.7 and 16.3 mg kg^{-1} corresponding to values of initial concentrations 3.0, 6.0, 8.0 and 12.0 mg l^{-1} respectively.

It was generally observed that there was an increased pattern in the amount of metal adsorbed with an increase in initial concentrations, except for Cr which was nearly constant from 1.0 to 3.0 mg l^{-1}. At comparative initial concentration levels, the amount of Pb (16.3 mg kg^{-1}) adsorbed was higher than those of Ni (4.3 m kg^{-1}) and Cr (6.2 mg kg^{-1}).

Adsorption percentage (%) of Ni, Cr and Pb as a function of their initial concentration at equilibrium is presented in figure 5. The adsorption percentage was relatively increasing with the concentration levels for Pb (47.2 – 56.2%) while it was in contrast decreasing for Ni (45.5 to 34.5%) and Cr (54.7 to 44.2%) from low concentrations to high concentrations. The adsorption efficiencies recorded in this study for lead on gravel are similar to the findings of Chen et *al.* (2009) who showed that the removal efficiencies of Pb on gravel material were within 95-99%. In contrast, Allende et *al.* (2012) showed a modest capacity for arsenic removal (43%) in gravel columns. In agreement with the present study, Aziz et *al.* (2008) showed the adsorption percentage of Ni, Cr and Pb on gravel were 44-73%, 42-72% and 42-78% respectively.

According to Holmberg (2006), adsorption processes include physical (physical sorption) or chemical (chemisorption). The forces of attraction of ions to the solid surface control the type of bonds and their classification. There is no transfer of electron in the physical adsorption. It is fundamentally reversible (can return to liquid phase through desorption process) and has a small degree of specificity such adsorbed ions are able to cover the whole surface of the gravel. In contrast, chemical sorption consists of chemical bonds and is basically irreversible. Metal ions adsorbed are fixed at specific sites at the surface of the adsorbent.

Adsorption Isotherm Models

Numerous adsorption models initially used for liquid phase adsorption are accessible and approved to correlate adsorption equilibriums in the case of heavy metals sorption (Rashid et *al.*, 2015). Freundlich and Langmuir models are among the most commonly used. The application of these models on gravel materials assisted heavy metal removal from the supernatant.

The values of different constants and parameters used in Freundlich and Langmuir adsorption isotherm for Ni, Cr and Pb, equations generated and the coefficients of determination (R^2) of the adsorption data are presented in Table 4 and illustrated in Figure 6. These models described the adsorption of these metal ions on gravel which is a function of increasing metal ions concentration in solution after equilibrium.

The Freundlich and Langmuir isotherms generally fitted data for all metal ions studied. More precisely, Langmuir models fitted better for Pb adsorption ($R^2 = 0.70$), except for Ni and Cr ($R^2<0.5$), while Freundlich model was favourable for Ni and Cr ($R^2 > 0.90$), except Pb ($R^2 = 0.54$).

The adsorption capacity is very important for describing an adsorbent (Febrianto et al., 2009). Freundlich parameters (k_F and n) specify whether the adsorption process is favourable or unfavourable (Fifi et *al.*, 2013). The adsorption Freundlich coefficients (k_F) were k_F= 8.4 mg/kg for Ni and 3.6 mg kg-1 for Cr. These values showed that gravel had good adsorption capacity (k_F> 2) while the adsorption intensity *n* was relatively low. The results also showed little preferential adsorption of Ni over Cr. This was in agreement with the variation of the adsorption percentages at equilibrium obtained in the previous section, which ranged between (34.5 – 45.5%) for Ni and (43.0 – 52.6%) for Cr. The values of *n* obtained were 0.59 for Cr and 0.58 for Ni. This indicated that the adsorption intensity was favourable (47.0 to 57.0%) overall concentrations (higher concentrations and lower concentrations) for Ni and Cr (n < 1) (Huang et *al.*, 2015). According to Asku et *al.* (1991), the low values of k_F and *n* reveal the low affinity between metal ions and the matrix. Hence, irrespective to the metal ion studied, the values of *n* were generally low ($n < 1$) for Ni ($n = 0.58$) and Cr ($n = 0.59$), indicating that the adsorption observed was physical adsorption that is due to weak bonds between the metal ions and the matrix with very low energy (Ho and Ofomaja, 2006). Physical adsorption is caused by van der Waals forces of attraction and does not develop strong bonds as is the case for chemical adsorption (Marchand et *al.,* 2010; Tripathi and Ranjan, 2015).

As explained in the previous section, the good adsorption capacity observed for gravel confirm the association of heavy metals with organic matter and/or Fe/oxydes coatings on solid surfaces while the low adsorption intensity (n<1) highlights the limited amount of organic matter and Fe/oxydes in 50 g of gravel collected from CWs for this experiment compare to the all constructed wetland systems, that is essentially due to the combination of gravel with macrophytes. In the constructed wetlands, plants release oxygen, organic matter and iron-plaques to gravel through shoot-to-root system. Iron-plaques developments on roots have been discussed for a number of macrophytes used in constructed wetlands (Colleen and Scott, 2001; Marchand et al., 2010). Iron plaque accumulation can account for up to 10% root dry weight into the rhizosphere of the foremost macrophytes (Colleen and Scott, 2001).

This type of adsorption is reversible for long term observation, leading to the release of metal ions (previously adsorbed) in the solution over a reversible process called desorption (Aziz et *al.,* 2008; Rashid et *al.,* 2015). Other studies also supported similar observations in the case of heavy metal adsorption using Freundlich isotherm (Tellan and Owalude, 2007; Fifi et *al.,* 2013).

Langmuir models fitted Pb metal ions adsorption on gravel. Regarding the parameters of the Langmuir isotherm, the higher the values of Q_{max}, b and R^2, best is the adsorbent studied. For Pb, $Q_{max} = 28.2$ mg kg^{-1} at equilibrium concentration (0.1 mg l^{-1}), b= 6443, indicated a low affinity of Pb to gravel compare to some bio sorbents with b > 10 000. Chitana (2002) reported that chitosan (a bio sorbent) could adsorb 59000 mg kg^{-1} of Pb at equilibrium concentration (0.9 mg l^{-1}) and b (Langmuir coefficient) was 60360 for Pb, showing the high affinity between Pb and chitosan. The important feature of Langmuir isotherm was described by a dimensionless constant factor R_L (Huang et *al.,* 2015). The range of values of this constant are as follows: $0<R_L<1$ for favourable adsorption; $R_L>1$ for unfavourable adsorption, $R_L=1$ for linear adsorption and $R_L=0$ for irreversible adsorption (Caliskan et *al.,* 2011; Huang et *al.,* 2015). Results showed that $R_L= 0.00032$ for Pb showing a favourable and reversible adsorption of Pb on this sorbent. Similar to Ni and Cr, this indicates that the process observed for Pb was physical adsorption that is due to weak bonds between the metal ions and the matrix with very low energy (Ho and Ofomaja, 2006).

Though the adsorption intensity was low with a value ($n = 0.58$ for Ni, 0.59 for Cr), the gravel was found to have a very good adsorption capacity (k>2 mg kg^{-1}) regardless of the metal species for Freundlich model. Langmuir model showed that gravel had a favourable adsorption of Pb indicated by $R_L=0.0003$ and b>6000).

Adsorption of metals onto gravel materials is an effective mechanism of removal of these pollutant in waste water using wetland plots. These process is favourable due to organic matter coated at the surface of gravel materials. Thus it is advantageous for engineers to use gravel materials as substrate in constructed wetlands. The evaluation of the performance of gravel in the removal of metal in constructed wetland can be easily performed using Freundlich isotherm and pseudo-second order models.

FUTURE RESEARCH DIRECTIONS

The investigation of wetland processes and their modelling need to be keeping forwards into a long term basis. Besides the adsorption processes analysed in this study, additional processes involved in heavy metal dynamics in constructed wetlands such as sedimentation, complexation, precipitation/co-precipitation, microbial activities should be carefully investigated and understood for all existing metal species and different wetland technology in relation with the environment. This results will help to capture the behaviour of wetland technology over extended period of time at low cost using modelling and simulation approach.

CONCLUSION

The adsorption equilibrium of Pb was achieved in 6 hours compared to 12 hours for Ni and Cr indicating a better affinity between Pb and gravel. The value $n<1$ for Ni and Cr showed that the sorption intensity was low but favourable. Values of Freundlich coefficient ($k_F > 2$ mg kg^{-1}) again showed that the gravels had a very good adsorption capacity regardless of the metal species. Evaluation of adsorption isotherms (Langmuir and Freundlich) showed that physical adsorption characterised by weak bonds (Van Der Waals bonds) was mainly responsible for metal adsorption on gravels. The adsorption of Pb was described better by Langmuir isotherms whereas Ni and Cr adsorption fitted better in Freundlich isotherms. The metal adsorption affinity with gravels was in the order Pb >Cr >Ni. Similarly pseudo-second-order which had higher values of coefficient of determination (R^2) compared to pseudo-first order kinetic model showed that pseudo-second-order model was better for predicting the adsorption rates of Ni, Cr and Pb on the gravel. Gravels were found to efficient for adsorption of Ni, Cr and Pb. Combining the contribution of gravels with plant uptake and microbial activities of constructed wetlands will further improve the pollutant removal efficiency of constructed wetlands.

REFERENCES

Abdić, Š., Memić, M., & Šabanović, E. (2018). Adsorptive removal of eight heavy metals from aqueous solution by unmodified and modified agricultural waste: Tangerine peel. *International Journal of Environmental Science and Technology*. doi:10.100713762-018-1645-7

Abdulfatah, A. Y., El Hamalawi, A., Wheatley, A. D., & Marto, A. (2015). *The Behaviour of Metal Contaminants in Silty Sand and Gravel*. Int'l Journal of Research in Chemical, Metallurgical and Civil Eng. doi:10.15242/IJRCMCE.E091501

Allende, L. K., Fletcher, T. D., & Sun, G. (2012). The effect of substrate media on the removal of the removal of Arsenic, boron and iron from an acidic wastewater in planted column reactors. *Chemical Engineering Journal*, *179*, 119–130. doi:10.1016/j.cej.2011.10.069

Annadurai, G., Juang, R. S., & Lee, D. J. (2003). Adsorption of heavy metals from water using banana and orange peels. *Water Science and Technology*, *47*(1), 185–190. doi:10.2166/wst.2003.0049 PMID:12578193

Appenroth, K-J. (2010). *Definition of "Heavy Metals" and Their Role in Biological Systems*. doi. doi:10.1007/978-3-642-02436-8_2

Aziz, H. A., Mohd, N. A. K., & Ariffin, S. (2008). Heavy metals (Cd, Pb, Zn, Ni, Cu and Cr (III)) removal from waterin Malaysia: Post treatment by high quality limestone. *Bioresource Technology*, *99*(6), 1578–1583. doi:10.1016/j.biortech.2007.04.007 PMID:17540556

Caliskan, N., Kul, A. R., Alkan, S., Sogut, E. G., & Alacabey, I. (2011). Adsorption of ZincII on diatomite and manganese-oxide-modified diatomite: A kinetic and equilibrium study. *Journal of Hazardous Materials*, *193*, 27–36. doi:10.1016/j.jhazmat.2011.06.058 PMID:21764214

Chitana, S. (2002). *Removal of heavy metals from wastewater using chitosan* (MSc thesis). Suranaree University of Technology.

Chong, H. L. H., Chia, P. S., & Ahmad, M. N. (2013). The adsorption of heavy metal by Bornean oil palm shell and its potential application as constructed wetland media. *Bioresource Technology*, *130*, 181–186. doi:10.1016/j.biortech.2012.11.136 PMID:23306127

Colleen, M. H. (2001). Fendorf characterization of fe plaque and associated metals on the roots of mine-waste impacted aquatic plants. *Environmental Science & Technology*, *35*(19), 3863–3868. doi:10.1021/es0105459 PMID:11642445

Defo, C., Kaur, R., Anshu, B., Khajanchi, L., & Paritosh, K. (2017). Modelling approaches for simulating wetland pollutant dynamics. *Critical Reviews in Environmental Science and Technology*, *47*(15), 1371–1408. doi:10.1080/10643389.2017.1350126

Fifi, U., Winiarski, T., & Evens, E. (2013). *Assessing the Mobility of Lead, Copper and Cadmium in a Calcareous Soil of Port-au-Prince*. International Journal of Environmental Resources and Public Health. doi:10.3390/ijerph10115830

Gunawardana, C., Goonetilleke, A., & Egodawatta, P. (2013). Adsorption of heavy metals by road deposited solids. *Water Science and Technology*, *67*(11), 2622–2629. doi:10.2166/wst.2013.171 PMID:23752398

Guo, X., Zhang, S., & Shan, X. (2008). Adsorption of metal ions on lignin. *Journal of Hazardous Materials*, *151*(1), 134–142. doi:10.1016/j.jhazmat.2007.05.065 PMID:17587495

Ho, Y., & Ofomaja, A. E. (2006). Biosorption thermodynamics of cadmium on coconut copra meal as bio sorbent. *Biochemical Engineering Journal*, *30*(2), 117–123. doi:10.1016/j.bej.2006.02.012

Holmberg, J. P. (2006). *Competitive adsorption and displacement behaviour of heavy metals on peat* (Master dissertation). Chalmers University of Technology, Goteborg, Sweden.

Huang, D. L., Min-Chen, G., Zeng, G. M., Lai, G., Zhang, C., Li, N. J., ... He, Y. (2015). Synthesis and application of modified zero-valent iron nanoparticle for removal of hexavalent chromium from wastewater. *Water, Air, and Soil Pollution*, *226*(11), 375. doi:10.100711270-015-2583-3

Kabbashi, N. A., Atieh, M. A., Al-Mamun, A., Mirghami, M. E. S., Alam, M. D. Z., & Yahya, N. (2009). Kinetic adsorption of application of carbon nanotubes for Pb(II) removal from aqueous solution. *Journal of Environmental Sciences (China)*, *21*(4), 539–544. doi:10.1016/S1001-0742(08)62305-0 PMID:19634432

Khajanchi, L., Kaur, R., Rosin, K. G., & Neelam, P. (2016). *Low-Cost Remediation and On-Farm Management Approaches for Safe Use of Wastewater in Agriculture*. Innovative Saline Agriculture. doi:10.1007/978-81-322-2770-0_12

Kivaisi, A. K. (2001). The potential for constructed wetlands for wastewater treatment and reuse in developing countries: A review. *Ecological Engineering*, *16*(4), 545–560. doi:10.1016/S0925-8574(00)00113-0

Maiga, Y., von Sperling, M., & Mihelcic, J. (2017). *Constructed Wetlands*. Global Water Pathogens Project. Retrieved from http://www.waterpathogens.org

Marchand, L., Mench, M., Jacob, D. L., & Otte, M. L. (2010). Metal and metalloid removal in constructed wetlands, with emphasis on the importance of plants and standardized measurements: A review. *Environmental Pollution*, *158*(12), 3447–3461. doi:10.1016/j.envpol.2010.08.018 PMID:20869144

Mautner, A., Maples, H. A., Kobkeatthawin, T., Kokol, V., Karim, Z., Li, K., & Bismarck, A. (2016). I Phosphorylated nanocellulose papers for copper adsorption from aqueous solutions. *International Journal of Environmental Science and Technology*, *13*(8), 1861–1872. doi:10.100713762-016-1026-z

Mohammad, M., Maitra, S., & Dutta, B. K. (2018). Comparison of Activated Carbon and Physic Seed Hull for the Removal of Malachite Green Dye from Aqueous Solution. *Water, Air, and Soil Pollution*, *229*(2), 45. doi:10.100711270-018-3686-4

Morari, F., Ferro, N. D., & Coco, E. (2015). Municipal wastewater treatment with *phragmites australis L.* and *Typha L.* for irrigation reuse. Boron and heavy metals. *Water, Air, and Soil Pollution*, *226*(3), 56. doi:10.100711270-015-2336-3

Mushtaq, M., Haq, N. B., Munawar, I., & Saima, N. (2016). Eriobotrya japonica seed biocomposite efficiency for copper adsorption: Isotherms, kinetics, thermodynamic and desorption studies. *Journal of Environmental Management*, *12*, 17–22. PMID:27039361

Rashid, H., Bhattia, N., Iqbal, M., & Noreen, S. (2015). Fungal biomass composite with bentonite efficiency for nickel andzinc adsorption: A mechanistic study. *Ecological Engineering*, *91*, 459–471. doi:10.1016/j.ecoleng.2016.03.014

Sahu, O. (2014). Reduction of heavy metals from wastewater by wetland. *International Letters of Natural Sciences*, *12*, 35–43. doi:10.18052/www.scipress.com/ILNS.12.35

Shakunthala, B., Shivanna, S., & Doddaiah, S. (2010). Urban Wastewater Characteristic and its Management in Urban Areas—A Case Study of Mysore City, Karnataka, India. *Journal of Water Resource and Protection*, *2*(08), 717–726. doi:10.4236/jwarp.2010.28082

Sizirici, B., & Yildiz. (2018). Adsorption capacity of iron oxide-coated gravel for landfill leachate: Simultaneousstudy. *I. International Journal of Environmental Science and Technology*. doi:10.100713762-018-1644-8

Tellan, A. C., & Owalude, S. O. (2007). Some Langmuir and Freundlich parameters of adsorption studies of chlorpheniramine maleate. *Res. J. Appl. Sci.*, *2*, 875–878.

Tripathi, A., & Ranjan, M. R. (2015). *Heavy Metal Removal from Wastewater Using Low Cost Adsorbents*. J Bioremed Biodeg. doi:10.4172/2155-6199.1000315

Tsibranska, I., & Hristova, E. (2011). Comparison of different kinetic models for adsorption of heavy metals onto activated carbon from apricot stones. *Izvestiia po Himiia*, *43*(3), 370–377.

Ullah, I., Nadeem, R., Iqbal, M., & Manzoor, Q. (2013). Biosorption of chromium onto native and immobilized sugarcane bagasse waste biomass. *Ecological Engineering*, *60*, 99–107. doi:10.1016/j.ecoleng.2013.07.028

Vymazal, J. (2010). Constructed wetlands for wastewater treatment. *Water (Basel)*, *2*(3), 530–549. doi:10.3390/w2030530 PMID:20795704

Water Technology Centre. (2014). Engineered Wetland Technology based. Eco-friendly Wastewater Treatment and Reuse. An initiative of Indian Agricultural Research Institute. WTC-IARI.

Yadav, A. K., Naresh, K., Sreerishnan, T. R., Santosh, S., & Narsi, R. B. (2010). Removal of Cr and Ni from acqueous solution in constructed wetland: Mass balance, adsorption-desorption and FTIR study. *Chemical Engineering Journal*, *160*(1), 122–128. doi:10.1016/j.cej.2010.03.019

KEY TERMS AND DEFINITIONS

Adsorbate: A substance that is adsorbed.

Adsorbent: A substance that adsorbs another material.

Adsorption: Removal process in which there is adhesion of atoms, ions or molecules from a gas, liquid or dissolved solid to a surface. This process creates a film of the adsorbate on the surface of the adsorbent.

Adsorption Capacity: The amount of adsorbate taken up by the adsorbent per unit mass (or volume) of the adsorbent. The adsorption capacity of a solid desiccant for water is expressed as the mass of water adsorbed per mass of desiccant.

Adsorption Isotherm: When an adsorbent and adsorbate are in contact for long period of time, an equilibrium is established between the amount of adsorbate adsorbed and the amount of adsorbate in solution. The equilibrium relationship is described by adsorption isotherms.

Constructed Wetlands: Constructed wetlands are a sanitation technology that utilize natural removal mechanisms provided by plant vegetation, substrate (soil, gravel, sand), and associated microbial populations (Maiga et al., 2017; Defo et al., 2017).

Heavy Metal: Any metal with a specific gravity of 5.0 g/cm^3 or greater, especially one that is toxic to organisms. “Heavy metals” would then refer to (1) transition elements; (2) rare earth elements, which can be subdivided into the lanthanides and the actinides, including La and Ac themselves; (3) a heterogeneous group including the metal Bi, the elements that form amphoteric oxides (Al, Ga, In, Tl, Sn, Pb, Sb and Po), and the metalloids Ge, As and Te (Appenroth, 2010).

Kinetics: The branch of chemistry or biochemistry concerned with measuring and studying the rates of reactions.

Chapter 12
Evaluation of Selected Reaeration Models for Water Quality Management

Lukman Salihu
University of Hafr Al-Batin, Saudi Arabia

Adekunbi E. Adedayo
Adeyemi Colloege of Education, Nigeria

Babajide Jelili Olalekan
Adeyemi Colloege of Education, Nigeria

Asani M. Afolabi
Ladoke Akintola University of Technology, Ogbomoso, Nigeria

Idi Dansuleiman Mohammed
Land, Air, and Water Consulting Engineers, Nigeria

Isaiah Adesola Oke
https://orcid.org/0000-0002-7082-7682
Obafemi Awolowo University, Nigeria

Obijole O. Adewale
Department of Ecology and Resources Management, School of Environmental Sciences, University of Venda, South Africa

ABSTRACT

In this chapter, a new proposed model was compared with selected standard models and evaluated statistically (model of selection criterion [MSC] and Akaike information criterion [AIC]). Suspended concentration and calculated reaeration rate were used to predict concentration of EPs removable by the

DOI: 10.4018/978-1-7998-1871-7.ch012

aeration and self-purification of the stream. The study revealed that MSC for the new proposed model were 0.75, - 0.44, - 0.32, - 0.45, and - 0.45 respectively. AIC for both dry and wet seasons were 11.85, 42.17, 41.37, 42.17, and 42.25 for the new proposed model, respectively. It was concluded the proposed model performed better than some of the standard models.

INTRODUCTION

Dissolved oxygen (DO) concentration is an important water quality index. A lower value of DO usually causes a serious environmental issue for a body of water, which include decrease in the capacity of creatures living within it to grow, reproduce, and survive. Efforts to increase DO concentration in lakes and rivers and prevent water quality deterioration involve utilization of aeration. Aeration is important to provide enough DO for aerobics organisms in Biochemical Oxygen Demand (BOD) removal and nitrification in activated sludge plants and retain contaminants suspension biomass. There are several hyraulic structure used in aeration process such as water jet aeration with circular, water jet aeration with venturi nozzles, pipe aeration with venturi tube, high-head conduit aeration, weir aeration and free-surface conduit aeration. Three methods that commonly used for aeration: gravity aeration, mechanical aeration and water diffused aeration. Utilizing a simple weir can makes gravity aeration, an inclined corrugated sheet or stepped cascade. The efficiency of aeration depends on the amount of surface contact between air and water, which is controlled primarily by the size of the water drop or air bubble. It has been reported that significant drop in DO below the standard due to a heavy pollutant load discharged at upstream can eliminate aquatic life. At this point artificial aeration augmentation can be used to improve the level of DO to a higher level. The problem of artificial aeration has previously been studied by various researchers such as Tarassov *et al.* (1969), Yu (1970), Hunter and Whipple (1970), Whipple *et al.* (1970), Davidson and Bradshaw (1970), Whipple and Yu (1971), Chang and Yeh (1973), Rinaldi and Soncini-Sessa (1978) and Greenop *et al.* (2001). It has been reported that useful models are not necessarily complicated ones. While the complicated models can simulate more conditions with more parameter inputs, the relative efforts, time and budget may be enormous. In a river system, the required water quality model is a one dimensional steady-state model that can be quickly solved using mathematical software and at the same time satisfies accuracy requirements.

BACKGROUND

Pollution is among severe and serious problem that urgently requires models for monitoring and implementation plans deriving solutions. Daily, domestic, institutional, industrial and agricultural wastes are discharged into the receiving water bodies (UN WWAP, 2003; Geissen *et al.,* 2015). Equivalent amount of these pollutants from human activities get into human and ecosystems. The United Nation (UN) estimates that the amount of wastewater produced annually is about 1500 km^3, which six times more water than exists in all the rivers of the world (UN WWAP, 2003). It was reported that lack of adequate sanitation practice is one of the most significant causes of water pollution. Worldwide, 2.5 billion people live without proper sanitation (UNICEF, 2009; Geissen *et al.2015,*). In some regions of the world, more than 50% of native freshwater fish species are at risk of extinction, and this is also the case for nearly one-third of the world's amphibians (Vié, Hilton-Taylor, and Stuart, 2009; Geissen *et al.,2015*).

In the last 70 years, there were several environmental pollution events which can be taken as historically events in ecotoxicology. The first events in ecotoxicology date back to the 1950s (Methylmercury in the food chain was found to cause cognition disorders and proceed) and a worldwide ecological movement in 1962 with the book Silent Spring. There are many main catastrophes as Tschernobyl or Sandoz, which contributed and strengthened the ecological thinking. Today, there are more than 1036 emerging pollutants, their metabolites and transformation products, are listed as present in the European aquatic environment (Geissen *et al.,2015 ;* NORMAN, 2019). Emerging pollutants (EPs) are defined as synthetic or naturally occurring chemicals that are not commonly monitored in the environment but which have the potential to enter the environment and cause known or suspected adverse ecological and human health effects. In some cases, release of emerging pollutants to the environment has likely occurred for a long time, but may not have been recognized until new detection methods were developed. In other cases, synthesis of new chemicals or changes in use and disposal of existing chemicals can create new sources of emerging pollutants (Geissen *et al., 2015*). EPs are currently not included in (inter)national routine monitoring programmes and their fate, behavior and ecotoxicological effects are often not well understood. EPs can be released from point pollution sources (wastewater treatment plants from urban or industrial areas, or from diffuse sources through atmospheric deposition or from crop and animal production.

EPs are categorized into more than 20 classes related to their origin (Geissen *et al.,2015*). The prominent classes are: pharmaceuticals (urban, stock farming), pesticides (agriculture), disinfection by-products (urban, industry), wood preservation and industrial chemicals (industry, Figure 1).

Mercedes (2017) reported that EPs can be classified by their occurrence, origin, use, molecular similarity, physical chemical characteristics, biochemical activity, and environmental and health effects. These classifications are as follows:

Figure 1.

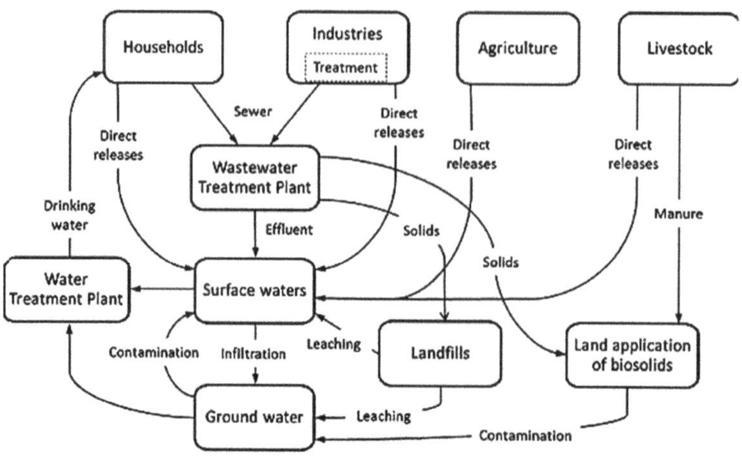

Classification According to Use

This classification allows associating to an activity a series of pollutants facilitating to focus in a smaller number of substances to investigate and regulate. The classification of EPs commonly used according to their use. Some compounds are overlap under several categories as follows:

Classification According to Simplifying Criteria: There is a great effort to find relationships that links physico-chemical properties of the substances with the most suitable removal technology for EPs. These criteria for this classification are as follows:

i. According Biological Transformation Rate (k_{biol}):

k_{biol} < 0.1 L/(gSS d) poor degradability
0.1< k_{biol} < 10 L/(gSS d) quite good biodegradability
k_{biol} > 10 L/(gSS d) very good degradability

ii. According To Partition Coefficient Octanol-Water (Log K_{ow}):

Log K_{ow} < 2.5 high hydrophilic compound
2.5 < Log K_{ow} < 4 moderate hydrophilic compound
Log K_{ow} > 4 high lipophilic compound

iii. According Partition Coefficient (Log K_d)

i. Log K_d < 2.7 low adsorption potential
ii. Log K_d > 2.7 high adsorption potential

Effects of emerging pollutants: The effects of emerging pollutants in the environment are partially known. As it is also little known the pathways of EP having into an ecosystem and the transformation process that suffer. EPs categories and associated health effects are as follows:

In light of the potential impact of these substances on aquatic life and human health, the lack of knowledge regarding their behaviour in the environment and the deficiency in analytical and sampling techniques, action is urgently required at multiple levels. Figure 2 provides various concept for management of emerging pollutants in the aquatic environment (Geissen *et al.,2015*).

Removal of EPs from Water and Wastewaters

Rashed *et al.* (2005) reported that Advanced Oxidation Technologies (AOTs) are the methods in use in removing EPs from both water and wastewaters. These technologies include Advanced Oxidation Process (AOPs) and other physicochemical conversion methods which are as follows:

a. Advanced Oxidation Process (AOPs),
b. Non-Thermal Plasmas (NTP) for air and wastewater treatment,
c. Electrohydraulic cavitation and sonolysis for wastewater treatment,
d. Electron beam and gamma-ray irradiation,

Class of Emerging	Pollutants Example	Definition
a) Antibiotics	Tetracycline, Erythromycin	Medications that fight bacterial infections, inhibiting or stopping bacterial growth.
b) Antimicrobials	Triclosan	Biochemicals that kill or inhibit the growth of microorganisms including bacteria and fungi.
c) Detergent metabolites	Nonylphenol	Chemical compounds formed when detergents are broken down by wastewater treatment or environmental degradation.
d) Disinfectants	Alcohols, Aldehydes and oxidizing agents	A chemical agent used on non-living surfaces to destroy, neutralize, or inhibit the growth of disease-causing and microorganisms. Disinfection by-products
e) Chloroform,	Nitrosodimethylamine (NDMA)	Chemical substances resulting from the interaction of organic matter in water with disinfection agents such as chlorine.
f) Estrogenic compounds	Estrone, Estradiol, Nonylphenol, Bisphenol	A Natural or synthetic chemicals that can elicit an estrogenic response.
g) Fire or flame retardants	Polybrominated Diphenyl Ethers (PBDEs)	Any of several materials or coatings that inhibit or resist the spread of fire.
h) Fragrances	Galaxolide	Chemical substances that impart a sweet or pleasant odour.
i) Insect repellents	DEET (N, N-diethyl-meta-tolumide)	Chemical substances applied to skin or other surfaces to discourage insects from coming in contact with the surface.
j) PAHs	Benzo(a)pyrene, Fluoranthene, Naphthalene	A large group of chemical substances usually found in the environment as a result of incomplete burning of carbon-containing materials like fossil fuels, wood, or garbage.
k) Personal Care Products	Para-hydroxybenzoate	Chemical substances used in a diverse group of personal items including toiletries and cosmetics.
l) Pesticides or Insecticides	Permethrin, Fenitrothion, Bacillus thuringiensis israelensis	Chemical substances or microbiological agents that kill, incapacitate or otherwise prevent pests from causing damage.
m) Pharmaceuticals	Fluoxetine (Prozac), Carbamazepine, Diphenhydramine	Chemical substances used in the prevention or treatment of physiological conditions.
n) Plasticizers	Dioctyl Phthalate (DOP)	Chemical additives that increase the plasticity or fluidity of a material.
o) Reproductive hormones	Dihydrotestosterone (DHT), Progesterone, Estrone, Estradiol	A group of chemical substances, usually steroids, whose purpose is to stimulate certain reproductive functions.
p) Solvents	Ethanol, Kerosene	Chemical solutions, other than water, capable of dissolving another substance.
q) Steroids	Cholesterol, Coprostanol, Estrone, Progesterone	A large group of fat-soluble organic compounds with a characteristic molecular structure, which includes many natural and synthetic hormones.
r) Surfactants	Sodium Lauryl Sulphate	Chemical substances that affect the surface of a liquid.

(Source: Petrie *et al.*, 2015, Mercedes, 2017)

Use Category	Suspected health effects from environmental exposure
a) Antibiotics	Antibiotic resistance in disease causing bacteria complicating treatment of infections
b) Disinfectants	Genotoxicity, cytotoxicity, carcinogenicity
c) Fire retardants	Endocrine disruption, indications of increased risk for cancer
d) Industrial additives	Can be toxic to animals, ecosystems, and humans
e) Life-style products	Can cause cellular stress, negative effects on reproductive activity (Caffeine, Nicotine) in animals
f) Non-prescription drugs	Unknown health effects
g) Other prescription drugs	Increased cancer rates, organ damage
h) Personal care products	Bacterial resistance, endocrine disruption
i) Pesticides	Endocrine disruption
j) Plasticizers	Endocrine disruption, increased risk of cancer
k) Reproductive hormones	Endocrine disruption
l) Solvents	Endocrine disruption, liver and kidney damage, respiratory impairment, cancer
m) Steroids	Endocrine disruption

(Source: Mercedes, 2017)

e. Catalytic oxidation for air treatment,
f. Wet air oxidation for wastewater treatment,
g. Supercritical water oxidation treatment,
h. Electrochemical redox reactions for wastewater treatment and
i. Photocatalysis for wastewater and air treatment.

It has been reported that physicochemical mechanisms EPs contributes to EP removal from wastewaters and surface waters. Literature has provided several removal and analysis processes to remove these EPs. Although, there are extensive researches on presence and removal of EPs, there is still a considerable lack of integrated and classified information to assess these documents. In addition, several empirical equations have been proposed in literature and some of these ones are widely applied in water quality studies, but recent investigation have demonstrated that they generally result in a poor fit with field data (Melching and Flores, 1999; Gualtieri and Gualtieri, 1999). This reason and many others form justification for the objective of this study, which proposed a general model for the prediction of K_2, determined the unknown parameter in the proposed model using field data, evaluated the model and compared its output with selected standard models that are in use. Therefore, in the current chapter, the removal of EPs from aqueous solutions specifically rivers using aeration process was presented.

Figure 2.

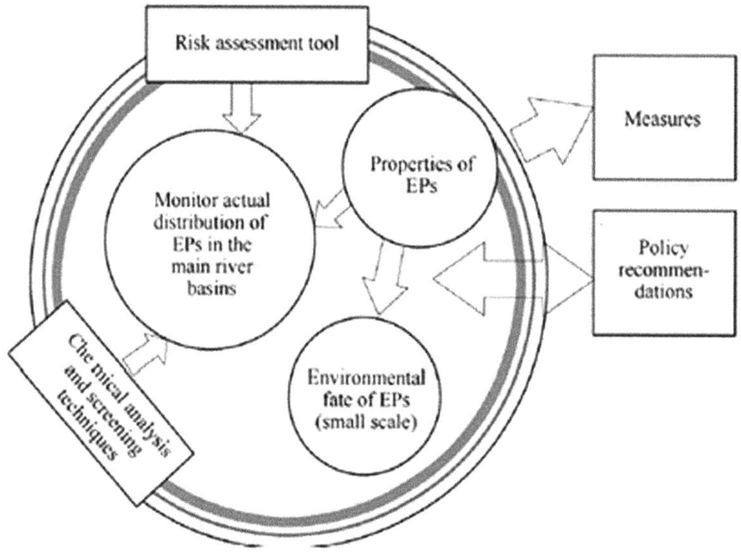

MATERIALS AND METHODS

This study is of two parts as follows:

a) In this first part of the study, survey of literature was conducted on the basis of accomplished information on EPs, presence and removal from environment using aeration in river. Information was collected from published in databases such as Google Scholar, PubMed, Elsevier, Scopus, Springer and Magiran using emerging pollutants, aeration, self - purification and aeration models as the keywords.
b) b) In this part information on Ogbe River and its catchment were obtained from relevant sources. Sampling points were established on the river at intervals of 500 m and the coordinate of each point was determined along Ogbe River. Table 1 shows the coordinates of the sampling points. Collections of samples from these points were conducted in accordance to standard methods (APHA, 2005), between February and March 2013 (dry season) and between June and July 2013 (wet season). Samples (110) were collected for both the dry and the wet seasons. Water quality based on suspend solid concentration, DO concentration, the depth and velocity of flow were monitored along the stream at all the sampling points. Computation of field K_2 was conducted as flows:

$$K_{2f} = \frac{1}{t}\ln\left(\frac{C_a}{C_b}\right) \tag{1}$$

Where, C_a and C_b are measured upstream and downstream DO concentrations.

A general non-linear regression model that represents generalization of simple formulae, which relates aeration coefficient to velocity of flow and depth of water, was proposed as follows:

$$K_2 = K\left(V^x\right)H^y \tag{2}$$

where, K_{2f} is the reaeration coefficient (per day); K, x and y are the constants parameters for the model, V is the velocity of flow and H is depth of water. The model was proposed and selected based on the simplicity and a minimum number of fixed parameters to be determined. The parameters in the proposed new numerical model (K, x and y) were determined using Microsoft Excel Solver and field data of K_{2f}. Microsoft Excel Solver was selected based on availability (at no additional cost) and accuracy in numerical solutions. Accuracy of model developed was evaluated and compared with standard models in use (Tsivogluo – Wallace, Streeter *et al*, Churchill *et al*, Oconnor and Dobbbins, IHP, Adedokun and Agunwambe) using statistical methods (analysis of variance (ANOVA), model of selection (MSC), Schwartz (SC) and Akaike Information Criterion (AIC)). Procedures employed in the computations of the parameters in the model are according to Oke *et al.*, (2016a) and evaluation are expressed by Equations (3) to (6), respectively:

a. Microsoft Excel Solver was added in on the toolbar of Microsoft Excel;
b. Target value was set as follows: $\left[\sum_{i=1}^{n} K_{2i} - \sum_{i=1}^{n} K_{2fi}\right]^2 = 0$; (3)
c. Changing cells of the iterations were selected, number of iterations, degree of accuracy and maximum time for the iteration were set for the software to meet the target;
d. and the iteration started through Microsoft Excel Solver (Figure 3).

$$MSC = \ln \frac{\sum_{i=1}^{n}\left(Y_{obsi} - \overline{Y}_{obs}\right)^2}{\sum_{i=1}^{n}\left(Y_{obsi} - Y_{cali}\right)^2} - \frac{2p}{n} \tag{4}$$

where, Y_{obsi} is the field data (K_{2f}); $\overline{Y}_{obs}$ is the average field data (K_{2f}); p is the total number of fixed parameters to be estimated in the equation; n is the total number of field data (K_{2f}), and Y_{cali} is the K_2 computed using models.

The Schwartz Criterion is defined by the formula

$$SC = n\ln\left(\sum_{i=1}^{n}\left(Y_{obsi} - Y_{cali}\right)^2\right) + p\ln(n) \tag{5}$$

The more appropriate model is the one with the *smaller* SC value.

The AIC was derived from the Information Criterion of Akaike (1976). It allows a direct comparison among models with a different number of parameters. The AIC presents the information on a given set of parameter estimates by relating the coefficient of determination to the number of parameters. The AIC values were computed using equation (6):

Figure 3.

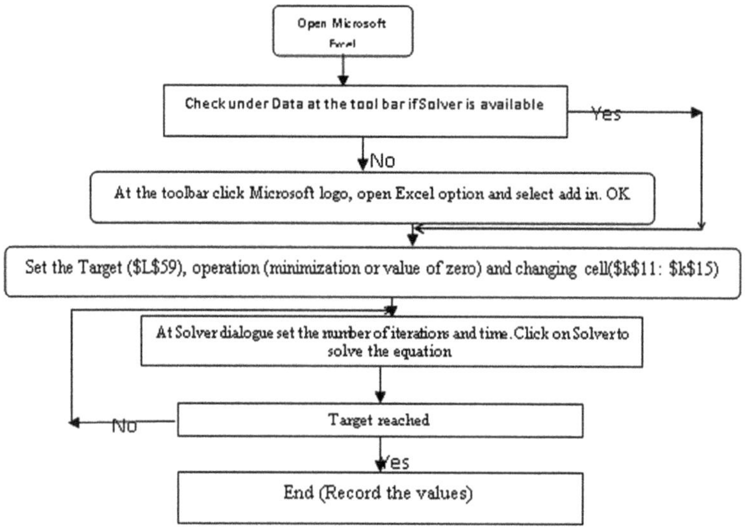

$$AIC = n\left(\ln\sum_{i=1}^{n}\left(Y_{obsi} - Y_{cali}\right)^{2}\right) + 2p \quad (6)$$

Suspended concentration and calculated reaeration rate were used to predict concentration of EPs removable by the aeration and self-purification of the stream. Predictions of concentrations (PEC) of EPs were determined as follows (Mercedes, 2017):

$$C_{st} = \frac{C_w}{D_l\left(1+\left(S_s x\, K_2 x 10^{-6}\right)\right)} \quad (7)$$

Where; C_{st} is the predicted EPs concentration removable, C_w is the concentration of EPs found in the water, S_s is the suspended solid concentration, Q is the flow rate from the stream and D_l is the dilution factor

$$D_l = \frac{K_2 + Q}{K_2} \quad (8)$$

Predictions of annual EPs' concentration were estimated as follows:

$$C_{s\tan} = \frac{C_{st}T_e}{365} \quad (9)$$

Where; C_{stan} is the annual predicted EPs concentration removable and T_e is the number of rainfall's days per year in Ile- Ife

Predictions of annual sediments from EPs' concentration were estimated as follows:

$$C_{sd\tan} = 1000\left(\frac{S_s C_s}{\rho_{ss}}\right) \quad (10)$$

Where; C_{sdtan} is the annual predicted sediment from EPs concentration, ρ_{ss} is the density of the suspended solid and C_s is the concentration of settleable solid

RESULTS AND DISCUSSION

Previous Studies on Removal of EPs by Aeration

Melanie *et al.* (2019) reported that the degradation of Ciprofloxacin, enoxacin, levofloxacin, lomefloxacin and norfloxacin by oxidation processes such as aeration can be described by first order kinetics. Equations 11 to 14 present expected reactions during oxidation processes

$$H_2O \ + \ h\upsilon \ \rightarrow \ H_2O^* \quad (11)$$

$$H_2O^* \rightarrow \ H^+ \ + \ OH^- \quad (12)$$

$$O_2 \ \rightarrow O^3P \ + \ O^1D \ \rightarrow \ 2^*OH \quad (13)$$

Where; O^3P is atomic oxygen in the ground energy state and O^1D is oxygen in the first excited energy state

$$H_2O \xrightarrow{e} \ H^\bullet \ + \ {}^\bullet OH \quad (14)$$

Sorption onto biomass during wastewater treatment, or into sediments when present in the riverine environment will result in removal from the aqueous phase (Petrie *et al.*, 2015). Ebele *et al.* (2017) stated that biodegradation, photodegradation and other abiotic processes such as hydrolysis transformation EPs (Blair *et al.*, 2004, Ebele *et al.* 2017), may reduce concentrations of EPs in the environment and result in partial loss and mineralization of these compounds (Alexy *et al.*, 2014, Ebele *et al.* 2017). The extent of photodegradation depends on the intensity of solar irradiation, water depth, organic matter composition, eutrophic conditions, latitude and seasonality.

It has been reported that antibiotics of loxacin and ciprofloxacin were removed by sorption during wastewater treatment due to their high affinity to solid organic matter (Petrie *et al.,* 2014c). EPs are susceptible to breakdown by photolysis. Photolysis has been shown to successfully degrade several EPs such as ketoprofen, propranolol, naproxen, E2, EE2, gemfibrozil and ibuprofen in river water (Lin and Reinhard, 2005). Half-lives ranged broadly from four minutes for ketoprofen to 15 hours for gemfibrozil and ibuprofen. This range of susceptibility to breakdown by photolysis observed is attributed to differences in their chemical structure. It can be postulated that the presence of dissolved organic matter of comparatively high concentration, as well as particulates in environmental waters will reduce EPs degradation kinetics by clouding sunlight intensity. West and Rowland (2012) found that humic acid (a small molecular weight charged species) slowed or increased degradation rate, dependant on the specific EP investigated.

Rashed *et al.* (2005), stated that photocatalysis is the combination of using a photocatalyst and UV or visible light for the treatment of wastewaters and gaseous pollutants. When illuminating a photocatalyst using UV or visible light, various organic compounds can be oxidized and mineralized at the photocatalyst surface or oxidized in solution ambient and atmospheric conditions. This is because strong oxidation and reduction sites are produced at the photocatalyst surface when this surface is illuminated with light of the appropriate wavelength. Radicals formed at the surface dissolve in solution and then react with pollutants. It has often been demonstrated that various organic pollutants can be oxidized photochemically under the influence of large amounts of free solar energy. Photocatalysis is one of the most important advanced oxidation technologies. It can be used not only for oxidative treatment of wastewater containing various organic and inorganic compounds, but also for reductive treatment such as reductive deposition of metals from wastewater.

Increased degradation in the presence of humic acid or nitrates can be attributed to indirect photolysis (Andreozzi et al., 2003). Wastewater effluents contain hydroxyl radicals and triplet excited state organic matter which facilitates indirect photolysis of some EPs (Ryan et al., 2011). Other environmental factors such as depth of river, shading from bankside vegetation, presence of particulate matter and season also require further investigation to assess their impact to EP photolysis in environmental conditions. Beyond removal at wastewater treatment plants, EPs may be removed by natural attenuation processes in the environment such as exposure to sunlight, aeration, biodegradation and adsorption onto sediments in watercourses. The removal efficiency of these processes is highly variable, depending on the properties of the contaminant and the environment. Natural attenuation is a very important mechanism, but is particularly challenging to assess because the process is site specific. It is generally documented that more compounds are greatly degraded faster and easier under aerobic (oxidation, aeration) conditions than anaerobic conditions. These reactions can be related to the different microbial activities of aeration and anaerobic microorganisms (Johnson et al., 1998; Sui et al., 2015). In sub- surface water and groundwater, microorganisms are lower numerous and diverse than those in soils. In redox conditions microorganisms' reactions and activities are generally poor (Sui et al., 2015). As such, EPs in groundwater may undergo incomplete degradation, potentially changing into hazardous metabolites, or even remain unchanged in groundwater for long periods. Redox controls were commonly observed in studies regarding EPs in groundwater. Burke et al. (2014) evaluated the redox-dependent removal of 27 wastewater-derived trace compounds by tank aeration experiments and found six compounds, including doxycycline, phenazone and propyphenazone, etc., were efficiently removed under anoxic conditions and could persist under anoxic conditions, while three antibiotics (roxithromycin, clarithromycin and clindamycin) were only removed under anoxic conditions (Sui et al., 2015). The article investigated the fate of several EPs in the hyporheic zone at two different temperatures simulating winter and summer conditions.

The results showed that a change in temperature resulted in a distinct difference of redox conditions and the attenuation of the tested EPs metoprolol, iopromide and diclofenac was more efficient in summer (Sui *et al.,* 2015). The diverse physical and chemical properties of different EPs may also result in different biodegradability. Although caffeine and paracetamol were more frequently used, they underwent more degradation during wastewater treatment and transport in the subsurface; as such they were generally detected less frequently, compared to other EPs such as carbamazepine and sulfamethoxazole, which were more resistant to degradation (Benotti and Brownawell, 2009; Sui *et al.,* 2015). Zhang *et al.* (2007) analyzed degradation of six sulfonamides in sandy soils and found that different stabilities and antibacterial activities of the sulfonamides might lead to the differences in degradation (Sui *et al.,* 2015). The environmental conditions exhibit different influences on the degradation of EPs with different properties. Sui *et al.*(2015) reported that degradation rates of chlortetracycline, tetracycline and tetracycline were significantly different at different pH and temperatures, while those of lincomycin and sulfonamides, such as sulfachlorpyridazine, sulfadimethoxine and sulfathiazole were less susceptible to changes in pH and temperature, indicating that pH-related reactions such as hydrolysis were not likely to be the removal mechanisms in ground water. Some of the equation

Modeling frameworks are poorly developed for EPs' fate – with the exception of pesticides – in the soil–water environment and require further development. The challenge is to model transport and fate of various EPs from all possible sources (urban, industrial, agriculture), through catchments to their outlet. EPs from diffuse sources are transported via the soil–water system to the sink (water body, McGuire and McDonnell, 2006). Depending on their properties, EPs can be transported by different processes, such as by runoff, erosion or leaching and enter into groundwater or surface water. They may be intercepted by the soil through adsorption or can be degraded during the transport and never reach the water bodies. Once water bodies are reached, further transport downstream in solution or attached to suspended material may occur. As EPs are "emerging", little experimental evidence is available on their transport and partitioning in catchments – and models to simulate this transport have not been parameterized for EPs except for pesticides. Little information is available on the dynamics of EPs in between the water column, sediments and the accumulation in the aquatic food chain, and on the loads from the agro environment through diffuse pollution or from urban and industrial areas through point pollution.

Amrita *et al.* (2014) reported that contamination of the urban water cycle (UWC) with a wide array of emerging organic compounds (EOCs) increases with urbanization and population density. Inorder to generate drinking water from the UWC requires close examination of their sources, occurrence, pathways, and health effects and the efficacy of wastewater treatment and natural attenuation processes that may occur in surface water bodies and groundwater. Amrita *et al.* (2014) provides in details the structure of the UWC and investigates the routes by which the water cycle is increasingly contaminated with compounds generated from various anthropogenic activities. Along with a thorough survey of chemicals representing compound classes such as hormones, antibiotics, surfactants, endocrine disruptors, human and veterinary pharmaceuticals, X-ray contrast media, pesticides and metabolites, disinfection-byproducts, algal toxins and taste-and-odour compounds, this paper provides a comprehensive and holistic review of the occurrence, fate, transport and potential health impact of the emerging organic contaminants of the UWC. Amrita *et al.* (2014) illustrated the widespread distribution of the emerging organic contaminants in the different aortas of the ecosystem and focuses on future research needs.

Edyta (2018a) compared the removal degrees of selected contaminants of emerging concern, such as H_2O_2, O_3, UV and UV/TiO_2, in water solutions during advanced oxidation processes (AOPs). The tested micropollutants belong to the following groups: pharmaceuticals, dyes, UV filters, hormones, pesticides and food additives. The highest removal rate of pharmaceutical compounds was observed during the UV/TiO_2 process. The decomposition of hormones in this process exceeded 96% and the concentration of the UV filter dioxybenzone was reduced by 75%. Toxicological analysis conducted in post-processed water samples indicated the generation of several oxidation by-products with a high toxic potential.

Edyta (2018b) compares the removal degrees of selected contaminants of emerging concern in water solutions during advanced oxidation processes (AOPs), such as H_2O_2, O_3, UV, UV/TiO_2, UV/H_2O_2, and UV/O_3. The tested micropollutants belong to the following groups: pharmaceuticals, dyes, UV filters, hormones, pesticides, and food additives. The highest removal rate of pharmaceutical compounds was observed during the UV/TiO_2 process. The decomposition of hormones in this process exceeded 96% and the concentration of the UV filter dioxybenzone was reduced by 75%. The pesticide triallat and the food additive butylated hydroxytoluene were most effectively oxidized by the UV process and their removal degrees exceeded 90%. The lowest removal degree in all examined processes was observed in the case of caffeine. Toxicological analysis conducted in post-processed water samples indicated the generation of several oxidation by-products with a high toxic potential. The presence of those compounds was confirmed by the GC-MS analysis. The performance of the UV/O_3 process leads to the increase of the toxicity of post-processed water solutions, especially solutions containing degradation by-products of carbamazepine, diclofenac sodium salt, acridine, trialatte, triclosan, and -estradiol were characterized by high toxicity.

Previous Studies on Models of Aeration

Oxidation with ozone, oxygen or hydrogen peroxide has been found to be an important alternative to chlorination, because the oxidation does not result in toxic chlorinated organic compounds. In streams and rivers dissolved oxygen and aeration rate are main sources of oxidation. Dissolved oxygen concentrations are the result of an interaction among some processes (sources and sinks, which affect the DO concentration). The first processes (atmospheric reaeration and photosynthetic production, tend to increase oxygen levels) and the second processes (oxidation of carbonaceous and nitrogenous waste material, respiration of aquatic plants, and sediments oxygen demand (SOD)), produce a decrease of DO concentration. Each of these phenomena can be expressed through a different kinetic expression to quantify their effects. Many theories have been proposed and a large number of techniques and equations have been used to estimate the reaeration coefficient (K_2). There are three basic techniques for measuring the reaeration coefficient of a stream. The techniques are as follows:

a) Dissolved Oxygen balance techniques
b) Distributed equilibrium technique
c) Tracer technique

A review of the estimation equations revealed that they can be subdivided into four general groups and represented as follows:

a. a. Group 1 (dimensional coefficient):

$$K_2 = K\left(V^x\right)H^y \tag{15}$$

where V is the stream velocity; x and y are constant exponents; and K is a dimensional coefficient.

b. b. Group 2 (dimensional coefficient):

$$K_2 = K\left(V^x\right)H^y S^z \tag{16}$$

where S is the stream slope, z is constant exponent.

c. c. Group 3 (dimensionless coefficient):

$$K_2 = 2.911\,x\,10^{-2}\left(V_s^3\right)V^{-2}H^{-1} \tag{17}$$

$$K_2 = 2.88\,x\,10^{-4}\left(\left(1+F\right)V_s\right)H^{-1} \tag{18}$$

where F is the Froude number.

d. d. Group 4 (dimensionless coefficient):

$$K_2 = g^{0.333}D_f^{0.667}R_m^{-0.667}\left(S^{0.333}\right)V^{0.333}\left(\left(2g_s\right)^{0.333}LH\right)^{-1} \tag{19a}$$

$$\text{where } g_s = \rho^{-3}\mu^{-4}T_s^3 \tag{19b}$$

$$L = \rho\mu^2 T_s \tag{19c}$$

Where, D_f is the molecular diffusion coefficient, T_s is the surface tension, μ is the water kinematic viscosity and ρ is the density of water

Some of the models for estimation of reaeration coefficients are Expressed by equations (20) to (29) (King 1970, Adedokun and Agunwamba, 2013):

Chezy model (non- isotropic, Asani and Oke, 2018):

$$K_2 = 480\frac{D_m^{0.5}S^{0.25}}{H^{1.25}} \tag{20}$$

Chezy model (isotropic, Asani and Oke, 2018):

$$K_2 = 127\frac{D_m^{0.5}V^{0.5}}{H^{1.5}} \tag{21}$$

Bennett and Rathbun (1972) model:

$$K_2 = 0.12\frac{b_1 aH^{\left(\frac{3}{8}\right)}\,\mathrm{Coth}\left(\frac{\left(bH^{\left(\frac{3}{8}\right)}\right)}{Hc^{0.5}}\right)^{0.5}}{Hc^{1.5}} \tag{22}$$

Where

$$b_1 = 1 + F^2; \tag{22a}$$

$$c = 0.9 + F \tag{22b}$$

$$a = 9.68 + 0.054(T - 20) \tag{22c}$$

$$b = 0.976 + 0.0137(30 - T)^{1.5} \tag{22d}$$

$$H = 30SV \tag{22e}$$

$$F = \frac{V}{\sqrt{gH}} \tag{22f}$$

Krenkel – Orlob model (Asani and Oke, 2018):

$$K_2 = 7.235(VS)^{0.408} H^{-0.66} \tag{23}$$

Tsivoglou – Wallace model (Ellis, 1998):

$$K_2 = 556(VS) \tag{24}$$

Parkhurst – Pomeroy model (Asani and Oke, 2018):

$$K_2 = 0.959\left(1+0.17F^2\right)(VS)^{0.375} H^{-1.0} \tag{25}$$

Churchill *et al.* (1962) model (Asani and Oke, 2018):

$$K_2 = 0.0009\left(V^{2.695} S^{0.823}\right) H^{-3.085} \tag{26}$$

Owens *et al.* (1964) model (Ellis, 1998):

$$K_2 = 0.289(V)^{0.72} H^{-1.75} \tag{27}$$

Gameson *et al.* (1955) model (Ellis, 1998):

$$K_2 = 0.222(V)^{0.67} H^{-0.185} \tag{28}$$

Isaac and Gaudy (1968) model (Ellis, 1998):

$$K_2 = 0.198(V) H^{-1.5} \tag{29}$$

Previous Studies on Aeration as Self- Purification in Rivers

Water self-purification is a complex process involving physical, chemical and biological processes that occur simultaneously, allowing a river can recover its natural state over a certain distance. Generally, self-purification is a biochemical transformation of pollutants (containing mainly organic compounds) into simpler forms, often inorganic, that have microorganism activity; this is done at the expense of the consumption of oxygen from the air and water and through sedimentation processes that cause the physical elimination of water pollution. The subject of the self-purification of streams divides itself naturally into two parts: that relating to the disappearance or oxidation of certain chemical constituents of sewage, and that relating to the disappearance of the sewage bacteria. Depuration comprises different mechanisms such as dilution, sedimentation, reaeration, adsorption, absorption, and chemical and biological reactions, and can be evaluated by the same mathematical model. The most important factors influencing a river's self-purification ability include, among others, topography, soil type, vegetation, the hydraulic characteristics of the river, the retention time of water in the catchment, biodiversity and temperature.

Shimin *et al,* (2011) indicated that the Juma River has a great ability to purify the organic pollutants such as TN and TP, but little ability to the heavy metal ions. By analyzing the pollutant absorption ability of the three aquatic plants, Myriophyllum spicatum has greatest ability to absorb the organic pollutants. Ifabiyi (2008) reported that water chemical parameters of ogbe river vary in both relative and absolute terms. The extent of variability suggests a need for a close monitoring of the chemical parameters for the purposes of water resource management

Karthiga *et al.(*2017) stated that Self-Purification capacity of Bhavani River was tested. It was concluded that the turbidity was high due to over release of domestic and industrial waste in to the water. Yonik and Imas (2017) reported that almost all of the rivers in Indonesia have Biochemical Oxygen Demand (BOD) and Chemical Oxygen Demand (COD) concentrations ratio less than 0.2, indicating the no-biodegradation capability. Existence of metals can also inhibit the biodegradation process. The water quality data showed significant increase of metals (Copper and Zinc) in several rivers. Surfactant was also observed in extremely high concentration. Thus, most of the Jakarta's rivers have relatively low capability in biodegradation capacity and self purification capability. Although, some year data shows an improvement, yet the BOD/COD ratio is still in the range of slow of biodegradation capacity. Yonik *et al.* (2018) showed that the value of urban river deoxygenation rate is relatively low. The low value of the rate of deoxygenation led to the difficulty in rivers to purify themselves. Pollutants inhibiting the process include phenol, detergent, and heavy metals, which are contaminating the river over the maximum standard. Biologically, the decomposer consists in the river is few. It leads to a slow organic degradation rate. The condition is representing the low capacity of self-purification of the Cikapundung River which is caused mainly by chemical and biological parameters.

Mala and Maly (2009) stated that there is a strong inhibitory effect of heavy metals on self purification of river. Effect ranges from 50 to 80% was observed for Cd (from 0.2 to 0.7 $mg.l^{-1}$), Cu (from 0.4 to 1.0 $mg.l^{-1}$), Be (from 0.5 to 2.0 $mg.l^{-1}$), Zn (from 1.2 to 5.0 mg. $mg.l^{-1}$), SnII (from 2.2 to 4.2 $mg.l^{-1}$), and CrVI (above 4 $mg.l^{-1}$). AsIII, AsV, MnII, SeIV and SeVI did not cause any BOD inhibition to 2 $mg.l^{-1}$ (As) res. 5 $mg.l^{-1}$ (other metals). The inhibitory effect of metals decreased significantly or disappeared with increasing incubation time. Being observed with all metals with toxic effects except for Be, SnII, and CrIII, the decrease was more intense at the beginning of the incubation period. Evy et al (2017) reported that the highest rate of carbon deoxygenation in rivers in Malang, Indonesia was 0.63 / day and the smallest was 0.48/day. The carbon deoxygenation rate in Metro River is higher than Brantas River of Malang City, Upper Citarum River, Cikapundung River and Citepus River. This shows that Metro River has better self-purification capability so that its quality is better. Trang *et al.* (2015) stated that the main processes that dominate the self-purification of Thi Nai lagoon are materials exchange, sedimentation, dispersion, decompose, and photosynthesis, respectively. The pollution loads discharged everyday into the Thi Nai lagoon is high, reducing the self-purification of lagoon, therefore it need to be controlled restrict.

In Nigeria, Badejo *et al.* (2018) reported that de-oxygenation constant, re-aeration constant and autopurification factor of Ona river indicated that the river is a swift stream. A predicted critical DO deficit of 8.01 mg/L was calculated while the measured DO deficit was 3.61 mg/l occurring at the point of discharge. The river Ona was able to purify comparatively the pollution load discharged into it.

Present Study on River Ogbe

Ogbe River is located at Ile-Ife, Ogbe River is one of the tributes of Opa River, which contribute water to Opa Dam in Obafemi Awolowo University, others tributaries are esinmirin and amota. Figure 4 shows detail of the tributaries. Ogbe River is a perennial water body that runs through a highly polluted new market and Mokola street in Ile-Ife. The river serves as a source of water for irrigation agriculture along the river.

Table 2a presents quality of water in the river along the river in the wet and in the dry seasons. From the table concentration of total solids ranges from 0.04 to 0.45 mg/l in the wet season and 0.34 to 1.06 mg/l in the dry season. Total solid concentration is a measure of the dissolved solid, volatile and non-volatile solids, suspended solids in wastewater, effluent, or water bodies. The amount of solids in wastewater is frequently used to describe the strength of the waste. The more solids present in a particular wastewater, the stronger that wastewater will be. If the solids in wastewater are mostly organic, the impact on a treatment plant is greater than if the solids are mostly inorganic. Normal domestic wastewater contains a very smaller amount of solids when compared to the amount of water that carries it, generally less than 0.1% (10 mg/l). This can be misleading, however, because it may take only a very small amount of organic residue to create large pollution problems. The number and severity of pollution problems will depend on the type of solids that are involved. Also results of suspended solid concentrations are presented in Table 2a. From the table suspended solid concentrations varied from 0.02 mg/l to 0.08 mg/l during the wet season and 0.02 to 0.06 mg/l during the dry season. Figures 6, 7 and 8 present view of selected EPs material on stream.

Dissolved oxygen has been a major constituent of interest in water-quality investigations. It is significant in the protection of aesthetic qualities of water as well as for the maintenance of fish and other aquatic life. The aesthetic qualities of water require sufficient dissolved oxygen to prevent septic conditions and attendant malodorous emissions. Insufficient dissolved oxygen in the water column causes the anaerobic decomposition of any organic materials present in the wastewater, which tends to cause the formation of noxious gases. In the Ogbe River, DO concentrations were in the range of 0.80 mg/l to 6.0 mg/l for dry season. In the wet season higher DO concentrations were obtained.

At some locations, dissolved oxygen concentrations were higher than 2.0 mg/l, the minimum standard for any class of surface water (Hammett, 1990). These results agree with literature such as Longe and Omole (2008); Ifabiyi (2008); Emeaso (2011); Ismaila *et al.,* (2012), Yotinov *et al.* (2014), Omole *et al* (2012); Viswanathan *et al.* (2015). Reaeration (Reoxygenation) rate ranged from 0.043 to 0.564 d^{-1} in the dry season and from 0.058 to 2.121 d^{-1} during the wet season. The mean value of reoxygenation rate for the dry and the wet seasons are 0.290 and 0.590 (d^{-1}) respectively. This result indicates that Ogbe River can be classified as a large stream of low velocity during dry season and as a large stream of normal velocity during raining season according to David (2000). As reported by Punmia *et al.*(1998), speed plays a major role in self- purification, the result of the parameters indicate that, the self-purification was higher in the wet season than in the dry season due to the high velocity in the wet season with re-oxygenation rate of 0.590(d^{-1}).

Evaluation of Aeration Models

Table 2b presents field reaeration rate and simulated reaeration rate for both wet and dry seasons respectively. From the tables, the highest and the lowest values of K_2 came from Tsivoglou and Wal-

Figure 4.

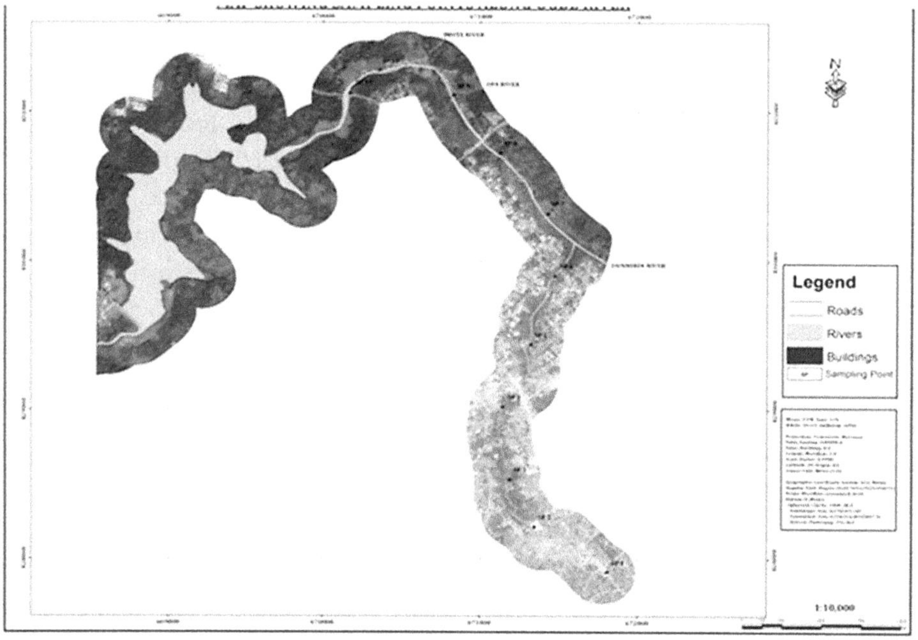

lace, and the new proposed models respectively. This result indicates that K_2 is a function of model of simulation and season. Statistical analysis of these K_2 revealed that sampling points, models and season have effects on K_2 (Tables 3a, b and c). Table 3a presents statistical evaluation of the reaeration rate in the wet season. Table 3b shows statistical evaluation of the reaeration rate in the dry season and Table 3c presents statistical evaluation of the reaeration rate in the combination of both dry and wet seasons. Table 4 presents effect of seasons on the values of K_2 (both simulated and field data). Further analysis using ANOVA shows that the models ($F_{7, 70} = 8.95$, $p = 7.84 \times 10^{-8}$ in the wet season, $F_{7, 70} = 14.84$, p

Figure 5 shows satellite image of Ogbe river

Figure 6.

= 1.05 x 10 $^{-11}$ in the dry season, $F_{7, 147}$ = 23. 63, p = 2.67 x 10 $^{-21}$ in the combination of both wet and dry seasons) and season ($F_{1, 7}$ = 9.05, p = 0.020) have significant effects on K_2 at 95% confidence level. Table 5 shows statistical evaluations (MSC, SC and AIC) of the models. The table revealed that MSC ranges from 0.597 to -13.815, -16.327 to – 0.657 and -14.337 to -0.573 for wet season, dry season and combination of both dry and wet seasons respectively. The values of SC calculated are between 12.164 and 170.697, -4.293 and 168. 080, and 26.712 and 178.115 for wet season, dry season and combination of dry and wet seasons, respectively. AIC is between 10.970 and 169.504, 25.519 and 176.921 for wet season, dry season and combination of both dry and wet seasons respectively. From the table new proposed models has the highest MSC, the lowest AIC and SC in all the cases (dry and wet season, combination of both wet and dry seasons). These results indicate that the order of accuracy is: the new

Figure 7.

Figure 8.

proposed model > Adedokun and Agunwamba > IHP > Oconnor and Dobbbins > Churchill *et al* = Streeter *et al* > Tsivogluo – Wallace based on the values of MSC, SC and AIC.

CONCLUSION

It can be concluded based on the study that:

i. the presence of EPs wastes had effects on the self- purification of the stream
ii. the volume of sediments as suspended and floating solid were in river ogbe more than necessary,
iii. the main processes that dominate the self-purification of rivers are materials exchange, sedimentation, dispersion, decompose, and photosynthesis, respectively.
iv. accuracy of the models in reaeration computation is in the order of new proposed model > Adedokun and Agunwambe > IHP > Oconnor and Dobbbins > Churchill *et al* = Streeter *et al* > Tsivogluo – Wallace based on the values of MSC, SC and AIC

FUTURE RESEARCH DIRECTIONS

In order to guide the development of rational water management practices that are protective of human and ecological health, a significant research effort is needed focusing on the following key issues.

a. Extensive and complete occurrence data on different EOCs in drink-ing water required to identify those that reach the drinking water at potentially significant concentrations. EOC behavior in different UWCs should be studied using consistent approaches.
b. Human health effects of continuing low level exposure to EOCs should be studied with emphasis on those that could enter the food chain indirectly through the consumption of water.
c. Long-term ecotoxicological effects of EOCs and EOC mixtures to wild-life and ecosystem structure and dynamics need to be researched.
d. EOC behavior needs to be monitored in the UWC. This should include compounds which have only been recently discovered, such as, steroidal drugs, anti-psychotic drugs and banned medications, and compounds with a potential of being a contaminant to the UWC inthe future such as illicit drugs, melamine or melamine resins or chemicals used in textile dyes.
e. Sources and entry points of EOCs into the UWC requires to be identified. Such studies should quantify mass flow, identify sinks, and removal mechanisms and transformation products
f. For products that are difficult to control, ecologically friendly alternatives should be found.
g. The UWC should be re-engineered to minimize input and maximize natural attenuation. This could include the incorporation of natural attenuation processes into the UWC, such as specially designed wetlands and reservoirs for the removal of EOCs.
h. Wastewater and drinking water treatment technologies need to be evaluated with respect to their removal of EOCs and technologies need to be developed that are capable of removing a broad spectrum of EOCs while minimizing the formation of treatment by-products.
i. Water quality models should include fate and transport models to evaluate relevant EOCs and their impacts on ecosystems and the environment. Development of such models requires an understanding of kinetics and transformation processes which is still lacking for many EOCs.

REFERENCES

Adedokun, T. C., & Agunwamba, J. C. (2013). Modelling the Effect of Industrial Effluents on Water Quality: "A Case Study of River Challawa in Nigeria. *International Journal of Modern Engineering Research.*, *3*(5), 2797–2803.

Akaike, H. (1976). An information criterion (AIC). *Math. Sci*, *14*, 1–5.

Alexy, Sch€oll, Kümpel, & Kümmerer. (2004). What do we know about antibiotics in the environment? In K. Kümmerer (Ed.), *Pharmaceuticals in the Environment, 2004*. Springer Berlin Heidelberg.

American Public Health Association (APHA). (1998). *Standard methods for Examination of water and waste water* (20th ed.). American Public Health Association Inc.

Amrita, P., Yiliang, H., Martin, J., Martin, R., & Karina, Y. (2014). Emerging contaminants of public health significance as water quality indicator compounds in the urban water cycle. *Environment International, 71*, 46–62. doi:10.1016/j.envint.2014.05.025 PMID:24972248

Andreozzi, R., Caprio, V., Ciniglia, C., De Champdore, M., Lo Giudice, R., Marotta, R., & Zuccato, E. (2004). Antibiotics in the environment: Occurrence in Italian STPs, fate, and preliminary assessment on algal toxicity of amoxicillin. *Environmental Science & Technology, 38*(24), 6832–6838. doi:10.1021/es049509a PMID:15669346

Badejo, A., Adekunle, A., Adeegbe, T., & Nwosu, G. (2018). Auto-Purification Response Of Ona River, Ibadan To Industrial And Domestic Effluent Discharge. FULafia Journal of Science & Technology, 4, 74 – 78.

Blair, B. D. J. P., Crago, J. P., Hedman, C. J., & Klaper, R. D. (2013). Pharmaceuticals and personal care products found in the Great Lakes above concentrations of environmental concern. *Chemosphere, 93*(9), 2116–2123. doi:10.1016/j.chemosphere.2013.07.057 PMID:23973285

Burke, V., Richter, V., Hass, D., Duennbier, U., Greskowiak, J., & Massmann, G. (2014). Redox-dependent removal of 27 organic trace pollutants: Compilation of results from tank aeration experiments. *Environmental Earth Sciences, 7*(8), 3685–3695. doi:10.100712665-013-2762-8

Chang, S., & Yeh, W. W.-G. (1973). Optimal allocation of artificial aeration along a polluted stream using dynamic programming. *Water Resources Bulletin, 9*(5), 985–997. doi:10.1111/j.1752-1688.1973.tb05824.x

David, A. C. (2000). *Water resource engineering*. Prentice-Hall.

Davidson, B., & Bradshaw, R. W. (1970). A steady state optimal design of artificial induced aeration in polluted streams by the use of Pontryagin's Minimum Principle. *Water Resources Research, 6*(2), 383–397. doi:10.1029/WR006i002p00383

Ebele, A. J., Mohamed, A. A., & Stuart, H. (2017). Pharmaceuticals and personal care products (PPCPs) in the freshwater aquatic environment. *Emerging Contaminants, 3*(1), 1–16. doi:10.1016/j.emcon.2016.12.004

Edyta, K. (2018). Decomposition of Contaminants of Emerging Concern in Advanced Oxidation Processes. *Proceedings, 2*, 180. doi:10.3390/ecws-2-04949

Emeaso, O. I. (2011). *Pollution and self- purification Studies of River Romi, Chikun Local government area Kaduna State* (Unpublished M.Sc. Thesis). Department of Water Resources and Environmental Engineering, Ahmadu Bello University, Zaria.

Evy, H., Candra, D. W., & Evelin, N. (2017). River Water Quality Performance From Carbondeoxygenation Rate. *International Journal of Engineering and Management, 1*(1), 27–34.

Geissena, V., Hans, M., Erwin, K., Günter, U., & Marti, N. (2015). Emerging pollutants in the environment: A challenge for water resource management. *International Soil and Water Conservation Research*, *3*(1), 57–65. doi:10.1016/j.iswcr.2015.03.002

Greenop, B., Lovatt, K., & Robb, M. (2001). The use of artificial oxygenation to reduce nutrient availability in the Canning River, Western Australia. *Water Science and Technology*, *43*(9), 133–144. doi:10.2166/wst.2001.0524 PMID:11419121

Gualtieri, C., & Gualtieri, P. (1997). *Hydrodynamic model for reaeration estimation in surface waters* (in Italian). AIMETA '97, Siena, Italy.

Gualtieri, C., & Gualtieri, P. (1999). Statistical analysis of reaeration rate in streams. *International Agricultural Engineering Conference (ICAE).*

Gualtieri, C., & Gualtieri, P. (2000). Field verification for a reaeration model in streams. *4th International Conference On Hydroscience & Engineering.*

Hunter, J.F., & Whipple, W., Jr. (1970). Evaluating instream aerators of polluted river. *Journal - Water Pollution Control Federation*, *42*, 249–262.

Ifabiyi, I. P. (2008). 'Self-purification of a Freshwater Stream in Ile-Ife'. Lessons for Water Management J. *Human Ecology*, *24*(2), 131–137. doi:10.1080/09709274.2008.11906109

Ismaila, S. M., Adie, D. B., Okuofu, C. A., & Ismail, A. (2012). Modelling of Water Quality of Bindare Stream, Zaria, Nigeria Using MATLAB. *Proceeding of the OAU Faculty of Technology Conference 2012.*

Jr, Whipple, W.,., Coughlan, F.P., & Yu, S.L. (1971). Instream aerations for polluted rivers. *Journal of the Sanitary Engineering Division*, *96*, 1153–1165.

Jr., & , Whipple, W.,Yu, S.L. (n.d.). Aeration system of large navigable rivers. *ASCE Journal of Sanitary Engineering Division, 97*, 883-901.

Karthiga, D. M., Mano, B. E., Geethamani, R., & Abinaya, S. (2017). Self-purification capacity of Bhavani River, India. *Research Journal of Engineering Sciences.*, *6*(3), 1–9.

King, D. L. (1970). *Reaeration of Streams and Reservoirs Analysis and Bibliography*. Hydraulics Branch Division of General Research Engineering and Research Center Denver.

Lin, A. Y.-C., & Reinhard, M. (2005). Photodegradation of common environmental pharmaceuticals and estrogens in river waters. *Environmental Toxicology and Chemistry*, *24*(6), 1303–1309. doi:10.1897/04-236R.1 PMID:16117104

Longe, E. O., & Omole, D. O. (2008). Analysis of Pollution Status of River Illo, Ota, Nigeria. *The Environmentalist*, *28*(4), 451–457. doi:10.100710669-008-9166-4

Mala, J., & Maly, J. (2018). Effect Of Heavy Metals On Self-Purification Processes In Rivers. *Applied Ecology and Environmental Research*, *7*(4), 333–340.

Map, G. (2014). *Imagery @ 2014 ones/spotimage, Digital Landsat' internet access*. Available at http://wn.Map.harvard.edu/maps/Ile-Ife

McGuire, K. J., & McDonnell, J. J. (2006). A review and evaluation of catchment transit time modelling. *Journal of Hydrology (Amsterdam)*, *330*(3–4), 543–563. doi:10.1016/j.jhydrol.2006.04.020

Melanie, V., Indra, B., Anna, N., & Martin, J. (2017). Comparison of UVA and UVC irradiation of FQs and their MICs for sustainable water treatment. *Conference: 2nd Green and Sustainable Chemistry Conference at Berlin.* Retrieved from https://www.researchgate.net/publication/316977230

Melching, C. S., & Flores, H. E. (1999). Reaeration equations derived from U.S. Geological Survey database. J. Environmental Engineering Division, 125(5), 407-414.

Mercedes, V. (2017). *Emerging Pollutants Removal in Wastewater Treatment Plants: A review and their implications in a river basin in Uruguay* (M.Sc thesis). UNESCO-IHE Institute for Water Education, Delft, the Netherlands.

NORMAN. (2019). *Network of reference laboraties, research centres and related organisations for Monitoring of emerging environmental substances.* Retrieved from https://www.norman-network.net/?q=Home

Oke, Ismail, Lukman, Ojo, Adeosun, & Nwude. (2016). Economics and Statistical Evaluations of Using Microsoft Excel Solver in Pipe Network Analysis. *J. Pipeline Syst. Eng.*, *7*(4).

Oke, I. A., Amele, S. O., Lukman, S., Ismail, A., & Foghi, P. O. (2016). Water quality and self purification properties of Ogbe River, Ile-Ife, Nigeria. *Journal of Environmental Design and Management*, *8*(1 & 2), 174–189.

Omole, D. O., Adewumi, I. K., Longe, E. O., & Ogbiya, A. S. (2012). Study of Auto-Purification Capacity of River Atuwara in Nigeria. *Environmental Research Journal*, *6*(3), 151-157.

Omole, D. O., & Longe, E. O. (2012). Re-aeration Coefficient Modeling: A Case Study of River Atuwara in Nigeria. *Research Journal of Applied Sciences, Engineering and Technology*, *4*(10), 1237–1243.

Petrie, B., Barden, R., & Kasprzyk-Hordern, B. (2015). A review on emerging contaminants in wastewaters and the environment: Current knowledge, understudied areas and recommendations for future monitoring. *Water Research*, *72*, 3–27. doi:10.1016/j.watres.2014.08.053 PMID:25267363

Petrie, B., McAdam, E. J., Hassard, F., Stephenson, T., Lester, J. N., & Cartmell, E. (2014a). Diagnostic investigation of steroid estrogen removal by activated sludge at varying solids retention time. *Chemosphere*, *113*, 101–108. doi:10.1016/j.chemosphere.2014.04.051 PMID:25065796

Petrie, B., McAdam, E. J., Lester, J. N., & Cartmell, E. (2014b). Assessing potential modifications to the activated sludge process to improve simultaneous removal of a diverse range of micropollutants. *Water Research*, *62*, 180–192. doi:10.1016/j.watres.2014.05.036 PMID:24956600

Petrie, B., McAdam, E.J., Lester, J.N., & Cartmell, E. (2014c). Obtaining process mass balances of pharmaceuticals and triclosan to determine their fate during wastewater treatment. *Sci. Total Environ.*, *497 – 498*, 560 -568.

Petrie, B., McAdam, E. J., Whelan, M. J., Lester, J. N., & Cartmell, E. (2013c). The determination of nonylphenol and its precursors in a trickling filter wastewater treatment process. *Analytical and Bioanalytical Chemistry*, *405*(10), 3243–3253. doi:10.100700216-013-6765-9 PMID:23388691

Punmia, B. C., Arun, K. J., & Ashok, K. J. (1998). *Waste water Engineering (including air pollution)* (2nd ed.). New Delhi: Laxmi Publication LTD.

Rashed, I. G., Hanna, M. A., El-Gamal, H. F., Al-Sarawy, A. A., & Wali, F. K. M. (2005). *Overview On Chemical Oxidation Technology In Wastewater Treatment*. Ninth International Water Technology Conference, IWTC9 2005, Sharm El-Sheikh, Egypt.

Rinaldi, S., & Soncini-Sessa, R. (1978). Optimal allocation of artificial instream aeration. Journal of the Environmental Engineering Division. *Proceedings of the American Society of Civil Engineers*, *104*, 147–160.

Ryan, C. C., Tan, D. T., & Arnold, W. A. (2011). Direct and indirect photolysis of sulfamethoxazole and trimethoprim in wastewater treatment plant effluent. *Water Research*, *45*(3), 1280–1286. doi:10.1016/j.watres.2010.10.005 PMID:21044793

Shimin, T., Zhaoyin, W., & Hongxia, S. (2011). *Study on the Self-purification of Juma River. Procedia Environmental Sciences*, *11*, 1328–1333.

Stefano, M., & Elisabetta, G. (2008). Water quality modelling for small river basins. *Environmental Modelling & Software*, *23*(4), 451–463. doi:10.1016/j.envsoft.2007.06.008

Sui, Q., Xuqi, C., Shuguang, L., Wentao, Z., Zhaofu, Q., & Gang, Y. (2015). Review article Occurrence, sources and fate of pharmaceuticals and personal care products in the groundwater: A review. *Emerging Contaminants*, *1*(1), 14–24. doi:10.1016/j.emcon.2015.07.001

Tarassov, V. J., Perlis, H. J., & Davidson, B. (1969, June). Optimization of a class of river aeration problems by the use of multivariable distributed parameter control theory. *Water Resources Research*, *5*(3), 563–573. doi:10.1029/WR005i003p00563

Trang, C., Thi, T., Dieu, L. V., Thanh, T. D., & Sinh, L. X. (2015). Assessment of Self-Purification Process of Thi Nai lagoon (Binh Dinh Province, Viet Nam). *Environment and Natural Resources Research*, *5*(3), 19–27.

UN WWAP. (2003). *Water for people, water for life (2003): 3rd World Water Forum in Kyoto, Japan*. Retrieved from www.norman-network.net

Vié, J.-C., Hilton-Taylor, C., & Stuart, S. N. (2009). *Wildlife in a changing world–an analysis of the 2008 IUCN Red List of Threatened Species*. Gland.

Viswanathan, Anitha, Amuthan, Rajesh, Veilumuthu, & Narayanan. (2015). Studies on the Role of Bacteria in self- purification of the River Tamirabarani. *Europ. J. Appl. Sci.*, *7*(1), 1-8.

West, C. E., & Rowland, S. J. (2012). Aqueous photo-transformation of diazepam and related human metabolites under simulated sunlight. *Environmental Science & Technology*, *46*(9), 4749–4756. doi:10.1021/es203529z PMID:22489744

Yonik, M. Y., & Imas, K. (2017). Investigation on the biodegradation capacity of urban rivers in Jakarta, Indonesia. *International Journal of GEOMATE, 12*(34), 45–50.

Yonik, M. Y., Mia, N., Neneng, S., & Annisa, N. (2018). Influencing parameter of self purification process in the urban area of Cikapundung River, Indonesia. *International Journal of GEOMATE, 14*(43), 50–54.

Yotinov, I., Lincheva, S., & Topalova, Y. (2014). Vertical fl uctuations of key parameters of the self-purification in the "Middle Iskar" cascade's reservoir Lakatnik. *Bulgarian Journal of Agricultural Science*, (Supplement 1), 113–117.

Yu, S. L. (1970). Aeration performance in natural streams. *Journal of the Sanitary Engineering Division, 96*, 1099–1113.

Zhang, C., Liang, Y., & Wang, F. (2007). Microbial degradation of sulfonamides in soils. *Nongye Huanjing Kexue Xuebao, 26*, 1658–1662.

ADDITIONAL READING

Alp, E., & Melching, C. S. (2011). Allocation of supplementary aeration stations in the Chicago waterway system for dissolved oxygen improvement. *Journal of Environmental Management, 92*(6), 1577–1583. doi:10.1016/j.jenvman.2011.01.014 PMID:21310524

Harvey, J. (1995). Mathematical Programming Models for Environmental Quality Control. *Operations Research, 43*(4), 578–622. doi:10.1287/opre.43.4.578

Leu, S.-Y., Chan, L., & Stenstrom, M. K. (2012). Toward Long Solids Retention Time of Activated Sludge Processes: Benefits in EnergySaving, Effluent Quality, and Stability. *Water Environment Research, 84*(1), 42–53. doi:10.2175/106143011X12989211841052 PMID:22368826

Miguel, A. (2016). Removal of micropollutants through a biological wastewater treatment plant in a subtropical climate, Queensland-Australia. *Journal of Environmental Health Science & Engineering, 14*(1), 14. doi:10.118640201-016-0257-8 PMID:27822379

Visanji, Z., Sadr, S. M. K., & Memon, F. A. (2018). An Implementation of a Decision Support Tool to Assess Treatment of Emerging Contaminants in India. *Journal of Water Resource and Protection, 10*(04), 422–440. doi:10.4236/jwarp.2018.104023

KEY TERMS AND DEFINITIONS

AIC: Akaike information criterion.
AOTs: Advanced oxidation technologies.
BOD: Biochemical oxygen demand.
COD: Chemical oxygen demand.
DHT: Dihydrotestosterone.
DO: Dissolved oxygen.
DOP: Dioctyl phthalate.
EOCs: Emerging organic compounds.
EPs: Emerging pollutants.
MSC: Model of selection criterion.
NTP: Non-thermal plasmas.
PBDEs: Polybrominated diphenyl ethers.
TN: Total nitrogen.
TP: Total phosphorous.
UWC: Urban water cycle.

Chapter 13
Fabrication of Metal Oxide–Biopolymer Nanocomposite for Water Defluoridation

Dickson Mubera Andala
https://orcid.org/0000-0002-3311-2469
Multimedia University of Kenya, Kenya

Erick Mobegi
Kenyatta University, Kenya

Mildred Nawiri
Kenyatta University, Kenya

Geoffrey Otieno
Technical University of Kenya, Kenya

ABSTRACT

Fluoride contamination in groundwater affects about 150 million people worldwide. In this study, the authors focused on synthesizing biopolymer metal oxide nanocomposite for fluoride removal. Nanocomposite material was done using SEM. As(V), Al, Ti, Zr, and Fe water samples were analysed by ICP-MS (inductively coupled plasma-mass spectrometry). Fluoride level was determined using the standard method – Ion-Selective Electrode method. Preliminary results indicate arsenic (V) removal was below the 10 ppb and fluoride less than 1.5 ppm as prescribed by WHO. The removal efficiency was after 60-70 minutes with recyclability of 11 cycles. The nanocomposite worked well in all pH ranges 6.5-8.5. A filter cartridge biopolymer metal oxide nanocomposite constituting of template aluminium homogenized in the aggregated network of chitosan was developed as an adsorbent for fluoride from the water with better adsorption limit.

DOI: 10.4018/978-1-7998-1871-7.ch013

INTRODUCTION

Fluoride is beneficial to human health. However, levels exceeding 1.5 mg/L in drinking water is detrimental to human health. For example, occurrence of fluorosis. In addition, exposure to high concentration of fluoride for a long period of time may lead to kidney, brain and liver damage. Drinking water with elevated levels of fluoride exceeding 1.5 ppm is a serious health challenge globally. Fluoride has been associated with fluorosis and bone diseases.

Interventions such as reverse osmosis and adsorption have been made to reduce fluoride contamination in drinking water. However, some of these solutions are expensive for the poor and cannot be used in areas with no electrical power. With the emerging field of nanotechnology, nano-adsorbents have received considerable attention as an alternative solution to water purification. However, some of the nanomaterials are either powders or fine particles that render separation difficult due to blockage in columns, poor chemical and mechanical stability. This renders their application in point-of-use devices ineffective.

Therefore, the main aims of this study was to synthesize and investigate the potential of a nano-sized granulated biopolymer-mixed metal oxide nanocomposite in reduction of fluoride in drinking water and fabricate a point-of-use water filter device for house-hold and commercial scale. The specific objectives of this study were:

i. To synthesize and characterize chitosan – Aluminum, iron and titanium and zirconium nanocomposite.
ii. To evaluate the reduction efficiency of the nanocomposite for fluoride in drinking water.
iii. To investigate the influence of pH, competing ions, adsorbent dose, initial concentration contact time on the nanocomposite performance.

BACKGROUND

Occurrence of Fluoride

Fluoride is a compound derived from fluorine. Fluorine does not exist in elemental form but acquires the negative charge to form fluoride, which exists in many rocks and minerals (Jadhav and Jadhav, 2014). It is the 13th most abundant element in the earth's crust and in the halogen family, it is the most electronegative and reactive (Waghmare *et al*, 2015). Fluoride in the earth's crust exists in granite, limestone and sandstone rocks. Fluorspar (CaF_2), sellaite (MgF_2), fluoroapatite ($Ca_5(PO_4)_3F$) and cryolite (Na_3AlF_6) are the main forms in which fluoride occurs (Tomar and Kumar, 2013). In cryolite form it is present in igneous rock whereas in fluorspar it is mainly found in sedimentary rocks (Mohapatra *et al.,* 2009). Other forms in which fluoride can exist are topaz ($Al_2F_2SiO_4$), amblygonite (Li(F, OH) $AlPO_4$) and apophylite ($KFCa_4(Si_8O_{20})$ $8H_2O$ (Dongre *et al.,*2012). Table 1 below summarizes percentage fluorine in different fluoride bearing minerals.

Distribution of Fluoride in Water

Fluoride in drinking water is a public health concern worldwide. The East African Rift Valley, a region that runs from Jordan valley down through Sudan, Ethiopia, Uganda, Kenya and Tanzania, is reported to

Table 1. Fluoride Bearing Minerals

Mineral	Chemical Formula	% Fluorine
Sellaite	MgF_2	61
Villiamite	NaF	55
Fluorite (Fluorspar)	CaF_2	49
Cryolite	Na_3AlF_6	45
Fluoroapatite	[3Ca3(PO4)2Ca(F,Cl2)]	34

Source: (Awasti *et al.*, 2002)

be a high fluoride Zone. Fluoride occurrence in both ground water and surface water is ascribed to the geochemical reaction in the aquifer and in some cases anthropogenic influence (Habuda-Stanić *et al.*, 2014: Adak *et al.,* 2016). During the process of weathering these minerals in conjunction with volcanic processes increase the concentration of fluoride in ground water through slow dissolution (Josh *et al.,* 2012; Adak *et al.,* 2016).

Fluoride occurrence and levels in groundwater and surface water depends contributing factors such as pH, alkalinity, total dissolved solids (TDS), hardness and the geochemical composition of aquifers (Habuda-Stanić *et al.,* 2014). Biswas, *et al* (2007) have proposed a mechanism on how fluoride leaches into the water as it percolates through fluoride minerals. This is described by the chemical equations below.

$CaF_2 + 2HCO_3^- \rightarrow CaCO_3 + 2F^- + H_2CO_3$

$CaF_2 + H_2CO_3 + 2OH^- \rightarrow CaCO_3 + 2F^- + 2H_2O$

Further, when water contaminated with fluoride percolates through $CaCO_3$ (calcite) rich soils or rocks it is converted to CaF_2 (fluorite) mineral as shown below.

$CaCO_3 + 2F^- + 2H^+ + \rightarrow CaF_2 + H_2CO_3$

Besides natural sources in ground water, fluoride can be discharged into water sources through discharges from industrials and agricultural activities during the manufacturing of steel, glass, electronics, ceramics, tiles, aluminium, pesticides and fertilizer and power stations fired by coal (Josh *et al.,* 2012; Waghmare *et al*, 2015).

Health Effects of Fluoride

Fluoride is an essential element for human beings as well as animals. This is related to the function it plays in mineralization of dentin and bones. Consumption of trace amounts of fluoride within the allowed limits of 0.5-1.0 mg/L, is beneficial to human health, particularly in children, on the rate of occurrence of dental carries (Loganathan *et al.,* 2013). However, according to WHO fluoride concentration in drinking water should not exceed 1.5 mg/L (WHO, 2006). According to Bhatnagar *et al.* (2011) about 200 million people globally depend on water, especially for drinking, that has goes beyond the WHO limit.

Exposure to high levels of fluoride has got detrimental health effects. It can lead to permanent demineralization of bones and teeth, a condition referred as fluorosis. Fluorosis is classified as dental and skeletal. Dental fluorosis refers to the disturbance that occurs to the dental enamel during tooth development. Skeletal fluorosis is an advanced form of fluorosis in the bones. It causes pain and damage to the bones and joints (Zhao *et al.,* 2010). This is largely due to the fact that Fluoride ion is electronegative and gets easily attracted to the positively charged calcium ion in bones and teeth (Tomar and Kumar, 2013).

Further studies have been conducted and it has been established that exposure to elevated fluoride levels can lead to interference towards carbohydrates, proteins, vitamins and e effects on DNA synthesis (Zhou *et al.,* 2004; Islam and Patel, 2011). In line with this, kidney diseases, interference of pineal glands, infertility, arthritis, Alzheimer's disease, and cancer have been attributed to intake of excessive fluoride (Kumar *et al.,* 2009; Khichar and Kumbhat, 2015; Waghmare *et al.,*2015).

Excessive exposure has also been linked to neurological disorders (Prasad *et al.*, 2014). Long-term consumption of water with high levels of fluoride can lead to effects of brain development in infants. This results to reduced intelligence quotient (IQ) in children. Additionally, long-term exposure may lead to damage of the thyroid and liver (Adak *et al.,* 2016). Table 2 illustrates the effects of fluoride on health.

Fluoride Levels in Kenya

Fluoride water contamination is a serious issue in Kenya. Presence of high fluoride levels in drinking water in some parts of the country specifically Rift valley, Nairobi, Central and Eastern has been reported. Actually, endemic fluorosis is and has been a serious health concern among the residents of these areas. In their studies on the occurrence and distribution of fluoride in ground waters of Kenya, Nair and co-workers ranked provinces (however now we have counties) according to the maximum fluoride levels. It was found out that Rift Valley, Central, Eastern and Nairobi provinces had the highest levels of fluoride whereas Western, Coast and Nyanza recorded minimal levels as depicted in Table 3 (Nair *et al.,* 1984). A number of studies in Kenya have pointed out high fluoride levels that exceed the 1.5 mg/L WHO standard exist in the following regions: Elementaita and Gilgil (Wambu *et al.*, 2011), Nakuru (Naslund *et al.*, 2005) and Baringo (Naslund *et al.*, 2005; CCEFW, 2010).

Approaches to Fluoride Removal

Defluoridation refers to the mechanism of fluoride removal from water (Khichar and Kumbhat, 2015). Various technologies have been applied in reducing fluoride levels in drinking water. These include ion exchange, adsorption, coagulation and precipitation, and membrane (reverse osmosis, nanofiltration membrane and dialysis) and electrochemical methods (Tomar & Kumar, 2013; Flanagan and Habuda-Stanić, 2014; Waghmare and Arfin, 2015). Each of these techniques has got both its advantages and disadvantages. However, adsorption process is more preferred due to its simplicity in operation and the easy of availability of adsorbents, as compared to other techniques due to drawbacks associated with many of these technologies (Tomar & Kumar, 2013: Waghmare and Arfin, 2015).

Table 2. Effects of Fluoride on Health

Concentration of fluoride	Impact on health
Nil	Limited growth and fertility
0.0 – 0.5 mg/L	Dental Caries
0.5 – 1.5 mg/L	Promotes dental health, prevents tooth decay
1.5 – 4.0 mg/	Dental fluorosis (mottling of teeth)
4.0 – 10.0 mg/L	Dental fluorosis, skeletal fluorosis
>10.0 mg/L	Crippling fluorosis

Table 3. Maximum fluoride ion concentrations (ppm) recorded for water samples taken in each province

Province	District	Fluoride Concentration
Nairobi	Nairobi	30.2
Central	Murang'a	22.0
Coast	Taita Taveta	15.0
Eastern	Machakos	19.3
North Eastern	Wajir	38.2
Nyanza	Kisumu	10.4
Rift Valley	Nakuru	57.0
Western	Bungoma	7.1

Source: (Nair *et al.*, 1984)

Nalgonda Technique (Precipitation)

Nalgonda technique is a common defluoridation techniques applied in countries like Kenya and Tanzania. It involves mixing water with calculated amount of alum, lime and bleaching powder. There after the water is taken through normal treatment process of flocculation, sedimentation, filtration and disinfection. The main purpose of sodium aluminate (alum) is to increase the rate of precipitate settlement. The bleaching powder is for disinfection. Using this method, fluoride is reduced by charge neutralization and adsorption onto $Al(OH)_3$. The major drawback of this technique is that the treated water will have residual aluminium of about 2-7 ppm way above the recommended 0.2 ppm by the WHO (Tomar & Kumar, 2013).

Adsorption

Adsorption can be defined as the change in the concentration of a substance in the surface layer compared to the phase in bulk with relation to the unit surface area. The component adsorbed is referred as the adsorbate while the adsorbing is the adsorbent. Fluoride is classified as a hard base due to its small ionic size and high electro negativity. Thus it has got strong attraction (adsorption) towards electropositive metals such as Zirconium(iv), iron(iii), calcium and Aluminium (iii) (Qiusheng *et al.,*2015). The ability of such metals to adsorb Fluoride makes them good candidates for the synthesis of fluoride adsorbent materials. Several adsorbents have been applied in water defluoridation including nano-based materials as discussed below.

The process of defluoridation using adsorption involves three phases. One, external mass transfers in which fluoride ions move through the layer surrounding the adsorbent particle. Two, fluoride ions are adsorbed on to particle surfaces. Lastly, there is an exchange of ions inside the adsorbent particles depending on the chemistry of the adsorbent (Flanagan and Habuda-Stanić, 2014).

Theory of Adsorption

Adsorption can be categorized into physical (physisorption) or chemical (chemisorption). In physical adsorption, the principle of adsorption is based on the bonding of weak intermolecular forces, dipole interactions and hydrogen binding between the adsorbate in solution and adsorbent. In addition to that, there is no exchange of electron between the adsorbent and the adsorbate. This is due to the fact that activation energy is not a prerequisite for physical adsorption. On the other hand, chemical adsorption occurs due to the existence of chemical bond between the adsorbent and adsorbate (Mohapatra *et al.*, 2009). Other differences are summarized in Table 4.

Adsorption Isotherms

The adsorption process on the surface of the adsorbent is characterized by an adsorption isotherm. Adsorption isotherm refers to the relationship between the amounts of adsorbate adsorbed on the surface the adsorbent and the equilibrium concentration of the adsorbate at a certain temperature. Adsorption isotherms help scientists to evaluate adsorbent performance. Adsorption efficiency is influenced by a number of factors including surface area, size of the particles, pH and degree of ionization of the adsorbate. The size of the particles minimizes mass transfer of the adsorbate into the adsorbent and reduces internal diffusion. Freundlich and Langmuir adsorption isotherms are the commonly employed isotherms in studying adsorption capacity form liquid to solid systems (Qiusheng *et al.*, 2015)

Langmuir Isotherm

Langmuir isotherm predicates monolayer adsorption onto homogeneous surface (Prabhu and Meenakshi, 2013). Equilibrium distribution of sorbate ions in solid and liquid phases is well represented by Langmuir. The Isotherm is predicated on the following assumptions. One, adsorbent surface is uniform. Two, there is equality in all the adsorption sites. Three, there is no interaction between the adsorbed molecules. Next, adsorption proceeds in a similar mechanism. Finally, a monolayer is the only one formed at maximum adsorption (Kumar *et al.*, 2011).

The Langmuir adsorption isotherm is expressed using the equation illustrated below (Qiusheng *et al.*,2015)

$$q_e = \frac{Q_{max} b C_e}{1 + b C_e} (\text{nonlinear})$$

Where:
qe = equilibrium adsorption capacity (mg/g)

Table 4. Differences in chemical and physical adsorption

Chemisorptions	Physical adsorption
Irreversible	Reversible
Adsorption increases with temperature	Adsorption decreases with temperature
Chemical bonds are involved	Weak Van Der Waals forces are involved
Dissociation may be encountered	No dissociation
Heat of adsorptions are more than 40 kJ/Mol	Heat of adsorptions are less 40 kJ/Mol

Ce = equilibrium concentration of fluoride in solution (mg/L),
Q_{max} = maximum fluoride adsorption capacity (mg/g),
b = *Langmuir* isotherm coefficient which is related to the strength of adsorption.

Freundlich Isotherm

Freundlich adsorption isotherm assumes heterogeneous adsorption on the adsorbent surface (Ramdani *et al.*, 2010; Qiusheng *et al.*, 2015).The Freundlich isotherm model may be expressed as:

$$\ln q_e = \ln K_F + 1/nlnCe(\text{linear})$$

$$q_e = K_F C_e^{1/n}(\text{nonlinear})$$

Where K_F and *1/n* are the constants that are related to the adsorption capacity and the adsorption intensity. *1/n* represents heterogeneity factor. This factor gives an indication of the favourability and capacity of the adsorption. Values of K_F and *1/n* are calculated from the slope and the linear plot of ln*qe* versus ln*Ce*. K_F and *1/n* values are calculated using the slope and the linear plot of ln*qe* versus ln*Ce*. The values of $n>1$ demonstrate beneficial adsorption

Adsorbents Used for Water Defluoridation

A number of adsorbents prepared from different materials have been applied for defluoridation. These include oxides, oxyhydroxides, biosorbents, geomaterials, and carbonaceous derived materials. These classes of materials are discussed below.

Oxides and Hydroxides/Oxyhydroxides

Metal oxides and hydroxides have got surface oxygen in which the number of coordinating metal ions that influences the adsorption of various anions and cations. This unique property is key for fluoride adsorption. Various metal oxides and oxyhydroxides have been applied in the development of defluoridation process. Lv *et al.* (2013) used Zirconium oxyhydroxide to modify red mud for fluoride removal. In their study it was observed that Zirconium modified red mud had an adsorption capacity of 0.6 mg/g and optimum pH for fluoride removal ranged between 3 and 4.

Bhaumik *et al.* (2011) in their work have studied the uptake capacity of fluoride from aqueous solution using polypyrrole/Fe_3O_4 magnetic nanocomposite. Fluoride adsorption was found to increase with the increase in adsorbent dosage and with increase in the pH from 2 to 6. Further, it was observed that there was negligible effect of nitrate and chloride ions with respect to defluoridation capacity. However, phosphate and sulphate ions were found to form complexes which in turn increased competition for the active sites of adsorption.

According to (Sujana and Anand, 2013) in their study on ferric hydroxide capability to remove fluoride it was observed that pH is a critical parameter in the absorption of fluoride on amorphous ferric chloride and that fluoride removal mechanism involved ion exchange and complexation at the surface of the adsorbent. Tests involving ground water under prevalent conditions showed that ferric hydroxide could reduce fluoride concentrations to less than 1.0 mg/L. A related study using granular ferric hydroxide reported a maximum sorption capacity of 7.0 mg/g and their data fitted well with the Langmuir model (Kumar *et al.*, 2009).

Jahin (2014) worked with the removal of fluoride from water using nanoscale zero-valent iron. He observed that increasing the dosage of the adsorbent and time of reaction, fluoride removal efficiency increased. On the other hand increase on the initial concentration of the adsorbate and solution pH decreased the removal efficiency. The main conclusion from this study was that nanoscale zero-valent iron could be a promising adsorbent for defluoridation of contaminated water.

Patnaik and co-workers studied the defluoridation capacity using chitosan- Fe^{3+} complex and recorded a maximum adsorption capacity of 2.34 mg/g of F^- at an initial concentration of 50 mg/L of F- and adsorbent dosage of 10 g/L. In the pH range of 3.0-10.0 there was insignificant variation in efficiency removal of fluoride. The effect of competing ions in the removal of fluoride was noted significantly with sulphate followed by bicarbonate and nitrate. However, the presence of chloride did not adversely affect the removal efficiency. Data fitted well with D-R and Langmuir-Freundlich models.

Biosorbents

This is a class of adsorption commonly referred to as biosorption. It involves the use of biomaterial such as biopolymers like chitin, chitosan (discussed in detail later in this section). Biomass from algae which has been pre-treated with calcium ions has been studied for fluoride removal. Other biomaterials that have been used for drinking water defluoridation include collagen fiber and alginate. All these materials have found this application due to the availability of abundant functional groups such as hydroxyl and amino groups (Jagtap *et al.*, 2011).

Biomass materials such as rice husk and egg shells have been applied for defluoridation. However, such materials have limited adsorption capacities as discovered from literature (Bhaumik *et al.*, 2011). Christina and Viswanathan (2015) used saponified orange peel residue for defluoridation. They observed that the maximum adsorption capacity was mg/g with batch equilibrium studies following Langmuir isotherm model. FTIR studies in the material after adsorption confirmed the involvement of the hydroxyl and carbonyl groups in the adsorption of fluoride.

Geomaterials

Geomaterials such as layered double hydroxides, zeolites, pumice stone, clay and Goethite have also been explored for defluoridation applications (Flanagan and Habuda-Stanić, 2014). Clays for instance, has been characterized with high surface area and ion exchange capacities hence high adsorption capacity. Guo and Tian, (2013) utilizing anion clay, hydrocalumite exhibited potential to remove fluoride and arsenic. Similar study by Guo and Reardon showed that anion clay – meixnerite has the capacity to remove over 95% of fluoride from water (Guo and Reardon, 2012).

Sepehr *et al.* (2013) have employed modified pumice for defluoridation. The authors found out that fluoride sorption capacity was 65% and 68% for unmodified and modified pumice with maximum capacity at 11.765 mmg/g. Zeolites have been as adsorbents due to their cation exchange properties. However, zeolite has minimal applications in fluoride adsorption due to the negative surface charges at all pH. This limitation favors high adsorption of cations compared to anions. To overcome this challenge researchers modify the surface of the Zeolite in order to increase its uptake capacity. For example, work done by Sun and co-authors reported that modified natural stilbite zeolite had an increased efficiency of fluoride removal with a capacity of 92% (Sun *et al.,* 2011)

Lanthanum is one of such geomaterial that has found wide use in defluoridation. Srimurali and Vardhan, (2016) prepared lanthanum- impregnated bauxite for defluoridation. The results showed that a dose of 2 g/L of the adsorbent had a reduction capacity of 99% of fluoride from an initial concentration of 20 mg F/L with a maximum sorption capacity of 18.18 mg/g. Equilibrium studies followed Langmuir isotherm model. They further noted that natural water pH (6.5-8.5) was favourable for adsorption.

Carbonaceous Materials

The typical carbonaceous materials used for defluoridation are bone char and Charcoal. These materials have the capacity to reduce fluoride concentration from levels ranging from 2 ppm to less than 0.2 ppm. However, while activated charcoal has high fluoride uptake it is highly influenced by the pH. It works well in pH of 5 and very low uptake at pH of 7. This means that its application in the pH ranges of natural drinking water (6.5 – 8.5) is quite limited. Bone char has been extensively used for drinking water defluoridation. Bone char is rich in apatite ($Ca_9(PO_4)6.CaCO_3$). The removal mechanism involves ion exchange between fluoride and apatite to form insoluble fluoroapatite as shown in the equation below:

$$Ca_9(PO_4)_6.CaCO_3 + 2\,F^- \rightarrow Ca_9(PO_4)_6.CaF_2 + CO_2$$

The local availability, easy and cheap processing facilities for bone char has made it more appropriate. However, use of bone char has its own disadvantages such as: obnoxious smell, distasteful and imparted yellow color in the treated water.

Comparison of Various Water Defluoridation Technologies

Table 5 below provides a summary of the various methods used in drinking water defluoridation with the main emphasis being on the merits and demerits.

Nanotechnology as an Emerging Water Defluoridation Technology

Nanomaterials (nanoparticles) refer to particles whose dimension range from 1 to 100 nm. Their small particle sizes and large surface area to volume ratio are responsible for their high reactivity and easy delivery through small spaces in the subsurface to contaminated sites. Nanomaterials are preferred because of time and cost effectiveness (U.S. EPA, 2008).

Nanotechnology holds immense potential in advancing water treatment and purification (Bootharaju and Pradeep, 2010). Through nanotechnology, concentrations of some contaminants can be reduced to near zero (Cloete *et al.,* 2010). Smith (2006) argues that nanotechnology for water remediation will play a crucial role in water security and consequently the food security of the world and more particularly developing countries.

Another class of material that has attracted significant research interest in nanotechnology is, biopolymers. Biopolymers are inexpensive and non-toxic. The chemical and physical parameters of these materials enhance the adsorption of both organic and inorganic contaminants in water. Biopolymers are categorized into natural or synthetic. This depends on the manufacturing process. Natural polymers include polysaccharides such as chitosan, cellulose, agar, starch and alginate. Synthetics polymers include

Table 5.Comparison of fluoride removal technologies

Technique	Advantage(s)	Disadvantage(s)
Adsorptive materials	• Easy of operation • Less expensive • simple operation, • availability of wide range of adsorbents	• Interference by co-ions •
Membrane filtration (Reverse Osmosis)	• High efficiency	• Expensive • Fouling • Imbalance of minerals in treated water
Ion-exchange	• High efficiency	• Expensive, • Interference by co-ions • Frequent regeneration • pH-dependent
Electrochemical treatments (electro-dialysis)	• High efficiency • high selectivity	• High cost of installation • Expensive maintenance
Coagulation/precipitation (aluminum hydroxide)	• High efficiency	• Expensive, • pH dependent • Interference by co-ions in water, • Increase in Al levels,

Source: Meenakshi and Maheshwari, 2006; El-Gohary *et al.*,2010; Bhatnagar *et al.*,2011; Chakrabortty *et al.*,2013; Tomar *et al.*, 2013)

polyvinyl alcohol, poly glycolic acid and polyhydroxy alkonates (Rhim *et al.*,2012). The hydrophilic nature of biopolymers makes them good adsorbents for removing pollutants from water (Averous and Pollet, 2012).

Smith (2006) argues that nanotechnology for water remediation will play a crucial role in water security more particularly developing countries. He further points out that application of nanotechnology in the remediation of contaminated water can either be through nanoscale filtration techniques, adsorption of pollutant on the nanoparticles or the breakdown of pollutants by nanoparticles catalysis. In general, nanoparticles remove pollutants from the environment through adsorption, redox catalytic reactions, and/or photocatalytic reactions (Sanchez *et al.*, 2011).

Adsorption process involves removal of organic and inorganic contaminants in water and wastewater treatment using adsorbents. It is important to note that nano-adsorbents have superior adsorption efficiency compared to conventional adsorbents because of improved surface area or active sites, selectivity, and adsorption kinetics (Qu *et al.*, 2013)

Nanotechnology

Nanotechnology involves the deliberate manipulation of matter at size scales of less than 100 nm. It holds the promise of creating new materials and devices which take advantage of unique phenomena realized at those length scales. Nanotechnology gives the ability to observe and control individual atoms and molecules (Chattopadhyay *et al.*, 2009).

Nanotechnology refers to the research, and technology development at the atomic, molecular, or macromolecular levels, in the length scale of approximately 1–100 nanometer range, to provide a fundamental understanding of phenomena, and materials at the nanoscale; and to create, and use structures, devices, and systems that have novel properties, and functions because of their small, and/or intermediate size." At this level, the physical, chemical, and biological properties of materials differ in fundamental, valuable ways from the properties of individual atoms, molecules, or bulk matter. This discipline cuts across chemistry, physics, engineering and biology (Roco, 2001).

Nanoparticles, by the fact that they are nano-sized, exhibit advanced properties in terms of size, morphology, surface area, high reactivity, fast dissolution, and strong sorption (Qu *et al.,* 2013). Noble nanoparticles have properties of antimicrobial activity and reduction of halogenated pesticides (Bootharaju and Pradeep, 2010). In addition to nanoparticles, nanocomposites applications have grown exponentially due to their superior properties compared to monolithics and microcomposites (Armelao *et al.,* 2006).

Synthesis Methods of Nanoparticles

There are two main approaches for synthesizing nanoparticles namely bottom-up and top-down. In top-down approach, nanomaterial is created from macro-scale to nano-scale. A good example of this approach is ball milling and laser ablation. The Bottom-up approach involves production on nanomaterials from molecular or atomic scale to nano-scale such as sol-gel and precipitation. Nanoparticles may be synthesized using various starting materials by different physical and chemical methods, with the produced particles differing in elemental composition, shape, size, and chemical or physical properties.

The methods can be divided to physical and chemical methods. Physical methods include: laser ablation, inert gas condensation, high energy ball milling, chemical vapour deposition, and electro-deposition. Chemical methods include: hydrothermal method, sol-gel method, sono-chemical methods, co-precipitation and Electrospinning method. However, the most commonly used methods for the synthesis of titanium oxide nanoparticles are sol-gel, chemical vapour deposition and hydrothermal.

Sol-Gel Method

Sol-gel method involves the formation of a concentrated suspension of a metallic oxide or hydroxide (sol), which is subsequently dehydrated by evaporation or solvent extraction, resulting in a semi-rigid mass (gel). A wide range of pure and mixed oxides can be produced on controlled heating of this gellated material. In a typical sol-gel method, a colloidal suspension or a sol is formed from hydrolysis and polymerization reactions of precursors, which are usually inorganic metal salts or metal organic compounds such as metal alkoxide.

Hydrothermal Synthesis

Hydrothermal method can be defined as a method of synthesis of single crystals which depends on the solubility of minerals in hot water under high pressure. Since ionic product of water (Kw) has a maximum value of around 250-300°C, hydrothermal synthesis is usually carried out below 300°C. The crystal growth is performed in an apparatus consisting of a steel pressure vessel called autoclave, in which a nutrient is supplied along with water. The formation mechanism of metal oxide particles from metal salt solution is as follows: first hydrated metal ions are hydrolyzed to metal hydroxide. Then, metal hydroxides proceed to precipitate as metal oxides through dehydration.

Chemical Vapour Deposition

Chemical vapour deposition is a process in which a substrate is exposed to one or more volatile precursors, which react and/or decompose on the substrate surface to produce a non-volatile solid deposit or thin film. The solid material is obtained as a coating, a powder, or as single crystals. The process is carried out in autoclaves under controlled temperature and pressure. It allows the use of temperatures above the boiling point of water/organic solution.

Factors Influencing Performance of Water Defluoridation Adsorbents

Effect of Contact Time

Contact time studies, in adsorption studies, help in establishing the minimum time required for maximum adsorption of a given contaminant from water (solution). In most cases, the sorption capacity increases with time until stability is achieved whereby there would be no significant difference in sorption capacity thereafter. It is assumed that all sites in the adsorbent are free but as sorption proceeds adsorptions sites decrease significantly.

Effect of Adsorbent Dose

The amount or dose of an adsorbent is one parameter that is of great significance that influences the uptake capacity of a sorbate (Saifuddin *et al.,* 2011). Generally, it is presumed that the higher the amount of the adsorbent the higher the uptake capacity of a given contaminant. However, at equilibrium, increase in the amount of dosage does not have significant effect on absorption. Increase in adsorbent dosage enhances the surface area and consequently boosts adsorption. On the contrary, increasing dosage beyond the optimum amount only serves to a negligible improvement in sorption but adsorption capacity decreases (Tang *et al.,* 2009**;** Josh *et al.,* 2013).

Thakre and co-authors observed that uptake of fluoride increased from 7.66–94.02% by increasing the adsorbent dosage from 0.2 to 6 g/L. This was attributed to high number of active sites for fluoride adsorption. On the other hand, they also noted that after a 6 g/L dosage there was no significant change in removal efficiency. This could have been due to lower equilibrium fluoride ion concentration per active site of adsorbent for adsorption (Thakre *et al*., 2010)

Christina and Viswanathan (2015) have investigated adsorptive fluoride removal from aqueous solution by using saponified orange peel residue immobilized sorbent. They observed that when the amount of the sorbent was increased from 0.25 to 1 g/L, the adsorption capacity increased from 4.13 mg/g to a maximum of 7 mg/g. It may be due to the large amounts of adsorption that are not yet saturated .However, decreased sorption capacity was noted beyond 1 g/L. this phenomenon can be attributed to the occupation of the F- ions in the free active sites.

Effect of pH

pH is a critical parameter that greatly influences the adsorption process in solid liquid interfaces (Mahdavi et *al.,* 2014). The performance of an adsorbent towards contaminant removal mainly depends on the initial pH of the solution (Garg and Chaudhari, 2012: Adak *et al.,* 2017). Further, the pH of a solution determines the surface charge of the adsorbent and speciation and degree of ionization of the adsorbate. For water related application, the range of pH study varies from 3 -11 but for drinking water should be between 6.5 and 8.5. For a good adsorbent the percentage uptake of the contaminant should not significantly vary with pH.

Thakre *et al.* (2010) have investigated defluoridation of water using chitosan based mesoporous Ti–Al binary metal oxide supported beads. They found out that there was no significant difference in the removal efficiency of fluoride in the pH range of 3-11. This outcome is explained by aluminium and titanium oxides forming aqua complex with water. This results to formation of surface charges through amphoteric dissociation. Hence, in acidic pH range, the positively charged surfaces created attract the negatively charged fluoride ion electrostatically. Consequently, resulting to increased uptake of fluoride by the adsorbent in acidic pH.

At neutral pH, ligand exchange interaction occurs between hydroxyl and fluoride ions. However, pH above 10, removal efficiency decreases by almost 52%. This is due to the competition of the active sites for both hydroxyl and fluoride ions. According to Tembhurkar and Dongre, (2006), they also recorded the same observation with pH values greater than 10. Decline in fluoride removal efficiency could be due to the high volume of OH^- ions, which tend to hinder the diffusion of F^- ions. Similar trend has been reported in other related studies (Liao and Shi, 2005; Kumar *et al.,*2009; Chen *et al.,* 2010; Zhao *et al.,* 2010**;** Li *et al.,*2010)

Christina and Viswanathan, (2015) in their study on adsorptive fluoride removal from aqueous solution by using saponified orange peel residue immobilized sorbent, reported that the adsorption capacity increased from 0.29 mg/g at pH 3 to 5.02 mg/g at neutral pH and decreased when pH was increased any further. This observation could be due to the fact that beyond neutral pH, the abundance of ^-OH groups compete for F- ions in solution hence decreased fluoride uptake capacity. Similar observation has been documented by Chen and co-workers in their study of fluoride removal from water by granular ceramic adsorption (Chen *et al.,* 2010).

Mahdavi *et al.* (2014) have utilized Al_2O_3 nanoparticles modified with natural and chemical modifiers for heavy metal removal. They observed that when the pH solution was increased from 4 to 8, the adsorption efficiency increased slightly. At low pH, there is a likelihood of competition between metal ions and the protons on the surface groups leading to reduced metal uptake by the material.

Effect of Sorbet's Initial Concentration

The study of the effect of initial concentration of the target water contaminant enables us to determine the minimum concentration of the Adsorbate (such fluoride and heavy metals) required for maximum sorption. Thakre *et al.* (2010) in their study of defluoridation using chitosan, titanium and aluminium, observed that increase in fluoride concentration while keeping the dosage of adsorbent constant at 4 mg resulted to decrease in fluoride uptake capacity. This was due to the fact that the number of active sites on the surface of the adsorbent was limited to uptake the fluoride ions. Similar phenomenon was also been reported in other studies (Tomar *et al.*, 2013; Christina and Viswanathan, 2015)

Effect of Competing Ions

In real time applications of adsorbents in filters, it is imperative to take into account of the effects of competing (co-existing ions). These are ions which are present in water along with the target pollutant such as fluoride. Examples include such as bicarbonate, carbonate, chloride, sulfate, nitrate and phosphate. Co-existing ions tend to scavenge for the active sites (Garg and Chaudhari, 2012)

Zhu *et al.* (2015) studied a facile precipitation synthesis of mesoporous 2-line ferrihydrite with good fluoride removal properties established that the presence of nitrate and chloride had minimal effect on the fluoride adsorption. However, existence of phosphate, sulfate, carbonate and bicarbonate had significant effect on the adsorption capacity and more especially carbonate at high concentrations. Comparable phenomenon has also been reported (Zhang *et al.,* 2015)

Challenges of Conventional Defluoridation Technologies

Defluoridation techniques available for drinking water treatment include chemical coagulation, precipitation, ion-exchange, adsorption and membrane technology. However, these techniques are characterized by certain limitations. Conventional water treatment methods such as ion exchange, reverse osmosis are chemically, energetically, operationally and financially intensive and beyond reach for the poor people and rural communities.

The above limitations restrict their effective utilization. However, among these techniques, adsorption is considered to be the cheapest in terms of cost, maintenance and ease of operation. Domestic water purification systems available in the market are not suitable for rural communities. For instance lack of electricity to operate them. Similarly, some traditional adsorbents used for water purification are expensive and difficult to regenerate once they are saturated with contaminants.

To overcome these challenges natural biopolymers such as chitosan and cellulose and metal oxides, engineered using nanotechnology, can be utilized for water purification. Nano-based composite of Chitosan and oxides of Titanium, Iron, Aluminium, Zirconium, Tin and Manganese have been proven to have application in water defluoridation. Chitosan is a good adsorbent for metal ions, has antimicrobial properties and can degrade pesticides. Titanium has got antibacterial activity and adsorption of fluoride and metal ions. Aluminium and Zirconium have good adsorbent for Fluoride ions.

In view of the above challenges of modern technologies for water purification and demerits of traditional adsorbents, nanotechnology proves to be an alternative application for water treatment. However, from literature many of the nanomaterials have low adsorption capacity and cannot be used in granular form as most are in powder form which renders filtration difficult. Therefore, there exists a huge gap in terms of developing nanomaterial for water purification that have high adsorption capacity, cheap, low or nil maintenance costs.

Review of Biopolymer and Metal Oxides Used in this Study

This section seeks to explain the chemistry of the biopolymer and metals from which the various oxides and hydroxides used in this study are derived from. This will build up on the synergistic properties of these materials which are tapped for removal of fluoride from drinking water.

Chitosan

Chitin, poly (b-(1-4)-N-acetyl-D-glucosamine) is found in the shells of many sea organisms. When chitin is deacetylated, it forms chitin derivatives, one of which is chitosan (Homayoni *et al.,* 2009). Figure 1 shows structures of chitin and chitosan, respectively.

Chitosan is a type of copolymer which consists of two repeating units i.e. N-acetyl-2-amino-2-d-glucopyranose and 2-amino-2-deoxy-d-glucopyranose which are linked by β -(1→ 4)-glycosidic bond.

Chitosan has received greater attention in water purification due to presence of large number of $–NH_2$ and OH^- groups. The free amine function of chitosan gives it a better ability to chelate ions and it also has the potential to bind antagonistically with microbial and mammalian cells (Sudheesh *et al.,* 2013). These chelating properties are of great interest for water treatment and particularly to recover metals present in contaminated water. (Koshijima *et al.,* 1973). Chitosan and modified chitosan can chelate toxic pollutant like heavy metals and dyes. On the basis of such properties chitosan nowadays has been identified as a promising adsorbent for the removal of toxic pollutant from waste water (Kyzas *et al.,* 2009). Additionally, Chitosan presents a large capacity to fix molecules such as pesticides (Van Daele and Thome, 1986; Thome *et al.,* 1997).

Chitosan and chitin have been shown to remove metals and dyes hence can be used to clean water. This is due to the adsorption of the metals and dyes to the cationic amine and hydroxyl functional groups on chitosan (Crini, 2005; Prabhu and Meenakshi, 2016). IN addition, chitosan also has anti-viral properties (Bai, 2012) and antimicrobial activity (Hwang *et al.,* 1999). It is generally accepted that the amine group of chitosan can react with the anionic groups on the bacteria cell surface and this interaction brings extensive change to the cell surface and the cell permeability (Sudardshan *et al.*, 1992, Fang *et al.,* 1994, Hwang *et al.,* 1999).

Aluminium

Activated alumina for a long time has been effective in removal of fluoride from water. Amorphous $Al(OH)_3$ and alumina or gibbsite Al_2OH_3 have been used for this purpose. From literature, this type of adsorbents react with fluoride at pH ranging from 3 -8 and the optimum pH for fluoride uptake is found to be in the range of 5.5 -6.5. Fluoride uptake usually decreases with lower pH. This is as a result of the preferential formation of ALF_X. However, increasing the pH hydroxide ion displaces the fluoride ion from the solid $Al(OH)_3$ (Tomar & Kumar, 2013).

Modification, through impregnating with other metals such as La(III) has been done in order to improve the adsorption capacity of fluoride. For instance, the defluoridation capacity of alumina impregnated with lanthanum hydroxide increased from 25 mg/g (for the original alumina) to 48 mg/g. The increase of adsorption capacity was ascribed to the exchange of ion between the OH on the surface of the material and the anion (Wasay, *et al.*, 1996).

Figure 1. Structure of chitin and chitosan (Bai, 2012)

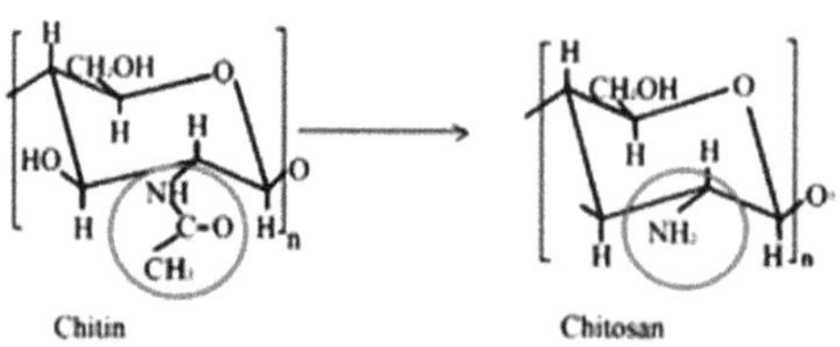

Titanium

In the Photodegradation of organic contaminants, photocatalysis has proven to have huge potential as a low-priced and sustainable water treatment technology. It is based on the intense contact between the pollutant and the surface of a catalyst material (for instance semiconductor) susceptible to generate oxidizing species upon photon absorption. Nanosized semiconductors such as TiO_2, has been used as a photocatalytic nanoparticles in water and wastewater treatment. TiO_2 is mostly preferred one due to its low toxicity, chemical stability, low cost, and large availability as a raw material (Qu *et al.*, 2013).

Zirconium

Zirconium oxide is transition metal that is chemically inert, non-toxic, biocompatible, insoluble in water and stable thermally. Additionally, Zirconium (IV) has been found to have high affinity towards fluoride through metal-ligand interactions (Hang *et al.*, 2012; Prabhu and Meenakshi 2013; Swain *et al.*, 2010). Previous studies have indicated that Zirconium has got high affinity towards fluoride. The chemical stability of Zr in relation to fluoride adsoprtion is due to its high affinity, minimal leaching effect and strong binding tendency (Prabhu and Meenakshi, 2016). Prabhu and Meenakshi (2013) in their work involving synthesis Zr(IV) encapsulated silica gel/chitosan (Zr SGCS) composite for fluoride removal, they established that their material had an adsorption capacity of 4,530 mgF−/kg compared to silica gel-chitosan composite, which had a defluoridation capacity of 3402 mgF−/kg.

Tomar *et al.* (2013) explored the effective adsorption of fluoride Zr-Mn composite material. They observed that under optimum condition of pH 7, contact time of 145 minutes and adsorbent dosage of 1.2g/50 mL, percentage removal was up to 90%. Qiusheng *et al* (2015) in their study of porous zirconium alginate beads adsorbent for fluoride adsorption observed that the presence of HCO_3^-, PO_4^{3-} and SO_4^{2-} reduced the adsorption of fluoride significantly. The adsorption kinetics obeyed the pseudo-second-order with correlation of 0.9953 and maximum adsorption capacity was 37.797 mg/g. Prasad, *et al.* (2014) reported the adsorption of fluoride on to zirconium-chitosan composite, where 99% of fluoride was found to be adsorbed on adsorbent at pH 7.0.

Swain *et al.*, (2010) prepared inorganic – organic hybrid exchanger using Zr(IV) and propanolamine for fluoride adsorption. The material had 99% capacity of fluoride removal at pH 3. The presence of counter ions NO_3-, Cl-, and SO4 did not have any noticeable effect on the fluoride adsorption. Lv *et al.*, (2013) reported that porous material prepared from red mud modified with Zirconium hydroxide had an optimum fluoride removal between pH range of 3 and 4 and adsorption capacity of 0.6 mg/g.

Adak *et al.* (2016) compared fluoride removal using Nano-sized Alumina and Zirconia Modified Alumina. The results indicated a more than 98% fluoride removal capacity was achieved using 0.4g/100 mL of each of the adsorbent at 5 ppm concentration of Fluoride at pH of 6.7. Increase in pH led to decrease in fluoride removal capacity. Further they proposed a defluoridation mechanism in which the Zr^{4+} and the free amine groups from chitosan electrostatically attract the negatively charged F^- ion.

MAIN FOCUS OF THE CHAPTER

Studies have pointed that almost 80% of diseases in the world emanate from drinking contaminated water whereas 65% of endemic fluorosis is due to consumption of fluoride contaminated water (Tomar

& Kumar, 2013). Conventional methods are often chemically, energetically and operationally intensive, focused on large systems, and thus require considerable infusion of capital, engineering expertise and infrastructure, all of which precludes their use in much of the world, specifically, developing world with large population living below the poverty line.

Techniques available for drinking water treatment include chemical coagulation, precipitation, ion-exchange, adsorption and membrane technology. However, these techniques are characterized by certain limitations. These limitations restrict their effective utilization. However, among these techniques, adsorption is considered to be the cheapest in terms of cost, maintenance and ease of operation.

Nanoscale engineered adsorbents have gained great attention in the application in water purification compared to existing bulk adsorbents. However, this class of nanomaterials cannot be used in field applications due to challenges in particle separation hence a potential danger in view of their leaching into the purified water. In line with this demerit, is that most of these nanoadsorbents have poor hydraulic conductivity making their application in filtration devices impossible (Kumar *et al.,* 2017).

From literature many of the nanomaterials used for defluoridation and arsenic reduction have low adsorption capacity and cannot be used in granular form as most are in powder form which renders filtration difficult. Further, many of these adsorbents do not have satisfactory wet strength to stay intact as a granular composition which characterizes them with poor hydraulic conductivity which leads to excessive pressure drop in the water purification cartridge. On the other hand many of these adsorbents are economically non-viable making them ineffective in commercial applications.

Therefore, there exists a huge gap in terms of developing nanomaterial for water purification that have high adsorption capacity, cheap, low or nil maintenance costs. It is worth noting that most of the fluoride, studies are based at laboratory levels. This means that they are only of academic nature. Thus the present work reports on the synthesis of a new and novel nanocomposite material and development of a prototype of an innovation device that utilizes the nanomaterial in granular form to fabricate point-of-use water filters for use at community and household level.

HYPOTHESIS

Synergistic properties of biopolymer mixed metal oxide nanomaterial would significantly reduce fluoride, heavy metals and bacteria in drinking water.

MATERIALS AND METHODS

Study Design

The granular composite was synthesized using chitosan- metal oxyhydroxides nano-structures. Green synthetic process was adopted in the synthesis of the composite. Green synthesis route involves hydrolysis of a metal precursor and chitosan. Alkaline solution is used for the hydrolysis, there after the material is washed and dried in ambient air. This study involved the fabrication of chitosan impregnated with ternary oxides/oxyhydroxides of iron, aluminium, titanium and zirconium. Chitosan (Cs) was the only biopolymer unto which the oxides were anchored. Factors influencing adsorbent performance such as pH, contact time, adsorbent dosage, presence of competing ions and initial concentration of the contaminant were studied in order to establish how the sorbent behaves in different water conditions.

Chemicals

Chitosan (degree of deacetylation 80%) was purchased from (Pelican Biotech & Chemicals Labs Pvt. Ltd. India). Hydrochloric acid (HCl), Sodium hydroxide (NaOH), anhydrous sodium sulphate (Na_2SO_4), Ferric hexachloride ($FeCl_3.6H_2O$), Aluminium sulphate, sodium fluoride(NaF) were all obtained from Sigma Aldrich. Zirconium oxychloride octahydrate ($ZrOCl_2{\cdot}8H_2O$) and 5 Titanium(IV) oxysulfate were acquired from Rankem Glasswares and Chemicals, Pvt. Ltd. India.

Synthesis of the Nanoadsorbent

The granular composites, composed of aluminium sulphate, titanium (IV) oxysulfate, Zirconium oxychloride octahydrate and iron oxyhydroxide – chitosan nanostructures, were synthesized using green method, which as a rule includes the hydrolysis of a metal. 1.3 g of chitosan was dissolved in 100 mL of %1 HCl and incubated for 12 hours. 0.5M $Al_2(SO_4)_3.18H_2O$ was added to the incubating chitosan. Next, 4 mL of 0.25M titanium (IV) oxysulfate was added. After mixing for a short while 4 mL of 0.25M $ZrOCl_2{\cdot}8H_2O$ was added. Next, 7 mL of 1M $FeCl_3.6H_2O$ was added. About 3.6 g Na_2SO_4 was added in one step. The mixture was precipitated and the pH brought to of 6.5 by slow addition of 2M NaOH. The mixture was then left to incubate for 12 hours at ambient temperature. The subsequent gel was washed with copious amount of water in order to remove soluble salts. This step is essential for the removal of excess soluble salts. Finally, after drying the material was crushed into specific sizes (52 x 72 μm) and was ready for various adsorption tests.

Preparation of Fluoride Standard Solutions

Prior to the preparation of the various standards all apparatus glassware and plastic wares were thoroughly cleaned with nitric acid and rinsed repeatedly with deionized water. This was carried out to minimize any contamination that may affect the experiments.

A stock solution of 1000 ppm was prepared by dissolving 2.21g of anhydrous NaF in 1 L volumetric flask of deionized water. Prior to weighing, the solid NaF was dried at 100 °C for one hour. All other desired fluoride solution concentrations for adsorption studies were prepared by the dilution from the stock solution.

Preparation of Total Ionic Strength Adjustment Buffer (TISAB II)

500 mL of DI was measured into 1 L beaker. 57 mL of glacial acetic acid was added and the mixture homogenized for 15 minutes using magnetic stirrer. 58g of high purity NaCl was added and mixing continued for one hour. Next, 5M NaOH was added to the mixture drop wise to adjust the pH in the range of 5.0 – 5.5. The mixture was left to cool at room temperature for 30 minutes at room temperature. After cooling the mixture was transferred into 1 L flask and diluted to the mark using distilled water. In summary the primary function of TISAB II reagent during fluoride analysis is to buffer the pH of the solution which in turn minimizes interference from hydroxide ion. Next, it helps to regulate the ionic

strength of the samples. Further, TISAB II helps to break down any complexes that may be formed as a result of Fluoride reacting with hydrogen, iron, aluminium and silicon. The complexation of these polyvalent cations limits the availability of free fluoride concentration in solution. All these are likely to affect the operation of the fluoride probe (Chauhan, *et al.*, 2007; Thakre *et al.*, 2010).

Fluoride Ion Selective Electrode Meter (ISE)

Fluoride ion concentration in water samples was measured using fluoride ion selective electrode meter (Model number: 98185, Hanna Instruments, USA). The electrode is made of a sensing element that is bonded to an epoxy body. Once the sensing tip is in contact with fluoride ions in solution, an electrode potential develops. The concentration of fluoride ions in water influences the magnitude of the potential. This potential is then measured against a constant reference potential using ion meter or digital pH/mV meter. The potential that is measured corresponds to the concentration of F^- and is described by the Nernst equation.

$$E = Eo + S \log (A)$$

Where:

E = measured electrode potential.

Eo = reference potential.

A = fluoride ion activity level in solution.

S = electrode slope (about 57 mV per decade)

Calibration Procedure for Fluoride

Calibration of the fluoride ion selective electrode was carried out before the measurement of F- concentrations in samples. This was done with the view of obtaining an optimum slope of about 59.16 mV with an acceptable range of 56.0 to 60 mV. The measurement of the slope factor gives an indication on the performance the ion selective electrode. Calibration was carried out in the range of 1, 10, 100 ppm using standard fluoride solution. These concentrations were prepared using serial dilution of the 1000 ppm stock solution. Once the F^- meter was calibrated, the residual fluoride concentration in experiment samples was measured.

Testing Protocol for Fluoride

Fluoride, Arsenic and bacteria were carried out using batch experiment technique. This involved mixing the adsorbent nanocomposite with a certain fixed concentration of fluoride, arsenic and bacteria. The mixture was agitated at 250 RMP until equilibrium state was achieved. The solutions were then filtered separately using microfilter and then concentration of the separate filtrates were determined. Batch experiments were carried out in order to optimize the experimental conditions by varying pH, initial concentration of the contaminant, contact time and dosage of the adsorbent and effect of co-existing ions. All tests were carried out before (Input) and after adsorption (Output).

To test for fluoride adsorption 100 mL of both tap water was taken and transferred to polypropylene conical flasks. Of the 100 mL taken 1 mL was taken out and to it 1 mL of fluoride standard (1000 ppm) was added. This gave a 10 ppm fluoride concentration. Next, 100 mg of the nanocomposite was added separately. The mixture was shaken at RPM 250 for 2 hours using orbital shaker (Orbitek, India). The shaking was stopped and the mixture allowed settling for 20 minutes. 5 mL of the supernatant was taken using a micropipette and 5 mL of TISAB II buffer solution was added and mixed thoroughly. Fluoride measurement was carried out using fluoride Ion Selective Electrode (Eutech instrument Model PCD 650).

Batch Adsorption Studies

Batch adsorption experiments were conducted to evaluate factors that influence adsorption efficiency (discussed earlier) and to examine adsorption isotherms and kinetics. These experiments were carried out with aqueous solutions of Fluoride, As (III) and As (V) in 250 mL polypropylene conical flasks. All the experiments were done in triplicates.

Working volume of the solution was taken as 100 mL while that of adsorbent mass was 100 mg unless mentioned herein with. Calculated amount of F^- were added from the stock solution of 1000 ppm sodium fluoride. This was primarily done in order to get the predetermined concentrations. Required amount of adsorbent was added and samples were agitated at 210 RPM in an orbital shaker at room temperature.

Effect of Adsorbent Dose

The amount of fluoride absorbed in a given adsorbent is greatly determined by the amount (dosage) of the adsorbent. To investigate the influence of adsorbent dose, experiments were carried out in which the dosage was varied from 5 to 200 mg. The initial concentrations of fluoride were kept constant at 10 ppm.

Effect of Contact Time

The impact of contact time on the adsorption of fluoride on the adsorbent was considered in the range of 1– 180 min. This was particularly to evaluate the equilibrium time for maximum adsorption. A fixed 10 ppm of fluoride solution was used and residual F^- concentration was measured by taking samples at varying contact time (1– 180 min). Equilibrium contact time was considered to be the time at which there was no increase in fluoride reduction.

Effect of pH

To gain understanding on the effect of raw water pH, adsorption experiments were carried out in the pH range of 4, 5,6,7,8, 9 and 10. The pH adjustments were achieved using 0.01M HCl and 0.01 M NaOH. The pH was measured using pH meter (Eutech instrument Model PCD 650). Adsorbent dose was kept at a constant 50 mg, fluoride ion concentration at 10 ppm and the solution volume maintained at 100 mL using deionized water.

Effect of Competing Ions

The presence of other existing ions alongside with fluoride can influence the competition for the active site. In this investigation, typical co-existing ions such as Cl^-, NO_3^- SO_4^{2-}, Na^+, SiO_3^{2-}, K^+, Ca^{2+}, Mg^{2+}, HCO_3^-, CO_3^{2-} and HCO_3^- and CO_3^{2-} were evaluated with respect to the effect they have to the adsorption efficiency process. This was carried out by the addition of required amounts of the ions of salt (have table in discussion section) in separate flasks. The experimental concentration for initial fluoride was maintained at 10 ppm while varying the initial concentration of co-ions from 0.6 –100 ppm. The working volume of water and weight of adsorbent was 50 mg/50 mL.

Adsorption Isotherms

Adsorption isotherm studies were done to investigate the relationship between solution phase and solid interface at equilibrium condition and constant temperature. Aqueous solutions of fluoride of variable concentration ranging from 1 to 200 mg/L were used. 50 mg of the nanocomposite (adsorbent) was used with the solution volume maintained at 100 mL using deionized water. The mixture was incubated for 48 hours at agitation speed of 300 RPM. There after the filtrate was examined for residue fluoride and arsenic. The pH of the solution was adjusted to 7 by adding 1.0 M sodium hydroxide or 1.0 M hydrochloric acid solutions. The filtrate was analysed for residue fluoride and equilibrium data fitted on the Langmuir adsorption isotherm.

RESULTS AND DISCUSSION

Adsorbent Characterization

Surface morphology, elemental analysis and elemental mapping studies were done using a Scanning Electron Microscope (SEM) coupled with Energy Dispersive Analysis of X-rays (EDAX) (FEI Quanta 200). The material is largely amorphous as shown from the images. EDAX analysis exhibits the elements constituting the surface of the adsorbent. Further, it shows the presence of an extra fluoride peak. The new peak is as a result of the fluoride ion adsorption onto the material.

Effect of Contact Time

The minimum amount of time required for adsorption of F-particles from aqueous solution was investigated by studying the impact contact time on the maximum fluoride adsorption capacity of the adsorbent. An initial F^- concentration of 10 ppm was used and the adsorption capacity was evaluated at a specified time interval ranging from 1 min to 3 hr. It can be observed that within the first 5 mins the concentration of F^- ions dropped from 10 ppm to less than 3 ppm as shown in Figure 2. The sorption capacity increased from 20 min to about 100 min and remained constant thereafter. This could be attributed to the presence and absence of free reaction sites for F^- ions on the adsorbent surface, respectively.

Effect of pH

To study the effect of pH on the defluoridation capacity of the adsorbent, studies were carried out at various solution pH ranging from 4-10. The average initial concentration of F^- was 10 ppm. Figure 3 shows the effect of pH on fluoride adsorption. Results showed that reduction of fluoride was up to below 1.5 ppm from an initial concentration of 10 ppm in all pH ranges. However greater reduction was observed at pH 8.

Figure 2. Effect of contact time

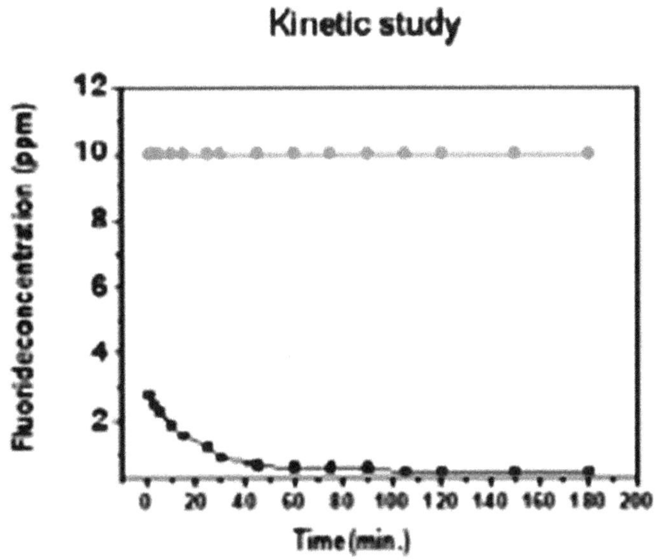

Effect of Counter Ions

Due to its nature, underground contains other dissolved ions that can interfere with the performance of the adsorbent with respect to competition of the adsorption active sites. Common ions include but not limited to bicarbonate, sulfate, chloride, nitrate, carbonate, and phosphate. Initial fluoride concentration was kept at 10 ppm while other parameters remained constant. The studies ions were added individually. The effect of the ions tested is depicted in Figure 4. It is observed that the removal percentage of fluoride was above 90% for all ions except chlorides, silicates, carbonate, bicarbonates and mixture of carbonate and bicarbonates. This can be ascribed to high competition rate for the active sites for these ions.

Figure 3. Effect of solution pH

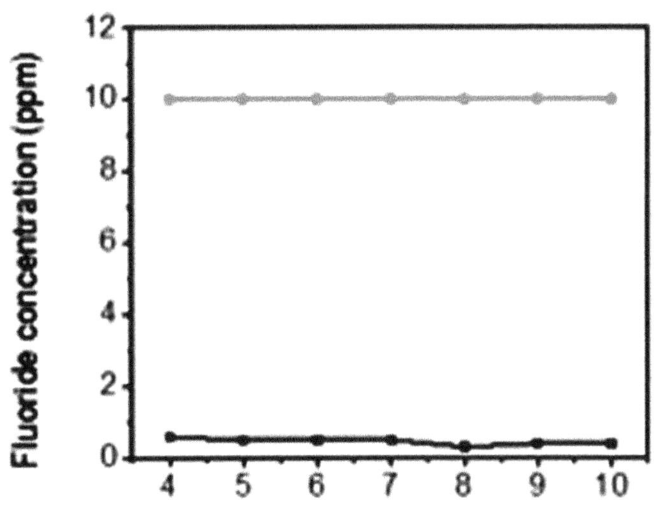

Effect of Dosage

The effect of adsorbent dosage on fluoride uptake was determined by varying the amount of the adsorbent ranging from 5 to 200 mg. The initial fluoride concentration was kept at 10 ppm. Figure 5 presents a reducing pattern in with the increase in the amount of the adsorbent. It is seen that Fluoride concentration dropped from 10 ppm to less than 2 ppm with approximately 50 mg of the adsorbent. This behaviour can be linked to the increase in the number of active adsorption sites. It can also be noted that sorption capacity after and above100 mg the adsorption capacity remained fairly constant (1.5 ppm). This could mean that the free adsorption sites are still available for F^- ion intake.

Adsorption Isotherm Studies

Maximum fluoride adsorption capacity was evaluated using Langmuir adsorption isotherm. The linear plot of Ce/q_e versus C_e in addition to high value of correlation coefficient provides that Langmuir isotherm provides a better fit for the equilibrium data. The Langmuir capacity (mg/g) of the nanomaterial for fluoride adsorption was calculated to be 62.5 mg/g.

CONCLUSION

From the present investigation the following main conclusions were drawn:

1. Chitosan templated aluminium-titanium-zirconium-iron oxyhydroxide composite was found to be an excellent adsorbent for the reduction of fluoride
2. The surface morphology of the nanocomposite as exhibited with the SEM was amorphous. Further, SEM-EDAX analysis of the elemental mapping confirmed the presence of all elements used in the synthesis of the adsorbent. After fluoride adsorption an extra peak of F^- is noted on the spectrum.
3. In relation to competing ions and their influence on the adsorption efficiency, it was recorded that there was a small decline in fluoride ion adsorption in the presence of sodium, sulfates, potassium,

Figure 4. Effect of counter-ions

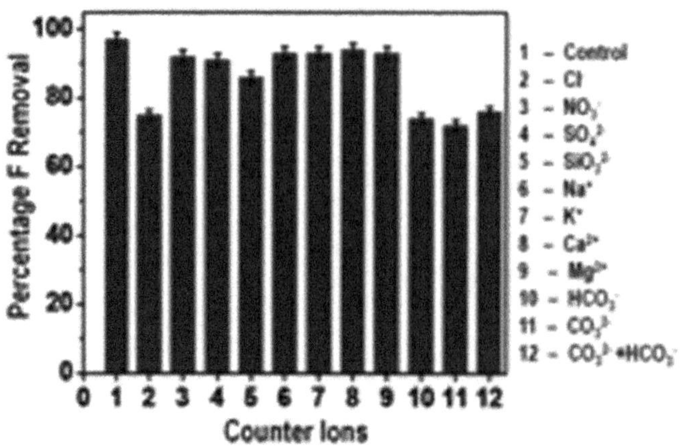

Figure 5. Effect of adsorbent dosage

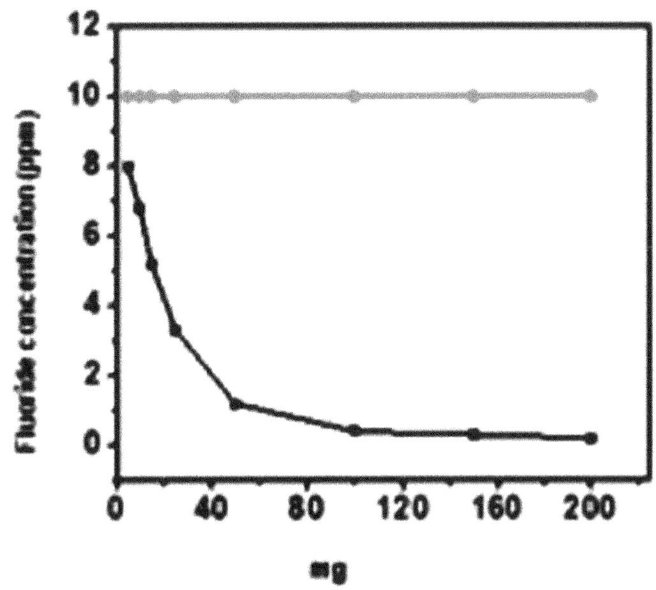

magnesium, calcium and nitrates ions. This could be linked to the increased competition for the adsorption sites.

4. The data fitted well in the Langmuir model (R^2=0.99931) compared to Freundlich model (R^2=0.99352) for fluoride equilibrium studies. Maximum adsorption capacity for fluoride stood at 63 mg/g. This is much better than those reported in literature especially those that can be up scaled.

FUTURE RESEARCH DIRECTIONS

Based on the results of this study the following areas should be considered for further research:

- Assess the efficiency of the prototype in field application for flouride reduction

REFERENCES

Adak, M.K., Mondal, B., Dhak, P., Sen, S., & Dhak, D. (2017). A comparative Study on Fluoride Removal Capacity from Drinking Water by Adsorption using Nano-sized Alumina and Zirconia Modified Alumina Prepared by Chemical Route. *Advances in Water Science and Technology*, *4*(1), 1-10.

Al-Badaii, F., Shuhaimi-Othman, M., & Gasim, M. B. (2013). Water quality assessment of the Semenyih river, Selangor, Malaysia. *Journal of Chemistry*, *2013*, 1–10. doi:10.1155/2013/871056

Armelao, L., Barreca, D., Bottaro, G., Gasparotto, A., Gross, S., Maragno, C., & Tondello, E. (2006). Recenttrends on nanocomposites based on Cu, Ag and Au clusters: A closer look. *Coordination Chemistry Reviews*, *250*(11-12), 1294–1314. doi:10.1016/j.ccr.2005.12.003

Averous, L., & Pollet, E. (2012). Green Nano-biocomposites. *Green. Energy. Technol.*

Bai, B. (2012). *Electrospun chitosan nanofibers for virus removal* (Master's Thesis). Michigan Technological University. Retrieved from http://digitalcommons.mtu.edu/etds/5

Bhatnagar, A., Kumar, E., & Sillanpaa, M. (2011). Fluoride removal from water by adsorption -A review. *Chemical Engineering Journal*, *171*(3), 811–840. doi:10.1016/j.cej.2011.05.028

Bhaumik, M., Leswifi, T. Y., Maity, A., Srinivasu, V. V., & Onyango, M. S. (2011). Removal of fluoride from aqueous solution by polypyrrole/Fe3O4 magnetic nanocomposite. *Journal of Hazardous Materials*, *186*(1), 150–159. doi:10.1016/j.jhazmat.2010.10.098 PMID:21112695

Biswas, K., Bandhoyapadhyay, D., & Ghosh, U. C. (2007). Adsorption kinetics of fluoride on iron(III)-zirconium(IV) hybrid oxide. *Adsorption*, *13*(1), 83–94. doi:10.100710450-007-9000-1

Bootharaju, M., & Pradeep, T. (2010). Uptake of Toxic Metal Ions from Water by Naked and Monolayer Protected Silver Nanoparticles: An X-ray Photoelectron Spectroscopic Investigation. *The Journal of Physical Chemistry C*, *114*(18), 8328–8336. doi:10.1021/jp101988h

CCEFW. (2010). *Excess fluoride in water. Report of the Consultative Committee Kenya Bureau of Standards*. Nairobi: KBS.

Chakrabortty, S., Roy, M., & Pal, P. (2013). Removal of fluoride from contaminated groundwater by cross flow nanofiltration: Transport modelling and economic evaluation. *Desalination*, *313*, 115–124. doi:10.1016/j.desal.2012.12.021

Chen, N., Zhang, Z., Feng, C., Sugiura, N., Li, M., & Chen, R. (2010). Fluoride removal from water by granular ceramicadsorption. *Journal of Colloid and Interface Science*, *348*(2), 579–584. doi:10.1016/j.jcis.2010.04.048 PMID:20510421

Christina, E., & Viswanathan, P. (2015). Development of a novel nano-biosorbent for the removal of fluoride from water. *Chinese Journal of Chemical Engineering*, *23*(6), 924–933. doi:10.1016/j.cjche.2014.05.024

Cloete, E., Michele, D., Marelize, B., & Manuel, L. (2010). *Nanotechnology in Water Treatment Applications*. Caister Academic Press.

Crini, G. (2005). Recent developments in polysaccharide-based materials used as adsorbents in wastewater treatment. *Progress in Polymer Science*, *30*(1), 38–70. doi:10.1016/j.progpolymsci.2004.11.002

Dongre, R., Ghugal, D., Meshram, J., & Ramteke, D. (2012). Fluoride removal from water by zirconium (IV) doped chitosan bio-composite. *African Journal of Environmental Science and Technology*, *6*(2). doi:10.5897/AJEST11.284

Dongre, S., & Tembhurkar, A.R. (2006). Studies on fluoride removal using adsorption process. *J. Environ. Sci. Eng.*, *48*, 151–156.

El-Gohary, F., Tawfik, A., & Mahmoud, U. (2010). Comparative study between chemical coagulation/precipitation (C/P) *versus* coagulation/dissolved air flotation (D/DAF) for pre-treatment of personal care products (PCPs) wastewater. *Desalination*, *252*(1-3), 106–112. doi:10.1016/j.desal.2009.10.016

Fang, S. W., Li, C. F., & Shih, D. Y. C. (1994). Antifungal activity of chitosan and its preservative effect on low-sugar candied kumquat. *Journal of Food Protection*, *56*(2), 136–140. doi:10.4315/0362-028X-57.2.136 PMID:31113143

Garg, P., & Chaudhari, S. (2012). Adsorption of Fluoride from Drinking Water on Magnesium substituted Hydroxyapatite. *International Conference on Future Environment and Energy*, *28*, 180–185.

Guo, Q., & Reardon, E. J. (2012). Fluoride removal from water by meixnerite and its calcination product. *Applied Clay Science*, *56*, 7–15. doi:10.1016/j.clay.2011.11.013

Guo, Q., & Tian, J. (2013). Removal of fluoride and arsenate from aqueous solution by hydrocalumite via precipitation and anion exchange. *Chemical Engineering Journal*, *231*, 121–131. doi:10.1016/j.cej.2013.07.025

Habuda-Stanić, M., Ravančić, M., & Flanagan, A. (2014). A Review on Adsorption of Fluoride from Aqueous Solution. *Materials (Basel)*, *7*(9), 6317–6366. doi:10.3390/ma7096317 PMID:28788194

Hang, C., Li, Q., Gao, S., & Shang, J. K. (2012). As(III) and as(V) Adsorption by hydrous Zirconium oxide Nanoparticles synthesized by a Hydrothermal process followed with heat treatment. *Industrial & Engineering Chemistry Research*, *51*(1), 353–361. doi:10.1021/ie202260g

Homayoni, H., Ravandi, H., & Valizadeh, M. (2009). Electrospinning of chitosan nanofibers: Processing optimization. *Carbohydrate Polymers*, *77*(3), 656–661. doi:10.1016/j.carbpol.2009.02.008

Hwang, K., Kim, J., Yoon, J., & Pyun, R. (1999). Bactericidal activity of chitosan on E. coli. *Advances in Chitin Science*, *62*, 239–243.

Islam, M., & Patel, R. (2011). Thermal activation of basic oxygen furnace slag andevaluation of its fluoride removal efficiency. *Chemical Engineering Journal*, *169*(1-3), 68–77. doi:10.1016/j.cej.2011.02.054

Jadhav, A.S., & Jadhav, M.V. (2014). Use of Maize Husk Fly Ash as an Adsorbent for Removal of Fluoride from Water. *International Journal of Recent Development in Engineering and Technology, 2.*

Jagpat, A. I., Yenkie, M. K., Das, S., & Rayalu, S. (2011). Synthesis and Characterization of Lanthanum impregnated Chitosan Flakes for Fluoride removal in Water. *Desalination*, *273*(2-3), 267–275. doi:10.1016/j.desal.2010.12.032

Jagtap, S., Thakre, D., Sakhare, N., Labhsetwar, N., Meshram, S., & Rayalu, S. (2010). Chitosan based mesoporous Ti–Al binary metal oxide supported beads for defluoridation of water. *Chemical Engineering Journal*, *158*(2), 315–324. doi:10.1016/j.cej.2010.01.008

Jahin, H. S. (2014). Fluoride Removal from Water Using Nanoscale Zero- Valent Iron (NZVI). *International Water Technology Journal*, *4*(3), 173–182.

Joshi, S., Adhikari, M., and Pradhananga, R. (2012). Adsorption of Fluoride Ion ontoZirconyl–Impregnated Activated Carbon Prepared from Lapsi Seed Stone. *J. Nepal Chem. Soc.,* (30), 13-23.

Joshi, S., Adhikari, M., & Pradhananga, R. (2013). Adsorption of Fluoride Ion onto Zirconyl-Impregnated Activated Carbon Prepared from Lapsi Seed Stone. *Journal of Nepal Chemical Society*, *30*(0).

Khichar, M., & Kumbhat, S. (2015). Defluoridation-A review of water from aluminium and alumina based compound. *International Journal of Chemical Studies*, *2*(5), 4–11.

Koshijima, T., Tanaka, R., Muraki, E., Yamada, A., & Yaku, F. (1973). Chelating polymers derived from cellulose and chitin. *Cellulose Chemistry and Technology*, *7*, 197–208.

Kumar, E., Bhatnagar, A., Ji, M., Jung, W., Lee, S., Kim, S., ... Jeon, B. (2009). Defluoridation from aqueous solutions by granular ferric hydroxide (GFH). *Water Research*, *43*(2), 490–498.

Kumar, E., Bhatnagar, A., Kumar, U., & Sillanpaa, M. (2011). Defluoridation from aqueous solution bynano –alumina: Characterization and Sorption studies. *Journal of Hazardous Materials*, *186*(2-3), 1042–1049. doi:10.1016/j.jhazmat.2010.11.102 PMID:21177029

Kumar, E., Bhatnagar, A., Minkyu, J., Jung, W., Lee, S., Kim, S., ... Jeon, B. (2009). Defluoridation from aqueous solutions by granularferric hydroxide (GFH). *Water Research*, *43*(2), 490–499. doi:10.1016/j.watres.2008.10.031 PMID:18995880

Kyzas, G. Z., Kostoglou, M., & Lazaridis, N. K. (2009). Copper and chromium(VI) removal by chitosan derivatives –Equilibrium and kinetic studies. *Chemical Engineering Journal*, *152*(2-3), 440–448. doi:10.1016/j.cej.2009.05.005

Li, Z., Deng, S., Zhang, X., Zhou, W., Huang, J., & Yu, G. (2010). Removal of fluoride from water using titanium-based adsorbents. *Frontiers of Environmental Science & Engineering in China*, *4*(4), 414–420. doi:10.100711783-010-0241-y

Liao, X., & Shi, B. (2005). Adsorption of Fluoride on Zirconium(IV)-Impregnated Collagen Fiber. *Environmental Science & Technology*, *39*(12), 4628–4632. doi:10.1021/es0479944 PMID:16047802

Loganathan, P., Vigneswaran, S., & Kandasamy, J. (2013). Enhanced removal of nitrate from water using surface modification of adsorbents. A review. *Journal of Environmental Management*, *131*, 363–374. doi:10.1016/j.jenvman.2013.09.034 PMID:24211565

Lv, G., Wu, L., Liao, L., Zhang, Y., & Li, Z. (2013). Preparation and characterization of red mud sintered porous materials for water defluoridation. *Applied Clay Science*, *74*, 95–101. doi:10.1016/j.clay.2012.10.004

Mahdavi, S., Jalali, M., & Afkhami, A. (2014). Heavy metals removal from aqueous solutions by Al2O3 nanoparticles modified with natural and chemical modifiers. *Clean Technologies and Environmental Policy*, *17*(1), 85–102. doi:10.100710098-014-0764-1

Meenakshi, G. (2010). Preparation and application of alumina/chitosan Biocomposite. *International Journal of Biological Macromolecules*, *47*(2), 146–154. doi:10.1016/j.ijbiomac.2010.05.008 PMID:20471415

Meenakshi, R. C., & Maheshwari, J. (2006). Fluoride in drinking water and its removal. *Journal of Hazardous Materials*, *137*(1), 456–463. doi:10.1016/j.jhazmat.2006.02.024 PMID:16600479

Memon, M., Soomro, M. S., Akhtar, M. S., & Memon, K. S. (2010). Drinking water quality assessment in southern Sindh (Pakistan). *Environmental Monitoring and Assessment*, *177*(1-4), 39–50. doi:10.100710661-010-1616-z PMID:20658360

Mohapatra, M., Anand, S., Mishra, B. K., Giles, D. E., & Singh, P. (2009). Review of fluoride removal from drinking water. *Journal of Environmental Management*, *91*(1), 67–77. doi:10.1016/j.jenvman.2009.08.015 PMID:19775804

Nair, K. R., Manji, F., & Gitonga, J. N. (1984). The occurrence and distribution of fluoride in groundwaters of Kenya. *East African Medical Journal*, *61*, 503–512. PMID:6545194

Naslund, J., & Snell, I. (2005). *GIS-Mapping of fluoride contaminated groundwater in Nakuru and Baringo District, Kenya* (MSc Thesis). Luleå University of Technology, Luleå, Sweden.

Patnaik, S., Mishra, P., Nayak, R., & Giri, A. (2016). Removal of fluoride from aqueous solution using Chitosan-Iron complex. *Journal of Analytical & Bioanalytical Techniques*, *7*(4).

Prabhu, S., & Meenakshi, S. (2016). Defluoridation of water using dicarboxylic acids mediated chitosan-polyaniline/zirconium biopolymeric complex. *International Journal of Biological Macromolecules*, *85*, 16–22. doi:10.1016/j.ijbiomac.2015.12.022 PMID:26706844

Prabhu, S. M., & Meenakshi, S. (2013). Defluoridation of water using synthesized Zr(IV) encapsulated silica gel/chitosan biocomposite: Adsorption isotherms and kinetic studies. *Desalination and Water Treatment*, *53*(13), 3592–3603. doi:10.1080/19443994.2013.871347

Prasad, K. S., Amin, Y., & Selvaraj, K. (2014). Defluoridation using biomimetically synthesized nano zirconium chitosan composite: Kinetic and equilibrium studies. *Journal of Hazardous Materials*, *276*, 232–24. doi:10.1016/j.jhazmat.2014.05.038 PMID:24887125

Qiusheng, Z., Xiaoyan, L., Jin, Q., Jing, W., & Xuegang, L. (2015). Porous zirconium alginate beads adsorbent for fluoride adsorption from aqueous solutions. *RSC Advances*, *5*(3), 2100–2112. doi:10.1039/C4RA12036A

Qu, X., Alvarez, J., & Li, Q. (2013). Applications of nanotechnology in water and wastewater treatment. *Water Research*, *47*(12), 3931–3946. doi:10.1016/j.watres.2012.09.058 PMID:23571110

Qu, X., Alvarez, P. J. J., & Li, Q. (2013). Applications of nanotechnology in water and wastewater treatment. *Water Research*, *47*(12), 3931–3946. doi:10.1016/j.watres.2012.09.058 PMID:23571110

Ramdani, A., Taleb, S., Benghalem, A., & Ghaffour, N. (2010). Removal of excess fluoride ions from Saharan brackish water by adsorption on natural materials. *Desalination*, *250*(1), 408–413. doi:10.1016/j.desal.2009.09.066

Rhim, J. W., Park, H. M., & Ha, C. S. (2013). Bio-nanocomposites for food packaging applications. *Progress in Polymer Science*, *38*(10-11), 1629–1652. doi:10.1016/j.progpolymsci.2013.05.008

Roco, M. C. (2001). International strategy for nanotechnology research and development. *Journal of Nanoparticle Research*, *3*(5/6), 353–360. doi:10.1023/A:1013248621015

Saifuddin, N., Nian, C.Y., Zhan, L.W., & Ning, K.X. (2011). Chitosan-silvernanoparticles composite as point-of-use drinking water filtrationsystem for household to remove pesticides in water. *Asian JBiochem*. doi:10.3923/ajb.2011

Sepehr, M. N., Sivasankar, V., Zarrabi, M., & Kumar, S. M. (2013). Surface modification of pumice enhancing its fluoride adsorption capacity: An insight into kinetic and thermodynamic studies. *Chemical Engineering Journal*, *228*, 192–204. doi:10.1016/j.cej.2013.04.089

Smith, A. (2006). Nanotech- the way forward for clean water. *Filtration & Separation*, *43*(8), 32–33. doi:10.1016/S0015-1882(06)70976-4

Sudardsha, N., Hoover, G., & Knorr, D. (1992). Antibacterial action of chitosan. *Food Biotechnology*, *6*(3), 257–272. doi:10.1080/08905439209549838

Sudheesh, K., Ajay, K., Omotayo, A., & Bhekie, B. (2013). Chitosan-base nanomaterials: A state-of-the-art review. *International Journal of Biological Macromolecules*, *59*, 46–58. doi:10.1016/j.ijbiomac.2013.04.043 PMID:23608103

Sujana, M. G., & Anand, S. (2013). Ferric hydroxide: Preparation, characterization and fluoride removal studies from water. *Desalination and Water Treatment*, *52*(34-36), 6453–6463. doi:10.1080/19443994.2013.822633

Sun, Y., Fang, Q., Dong, J., Cheng, X., & Xu, J. (2011). Removal of fluoride from drinking water by natural stilbite zeolite modified with Fe(III). *Desalination*, *277*(1-3), 121–127. doi:10.1016/j.desal.2011.04.013

Tang, T., Guan, X., Wang, J., Gao, N., McPhail, M., & Chusuei, C. (2009). Fluoride adsorption onto granular ferric hydroxide: Effects of ionic strength, pH, surface loading, and major co-existing anions. *Journal of Hazardous Materials*, *171*(1-3), 774–779. doi:10.1016/j.jhazmat.2009.06.079 PMID:19616377

Thome, J. P., Jeuniaux, C., & Weltrowski, M. (1997). Applications of chitosan for the elimination of organochlorine xenobiotics from wastewater. In M. F. A. Goosen (Ed.), *Applications of Chitin and Chitosan* (pp. 309–331). Lancaster, PA: Technomic.

Tomar, V., & Kumar, D. (2013). A critical study on efficiency of different materials for fluoride removal from aqueous media. *Chemistry Central Journal*, *7*(1), 51. doi:10.1186/1752-153X-7-51 PMID:23497619

U.S. EPA. (2008). *Nanotechnology for Site Remediation Fact Sheet*. EPA 542-08-009. Available: http://www.epa.gov/tio/download/remed/542-f-08-009.pdf

Van Daele, Y., & Thome, J. P. (1986). Purification of PCB contaminated water by chitosan: A biological test of efficiency using the common barbel, Barbus barbus. *Bulletin of Environmental Contamination and Toxicology*, *37*(1), 858–865. doi:10.1007/BF01607850 PMID:3098331

Viswanathan, P., & Christina, E. C. (2015). Adsorptive Fluoride Removal From Aqueous Solution By Using Saponified Orange Peel Residue Immobilized Sorbent. *International Journal of Pharma and Bio Sciences*, *6*(1), 577–588.

Vivek Vardhan, C., & Srimurali, M. (2016). Removal of fluoride from water using a novel sorbent lanthanum-impregnated bauxite. *SpringerPlus*, *5*(1), 1426. doi:10.118640064-016-3112-6 PMID:27625980

Waghmare, S.,Lataye, D.H., Arfin, T., & Rayalu, S.(2015). Defluoridation by Nano-Materials, Building Materials and Other Miscellaneous Materials:A Systematic Review. *International Journal of Innovative Research in Science, Engineering and Technology,* (4), 1198-1210.

Waghmare, S. S., & Arfin, T. (2015). Fluoride Removal from Water by various techniques: Review. *International Journal of Innovative Science. Engineering & Technology*, *9*(2), 560–571.

Wambu, E. W., & Muthakia, G. K. (2011). High fluoride water in the Gilgil area of Nakuru County, Kenya. *Journal of the International Society for Fluoride Research*, (1), 37-41.

Wasay, S. A., Tokunaga, S., & Park, S.-W. (1996). Removal of hazardous anions from aqueous solutions by la(lll)- and Y(lll)-lmpregnated alumina. *Separation Science and Technology*, *31*(10), 1501–1514. doi:10.1080/01496399608001409

WHO. (2006). *Guidelines for Drinking-water Quality: First Addendum to Volume 1: Recommendations.* Retrieved November 15, 2016, from http://comparamus.us

Zhang, J., Chen, N., Tang, Z., Yu, Y., Hu, Q., & Feng, C. (2015). A study of the mechanism of fluoride adsorption from aqueous solutions onto Fe-impregnated chitosan. *Physical Chemistry Chemical Physics*, *17*(18), 12041–12050. doi:10.1039/C5CP00817D PMID:25872764

Zhao, X., Wang, J., Wu, F., Wang, T., Cai, Y., Shi, Y., & Jiang, G. (2010). Removal of fluoride from aqueous media by Fe3O4@Al(OH)3 magnetic nanoparticles. *Journal of Hazardous Materials*, *173*(1-3), 102–109. doi:10.1016/j.jhazmat.2009.08.054 PMID:19747775

Zhou, Y., Yu, C., & Shan, Y. (2004). Adsorption of fluoride from aqueous solution on La3+ impregnated crosslinked gelatin. *Separation and Purification Technology*, *36*(2), 89–94. doi:10.1016/S1383-5866(03)00167-9

Zhu, B.-S., Jia, Y., Jin, Z., Sun, B., Luo, T., Kong, L.-T., & Liu, J.-H. (2015). A facile precipitation synthesis of mesoporous 2-line ferrihydrite with good fluoride removal properties. *RSC Advances*, *5*(103), 84389–84397. doi:10.1039/C5RA15619J

KEY TERMS AND DEFINITIONS

Defluoridation: The process of removing fluoride from water.
Fluorosis: Permanent demineralization of bones and teeth.

Compilation of References

Abargues, M. R., Ferrer, J., Bouzas, A., & Seco, A. (2013). Removal and Fate of Endocrine Disruptors Chemicals under Lab- Scale Postreatment Stage. Removal Assessment Using Light, Oxygen and Microalgae. *Bioresource Technology*, *149*, 142–148. doi:10.1016/j.biortech.2013.09.051 PMID:24096281

Abdel-Shafy, H. I., & Mansour, M. S. M. (2016). A review on polycyclic aromatic hydrocarbons: Source, environmental impact, the effect on human health and remediation. *Egyptian Journal of Petroleum*, *25*(1), 107–123. doi:10.1016/j.ejpe.2015.03.011

Abdić, Š., Memić, M., & Šabanović, E. (2018). Adsorptive removal of eight heavy metals from aqueous solution by unmodified and modified agricultural waste: Tangerine peel. *International Journal of Environmental Science and Technology*. doi:10.100713762-018-1645-7

Abdulfatah, A. Y., El Hamalawi, A., Wheatley, A. D., & Marto, A. (2015). *The Behaviour of Metal Contaminants in Silty Sand and Gravel*. Int'l Journal of Research in Chemical, Metallurgical and Civil Eng. doi:10.15242/IJRCMCE.E091501

Abong'o, D. A., Wandiga, S. O., & Jumba, I. O. (2018). Occurrence and distribution of organochlorine pesticide residue levels in water, sediment and aquatic weeds in the Nyando River catchment, Lake Victoria, Kenya. *African Journal of Aquatic Science*, *43*(3), 255–270. doi:10.2989/16085914.2018.1490244

Abong'o, D. A., Wandiga, S. O., Jumba, I. O., Van Den, B. P., Nazariwo, B. B., Madadi, V. O., ... Nkedi-Kizza, P. (2015). Organochlorine pesticide residue levels in soil from the Nyando River Catchment, Kenya. *Africa Journal of Physical Sciences*, *2*(1), 18–32.

Aboulhassan, M. A., Souabi, S., Yaacoubi, A., & Baudu, M. (2006). Removal of surfactant from industrial wastewaters by coagulation flocculation process. *International Journal of Environmental Science and Technology*, *3*(4), 327–332. doi:10.1007/BF03325941

Acero, J. L., Benitez, F. J., Teva, F., & Leal, A. I. (2010). Retention of emerging micropollutants from water and a municipal secondary effluent by ultrafiltration and nanofiltration. *Chemical Engineering Journal*, *163*(3), 264–272. doi:10.1016/j.cej.2010.07.060

Adachi, A., Ikeda, C., Takagi, S., Fukao, N., Yoshie, E., & Okano, T. (2001). Efficiency of Rice Bran for Removal of Organochlorine Compounds and Benzene from Industrial Wastewater. *Journal of Agricultural and Food Chemistry*, *49*(3), 1309–1314. doi:10.1021/jf001147c PMID:11312856

Adachi, A., Takagi, S., & Okano, T. (2001). Studies on Removal Efficiency of Rice Bran for Pesticides. *Journal of Health Science*, *47*(2), 94–98. doi:10.1248/jhs.47.94

Adak, M.K., Mondal, B., Dhak, P., Sen, S., & Dhak, D. (2017). A comparative Study on Fluoride Removal Capacity from Drinking Water by Adsorption using Nano-sized Alumina and Zirconia Modified Alumina Prepared by Chemical Route. *Advances in Water Science and Technology*, *4*(1), 1-10.

Adams, C., Asce, M., Wang, Y., Loftin, K., & Meyer, M. (2002). Removal of antibiotics from surface and distilled water in conventional water treatment processes. *Journal of Environmental Engineering*, *128*(3), 253–260. doi:10.1061/(ASCE)0733-9372(2002)128:3(253)

Adams, N. R. (1998). Natural and anthropogenic environmental oestrogens: The scientific basis for risk assessment. *Pure and Applied Chemistry*, *70*(9), 1855–1862. doi:10.1351/pac199870091855

Adedokun, T. C., & Agunwamba, J. C. (2013). Modelling the Effect of Industrial Effluents on Water Quality: "A Case Study of River Challawa in Nigeria. *International Journal of Modern Engineering Research.*, *3*(5), 2797–2803.

Adina, E. S., Cristina, O., Carmen, L, Paula, S., Paulina, V., Cornelia, B., & Ioan, G. (2013). *Waste Water Treatment Methods*. doi:10.5772/53755

Afifi, M., Alkaladi, A., Zinada, O. A. A., & Couderchet, M. (2017). Alteration in antioxidant genes expression in some fish caught from Jeddah and Yanbu coast as a bio-indicator of oil hydrocarbons pollution. *Saudi Journal of Biological Sciences*, *24*(7), 1580–1587. doi:10.1016/j.sjbs.2015.06.014 PMID:30294228

Agrafioti, E., Kalderis, D., & Diamadopoulos, E. (2014). Arsenic and chromium removal from water using biochars derived from rice husk, organic solid wastes and sewage sludge. *Journal of Environmental Management*, *133*, 309–314. doi:10.1016/j.jenvman.2013.12.007 PMID:24412594

Agrawal, N., Verma, P., & Shahi, S. K. (2018). Degradation of polycyclic aromatic hydrocarbons (phenanthrene and pyrene) by the ligninolytic fungi *Ganoderma lucidum* isolated from the hardwood stump. *Bioresources and Bioprocessing*, *5*(1), 11. doi:10.118640643-018-0197-5

Agüera, A., Perez Estrada, L. A., Ferrer, I., Thurman, E. M., Malato, S., & Fernandez-Alba, R. (2005). Application of time-of-flight mass spectrometry to the analysis of phototransformation products of diclofenac in water under natural sunlight. *Journal of Mass Spectrometry*, *40*(7), 908–915. doi:10.1002/jms.867 PMID:15934037

Ahmad Raston, N. H., Nguyen, V.-T., & Gu, M. B. (2016). A new lateral flow strip assay (LFSA) using a pair of aptamers for the detection of Vaspin. *Biosensors & Bioelectronics*. doi:10.1016/j.bios.2016.11.061 PMID:27916536

Ahmad, A. (2003). Policing Industrial Pollution in Nigeria. In *International Environmental Law and Policy in Africa* (pp. 131–150). Springer. doi:10.1007/978-94-017-0135-8_7

Ahmad, W., Al-Sibaai, A. A., Bashammakh, A. S., Alwael, H., & El-Shahawi, M. S. (2015). Recent advances in dispersive liquid-liquid microextraction for pesticide analysis. *Trends in Analytical Chemistry*, *72*, 181–192. doi:10.1016/j.trac.2015.04.022

Ahmed, M. B., Zhou, J. L., Ngo, H. H., Guo, W., & Chen, M. (2016). Progress in the preparation and application of modified biochar for improved contaminant removal from water and wastewater. *Bioresource Technology*, *214*, 836–851. doi:10.1016/j.biortech.2016.05.057 PMID:27241534

Ahmed, S. N., Shreeshivadasan, C., Zuriati, Z., & Saad, A. A. (2015). Treatment Performance of Textile Wastewater Using Electrocoagulation (EC) Process under Combined Electrical Connection of Electrodes Int. *Journal of the Electrochemical Society*, *10*, 5924–5941.

Aina, I., Inger, K., Anders, F., & Roland, M. (2002). High Prevalence of Vancomycin-Resistant Enterococci in Swedish Sewage. ASM. *Applied and Environmental Microbiology*, 2838–2842.

Akaike, H. (1976). An information criterion (AIC). *Math. Sci, 14*, 1–5.

Alani, R., Drouillard, K., Kehinde, O., & Alo, B. (2012). Bioaccumulation of polycyclic aromatic hydrocarbons in fish and invertebrates of Lagos Lagoon, Nigeria. *Journal of Emerging Trends in Engineering and Applied Sciences, 3*(2), 287–296.

Al-Badaii, F., Shuhaimi-Othman, M., & Gasim, M. B. (2013). Water quality assessment of the Semenyih river, Selangor, Malaysia. *Journal of Chemistry, 2013*, 1–10. doi:10.1155/2013/871056

Alder, A. C., McArdell, C. S., Golet, E. M., Molnar, E. M., Kuch, H. M., & Giger, W. (2003). Environmental exposure of antibacterial agents in hospital and municipal wastewater and river water in Switzerland (Abs.). *Proceedings of 3rd International Conference on Pharmaceuticals and Endocrine Disrupting Chemicals in Water*, 236–237.

Alexy, Sch€oll, Kümpel, & Kümmerer. (2004). What do we know about antibiotics in the environment? In K. Kümmerer (Ed.), *Pharmaceuticals in the Environment, 2004*. Springer Berlin Heidelberg.

Ali, I., Asim, M., & Khan, T. A. (2012). Low cost adsorbents for the removal of organic pollutants from wastewater. *Journal of Environmental Management, 113*, 170–183. doi:10.1016/j.jenvman.2012.08.028 PMID:23023039

Allende, L. K., Fletcher, T. D., & Sun, G. (2012). The effect of substrate media on the removal of the removal of Arsenic, boron and iron from an acidic wastewater in planted column reactors. *Chemical Engineering Journal, 179*, 119–130. doi:10.1016/j.cej.2011.10.069

Alvarez, D. A. (1999). *Development of an Integrative Sampling Device for Hydrophilic Organic Contaminants in Aquatic Environments*. Columbia, MO: University of Missouri-Columbia.

Amenu, D. (2014). Assessment of Disinfectant and Antibiotic Resistant Bacteria in Hospital Wastewater. *World Journal of Life Sciences Research., 1*(1), 1–7.

American Public Health Association (APHA). (1998). *Standard methods for Examination of water and waste water* (20th ed.). American Public Health Association Inc.

American Public Health Association (APHA). (1998). Standard methods for examination of water and wastewater. American Public Health Association, American Water Works Association and Water Pollution Control Federation.

Anastassiades, M., Lehotay, S. J., Stajnbaher, D., & Schenck, F. J. (2003). Fast and easy multiresidue method employing acetonitrile extraction/partitioning and "dispersive solid-phase extraction" for the determination of pesticide residues in produce. *Journal of AOAC International, 86*, 412. PMID:12723926

Anastopoulos, I., & Kyzas, G. Z. (2015a). Composts as Biosorbents for Decontamination of Various Pollutants: A Review. *Water, Air, and Soil Pollution, 226*(3), 61. doi:10.100711270-015-2345-2

Anastopoulos, I., & Kyzas, G. Z. (2015b). Progress in batch biosorption of heavy metals onto algae. *Journal of Molecular Liquids, 209*, 77–86. doi:10.1016/j.molliq.2015.05.023

Anderson, K. A., & Hillwalker, W. E. (2008). Bioavailability. Encyclopedia of Ecology, 348–357.

Andrady, A. L. (2011). Microplastics in the marine environment. *Marine Pollution Bulletin, 62*(8), 1596–1605. doi:10.1016/j.marpolbul.2011.05.030 PMID:21742351

Andreozzi, R., Canterino, M., Marotta, R., & Paxeus, N. (2005). Antibiotic removal from wastewaters: The ozonation of amoxicillin. *Journal of Hazardous Materials, 122*(3), 243–250. doi:10.1016/j.jhazmat.2005.03.004 PMID:15967280

Andreozzi, R., Caprio, V., Ciniglia, C., De Champdore, M., Lo Giudice, R., Marotta, R., & Zuccato, E. (2004). Antibiotics in the environment: Occurrence in Italian STPs, fate, and preliminary assessment on algal toxicity of amoxicillin. *Environmental Science & Technology, 38*(24), 6832–6838. doi:10.1021/es049509a PMID:15669346

Annadurai, G., Juang, R. S., & Lee, D. J. (2003). Adsorption of heavy metals from water using banana and orange peels. *Water Science and Technology*, *47*(1), 185–190. doi:10.2166/wst.2003.0049 PMID:12578193

Aparicio, I., Martín, J., Santos, J. L., Malvar, J. L., & Alonso, E. (2017). Stir bar sorptive extraction and liquid chromatography-tandem mass spectrometry determination of polar and non-polar emerging and priority pollutants in environmental waters. *Journal of Chromatography. A*, *1500*, 43–52. doi:10.1016/j.chroma.2017.04.007 PMID:28416215

Appenroth, K-J. (2010). *Definition of "Heavy Metals" and Their Role in Biological Systems*. doi. doi:10.1007/978-3-642-02436-8_2

Arboleda, C., Cabana, H., De Pril, E., Jones, J. P., Jimenez, G. A., & Mejia, A. I. (2013). *Elimination of Bisphenol A and Triclosan using the enzymatic system of autochthonous Colombian forest fungi*. Biotechnology. doi:10.5402/2013/968241

Arbulu, M., Sampedro, M. C., Unceta, N., Gomez-Caballero, A., Goicolea, M. A., & Barrio, R. J. (2011). A retention time locked gas chromatography-mass spectrometry method based on stir-bar sorptive extraction and thermal desorption for automated determination of synthetic musk fragrances in natural and wastewaters. *Journal of Chromatography. A*, *1218*(20), 3048–3055. doi:10.1016/j.chroma.2011.03.012 PMID:21444087

Arinaitwe K, Kiremire B.T., Muir D.C., Fellin P., Li H., Teixeira C., Mubiru D.N. (2016). Legacy and currently used pesticides in the atmospheric environment of Lake Victoria, East Africa. *Sci Total Environ.*, *543*(A), 9-18. doi:10.1016/j.scitotenv.2015.10.146

Armelao, L., Barreca, D., Bottaro, G., Gasparotto, A., Gross, S., Maragno, C., & Tondello, E. (2006). Recenttrends on nanocomposites based on Cu, Ag and Au clusters: A closer look. *Coordination Chemistry Reviews*, *250*(11-12), 1294–1314. doi:10.1016/j.ccr.2005.12.003

Arthur, C., & Baker, J. (Eds.). (2011). *Proceedings of the Second Research Workshop on Microplastic Debris*. NOAA Technical Memorandum NOS-OR&R-39.

Arthur, C., Baker, J., & Bamford, H. (2009). *Proceedings of the International Research Workshop on the Occurrence, Effects, and Fate of Microplastic Marine Debris*. Academic Press.

Arthur, C. L., & Pawliszyn, J. (1990). Solid phase microextraction with thermal desorption using fused silica optical fibers. *Analytical Chemistry*, *62*(19), 2145–2148. doi:10.1021/ac00218a019

Arukwe, A., Eggen, T., & Möder, M. (2012). Solid waste deposits as a significant source of contaminants of emerging concern to the aquatic and terrestrial environments—A developing country case study from Owerri, Nigeria. *The Science of the Total Environment*, *438*, 94–102. doi:10.1016/j.scitotenv.2012.08.039 PMID:22975307

Arvand, M., Gholizadeh, T. M., & Zanjanchi, M. A. (2012). MWCNTs/$Cu(OH)_2$ nanoparticles/IL nanocomposite modified glassy carbon electrode as a voltammetric sensor for determination of the non-steroidal anti-inflammatory drug diclofenac. *Materials Science & Engineering C-Materials for Biological Applications*, *32*(6), 1682–1689. doi:10.1016/j.msec.2012.04.066 PMID:24364977

Asgher, M., Shahid, M., Kamal, S., & Iqbal, H. M. N. (2014). Recent trends and valorization of immobilization strategies and ligninolytic enzymes by industrial biotechnology. *Journal of Molecular Catalysis. B, Enzymatic*, *101*, 56–66. doi:10.1016/j.molcatb.2013.12.016

Ashraf, S., Ali, Q., Zahir, Z. A., Ashraf, S., & Asghar, H. N. (2019). Phytoremediation: Environmentally sustainable way for reclamation of heavy metal polluted soils. *Ecotoxicology and Environmental Safety*, *174*, 714–727. doi:10.1016/j.ecoenv.2019.02.068 PMID:30878808

ATSDR. (1995). *Toxicological profile for otto fuels II, U.S. Department of Health and Human Services*. Atlanta, GA: U.S. Department of Health and Human Services, Public Health Service.

ATSDR. (2005). *Toxicology profile for polyaromatic hydrocarbons. ATSDR's Toxicological Profiles on CD-ROM*. Boca Raton, FL: CRC Press.

Averous, L., & Pollet, E. (2012). Green Nano-biocomposites. *Green. Energy. Technol.*

Aziz, H. A., Mohd, N. A. K., & Ariffin, S. (2008). Heavy metals (Cd, Pb, Zn, Ni, Cu and Cr (III)) removal from waterin Malaysia: Post treatment by high quality limestone. *Bioresource Technology*, *99*(6), 1578–1583. doi:10.1016/j.biortech.2007.04.007 PMID:17540556

Aznar, R., Sánchez-Brunete, C., Albero, B., Rodríguez, J. A., & Tadeo, J. L. (2014). Occurrence and analysis of selected pharmaceutical compounds in soil from Spanish agricultural fields. *Environmental Science and Pollution Research International*, *21*(6), 4772–4782. doi:10.100711356-013-2438-7 PMID:24363053

Badejo, A., Adekunle, A., Adeegbe, T., & Nwosu, G. (2018). Auto-Purification Response Of Ona River, Ibadan To Industrial And Domestic Effluent Discharge. FULafia Journal of Science & Technology, 4, 74 – 78.

Bai, B. (2012). *Electrospun chitosan nanofibers for virus removal* (Master's Thesis). Michigan Technological University. Retrieved from http://digitalcommons.mtu.edu/etds/5

Bailey, S. E., Olin, T. J., Bricka, R. M., & Adrian, D. D. (1999). A review of potentially low-cost sorbents for heavy metals. *Water Research*, *33*(11), 2469–2479. doi:10.1016/S0043-1354(98)00475-8

Bakir, A., Rowland, S. J., & Thompson, R. C. (2014). Enhanced desorption of persistent organic pollutants from microplastics under simulated physiological conditions. *Environ. Pollut.*, *185*, 16–23. doi:10.1016/j.envpol.2013.10.007 PMID:24212067

Baltussen, E., Sandra, P., David, F., & Cramers, C. (1999). Stir bar sorptive extraction (SBSE), a novel extraction technique for aqueous samples: Theory and principles. *The Journal of Microcolumn Separations*, *11*(10), 737–747. doi:10.1002/(SICI)1520-667X(1999)11:10<737::AID-MCS7>3.0.CO;2-4

Barnes, D. K. A., Galgani, F., Thompson, R. C., & Barlaz, M. (2009). Accumulation and fragmentation of plastic debris in global environments. *Philosophical Transactions of the Royal Society of London. Series B, Biological Sciences*, *364*(1526), 1985–1998. doi:10.1098/rstb.2008.0205 PMID:19528051

Barreca, S., Busetto, M., Colzani, L., Clerici, L., Daverio, D., Dellavedova, P., ... Ubaldi, V. (2019). Determination of estrogenic endocrine disruptors in water at sub-ng L−1 levels in compliance with Decision 2015/495/EU using offline-online solid phase extraction concentration coupled with high performance liquid chromatography-tandem mass spectrometry. *Microchemical Journal*, *147*, 1186–1191. doi:10.1016/j.microc.2019.04.030

Barroso, P. J., Santos, J. L., Martín, J., Aparicio, I., & Alonso, E. (2019). Emerging contaminants in the atmosphere: Analysis, occurrence and future challenges. *Critical Reviews in Environmental Science and Technology*, *49*(2), 104–171. doi:10.1080/10643389.2018.1540761

Barth, T., Calixto, L. A., Jabor, V. A. P., & Borges, K. B. (2015). Extração Líquido-Líquido. In Preparo de amostras para análise de compostos orgânicos. Rio de Janeiro: LTC.

Basheer, C., Vetrichelvan, M., Valiyaveettil, S., & Kee Lee, H. (2007). On-site polymer-coated hollow fiber membrane microextraction and gas chromatography-mass spectrometry of polychlorinated biphenyls and polybrominated diphenyl ethers. *Journal of Chromatography. A*, *139*(2), 157–164. doi:10.1016/j.chroma.2006.11.008 PMID:17118376

Batt, A. L., Kostich, M. S., & Lazorchak, J. M. (2008). Analysis of ecologically relevant pharmaceuticals in wastewater and surface water using selective solid-phase extraction and UPLC-MS/MS. *Analytical Chemistry*, *80*(13), 5021–5030. doi:10.1021/ac800066n PMID:18498179

Beckingham, B., & Ghosh, U. (2017). Differential bioavailability of polychlorinated biphenyls associated with environmental particles: Microplastic in comparison to wood, coal and biochar. *Environ. Pollut.*, *220*, 150–158. doi:10.1016/j.envpol.2016.09.033 PMID:27650963

Beita-Sandí, W., Ersan, M. S., Uzun, H., & Karanfil, T. (2016). Removal of N-nitrosodimethylamine precursors with powdered activated carbon adsorption. *Water Research*, *88*, 711–718. doi:10.1016/j.watres.2015.10.062 PMID:26584342

Benede, J. L., Chisvert, A., Salvador, A., Sanchez-Quiles, D., & Tovar-Sanchez, A. (2014). Determination of UV filters in both soluble and particulate fractions of seawaters by dispersive liquid–liquid microextraction followed by gas chromatography-mass spectrometry. *Analytica Chimica Acta*, *812*, 50–58. doi:10.1016/j.aca.2013.12.033 PMID:24491764

Benson, C. I., Sisipho, G., Larry, C. O., Uchechukwu, U. N., & Anthony, I. O. (2015). Antibiotic Susceptibilities of Enterococcus Species Isolated from Hospital and Domestic Wastewater Effluents in Alice, Eastern Cape Province of South Africa. *International Journal of Environmental Research and Public Health*, *12*(4), 4231–4246. doi:10.3390/ijerph120404231 PMID:25893999

Bera, R., Kundu, S., & Patra, A. (2015). 2D Hybrid Nanostructure of Reduced Graphene Oxide-CdS Nanosheet for Enhanced Photocatalysis. *ACS Applied Materials & Interfaces*, *7*(24), 13251–13259. doi:10.1021/acsami.5b03800 PMID:26029992

Berge, A. C. B., Dueger, E. L., & Sischo, W. M. (2006). Comparison of Salmonella enterica serovar distribution and antibiotic resistance patterns in wastewater at municipal water treatment plants in two California cities. *Journal of Applied Microbiology*.

Besseling, E., Wegner, A., Foekema, E. M., van den Heuvel-Greve, M. J., & Koelmans, A. A. (2012). Effects of microplastic on fitness and PCB bioaccumulation by the lugworm Arenicola marina (L.). *Environmental Science & Technology*, *47*(1), 593–600. doi:10.1021/es302763x PMID:23181424

Bhagat, J., Sarkar, A., & Ingole, B. S. (2016). DNA Damage and Oxidative Stress in Marine Gastropod *Morula granulata* Exposed to Phenanthrene. *Water, Air, and Soil Pollution*, *227*(4), 114. doi:10.100711270-016-2815-1

Bhatnagar, A., Kumar, E., & Sillanpaa, M. (2011). Fluoride removal from water by adsorption -A review. *Chemical Engineering Journal*, *171*(3), 811–840. doi:10.1016/j.cej.2011.05.028

Bhaumik, M., Leswifi, T. Y., Maity, A., Srinivasu, V. V., & Onyango, M. S. (2011). Removal of fluoride from aqueous solution by polypyrrole/Fe3O4 magnetic nanocomposite. *Journal of Hazardous Materials*, *186*(1), 150–159. doi:10.1016/j.jhazmat.2010.10.098 PMID:21112695

Bigus, P., Tobiszewski, M., & Namieśnik, J. (2014). Historical records of organic pollutants in sediment cores. *Marine Pollution Bulletin*, *78*(1-2), 26–42. doi:10.1016/j.marpolbul.2013.11.008 PMID:24300286

Birošová, L., Nagyová, K., Mackuľak, T., & Medveďová, A. (2015). The Occurrence of Antibiotic Resistant Escherichia Coli and Staphylococcus Aureus In Wastewater and Sludge From Treatment Plant In Czech Republic. *Proceedings of the 14th International Conference on Environmental Science and Technology*.

Bisesi, J. H. Jr, Sweet, L. E., van den Hurk, P., & Klaine, S. J. (2016). Effects of an antidepressant mixture on the brain serotonin and predation behavior of hybrid striped bass. *Environmental Toxicology and Chemistry*, *35*(4), 938–945. doi:10.1002/etc.3114 PMID:26076900

Biswas, K., Bandhoyapadhyay, D., & Ghosh, U. C. (2007). Adsorption kinetics of fluoride on iron(III)-zirconium(IV) hybrid oxide. *Adsorption*, *13*(1), 83–94. doi:10.100710450-007-9000-1

Blair, B. D. J. P., Crago, J. P., Hedman, C. J., & Klaper, R. D. (2013). Pharmaceuticals and personal care products found in the Great Lakes above concentrations of environmental concern. *Chemosphere*, *93*(9), 2116–2123. doi:10.1016/j.chemosphere.2013.07.057 PMID:23973285

Bloutsos, A. A., & Giannopoulos, P. C. (2011). Concentrations of selected toxic elements in airborne particulates of Patras, Greece. *Global NEST Journal*, *13*(2), 109–118.

Boehler, S., Lörracher, A. K., Schubert, J., & Braunbeck, T. (2018). Comparative live-imaging of in vivo EROD (ethoxyresorufin- O -deethylase) induction in zebrafish (Danio rerio) and fathead minnow (Pimephales promelas) embryos after exposure to PAHs and river sediment extracts. *The Science of the Total Environment*, *621*, 827–838. doi:10.1016/j.scitotenv.2017.11.101 PMID:29202294

Bolong, N., Ismail, A. F., Salim, M. R., & Matsuura, T. (2009). A review of the effects of emerging contaminants in wastewater and options for their removal. *Desalination*, *239*(1–3), 229–246. doi:10.1016/j.desal.2008.03.020

Book, C. (2016). *Germanium CAS#_ 7440-56-4*. Retrieved from http://www.chemicalbook.com/ProductChemicalPropertiesCB7733835_EN.htm#MSDSA

Bootharaju, M., & Pradeep, T. (2010). Uptake of Toxic Metal Ions from Water by Naked and Monolayer Protected Silver Nanoparticles: An X-ray Photoelectron Spectroscopic Investigation. *The Journal of Physical Chemistry C*, *114*(18), 8328–8336. doi:10.1021/jp101988h

Bound, J. P., & Voulvoulis, N. (2005). Household disposal of pharmaceuticals as a pathway for aquatic contamination in the United Kingdom. *Environmental Health Perspectives*, *113*(12), 1705–1711. doi:10.1289/ehp.8315 PMID:16330351

Bowa, K. O., Lalah, J. O., & Wandiga, S. O. (2009). Spatial and seasonal variations in concentrations of polycyclic aromatic hydrocarbons in water and sediment of Kisumu City Bay of Winam Gulf, Lake Victoria-Kenya. *Bulletin of Environmental Contamination and Toxicology*, *83*, 722–733.

Bozokalfa, G., Akbulut, H., Demir, B., Guler, E., Gumus, Z. P., Odaci Demirkol, D., ... Yagci, Y. (2016). Polypeptide Functional Surface for the Aptamer Immobilization: Electrochemical Cocaine Biosensing. *Analytical Chemistry*, *88*(7), 4161–4167. doi:10.1021/acs.analchem.6b00760 PMID:26928030

Brausch, J. M., & Rand, G. M. (2011). A review of personal care products in the aquatic environment: Environmental concentrations and toxicity. *Chemosphere*, *82*(11), 1518–1532. doi:10.1016/j.chemosphere.2010.11.018 PMID:21185057

Brennecke, D., Duarte, B., Paiva, F., Caçador, I., & Canning-Clode, J. (2016). Microplastics as vector for heavy metal contamination from the marine environment. *Estuarine, Coastal and Shelf Science*, *178*, 189–195. doi:10.1016/j.ecss.2015.12.003

Brette, F., Shiels, H. A., Galli, G. L. J., Cros, C., Incardona, J. P., Scholz, N. L., & Block, B. A. (2017). A Novel Cardiotoxic Mechanism for a Pervasive Global Pollutant. *Scientific Reports*, *7*(1), 41476. doi:10.1038rep41476 PMID:28139666

Brodin, T., Fick, J., Jonsson, M., & Klaminder, J. (2013). Dilute Concentrations of a Psychiatric Drug Alter Behavior of Fish from Natural Populations. *Science*, *339*(6121), 814–815. doi:10.1126cience.1226850 PMID:23413353

Brown, R. J., de Banate, M. A., & Rother, K. I. (2010). Artificial sweeteners: a systematic review of metabolic effects in youth. *International Journal of Pediatric Obesity*, *5*, 305–312.

Browne, M. A., Niven, S. J., Galloway, T. S., Rowland, S. J., & Thompson, R. C. (2013). Microplastic moves pollutants and additives to worms, reducing functions linked to health and biodiversity. *Current Bio.*, *23*(23), 2388–2392. doi:10.1016/j.cub.2013.10.012 PMID:24309271

Budavari, S., O'Neil, M. J., Smith, A., & Heckran, P. E. (1989). The Merck Index. In *An encyclopedia of chemicals, drugs, and biologicals* (11th ed.). Rahway, NJ: Merck.

Burke, V., Richter, V., Hass, D., Duennbier, U., Greskowiak, J., & Massmann, G. (2014). Redox-dependent removal of 27 organic trace pollutants: Compilation of results from tank aeration experiments. *Environmental Earth Sciences*, *7*(8), 3685–3695. doi:10.100712665-013-2762-8

Buszewski, B., & Szultka, M. (2012). Past, Present, and Future of Solid Phase Extraction: A Review. *Critical Reviews in Analytical Chemistry*, *42*(3), 198–213. doi:10.1080/07373937.2011.645413

Cabezaa, Y., Candelaa, L., & Ronenb, D. (2012). Monitoring the occurrence of emerging contaminants in treated wastewater and groundwater between 2008 and 2010. The Baix Llobregat. *Journal of Hazardous Materials*, *239*, 32–39. doi:10.1016/j.jhazmat.2012.07.032 PMID:22877748

Caliman, F. A., & Gavrilescu, M. (2009). Pharmaceuticals, personal care products and endocrine disrupting agents in the environment - A review. *Clean - Soil, Air. Water (Basel).*

Caliskan, N., Kul, A. R., Alkan, S., Sogut, E. G., & Alacabey, I. (2011). Adsorption of ZincII on diatomite and manganese-oxide-modified diatomite: A kinetic and equilibrium study. *Journal of Hazardous Materials*, *193*, 27–36. doi:10.1016/j.jhazmat.2011.06.058 PMID:21764214

Callén, M. S., López, J. M., Iturmendi, A., & Mastral, A. M. (2013). Nature and sources of particle associated polycyclic aromatic hydrocarbons (PAH) in the atmospheric environment of an urban area. *Environmental Pollution*, *183*, 166–174. doi:10.1016/j.envpol.2012.11.009 PMID:23245526

Carballa, M., Omil, F., & Lema, J. M. (2005). Removal of cosmetic ingredients and pharmaceuticals in sewage primary treatment. *Water Research*, *39*(19), 4790–4796. doi:10.1016/j.watres.2005.09.018 PMID:16246396

Carlsson, G., Örn, S., & Larsson, D. G. J. (2009). Effluent from bulk drug production is toxic to aquatic vertebrates. *Environmental Toxicology and Chemistry*, *28*(12), 2656. doi:10.1897/08-524.1 PMID:19186889

Carmalin, S. A., & Eder, C. L. (2018). Removal of emerging contaminants from the environment by adsorption. *Ecotoxicology and Environmental Safety*. PMID:29253687

Carpenter, E. J., Anderson, S. J., Harvey, G. R., Miklas, H. P., & Peck, B. B. (1972). Polystyrene spherules in coastal waters. *Sci.*, *178*(4062), 749–750. doi:10.1126cience.178.4062.749 PMID:4628343

Carpenter, E. J., & Smith, K. L. Jr. (1972). Plastics on the Sargasso Sea surface. *Sci.*, *175*(4027), 1240–1241. doi:10.1126cience.175.4027.1240 PMID:5061243

Carson, R. (1962). *Silent Spring* (1st ed.). Houghton Mifflin Company.

CCEFW. (2010). *Excess fluoride in water. Report of the Consultative Committee Kenya Bureau of Standards*. Nairobi: KBS.

Cerniglia, C. E., & Heitkamp, M. A. (1989). Microbial degradation of polycyclic aromatic hydrocarbons(PAH) in the aquatic environment. In U. Varanasi (Ed.), *Metabolism of polycyclic aromatic hydrocarbons in the aquatic environment* (pp. 41–68). Boca Raton, FL: CRC Press, Inc.

Chakrabortty, S., Roy, M., & Pal, P. (2013). Removal of fluoride from contaminated groundwater by cross flow nanofiltration: Transport modelling and economic evaluation. *Desalination*, *313*, 115–124. doi:10.1016/j.desal.2012.12.021

Chakraborty, S., & Kumar, R. N. (2016). Assessment of groundwater quality at MSW landfill site using standard and AHP based water quality index : A case study from Ranchi, Jharkhand, India. *Environmental Monitoring and Assessment*, *188*(335), 18. PMID:27155859

Chang, B.-V., Fan, S.-N., Tsai, Y.-C., Chung, Y.-L., Tu, P.-X., & Yang, C.-W. (2018). Removal of emerging contaminants using spent mushroom compost. *The Science of the Total Environment*, *634*, 922–933. doi:10.1016/j.scitotenv.2018.03.366 PMID:29660886

Chang, S., & Yeh, W. W.-G. (1973). Optimal allocation of artificial aeration along a polluted stream using dynamic programming. *Water Resources Bulletin*, *9*(5), 985–997. doi:10.1111/j.1752-1688.1973.tb05824.x

Chang, X., Meyer, M. T., Liu, X., Zhao, Q., Chen, H., Chen, J., ... Shu, W. (2010). Determination of antibiotics in sewage from hospitals, nursery and slaughter house, wastewater treatment plant and source water in Chongqing region of Three Gorge Reservoir in China. *Environ.Pollut.*, *158*(5), 1444–1450. doi:10.1016/j.envpol.2009.12.034 PMID:20096493

Chaudhary, P., Sharma, R., Singh, S. B., & Nain, L. (2011). Bioremediation of PAH by *Streptomyces* sp. *Bulletin of Environmental Contamination and Toxicology*, *86*(3), 268–271. doi:10.100700128-011-0211-5 PMID:21301805

Chaukura, N., Gwenzi, W., Tavengwa, N., & Manyuchi, M. M. (2016). Biosorbents for the removal of synthetic organics and emerging pollutants: Opportunities and challenges for developing countries. *Environmental Development*, *19*, 84–89. doi:10.1016/j.envdev.2016.05.002

Chelliapan, S., & Golar, S. (2011). Pre-treatment of Antibiotic Wastewater Using an Anaerobic Reactor. *Research Journal of Chemical Sciences*, *1*(6), 59–63.

Chelliapan, S., Wilby, T., & Sallis, P. J. (2006). Performance of an up-flow anaerobic stage reactor (UASR) in the treatment of pharmaceutical wastewater containing macrolide antibiotics. *Water Research*, *40*(3), 507–516. doi:10.1016/j.watres.2005.11.020 PMID:16387347

Chena, F., Hubera, C., May, R., & Schröder, P. (2016). Metabolism of oxybenzone in a hairy root culture: Perspectives for phytoremediation of a widely used sunscreen agent. *Journal of Hazardous Materials*, *306*, 230–236. doi:10.1016/j.jhazmat.2015.12.022 PMID:26736174

Chen, G. (2004). Electrochemical technologies in wastewater treatment. *Separation and Purification Technology*, *38*(1), 11–41. doi:10.1016/j.seppur.2003.10.006

Chen, G., Yu, Y., Zheng, K., Ding, T., Wang, W., Jiang, Y., & Yang, Q. (2015). Fabrication of Ultrathin Bi_2S_3 Nanosheets for High-Performance, Flexible, Visible-NIR Photodetectors. *Small*, *11*(24), 2848–2855. doi:10.1002mll.201403508 PMID:25703599

Chen, H., & Zhang, M. (2013). Effects of Advanced Treatment Systems on the Removal of Antibiotic Resistance Genes in Wastewater Treatment Plants from Hangzhou, China. *Environmental Science & Technology*, *47*(15), 8157–8163. PMID:23802698

Chen, N., Zhang, Z., Feng, C., Sugiura, N., Li, M., & Chen, R. (2010). Fluoride removal from water by granular ceramicadsorption. *Journal of Colloid and Interface Science*, *348*(2), 579–584. doi:10.1016/j.jcis.2010.04.048 PMID:20510421

Chen, Y., Hu, C., Hu, X., & Qu, J. (2009). Indirect Photodegradation of Amine Drugs in Aqueous Solution under Simulated Sunlight. *Environmental Science & Technology*, *43*(8), 2760–2765. doi:10.1021/es803325j PMID:19475946

Chiaradia, M. C., Collins, C. H., & Jardim, I. C. S. F. (2008). O estado da arte da cromatografia associada à espectrometria de massas acoplada à espectrometria de massas na análise de compostos tóxicos em alimentos. *Quimica Nova*, *31*(3), 623–636. doi:10.1590/S0100-40422008000300030

Chirikona, F., Filipovic, M., Ooko, S., & Orata, F. (2015). Perfluoroalkyl acids in selected wastewater treatment plants and their discharge load within the Lake Victoria basin in Kenya. *Environmental Monitoring and Assessment*, *187*(5), 238. doi:10.100710661-015-4425-6 PMID:25861900

Chitana, S. (2002). *Removal of heavy metals from wastewater using chitosan* (MSc thesis). Suranaree University of Technology.

Choi, K.-J., Son, H.-J., & Kim, S.-H. (2007). Ionic treatment for removal of sulphonamide and tetracycline classes of antibiotic. *The Science of the Total Environment*, *387*(1-3), 247–256. doi:10.1016/j.scitotenv.2007.07.024 PMID:17764725

Chokwe, T. B., Okonkwo, J. O., Sibali, L. L., & Ncube, E. J. (2015). Alkylphenol ethoxylates and brominated flame retardants in water, fish (Carp) and sediment samples from the Vaal River, South Africa. *Environmental Science and Pollution Research International*, *22*(15), 11922–11929. doi:10.100711356-015-4430-x PMID:25869432

Chong, H. L. H., Chia, P. S., & Ahmad, M. N. (2013). The adsorption of heavy metal by Bornean oil palm shell and its potential application as constructed wetland media. *Bioresource Technology*, *130*, 181–186. doi:10.1016/j.biortech.2012.11.136 PMID:23306127

Christina, E., & Viswanathan, P. (2015). Development of a novel nano-biosorbent for the removal of fluoride from water. *Chinese Journal of Chemical Engineering*, *23*(6), 924–933. doi:10.1016/j.cjche.2014.05.024

Clara, M., Strenn, B., Gans, O., Martinez, E., Kreuzinger, N., & Kroiss, H. (2005). Removal of selected pharmaceuticals, fragrances and endocrine disrupting compounds in a membrane bioreactor and conventional wastewater treatment plants. *Water Research*, *39*(19), 4797–4807. doi:10.1016/j.watres.2005.09.015 PMID:16242170

Cloete, E., Michele, D., Marelize, B., & Manuel, L. (2010). *Nanotechnology in Water Treatment Applications*. Caister Academic Press.

Coimbra, R., Escapa, C., Vázquez, N., Noriega-Hevia, G., Otero, M., Coimbra, R. N., & Otero, M. (2018). Utilization of Non-Living Microalgae Biomass from Two Different Strains for the Adsorptive Removal of Diclofenac from Water. *Water (Basel)*, *10*(10), 1401. doi:10.3390/w10101401

Collado, N., Buttiglieri, G., Marti, E., Ferrando-Climent, L., Rodriguez-Mozaz, S., Barcelo, D., ... Rodriguez-Roda, I. (2013). Effects on Activated Sludge Bacterial Community Exposed to Sulfamethoxazole. *Chemosphere*, *93*(1), 99–106. doi:10.1016/j.chemosphere.2013.04.094 PMID:23726012

Colleen, M. H. (2001). Fendorf characterization of fe plaque and associated metals on the roots of mine-waste impacted aquatic plants. *Environmental Science & Technology*, *35*(19), 3863–3868. doi:10.1021/es0105459 PMID:11642445

Collins, C.H., Braga, G.L., & Bonato, P.S. (2009). *Fundamentos de Cromatografia* (2nd ed.). Campinas: Unicamp.

Committee Report. (1999). Emerging Pathogens: Viruses, protozoa and Algal Toxins. *AWWA*, *91*(9), 110–121.

Cong, X., Fan, G.-C., Wang, X., Abdel-Halim, E. S., & Zhu, J.-J. (2016). Enhanced photoelectrochemical aptasensing platform amplified through the sensitization effect of CdTe@CdS core-shell quantum dots coupled with exonuclease-I assisted target recycling. *Journal of Materials Chemistry. B, Materials for Biology and Medicine*, *4*(36), 6117–6124. doi:10.1039/C6TB01807F

Connett, P. (1998). Municipal Waste Incineration: A Poor Solution for the Twenty First Century. *4th Annual International Management Conference*, 1-15.

Cooper, D. A., & Corcoran, P. L. (2010). Effects of mechanical and chemical processes on the degradation of plastic beach debris on the island of Kauai, Hawaii. *Marine Pollution Bulletin*, *60*(5), 650–654. doi:10.1016/j.marpolbul.2009.12.026 PMID:20106491

Covich, A. P., Palmer, M. A., & Crowl, T. A. (1999). The Role of Benthic Invertebrate Species in Freshwater Ecosystems. *Bioscience*, *49*(2), 119. doi:10.2307/1313537

Craggs, R., Sutherland, D., & Campbell, H. (2012). Hectare-scale demonstration of high rate algal ponds for enhanced wastewater treatment and biofuel production. *Journal of Applied Phycology*, *24*(3), 329–337. doi:10.100710811-012-9810-8

Crago, J., Bui, C., Grewal, S., & Schlenk, D. (2016). Age-dependent effects in fathead minnows from the anti-diabetic drug metformin. *General and Comparative Endocrinology*, *232*, 185–190. doi:10.1016/j.ygcen.2015.12.030 PMID:26752244

Crini, G. (2005). Recent developments in polysaccharide-based materials used as adsorbents in wastewater treatment. *Progress in Polymer Science*, *30*(1), 38–70. doi:10.1016/j.progpolymsci.2004.11.002

Curran, A. M., Rabin, S. I., Prada, P. A., & Furton, K. G. (2005). Comparison of the Volatile Organic Compounds Present in Human Odor Using SPME-GC/MS. *Journal of Chemical Ecology*, *31*(7), 1607–1619. doi:10.100710886-005-5801-4 PMID:16222796

Cypher, A. D., Consiglio, J., & Bagatto, B. (2017). Hypoxia exacerbates the cardiotoxic effect of the polycyclic aromatic hydrocarbon, phenanthrene in *Danio rerio*. *Chemosphere*, *183*, 574–581. doi:10.1016/j.chemosphere.2017.05.109 PMID:28570901

da Costa, J. P., Duarte, A. C., & Rocha-Santos, T. A. P. (2017). Chapter 1-microplastics - occurrence, fate and behavior in the environment. *Comprehensive Analytical Chemistry*, *75*, 1–24. doi:10.1016/bs.coac.2016.10.004

Dailianis, S., Tsarpali, V., Melas, K., Karapanagioti, H. K., & Manariotis, I. D. (2014). Aqueous phenanthrene toxicity after high-frequency ultrasound degradation. *Aquatic Toxicology (Amsterdam, Netherlands)*, *147*, 32–40. doi:10.1016/j.aquatox.2013.12.006 PMID:24373890

Dai, Y., Niu, J., Yin, L., Xu, J., & Xu, J. (2013). Laccase-carrying electrospun fibrous membrane for the removal of polycyclic aromatic hydrocarbons from contaminated water. *Separation and Purification Technology*, *104*, 1–8. doi:10.1016/j.seppur.2012.11.013

Danesh, N. M., Ramezani, M., Emrani, A. S., Abnous, K., & Taghdisi, S. M. (2016). A novel electrochemical aptasensor based on arch-shape structure of aptamer-complimentary strand conjugate and exonuclease I for sensitive detection of streptomycin. *Biosensors & Bioelectronics*, *75*, 123–128. doi:10.1016/j.bios.2015.08.017 PMID:26302000

Darnerud, P. O., Eriksen, G. S., Jóhannesson, T., Larsen, P. B., & Viluksela, M. (2001). Polybrominated diphenyl ethers: occurrence, dietary exposure, and toxicology. *Environmental Health Perspectives, 109*, 49–68.

Daughton, C. G. (2005). "Emerging" Chemicals as Pollutants in the Environment: A 21st Century Perspective. *Renewable Resources Journal*.

Daughton, C. G., & Ternes, T. A. (1999). Pharmaceuticals and personal care products in the environment: Agents of subtle change? *Environmental Health Perspectives*, *107*(suppl 6), 907–938. doi:10.1289/ehp.99107s6907 PMID:10592150

David, A. C. (2000). *Water resource engineering*. Prentice-Hall.

Davidson, B., & Bradshaw, R. W. (1970). A steady state optimal design of artificial induced aeration in polluted streams by the use of Pontryagin's Minimum Principle. *Water Resources Research*, *6*(2), 383–397. doi:10.1029/WR006i002p00383

De Campos, M. F., Lo Nostro, F. L., Da Cuña, R. H., & Moreira, R. G. (2018). Endocrine disruption of phenanthrene in the protogynous dusky grouper *Epinephelus marginatus* (Serranidae: Perciformes). *General and Comparative Endocrinology*, *257*, 255–263. doi:10.1016/j.ygcen.2017.06.020 PMID:28652135

De Gelder, S., Sæle, Ø., de Veen, B. T. H., Vos, J., Flik, G., Berntssen, M. H. G., & Klaren, P. H. M. (2017). The polycyclic aromatic hydrocarbons benzo[a]pyrene and phenanthrene inhibit intestinal lipase activity in rainbow trout (*Oncorhynchus mykiss*). *Comparative Biochemistry and Physiology Part C: Toxicology & Pharmacology*, *198*, 1–8. doi:10.1016/j.cbpc.2017.04.008 PMID:28476648

De Gusseme, B., Sintubin, L., Baert, L., Thibo, E., Hennebel, T., Vermeulen, G., ... Boon, N. (2010). Biogenic silver for disinfection of water contaminated with viruses. *Applied and Environmental Microbiology*, *76*(4), 1082–1087. doi:10.1128/AEM.02433-09 PMID:20038697

de Wilt, A., Butkovskyi, A., Tuantet, K., Leal, L. H., Fernandes, T. V., Langenhoff, A., & Zeeman, G. (2016). Micropollutant removal in an algal treatment system fed with source separated wastewater streams. *Journal of Hazardous Materials*, *304*, 84–92. doi:10.1016/j.jhazmat.2015.10.033 PMID:26546707

Dean, J. R. (2009). Classical Approaches for aqueous extraction. In Extraction Techniques in Analytical Sciences. West Sussex, UK: John Wiley and Sons. doi:10.1002/9780470682494.ch2

Defo, C., Kaur, R., Anshu, B., Khajanchi, L., & Paritosh, K. (2017). Modelling approaches for simulating wetland pollutant dynamics. *Critical Reviews in Environmental Science and Technology*, *47*(15), 1371–1408. doi:10.1080/10643389.2017.1350126

Dehghani, S., Moore, F., & Akhbarizadeh, R. (2017). Microplastic pollution in deposited urban dust, Tehran metropolis. Iran. *Environmental Science and Pollution Research International*, *24*(25), 20360–20371. doi:10.100711356-017-9674-1 PMID:28707239

Deng, S., & Zeng, D. (2017). Removal of phenanthrene in contaminated soil by combination of alfalfa, white-rot fungus, and earthworms. *Environmental Science and Pollution Research International*, *24*(8), 7565–7571. doi:10.100711356-017-8466-y PMID:28116628

Deniz, N., Angela, R., Dimitrios, B., & Viviane, Y. (2012). Removal of the antibiotic levofloxacin (LEVO) in water by ozonation and TiO_2 photocatalysis. *Chemical Engineering Journal, 189– 190*, 41– 48.

Derakhshan, Z., Mokhtari, M., Babaei, F., Roya, M. A., Mohammad, H. E., & Mohammad, F. (2016). Removal Methods of Antibiotic Compounds from Aqueous Environments–A Review. *J Environ Health Sustain Dev.*, *1*(1), 43–62.

Dhankhar, R., & Hooda, A. (2011). Fungal biosorption – an alternative to meet the challenges of heavy metal pollution in aqueous solutions. *Environmental Technology*, *32*(5), 467–491. doi:10.1080/09593330.2011.572922 PMID:21877528

Di Nisio, A., Sabovic, I., Valente, U., Tescari, S., Rocca, M. S., Guidolin, D., ... Plebani, M. (2018). Endocrine disruption of androgenic activity by perfluoroalkyl substances: Clinical and experimental evidence. *The Journal of Clinical Endocrinology and Metabolism*, *104*(4), 1259–1271. doi:10.1210/jc.2018-01855 PMID:30403786

Dimpe, K. M., & Nomngongo, P. N. (2016). Current sample preparation methodologies for analysis of emerging pollutants in different environmental matrices. *Trends in Analytical Chemistry*, *82*, 199–207. doi:10.1016/j.trac.2016.05.023

Dolphen, R., & Thiravetyan, P. (2015). Phytodegradation of Ethanolamines by *Cyperus alternifolius* : Effect of Molecular Size. *International Journal of Phytoremediation*, *17*(7), 686–692. doi:10.1080/15226514.2014.964839 PMID:25976882

Dongre, R., Ghugal, D., Meshram, J., & Ramteke, D. (2012). Fluoride removal from water by zirconium (IV) doped chitosan bio-composite. *African Journal of Environmental Science and Technology*, *6*(2). doi:10.5897/AJEST11.284

Dongre, S., & Tembhurkar, A.R. (2006). Studies on fluoride removal using adsorption process. *J. Environ. Sci. Eng.*, *48*, 151–156.

Doolette, C., McLaughlin, M., Kirby, J., Batstone, D., Harris, H., Ge, H., & Cornelis, G. (2013). Transformation of PVP Coated Silver Nanoparticles in a Simulated Wastewater Treatment Process and the Effect on Microbial Communities. *Chemistry Central Journal*, *7*(1), 1–18. doi:10.1186/1752-153X-7-46 PMID:23497481

Dorrington, T., Zanette, J., Zacchi, F. L., Stegeman, J. J., & Bainy, A. C. D. (2012). Basal and 3 methylcholanthrene-induced expression of cytochrome P450 1A, 1B and 1C genes in the Brazilian guppy, *Poecilia vivipara. Aquatic Toxicology (Amsterdam, Netherlands)*, *124-125*, 106–113. doi:10.1016/j.aquatox.2012.08.006 PMID:22940225

Dos Reis, I. M. M., Mattos, J. J., Garcez, R. C., Zacchi, F. L., Miguelão, T., Flores-Nunes, F., ... Bainy, A. C. D. (2015). Histological responses and localization of the cytochrome P450 (CYP2AU1) in *Crassostrea brasiliana* exposed to phenanthrene. *Aquatic Toxicology (Amsterdam, Netherlands)*, *169*, 79–89. doi:10.1016/j.aquatox.2015.10.011 PMID:26519834

Dubreuil, L., Devos, J., Neut, C., & Romond, C. (1984). Susceptibility of Anaerobic Bacteria fromSeveral French Hospitals to Three Major Antibiotics.ASM. *Antimicrobial Agents and Chemotherapy*, *25*(6), 764–766. doi:10.1128/AAC.25.6.764 PMID:6742819

Dulio, V., Bavel, B. V., Brorstrom-Lunden, E., Harmsen, J., Hollender, J., Schlabach, M., ... Koschorreck, J. (2018). Emerging pollutants in the EU: 10 years of NORMAN in support of environmental policies and regulations. *Environmental Sciences Europe*, *30*(1), 5. doi:10.118612302-018-0135-3 PMID:29568720

Dunne, P. W., Starkey, C. L., Gimeno-Fabra, M., & Lester, E. H. (2014). The rapid size- and shape-controlled continuous hydrothermal synthesis of metal sulphide nanomaterials. *Nanoscale*, *6*(4), 2406–2418. doi:10.1039/C3NR05749F PMID:24435800

Dyachenko, A., Mitchell, J., & Arsem, N. (2017). Extraction and identification of microplastic particles from secondary wastewater treatment plant (WWTP) effluent. *Analytical Methods*, *9*(9), 1412–1418. doi:10.1039/C6AY02397E

Ebele, A. J., Mohamed, A. A., & Stuart, H. (2017). Pharmaceuticals and personal care products (PPCPs) in the freshwater aquatic environment. *Emerging Contaminants*, *3*(1), 1–16. doi:10.1016/j.emcon.2016.12.004

Edyta, K. (2018). Decomposition of Contaminants of Emerging Concern in Advanced Oxidation Processes. *Proceedings*, *2*, 180. doi:10.3390/ecws-2-04949

Eerkes-Medrano, D., Thompson, R. C., & Aldridge, D. C. (2015). Microplastics in freshwater systems: A review of the emerging threats, identification of knowledge gaps and prioritisation of research needs. *Water Research*, *75*, 63–82. doi:10.1016/j.watres.2015.02.012 PMID:25746963

El-Gohary, F., Tawfik, A., & Mahmoud, U. (2010). Comparative study between chemical coagulation/precipitation (C/P) *versus* coagulation/dissolved air flotation (D/DAF) for pre-treatment of personal care products (PCPs) wastewater. *Desalination*, *252*(1-3), 106–112. doi:10.1016/j.desal.2009.10.016

Elmolla, E., & Chaudhuri, M. (2009a). Improvement of biodegradability of synthetic amoxicillin wastewater by photo-Fenton process. *World Applied Sciences Journal*, *5*, 53–58.

Elmolla, E., & Chaudhuri, M. (2009b). Optimization of Fenton process for treatment of amoxicillin, ampicillin and cloxacillin antibiotics in aqueous solution. *Journal of Hazardous Materials*, *170*(2-3), 666–672. doi:10.1016/j.jhazmat.2009.05.013 PMID:19515483

ELRI. (2011). A synopsis of laws and regulations on *environment*. ELRI.

Emad, S. E., & Malay, C. (2010). Photo-Fenton Treatment of Antibiotic Wastewater. Nature Environment and Pollution. *An International Quarterly Scientific Journal Technology.*, *9*(2), 365–370.

Emeaso, O. I. (2011). *Pollution and self- purification Studies of River Romi, Chikun Local government area Kaduna State* (Unpublished M.Sc. Thesis). Department of Water Resources and Environmental Engineering, Ahmadu Bello University, Zaria.

Emrani, A. S., Danesh, N. M., Lavaee, P., Ramezani, M., Abnous, K., & Taghdisi, S. M. (2016). Colorimetric and fluorescence quenching aptasensors for detection of streptomycin in blood serum and milk based on double-stranded DNA and gold nanoparticles. *Food Chemistry*, *190*, 115–121. doi:10.1016/j.foodchem.2015.05.079 PMID:26212949

EPA. (2010). Treating Contaminants of Emerging Concern A Literature Review Database. EPA.

Erdem, A., Eksin, E., & Muti, M. (2014). Chitosan-graphene oxide based aptasensor for the impedimetric detection of lysozyme. *Colloids and Surfaces. B, Biointerfaces*, *115*, 205–211. doi:10.1016/j.colsurfb.2013.11.037 PMID:24362059

Eskander, S., & Saleh, H. E.-D. (2017). Phytoremediation: An Overview. Environmental Science and Engineering Vol. 11: Soil Pollution and Phytoremediation. Studium Press LLC.

Evy, H., Candra, D. W., & Evelin, N. (2017). River Water Quality Performance From Carbondeoxygenation Rate. *International Journal of Engineering and Management*, *1*(1), 27–34.

Fang, S. W., Li, C. F., & Shih, D. Y. C. (1994). Antifungal activity of chitosan and its preservative effect on low-sugar candied kumquat. *Journal of Food Protection*, *56*(2), 136–140. doi:10.4315/0362-028X-57.2.136 PMID:31113143

FAO. (n.d.). *AQUASTAT*. Retrieved from http://www.fao.org/nr/water/aquastat/data/query/results.html

Fasnacht, M. P., & Blough, N. V. (2003). Mechanisms of the Aqueous Photodegradation of Polycyclic Aromatic Hydrocarbons. *Environmental Science & Technology*, *37*(24), 5767–5772. doi:10.1021/es034389c PMID:14717193

Fatta, K. D., Kalavrouziotis, I. K., Koukoulakis, P. H., & Vasquez, M. I. (2011). The risks associated with the wastewater reuse and xenobiotics in the agroecological environment. *The Science of the Total Environment*, *409*(19), 3555–3563. doi:10.1016/j.scitotenv.2010.03.036 PMID:20435343

Fawell, J. K., & Hunt, S. (1988). The polycyclic aromatic hydrocarbons. In J. K. Fawell & S. Hunt (Eds.), *Environmental toxicology: organic pollutants* (pp. 241–269). doi:10.1002/jat.2550090414

Fekete, S., Schappler, J., Veuthey, J.-L., & Guillarme, D. (2014). Current and future trends in UHPLC. *Trends in Analytical Chemistry*, *63*, 2–13. doi:10.1016/j.trac.2014.08.007

Feng, L., Yang, G., Zhu, L., Xu, J., Xu, X., & Chen, Y. (2016). Distribution and risk assessment of endocrine-disrupting pesticides in drinking water sources from agricultural watershed. *Water, Air, and Soil Pollution*, *227*(1), 23–32. doi:10.100711270-015-2704-z

Feng, N. X., Yu, J., Zhao, H. M., Cheng, Y. T., Mo, C. H., Cai, Q. Y., ... Wong, M. H. (2017). Efficient phytoremediation of organic contaminants in soils using plant–endophyte partnerships. *The Science of the Total Environment*, *583*, 352–368. doi:10.1016/j.scitotenv.2017.01.075 PMID:28117167

Fent, K., Weston, A. A., & Carminada, D. (2006). Ecotoxicology of human pharmaceuticals. *Aquatic Toxicology (Amsterdam, Netherlands)*, *76*(2), 122–159. doi:10.1016/j.aquatox.2005.09.009 PMID:16257063

Ferreira, L., Rosales, E., Danko, A. S., Sanromán, M. A., & Pazos, M. M. (2016). *Bacillus thuringiensis* a promising bacterium for degrading emerging pollutants. *Process Safety and Environmental Protection*, *101*, 19–26. doi:10.1016/j.psep.2015.05.003

FGN. (1999). *Constitution of the Federal Republic of Nigeria*. FGN.

FGN. (2004). *Water Resources Act*. FGN.

Fifi, U., Winiarski, T., & Evens, E. (2013). *Assessing the Mobility of Lead, Copper and Cadmium in a Calcareous Soil of Port-au-Prince*. International Journal of Environmental Resources and Public Health. doi:10.3390/ijerph10115830

Flores, C., Ventura, F., Martin-Alonso, J., & Caixach, J. (2013). Occurrence of Perfluorooctane Sulfonate (PFOS) and Perfluorooctanoate (PFOA) in N.E. Spanish Surface Waters and Their Removal in a Drinking Water Treatment Plant That Combines Conventional and Advanced Treatments in Parallel Lines. *The Science of the Total Environment*, *461–462*, 618–626. doi:10.1016/j.scitotenv.2013.05.026 PMID:23764674

FMWR. (2004). *National Water and Sanitation Policy*. FMWR.

FMWR. (2016). *Partnership for Expanded Water Supply, Sanitation & Hygiene*. PEWASH.

Forrez, I., Boon, N., & Verstraete, W. (2011). *Biodegradation of Micropollutants and Prospects for Water and Wastewater Biotreatment. In Comprehensive Biotechnology* (2nd ed.; Vol. 6, pp. 485–494). Elsevier.

Frank, W., Adair, S., Geftic, G., & Justus, G. (1969). Resistance of Pseudomonas to Quaternary Ammonium Compounds. ASM. *Applied Microbiology*, 299–302.

Free, C. M., Jensen, O. P., Mason, S. A., Eriksen, M., Williamson, N. J., & Boldgiv, B. (2014). High-levels of microplastic pollution in a large, remote, mountain lake. *Marine Pollution Bulletin*, *85*(1), 156–163. doi:10.1016/j.marpolbul.2014.06.001 PMID:24973278

García-Córcoles, M. T., Rodríguez-Gómez, R., de Alarcón-Gómez, B., Cipa, M., Martín-Pozo, L., Kauffmann, J.-M., & Zafra-Gómez, A. (2019). Chromatographic Methods for the Determination of Emerging Contaminants in Natural Water and Wastewater Samples: A Review. *Critical Reviews in Analytical Chemistry*, *49*(2), 160–186. doi:10.1080/10408347.2018.1496010 PMID:30285467

Garcia, N., Moreno, J., Cartmell, E., Rodriguez-Roda, I., & Judd, S. (2013). The application of microfiltration-reverse osmosis/nanofiltration to trace organics removal for municipal wastewater reuse. *Environmental Technology*, *34*(24), 3183–3189. doi:10.1080/09593330.2013.808244 PMID:24617078

Garg, P., & Chaudhari, S. (2012). Adsorption of Fluoride from Drinking Water on Magnesium substituted Hydroxyapatite. *International Conference on Future Environment and Energy*, *28*, 180–185.

Gauthier, P. T., Norwood, W. P., Prepas, E. E., & Pyle, G. G. (2016). Behavioural alterations from exposure to Cu, phenanthrene, and Cu-phenanthrene mixtures: Linking behaviour to acute toxic mechanisms in the aquatic amphipod, *Hyalella azteca. Aquatic Toxicology (Amsterdam, Netherlands)*, *170*, 377–383. doi:10.1016/j.aquatox.2015.10.019 PMID:26596825

Gavrilescu, M., Demnerová, K., Aamand, J., Agathos, S., & Fava, F. (2015). Emerging pollutants in the environment: Present and future challenges in biomonitoring, ecological risks, and bioremediation. *New Biotechnology*, *32*(1), 147–156. doi:10.1016/j.nbt.2014.01.001 PMID:24462777

Geier, M. C., James Minick, D., Truong, L., Tilton, S., Pande, P., Anderson, K. A., ... Tanguay, R. L. (2018). Systematic developmental neurotoxicity assessment of a representative PAH Superfund mixture using zebrafish. *Toxicology and Applied Pharmacology*, *354*, 115–125. doi:10.1016/j.taap.2018.03.029 PMID:29630969

Geissena, V., Hans, M., Erwin, K., Günter, U., & Marti, N. (2015). Emerging pollutants in the environment: A challenge for water resource management. *International Soil and Water Conservation Research*, *3*(1), 57–65. doi:10.1016/j.iswcr.2015.03.002

Ge, L., Deng, H., Wu, F., & Deng, N. (2009). Microalgae-promoted photodegradation of two endocrine disrupters in aqueous solutions. *Journal of Chemical Technology and Biotechnology (Oxford, Oxfordshire)*, *84*(3), 331–336. doi:10.1002/jctb.2043

Gerald, M., & Russell, A. D. (1999). Antiseptics and Disinfectants: Activity, Action, and Resistance.ASM. *Clinical Microbiology Reviews*, *12*(1), 147–179. doi:10.1128/CMR.12.1.147 PMID:9880479

GESAMP. (2015). *Sources, Fate and Effects of Microplastics in the Marine Environment: a Global Assessment (IMO/FAO/UNESCO-IOC/UNIDO/WMO/IAEA/UN/UNEP/UNDP Joint Group of Experts on the Scientific Aspects of Marine Environmental Protection).* Rep. Stud. GESAMP No. 90 Available from: http://ec.europa.eu/environment/marine/goodenvironmental-status/descriptor10/pdf/GESAMP_microplastics%20full%20study.pdf

Getenga, Z. M., Kengara, F. O., & Wandiga, S. O. (2004). Determination of organochlorine pesticides in soil and water from river Nyando drainage system within Lake Victoria Basin, Kenya. *Bulletin of Environmental Contamination and Toxicology*, *72*(2), 335–342. doi:10.100700128-003-9107-3 PMID:15106770

Ghauch, A., Tuqan, A., & Assi, H. A. (2009). Antibiotic removal from water: Elimination of amoxicillin and ampicillin by microscale and nanoscale iron particles. *Environ. Pollution.*, *157*(5), 1626–1635. doi:10.1016/j.envpol.2008.12.024 PMID:19168269

Gibbs, P. E., Bryan, G. W., & Pascoe, P. L. (1991). TBT-induced imposex in the dogwhelk, Nucella lapillus: Geographical uniformity of the response and effects. *Marine Environmental Research*, *32*(1–4), 79–87. doi:10.1016/0141-1136(91)90035-7

Gogoi, A., Mazumder, P., Tyagi, V. K., Chaminda, G. G. T., An, A. K., & Kumar, M. (2017). Occurrence and Fate of Emerging Contaminants in Water Environment: A Review. *Groundwater for Sustainable Development*, 37.

Golet, E. M., Alder, A. C., & Giger, W. (2002). Environmental exposure and risk assessment of fluoroquinolone antibacterial agents in wastewater and river water of the Glatt Valley Watershed. Switzerland. *Environmental Science & Technology*, *36*(17), 3645–3651. doi:10.1021/es0256212 PMID:12322733

Golet, E. M., Xifra, I., Siegrist, H., Alder, A. C., & Giger, W. (2003). Environmental exposure assessment of fluoroquinolone antibacterial agents from sewage to soil. *Environmental Science & Technology*, *37*(15), 3243–3249. doi:10.1021/es0264448 PMID:12966965

Golfinopoulos, S. K., Nikolaou, A. D., & Lekkas, T. D. (2003). The occurrence of disinfection by-products in the drinking water of Athens, Greece. *Environmental Science and Pollution Research International*, *10*(6), 368–372. doi:10.1065/espr2002.06.124 PMID:14690026

González-Naranjo, V., Boltes, K., & Biel, M. (2013). Mobility of ibuprofen, a persistent active drug, in soils irrigated with reclaimed water. *Plant, Soil and Environment*, *59*(2), 68–73. doi:10.17221/590/2012-PSE

Gopalan, A. I., Muthuchamy, N., & Lee, K. P. (2017). A novel bismuth oxychloride-graphene hybrid nanosheets based non-enzymatic photoelectrochemical glucose sensing platform for high performances. *Biosensors & Bioelectronics*, *89*, 352–360. doi:10.1016/j.bios.2016.07.017 PMID:27453439

Gowen, A. A., Tsenkova, R., Bruen, M., & O'donell, C. (2012). Vibrational Spectroscopy for Analysis of Water for Human Use and in Aquatic Ecosystems. *Critical Reviews in Environmental Science and Technology*, *42*(23), 2546–2573. doi:10.1080/10643389.2011.592758

Greenop, B., Lovatt, K., & Robb, M. (2001). The use of artificial oxygenation to reduce nutrient availability in the Canning River, Western Australia. *Water Science and Technology*, *43*(9), 133–144. doi:10.2166/wst.2001.0524 PMID:11419121

Gross, J. H. (2004). Mass spectrometry. A textbook. Springer.

Grover, D. P., Zhou, J. L., Frickers, P. E., & Readman, J. W. (2011). Improved removal of estrogenic and pharmaceutical compounds in sewage effluent by full scale granular activated carbon: Impact on receiving river water. *Journal of Hazardous Materials*, *185*(2), 1005–1011. doi:10.1016/j.jhazmat.2010.10.005 PMID:21035257

Gualtieri, C., & Gualtieri, P. (1997). *Hydrodynamic model for reaeration estimation in surface waters* (in Italian). AIMETA '97, Siena, Italy.

Gualtieri, C., & Gualtieri, P. (1999). Statistical analysis of reaeration rate in streams. *International Agricultural Engineering Conference (ICAE).*

Gualtieri, C., & Gualtieri, P. (2000). Field verification for a reaeration model in streams. *4th International Conference On Hydroscience & Engineering*.

Guillette, L. J. Jr, Gross, T. S., Masson, G. R., Matter, J. M., Percival, H. F., & Woodward, A. R. (1994). Developmental abnormalities of the gonad and abnormal sex hormone concentrations in juvenile alligators from contaminated and control lakes in Florida. *Environmental Health Perspectives*, *102*(8), 680–688. doi:10.1289/ehp.94102680 PMID:7895709

Gu, M. B., Mitchell, R. J., & Kim, B. C. (2004). Whole-cell-based biosensors for environmental biomonitoring and application. *Advances in Biochemical Engineering/Biotechnology*, *87*, 269–305. doi:10.1007/b13533 PMID:15217109

Gunawardana, C., Goonetilleke, A., & Egodawatta, P. (2013). Adsorption of heavy metals by road deposited solids. *Water Science and Technology*, *67*(11), 2622–2629. doi:10.2166/wst.2013.171 PMID:23752398

Guo, Q., & Reardon, E. J. (2012). Fluoride removal from water by meixnerite and its calcination product. *Applied Clay Science*, *56*, 7–15. doi:10.1016/j.clay.2011.11.013

Guo, Q., & Tian, J. (2013). Removal of fluoride and arsenate from aqueous solution by hydrocalumite via precipitation and anion exchange. *Chemical Engineering Journal*, *231*, 121–131. doi:10.1016/j.cej.2013.07.025

Guo, X., Zhang, S., & Shan, X. (2008). Adsorption of metal ions on lignin. *Journal of Hazardous Materials*, *151*(1), 134–142. doi:10.1016/j.jhazmat.2007.05.065 PMID:17587495

Gupta, S., & Babu, B. V. (2008). Economic feasibility analysis of low cost adsorbents for the removal of Cr (VI) from waste water. In *Proceedings of International Convention on Water Resources Development and Management (ICWRDM), BITS Pilani*. Citeseer.

Gupta, A. K., Dubey, R. S., & Malik, J. K. (2013). Application of Modern Electroanalytical techniques: Recent trend in Pharmaceutical and Drug Analysis. *International Journal of Pharmaceutical Sciences and Research*, *4*, 2450–2458.

Gupta, A., & Thakur, I. S. (2017). Treatment of Organic Recalcitrant Contaminants in Wastewater. In *Biological Wastewater Treatment and Resource Recovery*. InTech. doi:10.5772/66346

Gupta, V. K., Gupta, B., Rastogi, A., Agarwal, S., & Nayak, A. (2011). Pesticides removal from waste water by activated carbon prepared from waste rubber tire. *Water Research*, *45*(13), 4047–4055. doi:10.1016/j.watres.2011.05.016 PMID:21664639

Habuda-Stanić, M., Ravančić, M., & Flanagan, A. (2014). A Review on Adsorption of Fluoride from Aqueous Solution. *Materials (Basel)*, *7*(9), 6317–6366. doi:10.3390/ma7096317 PMID:28788194

Haggarty, J., & Burgess, K. E. (2017). Recent advances in liquid and gas chromatography methodology for extending coverage of the metabolome. *Current Opinion in Biotechnology*, *43*, 77–85. doi:10.1016/j.copbio.2016.09.006 PMID:27771607

Halden, R. U. (2015). Epistemology of contaminants of emerging concern and literature meta-analysis. *Journal of Hazardous Materials*, *282*, 2–9. doi:10.1016/j.jhazmat.2014.08.074 PMID:25294779

Han, B., Li, Y., Qian, B., He, Y., Peng, L., & Yu, H. (2018). Adsorption and determination of polycyclic aromatic hydrocarbons in water through the aggregation of graphene oxide. *Open Chemistry*, *16*(1), 716–725. doi:10.1515/chem-2018-0078

Han, C., Yang, M.-Q., Zhang, N., & Xu, Y.-J. (2014). Enhancing the visible light photocatalytic performance of ternary CdS-(graphene-Pd) nanocomposites via a facile interfacial mediator and co-catalyst strategy. *Journal of Materials Chemistry. A, Materials for Energy and Sustainability*, *2*(45), 19156–19166. doi:10.1039/C4TA04151H

Hang, C., Li, Q., Gao, S., & Shang, J. K. (2012). As(III) and as(V) Adsorption by hydrous Zirconium oxide Nanoparticles synthesized by a Hydrothermal process followed with heat treatment. *Industrial & Engineering Chemistry Research*, *51*(1), 353–361. doi:10.1021/ie202260g

Han, M. J., Choi, H. T., & Song, H. G. (2004). Degradation of phenanthrene by Trametes versicolor and its laccase. *Journal of Microbiology (Seoul, Korea)*, *42*(2), 94–98. PMID:15357301

Hans, P. H. A. (2012). Emerging Decontaminants. *Environmental Science & Technology*, *46*(8), 4259–4260. doi:10.1021/es301074u PMID:22462768

Hashkavayi, A. B., Raoof, J. B., Ojani, R., & Kavoosian, S. (2017). Ultrasensitive electrochemical aptasensor based on sandwich architecture for selective label-free detection of colorectal cancer (CT26) cells. *Biosensors & Bioelectronics*, *92*, 630–637. doi:10.1016/j.bios.2016.10.042 PMID:27829554

Hauser, B., & Popp, J. (2001). Membrane-assisted solvent extraction of organochlorine compounds in combination with large-volume injection/gas chromatography-electron capture detection. *Journal of Separation Science*, *24*(7), 551–560. doi:10.1002/1615-9314(20010801)24:7<551::AID-JSSC551>3.0.CO;2-2

He, J., Dai, J., Zhang, T., Sun, J., Xie, A., Tian, S., & Huo, P. (2016). *Preparation of highly porous carbon from sustainable a-cellulose for superior removal performance of tetracycline and sulfamethazine from water*. Academic Press.

Headley, J. V., & McMartin, D. W. (2004). A review of the occurrence and fate of naphthenic acids in aquatic environments. *Journal of Environmental Science and Health. Part A, Toxic/Hazardous Substances & Environmental Engineering*, *39*(8), 1989–2010. doi:10.1081/ESE-120039370 PMID:15332664

Hernández-Leal, L., Temmink, H., Zeeman, G., & Buisman, C. J. N. (2011). Removal of micropollutants from aerobically treated grey water via ozone and activated carbon. *Water Research*, *45*(9), 2887–2896. doi:10.1016/j.watres.2011.03.009 PMID:21453950

He, Y., & Chi, J. (2017). Pilot-scale demonstration of phytoremediation of PAH-contaminated sediments by *Hydrilla verticillata* and *Vallisneria spiralis*. *Environmental Technology*, *40*(5), 605–613. doi:10.1080/09593330.2017.1398783 PMID:29076392

Hidalgo-Ruz, V., Gutow, L., Thompson, R. C., & Thiel, M. (2012). Microplastics in the marine environment: A review of the methods used for identification and quantification. *Environmental Science & Technology*, *46*(6), 3060–3075. doi:10.1021/es2031505 PMID:22321064

Hidayat, A., & Yanto, D. H. Y. (2018). Biodegradation and metabolic pathway of phenanthrene by a new tropical fungus, *Trametes hirsuta* D7. *Journal of Environmental Chemical Engineering*, *6*(2), 2454–2460. doi:10.1016/j.jece.2018.03.051

Hirsch, R., Ternes, T., Haberer, K., & Kratz, K. L. (1999). Occurrence of antibiotics in the aquatic environment. *The Science of the Total Environment*, *225*(1-2), 109–118. doi:10.1016/S0048-9697(98)00337-4 PMID:10028708

Holmberg, J. P. (2006). *Competitive adsorption and displacement behaviour of heavy metals on peat* (Master dissertation). Chalmers University of Technology, Goteborg, Sweden.

Holm, J. V., Rügge, K., & Bjerg, P. L. (1995). Occurrence and distribution of pharmaceutical organic compounds in the groundwater downgradient of a landfill (Grindsted, Denmark). *Environmental Science & Technology*, *28*(5), 1415–1420. doi:10.1021/es00005a039 PMID:22192041

Homayoni, H., Ravandi, H., & Valizadeh, M. (2009). Electrospinning of chitosan nanofibers: Processing optimization. *Carbohydrate Polymers*, *77*(3), 656–661. doi:10.1016/j.carbpol.2009.02.008

Honarbakhsh, M. (2011). *Impact of Wastewater Effluent on the Structure and Antibiotic Resistance of Native Microbial Communities*. A Master of Science Graduate Program in Microbiology and Molecular Genetics thesis submitted to the Graduate School-New Brunswick Rutgers, The State University of New Jersey and The Graduate School of Biomedical Sciences University of Medicine and Dentistry.

Hong, A. D., Ngoc, H. P., Hoang, T. N., Thi, T. H., Hung, V. P., Van, C. P., ... Alfredo, C. A. (2008). Occurrence, fate and antibiotic resistance of fluoroquinolone antibacterials in hospital wastewaters in Hanoi, Vietnam. *Chemosphere*, *72*(6), 968–973. doi:10.1016/j.chemosphere.2008.03.009 PMID:18485444

Hongna, L., Zhiwei, S., & Changxiong, Z. (2014). Trends in Research on Electrochemical Oxidation. *Croatica Chemica Acta*, *87*(2), 185–194. doi:10.5562/cca2446

Hong, X., Qin, J., Chen, R., Yuan, L., Zha, J., Huang, C., ... Wang, Z. (2017). Phenanthrene-Induced Apoptosis and Its Underlying Mechanism. *Environmental Science & Technology*, *51*(24), 14397–14405. doi:10.1021/acs.est.7b04045 PMID:29161501

Hong, Y. W., Yuan, D. X., Lin, Q.-M., & Yang, T. L. (2008). Accumulation and biodegradation of phenanthrene and fluoranthene by the algae enriched from a mangrove aquatic ecosystem. *Marine Pollution Bulletin*, *56*(8), 1400–1405. doi:10.1016/j.marpolbul.2008.05.003 PMID:18597790

Hook, S. E., Mondon, J., Revill, A. T., Greenfield, P. A., Stephenson, S. A., Strzlecki, J., ... Barrett, S. (2018). Monitoring sublethal changes in fish physiology following exposure to a light, unweathered crude oil. *Aquatic Toxicology (Amsterdam, Netherlands)*, *204*, 27–45. doi:10.1016/j.aquatox.2018.08.013 PMID:30173120

Ho, Y., & Ofomaja, A. E. (2006). Biosorption thermodynamics of cadmium on coconut copra meal as bio sorbent. *Biochemical Engineering Journal*, *30*(2), 117–123. doi:10.1016/j.bej.2006.02.012

Huang, D. L., Min-Chen, G., Zeng, G. M., Lai, G., Zhang, C., Li, N. J., ... He, Y. (2015). Synthesis and application of modified zero-valent iron nanoparticle for removal of hexavalent chromium from wastewater. *Water, Air, and Soil Pollution*, *226*(11), 375. doi:10.100711270-015-2583-3

Huang, H. W., He, Y., Du, X., Chu, P. K., & Zhang, Y. H. (2015). A General and Facile Approach to Heterostructured Core/Shell $BiVO_4$/BiOI p-n Junction: Room-Temperature in Situ Assembly and Highly Boosted Visible-Light Photocatalysis. *ACS Sustainable Chemistry & Engineering*, *3*(12), 3262–3273. doi:10.1021/acssuschemeng.5b01038

Huan, H., Bin, H., Gen, F., Dan, X., Zhixiang, X., Xinhao, W., & Xuejun, P. (2018). Electrochemically modified dissolved organic matter accelerates the combining photodegradation and biodegradation of 17α- ethinylestradiol in natural aquatic environment. *Water Research*, *137*, 251–261. doi:10.1016/j.watres.2018.03.019 PMID:29550728

Huber, M. M., Göbel, A., Joss, A., Hermann, N., Löffler, D., McArdell, C. S., & Von Gunten, U. (2005). Oxidation of pharmaceuticals during ozonation of municipal wastewater effluents: A pilot study. *Environmental Science & Technology*, *39*(11), 4290–4299. doi:10.1021/es048396s PMID:15984812

Hu, D., Hu, W., Xie, L., Li, Y., Birch, L., & Prins, G. S. (2016). Actions of Estrogenic Endocrine Disrupting Chemicals on Human Prostate Stem/Progenitor Cells and Prostate Carcinogenesis. *The Open Biotechnology Journal*, *10*(77), 76–97. doi:10.2174/1874070701610010076

Huerta, B., Rodriguez-mozaz, S., Nannou, C., Nakis, L., Ruhí, A., Acuña, V., & Barcelo, D. (2015). Determination of a broad spectrum of pharmaceuticals and endocrine disruptors in the biofilm from a wastewater treatment plant-impacted river. *The Science of the Total Environment*. PMID:26087856

Huggins, T. M., Haeger, A., Biffinger, J. C., & Ren, Z. J. (2016). Granular biochar compared with activated carbon for wastewater treatment and resource recovery. *Water Research*, *94*, 225–232. doi:10.1016/j.watres.2016.02.059 PMID:26954576

Hu, H., Deng, Y., Fan, Y., Zhang, P., Sun, H., Gan, Z., & Yao, Y. (2016). Effects of artificial sweeteners on metal bioconcentration and toxicity on a green algae Scenedesmus obliquus. *Chemosphere*, *150*, 285–293. doi:10.1016/j.chemosphere.2016.02.043 PMID:26915590

Hunter, J.F., & Whipple, W., Jr. (1970). Evaluating instream aerators of polluted river. *Journal - Water Pollution Control Federation*, *42*, 249–262.

Huo, H., Xu, Z., Zhang, T., & Xu, C. (2015). Ni/CdS/TiO_2 nanotube array heterostructures for high performance photoelectrochemical biosensing. *Journal of Materials Chemistry. A, Materials for Energy and Sustainability*, *3*(11), 5882–5888. doi:10.1039/C4TA07190E

Hwang, K., Kim, J., Yoon, J., & Pyun, R. (1999). Bactericidal activity of chitosan on E. coli. *Advances in Chitin Science*, *62*, 239–243.

IARC. (1983). IARC Monographs on the Evaluation of Carcinogenic Risks to Humans. *Polynuclear aromatic compounds, Part 1, chemical, environmental and experimental data, 32*, 1–453.

IARC. (2010). IARC Monographs on the Evaluation of Carcinogenic Risks to Humans. *Some non-heterocyclic polycyclic aromatic hydrocarbons and some related exposures, 92*, 1–853.

Iatrou, E. I., Gatidou, G., Damalas, D., Thomaidis, N. S., & Stasinakis, A. S. (2017). Fate of antimicrobials in duckweed Lemna minor wastewater treatment systems. *Journal of Hazardous Materials*, *330*, 116–126. doi:10.1016/j.jhazmat.2017.02.005 PMID:28214647

Ifabiyi, I. P. (2008). 'Self-purification of a Freshwater Stream in Ile-Ife'. Lessons for Water Management J. *Human Ecology*, *24*(2), 131–137. doi:10.1080/09709274.2008.11906109

Ikonomou, M. G., Cai, S. S., Fernandez, M. P., Blair, J. D., & Fischer, M. (2008). Ultra-trace analysis of multiple endocrine-disrupting chemicals in municipal and bleached KraftMill effluents using gas chromatography-high-resolution mass spectrometry. *Environmental Toxicology and Chemistry*, *27*(2), 243–251. doi:10.1897/07-136R.1 PMID:18348620

Ikpesu, T. O. (2015). Assessment of occurrence and concentrations of paraquat dichloride in water, sediments and fish from Warri River Basin, Niger Delta, Nigeria. *Environmental Science and Pollution Research International*, *22*(11), 8517–8525. doi:10.100711356-014-3993-2 PMID:25548014

Ilgu, M., & Nilsen-Hamilton, M. (2016). Aptamers in analytics. *Analyst (London)*, *141*(5), 1551–1568. doi:10.1039/C5AN01824B PMID:26864075

Iliuk, A. B., Hu, L., & Tao, W. A. (2011). Aptamer in Bioanalytical Applications. *Analytical Chemistry*, *83*(12), 4440–4452. doi:10.1021/ac201057w PMID:21524128

Ilkhani, H., Sarparast, M., Noori, A., Bathaie, S. Z., & Mousavi, M. F. (2015). Electrochemical aptamer/antibody based sandwich immunosensor for the detection of EGFR, a cancer biomarker, using gold nanoparticles as a signaling probe. *Biosensors & Bioelectronics*, *74*, 491–497. doi:10.1016/j.bios.2015.06.063 PMID:26176209

Inam, E., Offiong, N.-A., Kang, S., Yang, P., & Essien, J. (2015). Assessment of the Occurrence and Risks of Emerging Organic Pollutants (EOPs) in Ikpa River Basin Freshwater Ecosystem, Niger Delta-Nigeria. *Bulletin of Environmental Contamination and Toxicology*, *95*(5), 624–631. doi:10.100700128-015-1639-9 PMID:26341253

IPCS. (1998). *Selected Non-heterocyclic Polycyclic Aromatic Hydrocarbons, Environmental Health Criteria 202*. Geneva: International Programme on Chemical Safety, World Health Organization.

Islam, M., & Patel, R. (2011). Thermal activation of basic oxygen furnace slag andevaluation of its fluoride removal efficiency. *Chemical Engineering Journal*, *169*(1-3), 68–77. doi:10.1016/j.cej.2011.02.054

Ismaila, S. M., Adie, D. B., Okuofu, C. A., & Ismail, A. (2012). Modelling of Water Quality of Bindare Stream, Zaria, Nigeria Using MATLAB. *Proceeding of the OAU Faculty of Technology Conference 2012*.

Ite, A. E., Harry, T. A., Obadimu, C. O., Asuaiko, E. R., & Inim, I. J. (2018). Petroleum hydrocarbons contamination of surface water and groundwater in the Niger Delta region of Nigeria. *Journal of Environment Pollution and Human Health*, *6*(2), 51–61. doi:10.12691/jephh-6-2-2

Ivan, S., Marin, M., Helena, K., Jakopovic, S. T., Josip, C., Ivan, M., & Marijan, A. (2011). Removal of antimicrobials using advanced wastewater treatment. *Journal of Hazardous Materials*, *192*, 319–328. PMID:21641718

Iwanowicz, L. R., Pinkney, A. E., Guy, C. P., Major, A. M., Munney, K., Blazer, V. S., ... Smith, D. R. (2019). Temporal evaluation of estrogenic endocrine disruption markers in smallmouth bass (Micropterus dolomieu) reveals seasonal variability in intersex. *The Science of the Total Environment*, *646*, 245–256. doi:10.1016/j.scitotenv.2018.07.167 PMID:30055487

Jacobson, J. L., & Jacobson, S. W. (1997). Evidence for PCBs as neurodevelopmental toxicants in humans. *Neurotoxicology*, *18*(2), 415–424. PMID:9291491

Jadhav, A.S., & Jadhav, M.V. (2014). Use of Maize Husk Fly Ash as an Adsorbent for Removal of Fluoride from Water. *International Journal of Recent Development in Engineering and Technology, 2*.

Jagpat, A. I., Yenkie, M. K., Das, S., & Rayalu, S. (2011). Synthesis and Characterization of Lanthanum impregnated Chitosan Flakes for Fluoride removal in Water. *Desalination*, *273*(2-3), 267–275. doi:10.1016/j.desal.2010.12.032

Jagtap, S., Thakre, D., Sakhare, N., Labhsetwar, N., Meshram, S., & Rayalu, S. (2010). Chitosan based mesoporous Ti–Al binary metal oxide supported beads for defluoridation of water. *Chemical Engineering Journal*, *158*(2), 315–324. doi:10.1016/j.cej.2010.01.008

Jahin, H. S. (2014). Fluoride Removal from Water Using Nanoscale Zero- Valent Iron (NZVI). *International Water Technology Journal*, *4*(3), 173–182.

Jaiswal, D. K., Kumar, A., & Yadav, R. R. (2011). Analytical Solution to the One-Dimensional Advection-Diffusion Equation with Temporally Dependent Coefficients. *Journal of Water Resource and Protection*, *2011*(January), 76–84. doi:10.4236/jwarp.2011.31009

Jambeck, J. R., Geyer, R., Wilcox, C., Siegler, T. R., Perryman, M., Andrady, A., ... Law, K. L. (2015). Plastic waste inputs from land into the ocean. *Sci. Magna.*, *347*, 768–771. PMID:25678662

Jandacek, R. J., & Genuis, S. J. (2013). An Assessment of the Intestinal Lumen as a Site for Intervention in Reducing Body Burdens of Organochlorine Compounds. *The Scientific World Journal*, *2013*, 1–10. doi:10.1155/2013/205621 PMID:23476122

Jara, C. C., Fino, D., Specchia, V., Saracco, G., & Spinelli, P. (2007). Electrochemical removal of antibiotics from wastewaters. *Applied Catalysis B: Environmental*, *70*(1-4), 479–487. doi:10.1016/j.apcatb.2005.11.035

Jardim, I. C. S. F. (2010). Extração em Fase Sólida: Fundamentos Teóricos e Novas Estratégias para Preparação de Fases Sólidas. *Scientia Chromatographica*, *2*, 13–25.

Jee, J. H., & Kang, J. C. (2004). Effect of phenanthrene on haematological parameters in olive flounder, *Paralichthys olivaceus* (Temminch et Schlegel). *Aquaculture Research*, *35*(14), 1310–1317. doi:10.1111/j.1365-2109.2004.01152.x

Jee, J. H., Kim, S. G., & Kang, J. C. (2004). Effects of phenanthrene on growth and basic physiological functions of the olive flounder, *Paralichtys olivaceus*. *Journal of Experimental Marine Biology and Ecology*, *304*(1), 123–136. doi:10.1016/j.jembe.2003.12.001

Jelic, A., Gros, M., Ginebra, A., Cespedes-Sanchez, R., Ventura, F., Petrovic, M., & Barcelo, D. (2011). Occurrence partition and removal of pharmaceutical in sewage water and sludge during wastewater treatment. *Water Research*, *45*(3), 1165–1176. doi:10.1016/j.watres.2010.11.010 PMID:21167546

Jiang, D., Du, X., Chen, D., Li, Y., Hao, N., Qian, J., ... Wang, K. (2016). Facile wet chemical method for fabricating p-type BiOBr/n-type nitrogen doped graphene composites: Efficient visible-excited charge separation, and high-performance photoelectrochemical sensing. *Carbon*, *102*, 10–17. doi:10.1016/j.carbon.2016.02.027

Jiang, L., Yuan, X., Zeng, G., Liang, J., Chen, X., Yu, H., ... Xiong, T. (2018). In-situ synthesis of direct solid-state dual Z-scheme WO 3 /g-C 3 N 4 /Bi 2 O 3 photocatalyst for the degradation of refractory pollutant. *Applied Catalysis B: Environmental*, *227*, 376–385. doi:10.1016/j.apcatb.2018.01.042

JICA. (2014). *Federal Republic of Nigeria the Project for Review and Update of Nigeria - National Water Resources Master Plan 2013 - MASTER PLAN PROGRESS REPORT - 2 SUMMARY*. JICA.

Ji, M.-K., Kabra, A. N., Choi, J., Hwang, J.-H., Kim, J. R., Abou-Shanab, R. A. I., & Jeon, B.-H. (2014). Biodegradation of bisphenol A by the freshwater microalgae Chlamydomonas mexicana and Chlorella vulgaris. *Ecological Engineering*, *73*, 260–269. doi:10.1016/j.ecoleng.2014.09.070

Jin, H., Capareda, S., Chang, Z., Gao, J., Xu, Y., & Zhang, J. (2014). Biochar pyrolytically produced from municipal solid wastes for aqueous As(V) removal: Adsorption property and its improvement with KOH activation. *Bioresource Technology*, *169*, 622–629. doi:10.1016/j.biortech.2014.06.103 PMID:25103038

Jose, M., Ramos, G., Agüera, A., & Fernández-alba, A. R. (2012). Occurrence and persistence of organic emerging contaminants and priority pollutants in five sewage treatment plants of Spain: Two years pilot survey monitoring. *Environmental Pollution*, *164*, 263–273. PMID:22387188

Joshi, S., Adhikari, M., and Pradhananga, R. (2012). Adsorption of Fluoride Ion ontoZirconyl–Impregnated Activated Carbon Prepared from Lapsi Seed Stone. *J. Nepal Chem. Soc.*, (30), 13-23.

Joshi, S., Adhikari, M., & Pradhananga, R. (2013). Adsorption of Fluoride Ion onto Zirconyl-Impregnated Activated Carbon Prepared from Lapsi Seed Stone. *Journal of Nepal Chemical Society*, *30*(0).

Jr., & , Whipple, W.,Yu, S.L. (n.d.). Aeration system of large navigable rivers. *ASCE Journal of Sanitary Engineering Division, 97*, 883-901.

Jr, Whipple, W.,., Coughlan, F.P., & Yu, S.L. (1971). Instream aerations for polluted rivers. *Journal of the Sanitary Engineering Division, 96*, 1153–1165.

Jurado, A., Vàzquez-Suñé, E., Carrera, J., López de Alda, M., Pujades, E., & Barceló, D. (2012). Emerging organic contaminants in groundwater in Spain: A review of sources, recent occurrence and fate in a European context. *The Science of the Total Environment*, *440*, 82–94. doi:10.1016/j.scitotenv.2012.08.029 PMID:22985674

Jury, K.L. Vancov, T., Stuetz, R.M., & Khan, S.J. (2010). Antibiotic resistance dissemination and sewage treatment plants. *Current Research, Technology and Education Topics in Applied Microbiology and Microbial Biotechnology*, 509 – 519.

Kabbashi, N. A., Atieh, M. A., Al-Mamun, A., Mirghami, M. E. S., Alam, M. D. Z., & Yahya, N. (2009). Kinetic adsorption of application of carbon nanotubes for Pb(II) removal from aqueous solution. *Journal of Environmental Sciences (China)*, *21*(4), 539–544. doi:10.1016/S1001-0742(08)62305-0 PMID:19634432

Kafilzadeh, F., Khosrobak, A., & Jamali, H. (2015). Isolation and Identification of Phenanthrene Degrading Bacteria from the Soil around Oil Company of Andimeshk and Investigation of Their Growth Kinetics. *Polycyclic Aromatic Compounds*, *36*(1), 58–71. doi:10.1080/10406638.2015.1037006

Kalantzi, O. I., & Alcock, R. A. (2012). Short-chain chlorinated paraffins in biota – levels and effects. *Global NEST Journal*, *14*(1), 66–71.

Kanangire, C. K., Matano, A., Dida, G. O., & Anyona, D. (2016). *A Systematic Review of Effects of Emerging pollutants on Human Health and Livelihoods of Population living in the Lake Victoria Basin of Kenya. Case study report*. UNESCO.

Karami, A., Romano, N., Hamzah, H., Simpson, S. L., & Yap, C. K. (2016). Acute phenanthrene toxicity to juvenile diploid and triploid African catfish (*Clarias gariepinus*): Molecular, biochemical, and histopathological alterations. *Environmental Pollution*, *212*, 155–165. doi:10.1016/j.envpol.2016.01.055 PMID:26845363

Karthiga, D. M., Mano, B. E., Geethamani, R., & Abinaya, S. (2017). Self-purification capacity of Bhavani River, India. *Research Journal of Engineering Sciences.*, *6*(3), 1–9.

Katouli, M., Thompson, J.M., Gündoğdu, A., & Stratton, H.M. (2012). Antibiotic Resistant Bacteria in Hospital Wastewaters and Sewage Treatment Plants. *Science Forum and Stakeholder Engagement: Building Linkages, Collaboration and Science Quality*, 225-229.

Kebede, T., Dube, S., Nindi, M., Kebede, T. G., Dube, S., & Nindi, M. M. (2019). Removal of Multi-Class Antibiotic Drugs from Wastewater Using Water-Soluble Protein of Moringa stenopetala Seeds. *Water (Basel)*, *11*(3), 595. doi:10.3390/w11030595

Keith, L. H. (2015). The Source of U.S. EPA's Sixteen PAH Priority Pollutants. *Polycyclic Aromatic Compounds*, *35*(2-4), 147–160. doi:10.1080/10406638.2014.892886

Kerrigan, J. F., Engstrom, D. R., Yee, D., Sueper, C., Erickson, P. R., Grandbois, M., ... Arnold, W. A. (2015). Quantification of hydroxylated polybrominated diphenyl ethers (OH-BDEs), triclosan, and related compounds in freshwater and coastal systems. *PLoS One*, *10*(10), e0138805. doi:10.1371/journal.pone.0138805 PMID:26466159

Khajanchi, L., Kaur, R., Rosin, K. G., & Neelam, P. (2016). *Low-Cost Remediation and On-Farm Management Approaches for Safe Use of Wastewater in Agriculture*. Innovative Saline Agriculture. doi:10.1007/978-81-322-2770-0_12

Khan, E. A., Bertotto, L., Dale, K., Lille-Langøy, R., Yadetie, F., Karlsen, O. A., ... Chen, Y. (2013). A battery of bioassays for the evaluation of phenanthrene biotoxicity in soil. *Archives of Environmental Contamination and Toxicology*, *65*(1), 47–55. doi:10.100700244-013-9879-3 PMID:23440446

Khan, M. U., Li, J., Zhang, G., & Naseem, R. (2016). Malik first insight into the levels and distribution of flame retardants in potable water in Pakistan: An underestimated problem with an associated health risk diagnosis. *The Science of the Total Environment*, *565*, 346–359. doi:10.1016/j.scitotenv.2016.04.173 PMID:27177141

Khan, Z., Roman, D., Kintz, T., delas Alas, M., Yap, R., & Doty, S. (2014). Degradation, Phytoprotection and Phytoremediation of Phenanthrene by Endophyte Pseudomonas putida, PD1. *Environmental Science & Technology*, *48*(20), 12221–12228. doi:10.1021/es503880t PMID:25275224

Khetan, S. K., & Collins, T. J. (2007). Human pharmaceuticals in the aquatic environment: A challenge to green chemistry. *Chemical Reviews*, *107*(6), 2319–2364. doi:10.1021/cr020441w PMID:17530905

Khichar, M., & Kumbhat, S. (2015). Defluoridation-A review of water from aluminium and alumina based compound. *International Journal of Chemical Studies*, *2*(5), 4–11.

Kiely, G. (1997). *Environmental engineering*. London: McGraw-Hill.

Kim, E., Jung, C., Han, J., Her, N., Park, C. M., Jang, M., & Yoon, Y. (2016). Sorptive removal of selected emerging contaminants using biochar in aqueous solution. *Journal of Industrial and Engineering Chemistry*, *36*, 364–371. doi:10.1016/j.jiec.2016.03.004

Kim, I., Yamashita, N., & Tanaka, H. (2009). Performance of UV and UV/H_2O_2 processes for the removal of pharmaceuticals detected in secondary effluent of a sewage treatment plant in Japan. *Journal of Hazardous Materials*, *166*(2), 1134–1140. doi:10.1016/j.jhazmat.2008.12.020 PMID:19147286

Kim, K.-H., Jahan, S. A., Kabir, E., & Brown, R. J. C. (2013). A review of airborne polycyclic aromatic hydrocarbons (PAHs) and their human health effects. *Environment International*, *60*, 71–80. doi:10.1016/j.envint.2013.07.019 PMID:24013021

Kim, K., Jeon, H. J., Choi, S. D., Tsang, D. C. W., Oleszczuk, P., Ok, Y. S., ... Lee, S. E. (2018). Combined toxicity of endosulfan and phenanthrene mixtures and induced molecular changes in adult Zebrafish (*Danio rerio*). *Chemosphere*, *194*, 30–41. doi:10.1016/j.chemosphere.2017.11.128 PMID:29197246

Kimosop, S. J., Getenga, Z. M., Orata, F., Okello, V. A., & Cheruiyot, J. K. (2016). Residue levels and discharge loads of antibiotics in wastewater treatment plants (WWTPs), hospital lagoons, and rivers within Lake Victoria Basin, Kenya. *Environmental Monitoring and Assessment*, *188*(9), 532. doi:10.100710661-016-5534-6 PMID:27553946

Kimosop, S. J., Orata, F., & Getenga, Z. (2012). The Influence of Filter Mud Compost and Tithonia diversifolia Leaves on the Dissipation of Diuron in Soils within the Nzoia River Drainage Basin, Kenya. *Bulletin of Environmental Contamination and Toxicology*, *89*(2), 328–333. doi:10.100700128-012-0685-9 PMID:22617949

Kimura, N., Kimura, T., Suzuki, M., & Totsukawa, K. (2006). Effect of gestational exposure to nonylphenol on the development and fertility of mouse offspring. *The Journal of Reproduction and Development*, *52*(6), 789–795. doi:10.1262/jrd.18007 PMID:17038804

Kim, Y. S., Raston, N. H. A., & Gu, M. B. (2016). Aptamer-based nanobiosensors. *Biosensors & Bioelectronics*, *76*, 2–19. doi:10.1016/j.bios.2015.06.040 PMID:26139320

King, D. L. (1970). *Reaeration of Streams and Reservoirs Analysis and Bibliography*. Hydraulics Branch Division of General Research Engineering and Research Center Denver.

Kivaisi, A. K. (2001). The potential for constructed wetlands for wastewater treatment and reuse in developing countries: A review. *Ecological Engineering*, *16*(4), 545–560. doi:10.1016/S0925-8574(00)00113-0

Klamerth, N., Malato, S., Maldonado, M. I., Agüera, A., & Fernández-Alba, A. R. (2010). Application of photo-fenton as a tertiary treatment of emerging contaminants in municipal wastewater. *Environmental Science & Technology*, *44*(5), 1792–1798. doi:10.1021/es903455p PMID:20131796

Koelmans, A. A., Besseling, A., & Foekema, E. M. (2014). Leaching of plastic additives to marine organisms. *Environ. Pollut.*, *187*, 49–54. doi:10.1016/j.envpol.2013.12.013 PMID:24440692

Kolpin, D. W., Furlong, E. T., Meyer, M. T., Thurman, E. M., Zaugg, S. D., Barber, L. B., & Buxton, H. T. (2002). Pharmaceuticals, hormones, and other organic wastewater contaminants in U.S. streams, 1999–2000: A national reconnaissance. *Environmental Science & Technology*, *36*(6), 1202–1211. doi:10.1021/es011055j PMID:11944670

Korsh, J., Shen, A., Aliano, K., & Davenport, T. (2015). Polycyclic Aromatic Hydrocarbons and Breast Cancer: A Review of the Literature. *Breast Care (Basel, Switzerland)*, *10*(5), 316–318. doi:10.1159/000436956 PMID:26688678

Koshijima, T., Tanaka, R., Muraki, E., Yamada, A., & Yaku, F. (1973). Chelating polymers derived from cellulose and chitin. *Cellulose Chemistry and Technology*, *7*, 197–208.

Kosma, C. I., Lambropoulou, D. A., & Albanis, T. A. (2010). Occurrence and removal of PPCPs in municipal and hospital wastewaters in Greece. *Journal of Hazardous Materials*, *179*(1-3), 804–817. doi:10.1016/j.jhazmat.2010.03.075 PMID:20395039

Kosutic, K., Dolar, D., Asperger, D., & Kunst, B. (2007). Removal of antibiotic from model wastewater by RO/NF membrane. *Separation and Purification Technology*, *53*(3), 244–249. doi:10.1016/j.seppur.2006.07.015

Kovalova, L., Siegrist, H., Von Gunten, U., Eugster, J., Hagenbuch, M., Wittmer, A., ... McArdell, C. S. (2013). Elimination of Micropollutants During Post- Treatment of Hospital Wastewater with Powdered Activated Carbon, Ozone, and UV. *Environmental Science & Technology*, *47*(14), 7899–7908. doi:10.1021/es400708w PMID:23758546

Kronenberg, M., Trably, E., Bernet, N., & Patureau, D. (2017). Biodegradation of polycyclic aromatic hydrocarbons: Using microbial bioelectrochemical systems to overcome an impasse. *Environmental Pollution*, *231*, 509–523. doi:10.1016/j.envpol.2017.08.048 PMID:28841503

Kuang, P. Y., Zheng, P. X., Liu, Z. Q., Lei, J. L., Wu, H., Li, N., & Ma, T. Y. (2016). Embedding Au Quantum Dots in Rimous Cadmium Sulfide Nanospheres for Enhanced Photocatalytic Hydrogen Evolution. *Small*, *12*(48), 6735–6744. doi:10.1002mll.201602870 PMID:27709776

Kumar, A., Chang, B., & Xagoraraki, I. (2010). Human health risk assessment of pharmaceuticals in water: Issues and challenges ahead. *International Journal of Environmental Research and Public Health*, *7*(11), 3929–3953. doi:10.3390/ijerph7113929 PMID:21139869

Kumar, E., Bhatnagar, A., Ji, M., Jung, W., Lee, S., Kim, S., ... Jeon, B. (2009). Defluoridation from aqueous solutions by granular ferric hydroxide (GFH). *Water Research*, *43*(2), 490–498.

Kumar, E., Bhatnagar, A., Kumar, U., & Sillanpaa, M. (2011). Defluoridation from aqueous solution bynano –alumina: Characterization and Sorption studies. *Journal of Hazardous Materials*, *186*(2-3), 1042–1049. doi:10.1016/j.jhazmat.2010.11.102 PMID:21177029

Kumar, E., Bhatnagar, A., Minkyu, J., Jung, W., Lee, S., Kim, S., ... Jeon, B. (2009). Defluoridation from aqueous solutions by granularferric hydroxide (GFH). *Water Research*, *43*(2), 490–499. doi:10.1016/j.watres.2008.10.031 PMID:18995880

Kumari, S., Regar, R. K., & Manickam, N. (2018). Improved polycyclic aromatic hydrocarbon degradation in a crude oil by individual and a consortium of bacteria. *Bioresource Technology*, *254*, 174–179. doi:10.1016/j.biortech.2018.01.075 PMID:29413920

Kurniawan, T. A., Chan, G. Y. S., Lo, W., & Babel, S. (2006). Comparisons of low-cost adsorbents for treating wastewaters laden with heavy metals. *The Science of the Total Environment*, *366*(2–3), 409–426. doi:10.1016/j.scitotenv.2005.10.001 PMID:16300818

Kuster, M., de Alda, M. J. L., Hernando, M. D., Petrovic, M., Martín-Alonso, J., & Barceló, D. (2008). Analysis and occurrence of pharmaceuticals, estrogens, progestogens and polar pesticides in sewage treatment plant effluents, river water and drinking water in the Llobregat River basin (Barcelona, Spain). *Journal of Hydrology (Amsterdam)*, *358*(1-2), 112–123. doi:10.1016/j.jhydrol.2008.05.030

Kyzas, G. Z., Kostoglou, M., & Lazaridis, N. K. (2009). Copper and chromium(VI) removal by chitosan derivatives –Equilibrium and kinetic studies. *Chemical Engineering Journal*, *152*(2-3), 440–448. doi:10.1016/j.cej.2009.05.005

Lamichhane, S., Bal Krishna, K. C., & Sarukkalige, R. (2017). Surfactant-enhanced remediation of polycyclic aromatic hydrocarbons: A review. *Journal of Environmental Management*, *199*, 46–61. doi:10.1016/j.jenvman.2017.05.037 PMID:28527375

Lamichhane, S., Krishna, K. C. B., & Sarukkalige, R. (2016). Polycyclic aromatic hydrocarbons (PAHs) removal by sorption: A review. *Chemosphere*, *148*, 336–353. doi:10.1016/j.chemosphere.2016.01.036 PMID:26820781

Lan, M. H., Liu, W. M., Ge, J. C., Wu, J. S., Sun, J. Y., Zhang, W. J., & Wang, P. F. (2015). A selective fluorescent and colorimetric dual-responses chemosensor for streptomycin based on polythiophene derivative. *Spectrochimica Acta Part a-Molecular and Biomolecular Spectroscopy, 136*, 871-874. doi:10.1016/j.saa.2014.09.106

Landrigan, P. J., Fuller, R., Acosta, N. J. R., Adeyi, O., Arnold, R., Basu, N. N., ... Zhong, M. (2018). The Lancet Commission on pollution and health. *Lancet*, *391*(10119), 462–512. doi:10.1016/S0140-6736(17)32345-0 PMID:29056410

Lapworth, D. J., Baran, N., Stuart, M. E., & Ward, R. S. (2012). Emerging organic contaminants in groundwater: A review of sources, fate and occurrence. *Environ. Pollut.*, *163*, 287–303. doi:10.1016/j.envpol.2011.12.034 PMID:22306910

Lawson, C., Gieske, M., Murdoch, B., Ye, P., Li, Y., Hassold, T., & Hunt, P. A. (2011). Gene expression in the fetal mouse ovary is altered by exposure to low doses of bisphenol A. *Biology of Reproduction*, *84*(1), 79–86. doi:10.1095/biolreprod.110.084814 PMID:20739668

Lee, S. H., & McCain, M. C. (2015). Wildlife Toxicity Assessment for Phenanthrene. In M. A. Williams, G. Reddy, M. J. Quinn Jr, & M. S. Johnson (Eds.), *Wildlife Toxicity Assessments for Chemicals of Military Concern* (pp. 591–598). Elsevier Inc. doi:10.1016/B978-0-12-800020-5.00031-4

Le, M. D., Duong, H. A., Nguyen, M. H., Sáiz, J., Pham, H. V., & Mai, T. D. (2016). Screening determination of pharmaceutical pollutants in different water matrices using dual-channel capillary electrophoresis coupled with contactless conductivity detection. *Talanta*, *160*, 512–520. doi:10.1016/j.talanta.2016.07.032 PMID:27591645

Lenz, R., Enders, K., Stedmon, C. A., Mackenzie, D. M. A., & Nielsen, T. G. (2015). A critical assessment of visual identification of marine microplastic using Raman spectroscopy for analysis improvement. *Marine Pollution Bulletin*, *100*(1), 82–91. doi:10.1016/j.marpolbul.2015.09.026 PMID:26455785

Leong, M. I., Fuh, M.-R., & Huang, S.-D. (2014). Beyond dispersive liquid-liquid microextraction. *Journal of Chromatography. A*, *1335*, 2–14. doi:10.1016/j.chroma.2014.02.021 PMID:24582396

Liao, X., & Shi, B. (2005). Adsorption of Fluoride on Zirconium(IV)-Impregnated Collagen Fiber. *Environmental Science & Technology*, *39*(12), 4628–4632. doi:10.1021/es0479944 PMID:16047802

Li, B., Sun, S. J., Huo, C. Y., Li, W. L., Zhu, N. Z., Qi, H., ... Ma, W. L. (2016). Occurrence and fate of PBDEs and novel brominated flame retardants in a wastewater treatment plant in Harbin, China. *Environmental Science and Pollution Research International*, *23*(19), 19246–19256. doi:10.100711356-016-7142-y PMID:27364485

Li, C. J., Wang, S. P., Wang, T., Wei, Y. J., Zhang, P., & Gong, J. L. (2014). Monoclinic Porous $BiVO_4$ Networks Decorated by Discrete g-C_3N_4 Nano-Islands with Tunable Coverage for Highly Efficient Photocatalysis. *Small*, *10*(14), 2783–2790. doi:10.1002mll.201400506 PMID:24740883

Lidia, C., Kaklamanos, G., Ioana, A., Adriana, A., & Linda, M. (2015). High sensitive multiresidue analysis of pharmaceuticals and antifungals in surface water using U-HPLC-Q-Exactive Orbitrap HRMS. Application to the Danube river basin on the Romanian territory. *The Science of the Total Environment*, *532*, 501–511. doi:10.1016/j.scitotenv.2015.06.010 PMID:26100729

Li, G., Chen, X., & Gao, G. (2014). Bi_2S_3 microspheres grown on graphene sheets as low-cost counter-electrode materials for dye-sensitized solar cells. *Nanoscale*, *6*(6), 3283–3288. doi:10.1039/C3NR06093D PMID:24509629

Li, H. L., Yu, K., Lei, X., Guo, B. J., Fu, H., & Zhu, Z. Q. (2015). Hydrothermal Synthesis of Novel MoS_2/$BiVO_4$ Hetero-Nanoflowers with Enhanced Photocatalytic Activity and a Mechanism Investigation. *The Journal of Physical Chemistry C*, *119*(39), 22681–22689. doi:10.1021/acs.jpcc.5b06729

Li, J., Tu, W., Li, H., Han, M., Lan, Y., Dai, Z., & Bao, J. (2014). In Situ-Generated Nano-Gold Plasmon-Enhanced Photoelectrochemical Aptasensing Based on Carboxylated Perylene-Functionalized Graphene. *Analytical Chemistry*, *86*(2), 1306–1312. doi:10.1021/ac404121c PMID:24377281

Li, L., Jiang, M., & Shen, X. (2016). Variability in antioxidant/detoxification enzymes of *Sinonovacula constricta* exposed to benzo[a]pyrene and phenanthrene. *Marine Pollution Bulletin*, *109*(1), 507–511. doi:10.1016/j.marpolbul.2016.05.013 PMID:27216044

Li, L., Zou, D., Xiao, Z., Zeng, X., Zhang, L., Jiang, L., & Liu, F. (2019). Biochar as a sorbent for emerging contaminants enables improvements in waste management and sustainable resource use. *Journal of Cleaner Production*, *210*, 1324–1342. doi:10.1016/j.jclepro.2018.11.087

Lin, A. Y.-C., & Reinhard, M. (2005). Photodegradation of common environmental pharmaceuticals and estrogens in river waters. *Environmental Toxicology and Chemistry*, *24*(6), 1303–1309. doi:10.1897/04-236R.1 PMID:16117104

Lisouza, F. A., Owuor, P. O., & Lalah, J. O. (2011). Variation in indoor levels of polycyclic aromatic hydrocarbons from burning various biomass types in the traditional grass-roofed households in Western Kenya. *Environ Pollut*, *159*(7), 1810–1815. doi:10.1016/j.envpol.2011.03.032 PMID:21501911

Little, C., Hepher, M. J., & El-Sharif, M. (2002). The sono-degradation of phenanthrene in an aqueous environment. *Ultrasonics*, *40*(1-8), 667–674. doi:10.10160041-624x(02)00196-8

Liu, J., Wang, R., Haung, B., Lin, C., Zhou, J., & Pan, X. (2011). Biological effects and bioaccumulation of steroidal and phenolic endocrine chemicals in highback Crucian carp exposed to wastewater treatment plant effluents. *Journal of Environmental Pollution, 162*, 325-331.

Liu, B. Q., Tang, D. P., Zhang, B., Que, X. H., Yang, H. H., & Chen, G. N. (2013). Au(III)-promoted magnetic molecularly imprinted polymer nanospheres for electrochemical determination of streptomycin residues in food. *Biosensors & Bioelectronics*, *41*, 551–556. doi:10.1016/j.bios.2012.09.021 PMID:23058661

Liu, B. Q., Zhang, B., Cui, Y. L., Chen, H. F., Gao, Z. Q., & Tang, D. P. (2011). Multifunctional Gold-Silica Nanostructures for Ultrasensitive Electrochemical Immunoassay of Streptomycin Residues. *ACS Applied Materials & Interfaces*, *3*(12), 4668–4676. doi:10.1021/am201087r PMID:22059488

Liu, H., Meng, F., Tong, Y., & Chi, J. (2014). Effect of plant density on phytoremediation of polycyclic aromatic hydrocarbons contaminated sediments with *Vallisneria spiralis. Ecological Engineering*, *73*, 380–385. doi:10.1016/j.ecoleng.2014.09.084

Liu, J., & Chen, W. (2015). Remediation of phenanthrene contaminated soils by nonionic-anionic surfactant washing coupled with activated carbon adsorption. *Water Science and Technology*, *72*(9), 1552–1560. doi:10.2166/wst.2015.357 PMID:26524446

Liu, M., Zhang, Y., Ding, R., Gao, Y., & Yang, M. (2013). Response of Activated Sludge to the Treatment of Oxytetracycline Production Waste Stream. *Applied Microbiology and Biotechnology*, *97*(19), 8805–8812. doi:10.100700253-012-4589-8 PMID:23188460

Liu, Q., Shi, J., & Jiang, G. (2012). Application of graphene in analytical sample preparation. *Trends in Analytical Chemistry*, *37*, 1–11. doi:10.1016/j.trac.2012.03.011

Liu, S. H., Zeng, Z. T., Niu, Q. Y., Xiao, R., Zeng, G. M., Liu, Y., ... Tao, J.-J. (2019). Influence of immobilization on phenanthrene degradation by *Bacillus* sp. P1 in the presence of Cd (II). *The Science of the Total Environment*, *655*, 1279–1287. doi:10.1016/j.scitotenv.2018.11.272 PMID:30577120

Liu, S., Wang, Y., Xu, W., Leng, X., Wang, H., Guo, Y., & Huang, J. (2017). A novel sandwich-type electrochemical aptasensor based on GR-3D Au and aptamer-AuNPs-HRP for sensitive detection of oxytetracycline. *Biosensors & Bioelectronics*, *88*, 181–187. doi:10.1016/j.bios.2016.08.019 PMID:27544787

Liu, Y., Ma, H., Zhang, Y., Pang, X., Fan, D., Wu, D., & Wei, Q. (2016). Visible light photoelectrochemical aptasensor for adenosine detection based on CdS/PPy/g-C_3N_4 nanocomposites. *Biosensors & Bioelectronics*, *86*, 439–445. doi:10.1016/j.bios.2016.06.089 PMID:27424261

Liu, Y., Yan, K., Okoth, O. K., & Zhang, J. (2015). A label-free photoelectrochemical aptasensor based on nitrogen-doped graphene quantum dots for chloramphenicol determination. *Biosensors & Bioelectronics*, *74*, 1016–1021. doi:10.1016/j.bios.2015.07.067 PMID:26264269

Li, Z., Deng, S., Zhang, X., Zhou, W., Huang, J., & Yu, G. (2010). Removal of fluoride from water using titanium-based adsorbents. *Frontiers of Environmental Science & Engineering in China*, *4*(4), 414–420. doi:10.100711783-010-0241-y

Li, Z., Wang, J., & Li, D. (2016). Applications of Raman spectroscopy in detection of water quality. *Applied Spectroscopy Reviews*, *51*(4), 333–357. doi:10.1080/05704928.2015.1131711

Llamas, A., Al-Lal, A. M., García-Martínez, M. J., Ortega, M. F., Llamas, J. F., Lapuerta, M., & Canoira, L. (2017). Polycyclic Aromatic Hydrocarbons (PAHs) produced in the combustion of fatty acid alkyl esters from different feedstocks: Quantification, statistical analysis and mechanisms of formation. *The Science of the Total Environment*, *586*, 446–456. doi:10.1016/j.scitotenv.2017.01.180 PMID:28215803

Loganathan, P., Vigneswaran, S., & Kandasamy, J. (2013). Enhanced removal of nitrate from water using surface modification of adsorbents. A review. *Journal of Environmental Management*, *131*, 363–374. doi:10.1016/j.jenvman.2013.09.034 PMID:24211565

Logan, B. E. (2001). Peer reviewed: Assessing the outlook for perchlorate remediation. *Environmental Science & Technology*, *35*(23), 482A–487A. doi:10.1021/es012564o PMID:11770780

Longe, E. O., & Omole, D. O. (2008). Analysis of Pollution Status of River Illo, Ota, Nigeria. *The Environmentalist*, *28*(4), 451–457. doi:10.100710669-008-9166-4

López-Darias, J., Germán-Hernández, M., Pino, V., & Afonso, A. M. (2010). Dispersive liquid–liquid microextraction versus singledrop microextraction for the determination of several endocrine-disrupting phenols from seawaters. *Talanta*, *80*(5), 1611–1618. doi:10.1016/j.talanta.2009.09.057 PMID:20152385

Lopez-Serna, R., Jurado, A., Vazquez-Sune, E., Carrera, J., Petrović, M., & Barceló, D. (2013). Occurrence of 95 pharmaceuticals and transformation products in urban groundwater underlying the metropolis of Barcelona, Spain. *Environ. Pollut.*, *174*, 305–315. doi:10.1016/j.envpol.2012.11.022 PMID:23302545

Lorgeoux, C., Moilleron, R., Gasperi, J., Ayrault, S., Bonté, P., Lefèvre, I., & Tassin, B. (2016). Temporal trends of persistent organic pollutants in dated sediment cores: Chemical fingerprinting of the anthropogenic impacts in the Seine River basin, Paris. *The Science of the Total Environment*, *541*, 1355–1363. doi:10.1016/j.scitotenv.2015.09.147 PMID:26479909

Loughery, J. R., Kidd, K. A., Mercer, A., & Martyniuk, C. J. (2018). Part B:Morphometric and transcriptomic responses to sub-chronic exposure to the polycyclic aromatic hydrocarbon phenanthrene in the fathead minnow (Pimephales promelas). *Aquatic Toxicology (Amsterdam, Netherlands)*, *199*, 77–89. doi:10.1016/j.aquatox.2018.03.026 PMID:29621673

Lozano, R., & Mouat, J. (2009). *Marine litter in the North-East Atlantic region: assessment and priorities for response.* KIMO Int.

Lozano, N., Rice, C. P., Ramirez, M., & Torrents, A. (2013). Fate of Triclocarban, Triclosan and Methyltriclosan During Wastewater and Biosolids Treatment Processes. *Water Research*, *47*(13), 4519–4527. doi:10.1016/j.watres.2013.05.015 PMID:23764601

Lubick, N. (2011). Tools for tracking antibiotic resistance. *Environmental Health Perspectives*, *119*(5), A214–A217. doi:10.1289/ehp.119-a214 PMID:21531665

Lüchmann, K. H., Clark, M. S., Bainy, A. C. D., Gilbert, J. A., Craft, J. A., Chipman, J. K., ... Schroeder, D. C. (2015). Key metabolic pathways involved in xenobiotic biotransformation and stress responses revealed by transcriptomics of the mangrove oyster *Crassostrea brasiliana*. *Aquatic Toxicology (Amsterdam, Netherlands)*, *166*, 10–20. doi:10.1016/j.aquatox.2015.06.012 PMID:26186662

Lu, J., Tian, S., Lv, X., Chen, Z., Chen, B., Zhu, X., & Cai, Z. (2018). TiO 2 nanoparticles in the marine environment: Impact on the toxicity of phenanthrene and Cd 2+ to marine zooplankton *Artemia salina*. *The Science of the Total Environment*, *615*, 375–380. doi:10.1016/j.scitotenv.2017.09.292 PMID:28988072

Luo, Y., Guo, W., Ngo, H. H., Nghiem, L. D., Hai, F. I., Zhang, J., & Wang, X. C. (2014). A review on the occurrence of micropollutants in the aquatic environment and their fate and removal during wastewater treatment. *The Science of the Total Environment*, *473–474*, 619–641. doi:10.1016/j.scitotenv.2013.12.065 PMID:24394371

Lusher, A. L., Burke, A., O'Connor, I., & Officer, R. (2014). Microplastic pollution in the Northeast Atlantic Ocean: Validated and opportunistic sampling. *Marine Pollution Bulletin*, *88*(1-2), 325–333. doi:10.1016/j.marpolbul.2014.08.023 PMID:25224764

Luz, R. A. S., Lost, R. M., & Crespilho, F. N. (2013). *Nanomaterials for biosensors and implantable biodevices. In From implantable biosensors to green power generation* (pp. 27–48). Springer.

Lv, G., Wu, L., Liao, L., Zhang, Y., & Li, Z. (2013). Preparation and characterization of red mud sintered porous materials for water defluoridation. *Applied Clay Science*, *74*, 95–101. doi:10.1016/j.clay.2012.10.004

Machado, A. A., Hoff, M. L., Klein, R. D., Cordeiro, G. J., Lencina Avila, J. M., Costa, P. G., & Bianchini, A. (2014). Oxidative stress and DNA damage responses to phenanthrene exposure in the estuarine guppy *Poecilia vivipara*. *Marine Environmental Research*, *98*, 96–105. doi:10.1016/j.marenvres.2014.03.013 PMID:24722053

MacLeod, S. L., McClure, E. L., & Wong, C. S. (2007). Laboratory calibration and field deployment of the polar organic chemical integrative sampler for pharmaceuticals and personal care products in wastewater and surface water. *Environmental Toxicology and Chemistry*, *26*(12), 2517–2529. doi:10.1897/07-238.1 PMID:18020693

Madadi, V. O. (2005). *Chemodynamics studies and assessment of pesticides residue levels in Lake Victoria catchment area for Rivers Sio and Nzoia* (MSc. Thesis). Department of Chemistry, University of Nairobi.

Mahdavi, S., Jalali, M., & Afkhami, A. (2014). Heavy metals removal from aqueous solutions by Al2O3 nanoparticles modified with natural and chemical modifiers. *Clean Technologies and Environmental Policy*, *17*(1), 85–102. doi:10.100710098-014-0764-1

Maiga, Y., von Sperling, M., & Mihelcic, J. (2017). *Constructed Wetlands*. Global Water Pathogens Project. Retrieved from http://www.waterpathogens.org

Majors, R. (2009). Sample Preparation. In Introduction to Modern Liquid Chromatography (3rd ed.; pp. 757-808). Hoboken, NJ: John Wiley and Sons.

Mala, J., & Maly, J. (2018). Effect Of Heavy Metals On Self-Purification Processes In Rivers. *Applied Ecology and Environmental Research*, *7*(4), 333–340.

Maldaner, L., & Jardim, I. C. S. F. (2012). UHPLC – Uma abordagem atual: Desenvolvimentos e desafios recentes. *Scientia Chromatographica*, *4*(3), 197–207. doi:10.4322c.2012.014

Maletz, S., Floehr, T., Beier, S., Klumper, C., Brouwer, A., Behnisch, P., ... Hollert, H. (2013). In Vitro Characterization of the Effectiveness of Enhanced Sewage Treatment Processes to Eliminate Endocrine Activity of Hospital Effluents. *Water Research*, *47*(4), 1545–1557. doi:10.1016/j.watres.2012.12.008 PMID:23305681

Maliszewska-Kordybach, B., Smreczak, B., & Klimkowicz-Pawlas, A. (2009). Effects of anthropopressure and soil properties on the accumulation of polycyclic aromatic hydrocarbons in the upper layer of soils in selected regions of Poland. *Applied Geochemistry*, *24*(10), 1918–1926. doi:10.1016/j.apgeochem.2009.07.005

Mallek, M., Chtourou, M., Portillo, M., Monclús, H., Walha, K., Salah, A., & Salvadó, V. (2018). Granulated cork as biosorbent for the removal of phenol derivatives and emerging contaminants. *Journal of Environmental Management*, *223*, 576–585. doi:10.1016/j.jenvman.2018.06.069 PMID:29975884

Manickum, T., John, W., & Terry, S. (2011). Determination of selected steroid estrogens in treated sewage effluent in the Umsuduzi (Duzi) Rier water catchment area. *Hydrology: Current Research*, *2*(3), 45–56.

Manisankar, P., Rani, C., & Viswanathan, S. (2004). Effect of halides in the electrochemical treatment of distillery effluent. *Chemosphere*, *57*(8), 961–966. doi:10.1016/j.chemosphere.2004.07.026 PMID:15488586

Map, G. (2014). *Imagery @ 2014 ones/spotimage, Digital Landsat' internet access*. Available at http://wn.Map.harvard.edu/maps/Ile-Ife

Marchand, L., Mench, M., Jacob, D. L., & Otte, M. L. (2010). Metal and metalloid removal in constructed wetlands, with emphasis on the importance of plants and standardized measurements: A review. *Environmental Pollution*, *158*(12), 3447–3461. doi:10.1016/j.envpol.2010.08.018 PMID:20869144

Margot, J., Kienle, C., Magnet, A., Weil, M., Rossi, L., de Alencastro, L. F., ... Barry, D. A. (2013). Treatment of Micropollutants in Municipal Wastewater: Ozone or Powdered Activated Carbon? *The Science of the Total Environment*, *461–462*, 480–498. doi:10.1016/j.scitotenv.2013.05.034 PMID:23751332

Marini, M., & Frapiccini, E. (2014). Do lagoon area sediments act as traps for polycyclic aromatic hydrocarbons? *Chemosphere*, *111*, 80–88. doi:10.1016/j.chemosphere.2014.03.037 PMID:24997903

Mark, C. M. (1982). Effect of UV Light Disinfection on Antibiotic-Resistant Coliforms in Wastewater Effluents.ASM. *Applied and Environmental Microbiology*, *43*(2), 371–377. doi:10.1128/AEM.43.2.371-377.1982 PMID:7059170

Martin, S. F., & Charles, H. D. (1966). Factors Influencing the Occurrence of High Numbers of Iodine-Resistant Bacteria in iodinated Swimming Pools.ASM. *Applied Microbiology*, *14*(4), 627–635. doi:10.1128/AEM.14.4.627-635.1966 PMID:5927040

Martins, M., Ferreira, A. M., Costa, M. H., & Costa, P. M. (2015). Comparing the genotoxicity of a potentially carcinogenic and a noncarcinogenic PAH, singly, and in binary combination, on peripheral blood cells of the European sea bass. *Environmental Toxicology*, *31*(11), 1307–1318. doi:10.1002/tox.22135 PMID:25728603

Masura, J., Baker, J., Foster, G., & Arthur, C. (2015). *Laboratory Methods for the Analysis of Microplastics in the Marine Environment: Recommendations for Quantifying Synthetic Particles in Waters and Sediments*. NOAA Technical Memorandum NOS-OR&R-48. Available from: https://marinedebris.noaa.gov/sites/default/files/publications-files/noaa_microplastics_methods_manual.pdf

Mat Lazim, Z., Hadibarata, T., Hafiz Puteh, M., & Yusop, Z. (2015). Adsorption Characteristics of Bisphenol A onto Low-Cost Modified Phyto-Waste Material in Aqueous Solution. *Water, Air, and Soil Pollution*, *226*(34).

Mato, Y., Isobe, T., Takada, H., Kanehiro, H., Ohtake, C., & Kaminuma, T. (2001). Plastic resin pellets as a transport medium for toxic chemicals in the marine environment. *Environmental Science & Technology*, *35*(2), 318–324. doi:10.1021/es0010498 PMID:11347604

Mautner, A., Maples, H. A., Kobkeatthawin, T., Kokol, V., Karim, Z., Li, K., & Bismarck, A. (2016). I Phosphorylated nanocellulose papers for copper adsorption from aqueous solutions. *International Journal of Environmental Science and Technology*, *13*(8), 1861–1872. doi:10.100713762-016-1026-z

Ma, Y. M., Pendlebury, S. R., Reynal, A., Le Formal, F., & Durrant, J. R. (2014). Dynamics of photogenerated holes in undoped $BiVO_4$ photoanodes for solar water oxidation. *Chemical Science (Cambridge)*, *5*(8), 2964–2973. doi:10.1039/C4SC00469H

McClusky, L. M., De Jager, C., & Bornman, M. S. (2006). Stage-related increase in the proportion of apoptotic germ cells and altered frequencies of stages in the spermatogenic cycle following gestational, lactational, and direct exposure of male rats to p-nonylphenol. *Toxicological Sciences*, *95*(1), 249–256. doi:10.1093/toxsci/kfl141 PMID:17065434

McGuire, K. J., & McDonnell, J. J. (2006). A review and evaluation of catchment transit time modelling. *Journal of Hydrology (Amsterdam)*, *330*(3–4), 543–563. doi:10.1016/j.jhydrol.2006.04.020

Meador, J. P., Stein, J. E., Reichert, W. L., & Varanasi, U. (1995). Bioaccumulation of polycyclic aromatic hydrocarbons by marine organisms. *Reviews of Environmental Contamination and Toxicology*, *143*, 79–165. doi:10.1007/978-1-4612-2542-3_4 PMID:7501868

Meenakshi, G. (2010). Preparation and application of alumina/chitosan Biocomposite. *International Journal of Biological Macromolecules*, *47*(2), 146–154. doi:10.1016/j.ijbiomac.2010.05.008 PMID:20471415

Meenakshi, R. C., & Maheshwari, J. (2006). Fluoride in drinking water and its removal. *Journal of Hazardous Materials*, *137*(1), 456–463. doi:10.1016/j.jhazmat.2006.02.024 PMID:16600479

Meffe, R., & de Bustamante, I. (2014). Emerging organic contaminants in surface water and groundwater: A first overview of the situation in Italy. *The Science of the Total Environment*, *481*, 280–295. doi:10.1016/j.scitotenv.2014.02.053 PMID:24602913

Meirinho, S. G., Dias, L. G., Peres, A. M., & Rodrigues, L. R. (2016). Voltammetric aptasensors for protein disease biomarkers detection: A review. *Biotechnology Advances*, *34*(5), 941–953. doi:10.1016/j.biotechadv.2016.05.006 PMID:27235188

Melanie, V., Indra, B., Anna, N., & Martin, J. (2017). Comparison of UVA and UVC irradiation of FQs and their MICs for sustainable water treatment. *Conference: 2nd Green and Sustainable Chemistry Conference at Berlin.* Retrieved from https://www.researchgate.net/publication/316977230

Melanie, V., Indra, B., Anna, N., & Martin, J. (2017). Comparison of UVA and UVC irradiation of FQs and their MICs for sustainable water treatment. *Conference: 2nd Green and Sustainable Chemistry Conference at Berlin.* Retrieved from https://www.researchgate.net/publication/316977230_

Melching, C. S., & Flores, H. E. (1999). Reaeration equations derived from U.S. Geological Survey database. J. Environmental Engineering Division, 125(5), 407-414.

Memon, M., Soomro, M. S., Akhtar, M. S., & Memon, K. S. (2010). Drinking water quality assessment in southern Sindh (Pakistan). *Environmental Monitoring and Assessment*, *177*(1-4), 39–50. doi:10.100710661-010-1616-z PMID:20658360

Mercedes, V. (2017). *Emerging Pollutants Removal in Wastewater Treatment Plants: A review and their implications in a river basin in Uruguay* (M.Sc thesis). UNESCO-IHE Institute for Water Education, Delft, the Netherlands.

Mesdaghinia, A. R., Naddafi, K., Nabizadeh, R., Saeedi, R., & Zamanzadeh, M. (2009). Wastewater Characteristics and Appropriate Method for Wastewater Management in the Hospitals. *Iranian J Pub. Health*, *38*(1), 34–40.

Michael, A. G., Mark, D. C., Kevin, B. A., & Jayne, C. B. (1985). Transfer of Plasmids pBR322 and pBR325 in Wastewater from Laboratory Strains of Escherichia coli to Bacteria Indigenous to the Waste Disposal System. *ASM. Applied and Environmental Microbiology*, 836–841.

Micheal, V. W., & John, W. V. (1985). Occurrence of Multiple-Antibiotic-Resistant Enteric Bacteria in Domestic Sewage and Oxidation Lagoons. *Applied and Environmental Microbiology*, 930–933. PMID:4083887

Miller, J. H., Novak, J. T., Knocke, W. R., Young, K., Hong, Y., Vikesland, P. J., ... Pruden, A. (2013). Effect of Silver Nanoparticles and Antibiotics on Antibiotic Resistance Genes in Anaerobic Digestion. *Water Environment Research*, *85*(5), 411–421. doi:10.2175/106143012X13373575831394 PMID:23789571

Miller, J. S., & Olejnik, D. (2001). Photolysis of polycyclic aromatic hydrocarbons in water. *Water Research*, *35*(1), 233–243. doi:10.1016/S0043-1354(00)00230-X PMID:11257878

Mills, G. (2015). *Active and Passive Sampling for Pollutants of Emerging Concern - Sheffield 4th March 2015. In Emerging Contaminants in Waters and Soils, Practical Considerations: Sampling*. Analysis and Consequences.

Mills, M. R., Arias-Salazar, K., Baynes, A., Shen, L. Q., Churchley, J., Beresford, N., ... Jobling, S. (2015). Removal of ecotoxicity of 17α-ethinylestradiol using TAML/peroxide water treatment. *Scientific Reports*, *5*(1), 10511. doi:10.1038rep10511 PMID:26068117

Miraji, Chande, & Ngassapa. (2018). The Evolutionary Engagement of Analytical Techniques in the Eco-toxicity of Emerging Contaminants. *International Journal for Research & Development in Technology*, *3*(3), 42–48.

Mishra, G. K., Sharma, A., & Bhand, S. (2015). Ultrasensitive detection of streptomycin using flow injection analysis-electrochemical quartz crystal nanobalance (FIA-EQCN) biosensor. *Biosensors & Bioelectronics*, *67*, 532–539. doi:10.1016/j.bios.2014.09.033 PMID:25266252

Moeder, M., Schrader, S., Winkler, U., & Rodil, R. (2010). At-line microextraction by packed sorbent-gas chromatography-mass spectrometry for the determination of UV filter and polycyclic musk compounds in water samples. *Journal of Chromatography. A*, *1217*(17), 2925–2932. doi:10.1016/j.chroma.2010.02.057 PMID:20334867

Mohammad, M., Maitra, S., & Dutta, B. K. (2018). Comparison of Activated Carbon and Physic Seed Hull for the Removal of Malachite Green Dye from Aqueous Solution. *Water, Air, and Soil Pollution*, *229*(2), 45. doi:10.100711270-018-3686-4

Mohapatra, M., Anand, S., Mishra, B. K., Giles, D. E., & Singh, P. (2009). Review of fluoride removal from drinking water. *Journal of Environmental Management*, *91*(1), 67–77. doi:10.1016/j.jenvman.2009.08.015 PMID:19775804

Moore, C., Lattin, G., & Zellers, A. (2011). Quantity and type of plastic debris flowing from two urban rivers to coastal waters and beaches of Southern California. Rev. Gestao Costeira Integr.-. *J. Integrated Coast. Zone Manag*, *11*(1).

Moraes, F. C., Rossi, B., Donatoni, M. C., de Oliveira, K. T., & Pereira, E. C. (2015). Sensitive determination of 17b-estradiol in river water using a graphene based electrochemical sensor. *Analytica Chimica Acta*, *881*, 37–43. doi:10.1016/j.aca.2015.04.043 PMID:26041518

Morais, G. dos S., Pesenti, E. C., Cestari, M. M., & Navarro-Silva, M. A. (2014). Genotoxic effect of Phenanthrene on *Chironomus sancticaroli* (Diptera: Chironomidae). *Zoologia*, *31*(4), 323–328. doi:10.1590/S1984-46702014000400003

Morari, F., Ferro, N. D., & Coco, E. (2015). Municipal wastewater treatment with *phragmites australis L.* and *Typha L.* for irrigation reuse. Boron and heavy metals. *Water, Air, and Soil Pollution*, *226*(3), 56. doi:10.100711270-015-2336-3

Morasch, B. (2013). Occurrence and dynamics of micropollutants in a karst aquifer. *Environ. Pollut.*, *173*, 133–137. doi:10.1016/j.envpol.2012.10.014 PMID:23202643

Morasch, B., Bonvin, F., Reiser, H., Grandjean, D., de Alencastro, L.F., & Perazzolo, C., Che vre, N., Kohn, T. (2010). Occurrence and fate of micropollutants in the Vidy Bay of Lake Geneva, Switzerland. Part II: Micropollutant removal between wastewater and raw drinking water. *Environmental Toxicology and Chemistry*, *29*(8), 1658–1668. PMID:20821617

Morin, N., Miège, C., Randon, J., & Coquery, M. (2012). Chemical calibration, performance, validation and applications of the polar organic chemical integrative sampler (POCIS) in aquatic environments. *Trends in Analytical Chemistry*, *36*, 144–175. doi:10.1016/j.trac.2012.01.007

Mottier, P., Parisod, V., & Turesky, R. J. (2000). Quantitative Determination of Polycyclic Aromatic Hydrocarbons in Barbecued Meat Sausages by Gas Chromatography Coupled to Mass Spectrometry. *Journal of Agricultural and Food Chemistry*, *48*(4), 1160–1166. doi:10.1021/jf991205y PMID:10775366

Moussavi, G., Hosseini, H., & Alahabadi, A. (2013). The investigation of diazinon pesticide removal from contaminated water by adsorption onto NH4Cl-induced activated carbon. *Chemical Engineering Journal*, *214*, 172–179. doi:10.1016/j.cej.2012.10.034

Mu, J., Wang, J., Jin, F., Wang, X., & Hong, H. (2014). Comparative embryotoxicity of phenanthrene and alkyl-phenanthrene to marine medaka (*Oryzias melastigma*). *Marine Pollution Bulletin*, *85*(2), 505–515. doi:10.1016/j.marpolbul.2014.01.040 PMID:24559736

Mu, L., Zhou, Q., Zhao, Y., Liu, X., & Hu, X. (2019). Graphene Oxide Quantum Dots Stimulate Indigenous Bacteria to Remove Oil Contamination. *Journal of Hazardous Materials*, *366*, 694–702. doi:10.1016/j.jhazmat.2018.12.044 PMID:30583239

Mule, S., Nguta, C., Kamau, G., Ongera, G., Wangila, P., Kiplimo, J., & Chirchir, D. (2015). Retention Efficiencies of Halogenated and Non-Halogenated Hydrocarbons in Selected Wetland Ecosystem in Lake Victoria Basin. *Intern. J. of Envir*, *4*(2), 38–52. doi:10.3126/ije.v4i2.12623

Munoz, R., Guieysse, B., & Mattiasson, B. (2003). Phenanthrene biodegradation by an algal bacterial consortium in two-phase partitioning bioreactors. *Applied Microbiology and Biotechnology*, *61*(3), 261–267. doi:10.100700253-003-1231-9 PMID:12698286

Mushtaq, M., Haq, N. B., Munawar, I., & Saima, N. (2016). Eriobotrya japonica seed biocomposite efficiency for copper adsorption: Isotherms, kinetics, thermodynamic and desorption studies. *Journal of Environmental Management*, *12*, 17–22. PMID:27039361

Nair, K. R., Manji, F., & Gitonga, J. N. (1984). The occurrence and distribution of fluoride in groundwaters of Kenya. *East African Medical Journal*, *61*, 503–512. PMID:6545194

Nair, V., Perkins, C. L., Lin, Q. Y., & Law, M. (2016). Textured nanoporous Mo:$BiVO_4$ photoanodes with high charge transport and charge transfer quantum efficiencies for oxygen evolution. *Energy & Environmental Science*, *9*(4), 1412–1429. doi:10.1039/C6EE00129G

Nam, T. H., Jeon, H. J., Mo, H., Cho, K., Ok, Y. S., & Lee, S. E. (2015). Determination of biomarkers for polycyclic aromatic hydrocarbons (PAHs) toxicity to earthworm (*Eisenia fetida*). *Environmental Geochemistry and Health*, *37*(6), 943–951. doi:10.100710653-015-9706-z PMID:25920560

Nanseu-Njiki, C. P., Dedzo, G. K., & Ngameni, E. (2010). Study of the removal of paraquat from aqueous solution by biosorption onto Ayous (Triplochiton schleroxylon) sawdust. *Journal of Hazardous Materials*, *179*(1), 63–71. doi:10.1016/j.jhazmat.2010.02.058 PMID:20334968

Napper, I. E., & Thompson, R. C. (2016). Release of synthetic microplastic plastic fibres from domestic washing machines: Effects of fabric type and washing conditions. *Marine Pollution Bulletin*, *112*(1-2), 39–45. doi:10.1016/j.marpolbul.2016.09.025 PMID:27686821

Nash, J. P., Kime, D. E., Van der Ven, L. T. M., Wester, P. W., Brion, F., Maack, G., ... Tyler, C. R. (2004). Long-Term Exposure to Environmental Concentrations of the Pharmaceutical Ethynylestradiol Causes Reproductive Failure in Fish. *Environmental Health Perspectives*, *112*(17), 1725–1733. doi:10.1289/ehp.7209 PMID:15579420

Naslund, J., & Snell, I. (2005). *GIS-Mapping of fluoride contaminated groundwater in Nakuru and Baringo District, Kenya* (MSc Thesis). Luleå University of Technology, Luleå, Sweden.

National Academy of Sciences. (1999). Historical Overview of Drinking Water Contaminants and Public Water Utilities. In Identifying Future Drinking Water Contaminants (p. 260). Author.

Naumczyk, J., Szpyrkowicz, L., & Zilio-Grandi, F. (1996). Electrochemical treatment of textile wastewater. *Water Science and Technology*, *34*(11), 17–24. doi:10.2166/wst.1996.0258

NBS. (2017). *Nigeria SDGs Indicators Baseline Report 2016 | UNDP in Nigeria.* National Bureau of Statistics.

Nduka, J. K., & Orisakwe, O. E. (2010). Water Quality Issues in the Niger Delta of Nigeria: Polyaromatic and Straight Chain Hydrocarbons in Some Selected Surface Waters. *Water Quality, Exposure, and Health*, *2*(2), 65–74. doi:10.100712403-010-0024-5

Ngeno, E. C., Orata, F., Baraza, L. D., Shikuku, V. O., & Kimosop, S. J. (2016). Adsorption of Caffeine and Ciprofloxacin onto Pyrolitically Derivedwater Hyacinth Biochar: Isothermal, Kinetic, and Thermodynamic Studies. *Journal of Chemistry and Chemical Engineering*. doi:10.17265/1934-7375/2016.04.006

Nie, G., Lu, X., Lei, J., Yang, L., & Wang, C. (2015). Facile and controlled synthesis of bismuth sulfide nanorods-reduced graphene oxide composites with enhanced supercapacitor performance. *Electrochimica Acta*, *154*, 24–30. doi:10.1016/j.electacta.2014.12.090

Niemuth, N. J., & Klaper, R. D. (2015). Emerging wastewater contaminant metformin causes inters ex and reduced fecundity in fish. *Chemosphere*, *135*, 38–45. doi:10.1016/j.chemosphere.2015.03.060 PMID:25898388

Nnamonu, L. A., & Onekutu, A. (2015). Green pesticides in Nigeria: An overview. *Journal of Biology. Agriculture and Healthcare*, *5*(9), 48–62.

Nogueira, D. J., Mattos, J. J., Dybas, P. R., Flores-Nunes, F., Sasaki, S. T., Taniguchi, S., ... Bainy, A. C. D. (2017). Effects of phenanthrene on early development of the Pacific oyster *Crassostrea gigas* (Thunberg, 1789). *Aquatic Toxicology (Amsterdam, Netherlands)*, *191*, 50–61. doi:10.1016/j.aquatox.2017.07.022 PMID:28800408

Nogueira, J. M. F. (2015). Stir-bar sorptive extraction: 15 years making sample preparation more environment friendly. *Trends in Analytical Chemistry*, *71*, 214–223. doi:10.1016/j.trac.2015.05.002

NORMAN. (2019). *Network of reference laboraties, research centres and related organisations for Monitoring of emerging environmental substances*. Retrieved from https://www.norman-network.net/?q=Home

NORMAN-Network. (2016). *Emerging Substances*. Retrieved from http://www.norman-network.net/sites/default/files/files/Emerging_substances_list_Feb_16/NORMAN%20list_2016_FINAL.XLSX

Nováková, L., & Vlcková, H. (2009). A review of current trends and advances in modern bio-analytical methods: Chromatography and sample preparation. *Analytica Chimica Acta*, *656*(1-2), 8–35. doi:10.1016/j.aca.2009.10.004 PMID:19932811

Nuñez, L., & Moretton, J. (2007). Disinfectant-resistant bacteria in Buenos Aires city hospital wastewater. *Brazilian Journal of Microbiology*, *38*(4), 123–132. doi:10.1590/S1517-83822007000400012

Núñez, O., Gallart-Ayala, H., Martins, C. P. B., & Lucci, P. (2012). New trends in fast liquid chromatography for food and environmental analysis. *Journal of Chromatography. A*, *1228*, 298–323. doi:10.1016/j.chroma.2011.10.091 PMID:22153282

Nyenje, P. M., Foppen, J. W., Uhlenbrook, S., Kulabako, R., & Muwanga, A. (2010). Eutrophication and nutrient release in urban areas of sub-Saharan Africa - A review. *The Science of the Total Environment*, *408*(3), 447–455. doi:10.1016/j.scitotenv.2009.10.020 PMID:19889445

Obst, U., Schwartz, T., & Volkmann, H. (2006). Antibiotic resistant pathogenic bacteria and their resistance genes in bacterial biofilms. *The International Journal of Artificial Organs*, *29*(4), 387–394. doi:10.1177/039139880602900408 PMID:16705607

Offiong, N.-A. O., Inam, E. J., & Edet, J. B. (2019). Preliminary Review of Sources, Fate, Analytical Challenges and Regulatory Status of Emerging Organic Contaminants in Aquatic Environments in Selected African Countries. *Chemistry Africa*, 1–13.

Ogbeide, O., Uhunamure, G., Okundaye, F., & Ejeomo, C. (2019). First report on probabilistic risk assessment of pesticide residues in a riverine ecosystem in South-South Nigeria. *Chemosphere*, *231*, 546–561. doi:10.1016/j.chemosphere.2019.05.105 PMID:31151015

Ogunah, J. A., Owuor, P. O., Kowenje, C. O., & Lalah, J. O. (2016). Adsorption and desorption of chlorsulfuron in agricultural soils of Mara River Basin, Kenya. *AJEA*, *12*(6), 1–10. doi:10.9734/AJEA/2016/26406

Oke, Ismail, Lukman, Ojo, Adeosun, & Nwude. (2016). Economics and Statistical Evaluations of Using Microsoft Excel Solver in Pipe Network Analysis. *J. Pipeline Syst. Eng.*, *7*(4).

Oke, I. A., Amele, S. O., Lukman, S., Ismail, A., & Foghi, P. O. (2016). Water quality and self purification properties of Ogbe River, Ile-Ife, Nigeria. *Journal of Environmental Design and Management*, *8*(1 & 2), 174–189.

Okereke, J. N., Ogidi, O. I., & Obasi, K. O. (2016). International Journal of Advanced Research in Biological Sciences Environmental and Health Impact of Industrial Wastewater Effluents in Nigeria-A Review. *Int. J. Adv. Res. Biol. Sci*, *3*(6), 55–67.

Oketola, A. A., & Fagbemigun, T. K. (2013). Determination of nonylphenol, octylphenol and bisphenol-A in water and sediments of two major rivers in Lagos, Nigeria. *Journal of Environmental Protection*, *4*(07), 38–45. doi:10.4236/jep.2013.47A005

Okoth, O. K., Yan, K., Feng, J., & Zhang, J. (2018). Label-free photoelectrochemical aptasensing of diclofenac based on gold nanoparticles and graphene-doped CdS. *Sensors and Actuators. B, Chemical*, *256*, 334–341. doi:10.1016/j.snb.2017.10.089

Okoth, O. K., Yan, K., Liu, Y., & Zhang, J. (2016). Graphene-doped Bi_2S_3 nanorods as visible-light photoelectrochemical aptasensing platform for sulfadimethoxine detection. *Biosensors & Bioelectronics*, *86*, 636–642. doi:10.1016/j.bios.2016.07.037 PMID:27471154

Okoth, O. K., Yan, K., & Zhang, J. (2017). Mo-doped BiVO4 and graphene nanocomposites with enhanced photoelectrochemical performance for aptasensing of streptomycin. *Carbon*, *120*, 194–202. doi:10.1016/j.carbon.2017.04.079

Okungu, J. O., Njoka, S., Abuodha, J. O. Z., & Hecky, R. E. (2005). *An Introduction to Lake Victoria Catchment, Water Quality, Physical Limnology and Ecosystem Status (Kenyan Sector)*. Lake Victoria Environment Management Project (LVEMP) Kisumu Kenya. Retrieved from http://hdl.handle.net/1834/7140

Oleszczuk, P., Hale, S. E., Lehmann, J., & Cornelissen, G. (2012). Activated carbon and biochar amendments decrease pore-water concentrations of polycyclic aromatic hydrocarbons (PAHs) in sewage sludge. *Bioresource Technology*, *111*, 84–91. doi:10.1016/j.biortech.2012.02.030 PMID:22391590

Ololade, I. A. (2014). Spatial distribution of perfluorooctane sulfonate (PFOS) in major rivers in southwest Nigeria. *Toxicological and Environmental Chemistry*, *96*(9), 1356–1365. doi:10.1080/02772248.2015.1028409

Ololade, I. A., Oladoja, N. A., Ololade, O. O., Oloye, F. F., Adeola, A. O., Alabi, A. B., ... Owolabi, M. B. (2018). Geographical distribution of perfluorooctanesulfonate and perfluorooctanoate in selected rivers from Nigeria. *Journal of Environmental Chemical Engineering*, *6*(4), 4061–4069. doi:10.1016/j.jece.2018.06.020

Olowe, O. A., Olayemi, A. B., Eniola, K. I. T., & Adeyeba, O. A. (2004). Antibacterial activity of some selected disinfectants regularly used in hospitals. *African Journal of Clinical and Experimental Microbiology*, *5*(1), 126–130. doi:10.4314/ajcem.v5i1.7365

Olya, S. K., Katherine, Y. B., Carla, C., Benjamin, J. F., Meagan, S. M., Austa, M. P., ... Holly, A. S. (2014). Emerging Pollutants – Part II: Treatment. *Water Environment Research*, *86*(10), 2036–2096. doi:10.1002/j.1554-7531.2014.tb00240.x

Omole, D. O., Adewumi, I. K., Longe, E. O., & Ogbiya, A. S. (2012). Study of Auto-Purification Capacity of River Atuwara in Nigeria. *Environmental Research Journal*, *6*(3), 151-157.

Omole, D. O., & Longe, E. O. (2012). Re-aeration Coefficient Modeling: A Case Study of River Atuwara in Nigeria. *Research Journal of Applied Sciences, Engineering and Technology*, *4*(10), 1237–1243.

Omwoma, S., Lalah, J. O., Virani, M., Schramm, K., & Henkelmann, B. (2015). Dioxin-like PCBs and PCDD/Fs in surface sediments near the shore of Winam Gulf, Lake Victoria. *Chemosphere*, *118*, 143–147. doi:10.1016/j.chemosphere.2014.07.062 PMID:25150826

Onunga, D. O., Kowino, I. O., Ngigi, A. N., Aggrey, O., Orata, F., Getenga, Z. M., & Were, H. (2015). Biodegradation of carbofuran in soils within Nzoia River Basin, Kenya. *Journal of Environmental Science and Health. Part. B, Pesticides, Food Contaminants, and Agricultural Wastes*, *50*(6), 387–397. doi:10.1080/03601234.2015.1011965 PMID:25844859

Orata F., Maes A., Werres F., Wilken D. F. (2011). Perflourinated compounds distribution and source identification in sediments of Lake Victoria Gulf Basin. *Soil and Sediment Contamination: An International Journal.* doi:10.1080/15320383.2011.546442

Orata F., Quinete N., Maes, A., Werres, F., & Wilken, R. (2008a). Perfluorooctanoic acid and perfluorooctane sulfonate in Nile Perch and tilapia from gulf of Lake Victoria. *African Journal of Pure and Applied Chemistry*, *2*(8), 75-79.

Orata, F. (2012). *Derivatization Reactions and Reagents for Gas Chromatography Analysis*. Advanced Gas Chromatography. doi:10.5772/33098

Orata, F. O. (2018). Conventional Wastewater Treatment Plants as a Discharge and Source Point for Biota Exposure to Micro-pollutants. In R. A. Hauser-Davis & T. E. Parente (Eds.), *Ecotoxicology, Perspectives on Key Issues*. doi:10.1201/b21896-11

Orata, F., Quinete, N., Werres, F., & Wilken, R. D. (2009). Determination of Perfluorooctanoic acid and Perfluorooctane sulfonate in Lake Victoria Gulf Water. *Bulletin of Environmental Contamination and Toxicology*, *82*(2), 218–222. doi:10.100700128-008-9543-1 PMID:18791652

Ötker, H. M., & Akmehmet-Balcioglu, I. (2005). Adsorption and degradation of enrofloxacin, a veterinary antibiotic on natural zeolite. *Journal of Hazardous Materials*, *122*(3), 251–258. doi:10.1016/j.jhazmat.2005.03.005 PMID:15967281

Ouvrard, S., Leglize, P., & Morel, J. L. (2014). PAH phytoremediation: Rhizodegradation or rhizoattenuation? *International Journal of Phytoremediation*, *16*(1), 46–61. doi:10.1080/15226514.2012.759527 PMID:24912214

Overmans, S., Nordborg, M., Díaz-Rúa, R., Brinkman, D. L., Negri, A. P., & Agustí, S. (2018). Phototoxic effects of PAH and UVA exposure on molecular responses and developmental success in coral larvae. *Aquatic Toxicology (Amsterdam, Netherlands)*, *198*, 165–174. doi:10.1016/j.aquatox.2018.03.008 PMID:29550714

Ozdemir, I., Şahin, M., Orhan, R., & Erdem, M. (2014). Preparation and characterization of activated carbon from grape stalk by zinc chloride activation. *Fuel Processing Technology*, *125*, 200–206. doi:10.1016/j.fuproc.2014.04.002

Pablo, C. F., Mart'ınez, C. J., Cristina, S., & Manuel, A. R. (2009). Technical and economic comparison of conventional and electrochemical coagulation processes. *Journal of Chemical Technology and Biotechnology (Oxford, Oxfordshire)*, *84*(5), 702–710. doi:10.1002/jctb.2102

Padrón, M. E. T., Ferrera, Z. S., & Rodríguez, J. J. S. (2009). Coupling of solid-phase microextraction with micellar desorption and high performance liquid chromatography for the determination of pharmaceutical residues in environmental liquid samples. *Biomedical Chromatography*, *23*(11), 1175–1185. doi:10.1002/bmc.1240 PMID:19444801

Pal, A., He, Y., Jekel, M., Reinhard, M., & Gin, K. Y. (2014). Emerging contaminants of public health significance as water quality indicator compounds in the urban water cycle Untreated water Sewer system. *Environment International*, *71*, 46–62. doi:10.1016/j.envint.2014.05.025 PMID:24972248

Panizza, M., & Cerisola, G. (2003). A comparative study on direct and indirect electrochemical oxidation of polyaromatic compounds. *Ann Chim.*, *93*(12), 977–984. PMID:14756253

Park, B. J., & Kidd, K. (2005). Effects of the synthetic estrogen ethinylestradiol on early life stages of mink frogs and green frogs in the wild and in situ. *Environmental Toxicology and Chemistry: An International Journal*, *24*(8), 2027–2036. doi:10.1897/04-227R.1 PMID:16152976

Park, H. S., Ha, H.-W., Ruoff, R. S., & Bard, A. J. (2014). On the improvement of photoelectrochemical performance and finite element analysis of reduced graphene oxide-$BiVO_4$ composite electrodes. *Journal of Electroanalytical Chemistry*, *716*, 8–15. doi:10.1016/j.jelechem.2013.08.036

Park, J.-W., Tatavarty, R., Kim, D. W., Jung, H.-T., & Gu, M. B. (2012). Immobilization-free screening of aptamers assisted by graphene oxide. *Chemical Communications*, *48*(15), 2071–2073. doi:10.1039/C2CC16473F PMID:22143382

Park, S., & Choi, K. (2008). Hazard assessment of commonly used agricultural antibiotics on aquatic ecosystems. *Ecotoxicology (London, England)*, *17*(6), 526–538. doi:10.100710646-008-0209-x PMID:18449638

Pasquini, L., Merlin, C., Hassenboehler, L., Munoz, J.-F., Pons, M.-N., & Görner, T. (2013). Impact of Certain Household Micropollutants on Bacterial Behavior. Toxicity Tests/Study of Extracellular Polymeric Substances in Sludge. *The Science of the Total Environment*, *463–464*, 355–365. doi:10.1016/j.scitotenv.2013.06.018 PMID:23827359

Pastore, M. C. (2015). Reworking the relation between sanitation and the city in Dar es Salaam, Tanzania. *Environment and Urbanization*, *27*(2), 473–488. doi:10.1177/0956247815592285

Patnaik, S., Mishra, P., Nayak, R., & Giri, A. (2016). Removal of fluoride from aqueous solution using Chitosan-Iron complex. *Journal of Analytical & Bioanalytical Techniques*, *7*(4).

Pattengale, B., Ludwig, J., & Huang, J. (2016). Atomic Insight into the W-Doping Effect on Carrier Dynamics and Photoelectrochemical Properties of $BiVO_4$ Photoanodes. *The Journal of Physical Chemistry C*, *120*(3), 1421–1427. doi:10.1021/acs.jpcc.5b11451

Pauwels, B., & Verstraete, W. (2006). The treatment of hospital wastewater: an appraisal The treatment of hospital wastewater: an appraisal. *Journal of Water and Health*, *04*(4), 406–416. doi:10.2166/wh.2006.0024 PMID:17176811

Pawliszyn, J. (1997). *Solid phase microextraction: theory and practice*. New York: Wiley-VHC.

Pearlman, R. S., Yalkowsky, S. H., & Banerjee, S. (1984). Water solubilities of polynuclear aromatic and heteroaromatic compounds. *Journal of Physical and Chemical Reference Data*, *103*(2), 555–562. doi:10.1063/1.555712

Pedro, A. S., Matthieu, F., Christian, G., & Sébastien, S. (2009). Review of the Occurrence of Anti-infectives in Contaminated Wastewaters and Natural and Drinking Waters. *Environmental Health Perspectives*, *117*(5), 675–684. doi:10.1289/ehp.11776 PMID:19479007

Pedrouzo, M., Borrull, F., Marce, R. M., & Pocurull, E. (2010). Stir-barsorptive extraction and ultra-high-performance liquid chromatography-tandem mass spectrometry for simultaneous analysis of UV filters and antimicrobial agents in water samples. *Analytical and Bioanalytical Chemistry*, *397*(7), 2833–2839. doi:10.100700216-010-3743-3 PMID:20428847

Peng, J., Wang, J., & Cai, L. (2017). Current understanding of microplastics in the environment: Occurrence, fate, risks, and what we should do. *Integrated Environmental Assessment and Management*, *13*(3), 476–482. doi:10.1002/ieam.1912 PMID:28440924

Peng, R.-H., Fu, X.-Y., Zhao, W., Tian, Y.-S., Zhu, B., Han, H.-J., ... Yao, Q.-H. (2014). Phytoremediation of Phenanthrene by Transgenic Plants Transformed with a Naphthalene Dioxygenase System from *Pseudomonas*. *Environmental Science & Technology*, *48*(21), 12824–12832. doi:10.1021/es5015357 PMID:25299803

Peng, X., Ou, W., Wang, C., Wang, Z., Huang, Q., Jin, J., & Tan, J. (2014). Occurrence and ecological potential of pharmaceuticals and personal care products in groundwater and reservoirs in the vicinity of municipal landfills in China. *The Science of the Total Environment*, *490*, 889–898. doi:10.1016/j.scitotenv.2014.05.068 PMID:24908648

Peng, X., Xiong, S., Ou, W., Wang, Z., Tan, J., Jin, J., ... Fan, Y. (2016). Persistence, temporal and spatial profiles of ultraviolet absorbents and phenolic personal care products in riverine and estuarine sediment of the Pearl River catchment, China. *Journal of Hazardous Materials*, *8*. doi:10.1016/j.jhazmat.2016.05.020 PMID:27209124

Peng, Y., Gautam, L., & Hall, S. W. (2019). The detection of drugs of abuse and pharmaceuticals in drinking water using solid-phase extraction and liquid chromatography-mass spectrometry. *Chemosphere*, *223*, 438–447. doi:10.1016/j.chemosphere.2019.02.040 PMID:30784750

Perra, G., Pozo, K., Guerranti, C., Lazzeri, D., Volpi, V., Corsolini, S., & Focardi, S. (2011). Levels and spatial distribution of polycyclic aromatic hydrocarbons (PAHs) in superficial sediment from 15 Italian marine protected areas (MPA). *Marine Pollution Bulletin*, *62*(4), 874–877. doi:10.1016/j.marpolbul.2011.01.023 PMID:21310440

Perreault, F., de Faria, A. F., & Elimelech, M. (2015). Environmental applications of graphene-based nanomaterials. *Chemical Society Reviews*, *44*(16), 5861–5896. doi:10.1039/C5CS00021A PMID:25812036

Pescod, M. B. (1992). *Wastewater treatment and use in agriculture*. Food and Agriculture Organization of the United Nations.

Petrie, B., McAdam, E.J., Lester, J.N., & Cartmell, E. (2014c). Obtaining process mass balances of pharmaceuticals and triclosan to determine their fate during wastewater treatment. *Sci. Total Environ., 497 – 498*, 560 -568.

Petrie, B., Barden, R., & Kasprzyk-Hordern, B. (2015). A review on emerging contaminants in wastewaters and the environment: Current knowledge, understudied areas and recommendations for future monitoring. *Water Research*, *72*, 3–27. doi:10.1016/j.watres.2014.08.053 PMID:25267363

Petrie, B., McAdam, E. J., Hassard, F., Stephenson, T., Lester, J. N., & Cartmell, E. (2014a). Diagnostic investigation of steroid estrogen removal by activated sludge at varying solids retention time. *Chemosphere*, *113*, 101–108. doi:10.1016/j.chemosphere.2014.04.051 PMID:25065796

Petrie, B., McAdam, E. J., Lester, J. N., & Cartmell, E. (2014b). Assessing potential modifications to the activated sludge process to improve simultaneous removal of a diverse range of micropollutants. *Water Research*, *62*, 180–192. doi:10.1016/j.watres.2014.05.036 PMID:24956600

Petrie, B., McAdam, E. J., Whelan, M. J., Lester, J. N., & Cartmell, E. (2013c). The determination of nonylphenol and its precursors in a trickling filter wastewater treatment process. *Analytical and Bioanalytical Chemistry*, *405*(10), 3243–3253. doi:10.100700216-013-6765-9 PMID:23388691

Piazza, R. S., Trevisan, R., Flores-Nunes, F., Toledo-Silva, G., Wendt, N., Mattos, J. J., ... Bainy, A. C. D. (2016). Exposure to phenanthrene and depuration: Changes on gene transcription, enzymatic activity and lipid peroxidation in gill of scallops *Nodipecten nodosus. Aquatic Toxicology (Amsterdam, Netherlands)*, *177*, 146–155. doi:10.1016/j.aquatox.2016.05.025 PMID:27286572

Pintado-Herrera, M. G., González-Mazo, E., & Lara-Martín, P. A. (2014). Atmospheric pressure gas chromatography-time-of-flight-mass spectrometry (APGC-ToF-MS) for the determination of regulated and emerging contaminants in aqueous samples after stir bar sorptive extraction (SBSE). *Analytica Chimica Acta*, *851*, 1–13. doi:10.1016/j.aca.2014.05.030 PMID:25440658

Piri-Moghadam, H., Ahmadi, F., & Pawliszyn, J. (2016). A critical review of solid phase microextraction for analysis of water samples. *Trends in Analytical Chemistry*, *85*, 133–143. doi:10.1016/j.trac.2016.05.029

Pomati, F., Castiglioni, S., Zuccato, E., Fanelli, R., Vigetti, D., Rossetti, C., & Davide, C. (2006). Effects of a complex mixture of therapeutic drugs at environmental levels on human embryonic cells. *Environmental Science & Technology*, *40*(7), 2442–2447. doi:10.1021/es051715a PMID:16646487

Porter, C. J. H., Trevaskis, N. L., & Charman, W. N. (2007). Lipids and lipid-based formulations: Optimizing the oral delivery of lipophilic drugs. *Nature Reviews. Drug Discovery*, *6*(3), 231–248. doi:10.1038/nrd2197 PMID:17330072

Portolés, T., Pitarch, E., Lopez, F., & Hernandez, J. F. (2011). Development and validation of a rapid and wide-scope qualitative screening method for detection and identification of organic pollutants in natural water and wastewater by gas chromatography time-of-flight mass spectrometry. *Journal of Chromatography. A*, *1218*(2), 303–315. doi:10.1016/j.chroma.2010.11.010 PMID:21134677

Pothitou, P., & Voutsa, D. (2008). Endocrine disrupting compounds in municipal and industrial wastewater treatment plants in Northern Greece. *Chemosphere*, *73*(11), 1716–1723. doi:10.1016/j.chemosphere.2008.09.037 PMID:18954890

Poynton, H. C., & Vulpe, C. D. (2009). Ecotoxicogenomics: Emerging technologies for emerging contaminants. *Journal of the American Water Resources Association*, *45*(1), 83–96. doi:10.1111/j.1752-1688.2008.00291.x

Prabhu, S. M., & Meenakshi, S. (2013). Defluoridation of water using synthesized Zr(IV) encapsulated silica gel/chitosan biocomposite: Adsorption isotherms and kinetic studies. *Desalination and Water Treatment*, *53*(13), 3592–3603. doi:10.1080/19443994.2013.871347

Prabhu, S., & Meenakshi, S. (2016). Defluoridation of water using dicarboxylic acids mediated chitosan-polyaniline/zirconium biopolymeric complex. *International Journal of Biological Macromolecules*, *85*, 16–22. doi:10.1016/j.ijbiomac.2015.12.022 PMID:26706844

Prasad, K. S., Amin, Y., & Selvaraj, K. (2014). Defluoridation using biomimetically synthesized nano zirconium chitosan composite: Kinetic and equilibrium studies. *Journal of Hazardous Materials*, *276*, 232–24. doi:10.1016/j.jhazmat.2014.05.038 PMID:24887125

Prieto, A., Basauri, O., Rodil, R., Usobiaga, A., Fernández, L. A., Etxebarria, N., & Zuloaga, O. (2010). Stir-bar sorptive extraction: A view on method optimisation, novel applications, limitations and potential solutions. *Journal of Chromatography. A*, *1217*(16), 2642–2666. doi:10.1016/j.chroma.2009.12.051 PMID:20083248

Prieto, A., Telleria, O., Etxebarria, N., Fernández, L. A., Usobiaga, A., & Zuloaga, O. (2008b). Simultaneous preconcentration of a wide variety of organic pollutants in water samples. Comparison of stir bar sorptive extraction and membrane-assisted solvent extraction. *Journal of Chromatography. A*, *1214*(1-2), 1–10. doi:10.1016/j.chroma.2008.10.060 PMID:18990398

Prieto, A., Zuloaga, O., Usobiaga, A., Etxebarria, N., & Fernandez, L. A. (2008a). Use of experimental design in the optimisation of stir bar sorptive extraction followed by thermal desorption for the determination of brominated flame retardants in water samples. *Analytical and Bioanalytical Chemistry*, *390*(2), 739–748. doi:10.100700216-007-1712-2 PMID:18038264

Prosser, C. M., Unger, M. A., & Vogelbein, W. K. (2011). Multistressor interactions in the zebrafish (*Danio rerio*): Concurrent phenanthrene exposure and *Mycobacterium marinum* infection. *Aquatic Toxicology (Amsterdam, Netherlands)*, *102*(3-4), 177–185. doi:10.1016/j.aquatox.2011.01.011 PMID:21356180

Pruden, A., Pei, R., Storteboom, H., & Carlson, K. H. (2006). Antibiotic resistance genes as emerging contaminants: Studies in Northern Colorado. *Environmental Science & Technology*, *40*(23), 7445–7450. doi:10.1021/es060413l PMID:17181002

Pulford, I., & Watson, C. (2003). Phytoremediation of heavy metal-contaminated land by trees—A review. *Environment International*, *29*(4), 529–540. doi:10.1016/S0160-4120(02)00152-6 PMID:12705950

Punmia, B. C., Arun, K. J., & Ashok, K. J. (1998). *Waste water Engineering (including air pollution)* (2nd ed.). New Delhi: Laxmi Publication LTD.

Putra, E. K., Pranowo, R., Sunarso, J., Indraswati, N., & Ismadji, S. (2009). Performance of activated carbon and bentonite for adsorption of amoxicillin from wastewater: Mechanism, isotherms and kinetics. *Water Research*, *43*(9), 2419–2430. doi:10.1016/j.watres.2009.02.039 PMID:19327813

Qamar, M., Saquib, M., & Muneer, M. (2005). Photocatalytic degradation of two selected dye derivatives, chromotrope 2B and amido black 10B, in aqueous suspensions of titanium dioxide. *Dyes and Pigments*, *65*(1), 1–9. doi:10.1016/j.dyepig.2004.06.006

Qian, S., Xuqi, C., Shuguang, L., Wentao, Z., Zhaofu, Q., & Gang, Y. (2015). Occurrence, sources and fate of pharmaceuticals and personal care products in the groundwater: A review. *Emerging Contaminants*, *1*(1), 14–24. doi:10.1016/j.emcon.2015.07.001

Qiao, M., Wang, C., Huang, S., Wang, D., & Wang, Z. (2006). Composition, sources, and potential toxicological significance of PAHs in the surface sediments of the Meiliang Bay, Taihu Lake, China. *Environment International*, *32*(1), 28–33. doi:10.1016/j.envint.2005.04.005 PMID:15996733

Qiao, Y., Li, J., Li, H., Fang, H., Fan, D., & Wang, W. (2016). A label-free photoelectrochemical aptasensor for bisphenol A based on surface plasmon resonance of gold nanoparticle-sensitized ZnO nanopencils. *Biosensors & Bioelectronics*, *86*, 315–320. doi:10.1016/j.bios.2016.06.062 PMID:27387262

Qiu, Q., Tan, Z., Wang, J., Peng, J., Li, M., & Zhan, Z. (2016). Extraction, enumeration and identification methods for monitoring microplastics in the environment. *Estuarine, Coastal and Shelf Science*, *176*, 102–109. doi:10.1016/j.ecss.2016.04.012

Qiusheng, Z., Xiaoyan, L., Jin, Q., Jing, W., & Xuegang, L. (2015). Porous zirconium alginate beads adsorbent for fluoride adsorption from aqueous solutions. *RSC Advances*, *5*(3), 2100–2112. doi:10.1039/C4RA12036A

Queiroz, S. C. N., Collins, C. H., & Jardim, I. C. S. F. (2001). Métodos de Extração e/ou Concentração de Compostos Encontrados em Fluidos Biológicos para Posterior Determinação Cromatográfica. *Quimica Nova*, *24*(1), 68–76. doi:10.1590/S0100-40422001000100013

Qu, X., Alvarez, J., & Li, Q. (2013). Applications of nanotechnology in water and wastewater treatment. *Water Research*, *47*(12), 3931–3946. doi:10.1016/j.watres.2012.09.058 PMID:23571110

Rabodonirina, S., Rasolomampianina, R., Krier, F., Drider, D., Merhaby, D., Net, S., & Ouddane, B. (2019). Degradation of fluorene and phenanthrene in PAHs-contaminated soil using *Pseudomonas* and *Bacillus* strains isolated from oil spill sites. *Journal of Environmental Management*, *232*, 1–7. doi:10.1016/j.jenvman.2018.11.005 PMID:30453222

Raftery, T. D., Jayasundara, N., & Di Giulio, R. T. (2017). A bioenergetics assay for studying the effects of environmental stressors on mitochondrial function in vivo in zebrafish larvae. *Comparative Biochemistry and Physiology Part C: Toxicology & Pharmacology*, *192*, 23–32. doi:10.1016/j.cbpc.2016.12.001 PMID:27939721

Raghav, M., Eden, S., Mitchell, K., & Witte, B. (2013). *Contaminants of Emerging Concern in Wastewater.* Water Resources Research Center.

Raghav, M., Eden, S., Mitchell, K., & Witte, B. (2013). *Contaminants of Emerging Concern in Water*. Tucson, AZ: Water Resources Research Center, College of Agriculture, University of Arizona. Retrieved from http://hdl.handle.net/10150/325905

Rahim, A., Mahnaz, N., Hossein, K., & Akbar, H. (2014). Monitoring and Comparison of Antibiotic Resistant Bacteria and Their Resistance Genes in Municipal and Hospital Wastewaters. *International Journal of Preventive Medicine*, *5*(7), 887–894. PMID:25105001

Rajapaksha, A. U., Vithanage, M., Ahmad, M., Seo, D.-C., Cho, J.-S., Lee, S.-E., ... Ok, Y. S. (2015). Enhanced sulfamethazine removal by steam-activated invasive plant-derived biochar. *Journal of Hazardous Materials*, *290*, 43–50. doi:10.1016/j.jhazmat.2015.02.046 PMID:25734533

Ramdani, A., Taleb, S., Benghalem, A., & Ghaffour, N. (2010). Removal of excess fluoride ions from Saharan brackish water by adsorption on natural materials. *Desalination*, *250*(1), 408–413. doi:10.1016/j.desal.2009.09.066

Ramos, L. (2012). Critical overview of selected contemporary sample preparation techniques. *Journal of Chromatography. A*, *1221*, 84–98. doi:10.1016/j.chroma.2011.11.011 PMID:22137129

Rana, P., Mohan, N., & Rajagopal, C. (2004). Electrochemical removal of chromium from wastewater by using carbon aerogel electrodes. *Water Research*, *38*(12), 2811–2820. doi:10.1016/j.watres.2004.02.029 PMID:15223274

Rao, M. A., Di Rauso Simeone, G., Scelza, R., & Conte, P. (2017). Biochar based remediation of water and soil contaminated by phenanthrene and pentachlorophenol. *Chemosphere*, *186*, 193–201. doi:10.1016/j.chemosphere.2017.07.125 PMID:28778017

Rao, P. M., Cai, L. L., Liu, C., Cho, I. S., Lee, C. H., Weisse, J. M., ... Zheng, X. L. (2014). Simultaneously Efficient Light Absorption and Charge Separation in $WO_3/BiVO_4$ Core/Shell Nanowire Photoanode for Photoelectrochemical Water Oxidation. *Nano Letters*, *14*(2), 1099–1105. doi:10.1021/nl500022z PMID:24437363

Rashed, I. G., Hanna, M. A., El-Gamal, H. F., Al-Sarawy, A. A., & Wali, F. K. M. (2005). *Overview On Chemical Oxidation Technology In Wastewater Treatment*. Ninth International Water Technology Conference, IWTC9 2005, Sharm El-Sheikh, Egypt.

Rashid, H., Bhattia, N., Iqbal, M., & Noreen, S. (2015). Fungal biomass composite with bentonite efficiency for nickel andzinc adsorption: A mechanistic study. *Ecological Engineering*, *91*, 459–471. doi:10.1016/j.ecoleng.2016.03.014

Raucher, R. S. (1996). Public Health and Regulatory Considerations of the Safe Drinking Water Act. *Annual Review of Public Health*, *17*(1), 179–202. doi:10.1146/annurev.pu.17.050196.001143 PMID:8724223

Rauf, A., Sher Shah, M. S. A., Choi, G. H., Humayoun, U. B., Yoon, D. H., Bae, J. W., ... Yoo, P. J. (2015). Facile Synthesis of Hierarchically Structured Bi_2S_3/Bi_2WO_6 Photocatalysts for Highly Efficient Reduction of Cr(VI). *ACS Sustainable Chemistry & Engineering*, *3*(11), 2847–2855. doi:10.1021/acssuschemeng.5b00783

Ravera, M., Ciccarelli, C., Gianotti, V., Scorza, S., & Osella, D. (2004). Electroassisted methods for waste destruction: Silver (II) and peroxydisulfate reagents in the electrochemically mediated oxidation of polyaromatic sulfonates. *Chemosphere*, *57*(7), 587–594. doi:10.1016/j.chemosphere.2004.08.035 PMID:15488920

Reddam, A., Mager, E. M., Grosell, M., & McDonald, M. D. (2017). The impact of acute PAH exposure on the toadfish glucocorticoid stress response. *Aquatic Toxicology (Amsterdam, Netherlands)*, *192*, 89–96. doi:10.1016/j.aquatox.2017.08.014 PMID:28942071

Regueiro, J., Becerril, E., Garcia-Jares, C., & Llompart, M. (2009a). Trace analysis of parabens, triclosan and related chlorophenols in water by headspace solid-phase microextraction with in situ derivatization and gas chromatography-tandem mass spectrometry. *Journal of Chromatography. A*, *1216*(23), 4693–4702. doi:10.1016/j.chroma.2009.04.025 PMID:19414177

Regueiro, J., Llompart, M., Psillakis, E., Garcia-Monteagudo, J. C., & Garcia-Jares, C. (2009b). Ultrasound-assisted emulsification-microextraction of phenolic preservatives in water. *Talanta*, *79*(5), 1387–1397. doi:10.1016/j.talanta.2009.06.015 PMID:19635375

Reynoldson, T. B., & Metcalfe-Smith, J. L. (1992). An overview of the assessment of aquatic ecosystem health using benthic invertebrates. *Journal of Aquatic Ecosystem Health*, *1*(4), 295–308. doi:10.1007/BF00044171

Rezaee, M., Assadi, Y., Hosseini, M.-R. M., Aghaee, E., Ahmadi, F., & Berijani, S. (2006). Determination of organic compounds in water using dispersive liquid–liquid microextraction. *Journal of Chromatography. A*, *1116*(1-2), 1–9. doi:10.1016/j.chroma.2006.03.007 PMID:16574135

Rezaee, M., Yamini, Y., & Faraji, M. (2010). Evolution of dispersive liquid–liquid microextraction method. *Journal of Chromatography. A*, *1217*(16), 2342–2357. doi:10.1016/j.chroma.2009.11.088 PMID:20005521

Rhim, J. W., Park, H. M., & Ha, C. S. (2013). Bio-nanocomposites for food packaging applications. *Progress in Polymer Science*, *38*(10-11), 1629–1652. doi:10.1016/j.progpolymsci.2013.05.008

Ribeiro-Claro, P., Nolasco, M. M., & Araújo, C. (2017). Chapter 5-characterization of microplastics by Raman spectroscopy. *Comprehensive Analytical Chemistry*, *75*, 119–151. doi:10.1016/bs.coac.2016.10.001

Richardi, V. S., Vicentini, M., Morais, G. S., Rebechi, D., da Silva, T. A., Fávaro, L. F., & Navarro-Silva, M. A. (2018). Effects of phenanthrene on different levels of biological organization in larvae of the sediment-dwelling invertebrate *Chironomus sancticaroli* (Diptera:Chironomidae). *Environmental Pollution*, *242*, 277–287. doi:10.1016/j.envpol.2018.06.091 PMID:29990935

Richardson, S. D. (2006). Environmental Mass Spectrometry: Emerging Contaminants and Current Issues. *Analytical Chemistry*, *78*(12), 4021–4046. doi:10.1021/ac060682u PMID:16771539

Richardson, S. D., & Ternes, T. A. (2017). Water Analysis: Emerging Contaminants and Current Issues. *America Chemical Society*, *90*, 398–428. PMID:29112806

Rinaldi, S., & Soncini-Sessa, R. (1978). Optimal allocation of artificial instream aeration. Journal of the Environmental Engineering Division. *Proceedings of the American Society of Civil Engineers*, *104*, 147–160.

Ríos, Á., & Zougagh, M. (2016). Recent advances in magnetic nano materials for improving analytical processes. *Trends in Analytical Chemistry*, *84*, 72–83. doi:10.1016/j.trac.2016.03.001

Roberts, D. (2006). DDT returns. *Nature Publishing Group*, *12*(8), 870–871. PMID:16892021

Rocha, M. J., Cruzeiro, C., Reis, M., Pardal, M. A., & Rocha, E. (2016). Pollution by oestrogenic endocrine disruptors and b-sitosterol in a South-Western European River (Mira, Portugal). *Environmental Monitoring and Assessment*, *188*(4), 240–254. doi:10.100710661-016-5236-0 PMID:27004433

Rocha-Santos, T., & Duarte, A. C. (2015). A critical overview of the analytical approaches to the occurrence, the fate and the behavior of microplastics in the environment. *Trends in Analytical Chemistry*, *65*, 47–53. doi:10.1016/j.trac.2014.10.011

Rochman, C. M., Cook, A. M., & Koelmans, A. A. (2016). Plastic debris and policy: Using current scientific understanding to invoke positive change. *Environmental Toxicology and Chemistry*, *35*(7), 1617–1626. doi:10.1002/etc.3408 PMID:27331654

Roco, M. C. (2001). International strategy for nanotechnology research and development. *Journal of Nanoparticle Research*, *3*(5/6), 353–360. doi:10.1023/A:1013248621015

Rodney, M. D., & William, J. C. B. (2002). Survival Mechanisms of Clinically Relevant Microorganisms.ASM. *Clinical Microbiology Reviews*, *15*(2), 167–193. doi:10.1128/CMR.15.2.167-193.2002 PMID:11932229

Rodrigues, E. M., Morais, D. K., Pylro, V. S., Redmile-Gordon, M., de Oliveira, J. A., Roesch, L. F. W., ... Tótola, M. R. (2017). Aliphatic Hydrocarbon Enhances Phenanthrene Degradation by Autochthonous Prokaryotic Communities from a Pristine Seawater. *Microbial Ecology*, *75*(3), 688–700. doi:10.100700248-017-1078-8 PMID:28971238

Rodriguez-Narvaez, O. M., Peralta-Hernandez, J. M., Goonetilleke, A., & Bandala, E. R. (2017). Treatment technologies for emerging contaminants in water: A review. *Chemical Engineering Journal*, *323*, 361–380. doi:10.1016/j.cej.2017.04.106

Román, I. P., Chisvert, A., & Canals, A. (2011). Dispersive solid-phase extraction based on oleic acid-coated magnetic nanoparticles followed by gas chromatography-mass spectrometry for UV-filter determination in water samples. *Journal of Chromatography. A*, *1218*(18), 2467–2475. doi:10.1016/j.chroma.2011.02.047 PMID:21411104

Rongjun, S., Guangshan, Z., Peng, W., Shixiong, L., Ryan, M. R., & Crittenden, J. C. (2015). Treatment of Antibiotic Pharmaceutical Wastewater Using a Rotating Biological Contactor. *Journal of Chemistry*, *2015*, 705275. doi:10.1155/2015/705275

Rosal, R., Rodrı'guez, A., Perdigo'n-Melo'n, J. A., Petre, A., Garcı'a-Calvo, E., Go'mez, M. J., ... Fernández-Alba, A. R. (2010). Occurrence of emerging pollutants in urban wastewater and their removal through biological treatment followed by ozonation. *Water Research*, *44*(2), 578–588. doi:10.1016/j.watres.2009.07.004 PMID:19628245

Rovani, S., Censi, M. T., Pedrotti, S. L., Lima, É. C., Cataluña, R., & Fernandes, A. N. (2014). Development of a new adsorbent from agro-industrial waste and its potential use in endocrine disruptor compound removal. *Journal of Hazardous Materials*, *271*, 311–320. doi:10.1016/j.jhazmat.2014.02.004 PMID:24647264

Ruane, J., Sonnino, A., & Agostini, A. (2010). Bioenergy and the potential contribution of agricultural biotechnologies in developing countries. *Biomass and Bioenergy*, *34*(10), 1427–1439. doi:10.1016/j.biombioe.2010.04.011

Ryan, C. C., Tan, D. T., & Arnold, W. A. (2011). Direct and indirect photolysis of sulfamethoxazole and trimethoprim in wastewater treatment plant effluent. *Water Research*, *45*(3), 1280–1286. doi:10.1016/j.watres.2010.10.005 PMID:21044793

Saba, B., Christy, A. D., & Jabeen, M. (2016). Kinetic and Enzymatic Decolorization of Industrial Dyes Utilizing Plant-Based Biosorbents: A Review. *Environmental Engineering Science*, *33*(9), 601–614. doi:10.1089/ees.2016.0038

Sahu, O. (2014). Reduction of heavy metals from wastewater by wetland. *International Letters of Natural Sciences*, *12*, 35–43. doi:10.18052/www.scipress.com/ILNS.12.35

Saichang, Z., Hong, C., & Jianan, L. (2013). Sources, distribution and potential risks of pharmaceuticals and personal care products in Qingshan Lake basin, Eastern China. *Ecotoxicology and Environmental Safety*, *96*, 154–159. doi:10.1016/j.ecoenv.2013.06.033 PMID:23871206

Saifuddin, N., Nian, C.Y., Zhan, L.W., & Ning, K.X. (2011). Chitosan-silvernanoparticles composite as point-of-use drinking water filtrationsystem for household to remove pesticides in water. *Asian JBiochem*. doi:10.3923/ajb.2011

Salgado, R., Marques, R., Noronha, J. P., Carvalho, G., Oehmen, A., & Reis, M. A. M. (2012). Assessing the removal of pharmaceuticals and personal care products in a full-scale activated sludge plant. *Environmental Science and Pollution Research International*, *19*(5), 1818–1827. doi:10.100711356-011-0693-z PMID:22207237

SalmaJ.RafikK. (2015). *Antibiotics*. Retrieved from https://www.researchgate.net/publication/272820065

Samanta, S. K., Chakraborti, A. K., & Jain, R. K. (1999). Degradation of phenanthrene by different bacteria: Evidence for novel transformation sequences involving the formation of 1-naphthol. *Applied Microbiology and Biotechnology*, *53*(1), 98–107. doi:10.1007002530051621 PMID:10645629

San Francisco Estuary Project. (2007). *Comprehensive Conservation And Management Plan: Pollution Prevention and Reduction*. San Francisco: Author.

Sánchez-Polo, M., Rivera-Utrilla, J., Prados-Joya, G., Ferro-García, M. A., & Bautista-Toledo, I. (2008). Removal of pharmaceutical compounds, nitroimidazoles, from waters by using the ozone/carbon systems. *Water Research*, *42*(15), 4163–4171. doi:10.1016/j.watres.2008.05.034 PMID:18762314

Sarantuyaa, Z., & Frank-Kamenetsky, D. (n.d.). *Emerging Pollutants in Water and Wastewater: UNESCO-HELCOM Case Study on Pharmaceuticals in the Aquatic and Marine Environment in the Baltic Sea Region*. Academic Press.

Schafer, A. K. S., & Kegley, S. E. (2002). Persistent Toxic Chemicals in the US Food Supply Persistent toxic chemicals in the US food supply. *Journal of Epidemiology and Community Health*, *56*(11), 813–817. doi:10.1136/jech.56.11.813 PMID:12388566

Schaider, L.A. Rudel, R.A., Ackerman, J.M. (2014). Pharmaceuticals, perfluorosurfactants, and other organic wastewater compounds in public drinking water wells in a shallow sand and gravel aquifer. *Sci. Total Environ., 468 – 469*, 384 – 393.

Schultz, M. M., Bartell, S. E., & Schoenfuss, H. L. (2012). Effects of Triclosan and Triclocarban, Two Ubiquitous Environmental Contaminants, on Anatomy, Physiology, and Behavior of the Fathead Minnow (Pimephales promelas). *Archives of Environmental Contamination and Toxicology*, *63*(1), 114–124. doi:10.100700244-011-9748-x PMID:22237462

Schwarz, A., Adetutu, E. M., Juhasz, A. L., Aburto-Medina, A., Ball, A. S., & Shahsavari, E. (2018). Microbial Degradation of Phenanthrene in Pristine and Contaminated Sandy Soils. *Microbial Ecology*, *75*(4), 888–902. doi:10.100700248-017-1094-8 PMID:29080101

Sebok, A., Vasanits-Zsigrai, A., Helenkar, A., Zaray, G., & Molnar-Perl, I. (2009). Multiresidue analysis of pollutants as their trimethylsilyl derivatives, by gas chromatography-mass spectrometry. *Journal of Chromatography. A*, *1216*(12), 2288–2301. doi:10.1016/j.chroma.2009.01.056 PMID:19201001

Sendzik, J., Shakibaei, M., Schafer-Korting, M., & Stahlmann, R. (2005). Fluoroquinolones cause changes in extracellular matrix, signalling proteins, metalloproteinases and caspase-3 in cultured human tendon cells. *Toxicology*, *212*(1), 24–36. doi:10.1016/j.tox.2005.04.002 PMID:15890441

Sepehr, M. N., Sivasankar, V., Zarrabi, M., & Kumar, S. M. (2013). Surface modification of pumice enhancing its fluoride adsorption capacity: An insight into kinetic and thermodynamic studies. *Chemical Engineering Journal*, *228*, 192–204. doi:10.1016/j.cej.2013.04.089

Servos, M. R., Bennie, D. T., Burnison, B. K., Jurkovic, A., McInnis, R., Neheli, T., & Ternes, T. A. (2005). Distribution of estrogens, 17β-estradiol and estrone, in Canadian municipal wastewater treatment plants. *The Science of the Total Environment*, *336*(1–3), 155–170. doi:10.1016/j.scitotenv.2004.05.025 PMID:15589256

Shakunthala, B., Shivanna, S., & Doddaiah, S. (2010). Urban Wastewater Characteristic and its Management in Urban Areas—A Case Study of Mysore City, Karnataka, India. *Journal of Water Resource and Protection*, *2*(08), 717–726. doi:10.4236/jwarp.2010.28082

Shala, L., & Foster, G. D. (2010). Surface Water Concentrations and Loading Budgets of Pharmaceuticals and Other Domestic-Use Chemicals in an Urban Watershed (Washington, DC, USA). *Archives of Environmental Contamination and Toxicology*, *58*(3), 551–561. doi:10.100700244-009-9463-z PMID:20098984

Shandilya, M., Rai, R., & Singh, J. (2016). Review: Hydrothermal technology for smart materials. *Advances in Applied Ceramics*, *115*(6), 354–376. doi:10.1080/17436753.2016.1157131

Shandost-fard, F., Salimi, A., Sharifi, E., & Korani, A. (2013). Fabrication of a highly sensitive adenosine aptasensor based on covalent attachment of aptamer onto chitosan-carbon nanotubes-ionic liquid nanocomposite. *Biosensors & Bioelectronics*, *48*, 100–107. doi:10.1016/j.bios.2013.03.060 PMID:23660341

Shanker, U., Jassal, V., & Rani, M. (2017). Degradation of toxic PAHs in water and soil using potassium zinc hexacyanoferrate nanocubes. *Journal of Environmental Management*, *204*, 337–348. doi:10.1016/j.jenvman.2017.09.015 PMID:28910732

Shan, L., Liu, Y., Ma, C., Dong, L., Liu, L., & Wu, Z. (2016). Enhanced Photocatalytic Performance in Ag^{+}-Induced $BiVO_4$/-Bi_2O_3 Heterojunctions. *European Journal of Inorganic Chemistry*, *2016*(2), 232–239. doi:10.1002/ejic.201500936

Sharma, D. R., Pradhan, B., & Mishra, S. K. (2010). Multiple drug resistance in bacterial isolates from liquid wastes generated in central hospitals of Nepal. *Kathmandu University Medical Journal, 8*(1), 29, 40-44.

Shen, B., Wu, J., & Zhao, Z. (2017). Organochlorine pesticides and polycyclic aromatic hydrocarbons in water and sediment of the Bosten Lake, Northwest China. *Journal of Arid Land*, *9*(2), 287–298. doi:10.100740333-017-0008-4

Shen, G., Guo, Y., Sun, X., & Wang, X. (2014). Electrochemical Aptasensor Based on Prussian Blue-Chitosan-Glutaraldehyde for the Sensitive Determination of Tetracycline. *Nano-Micro Letters*, *6*(2), 143–152. doi:10.1007/BF03353778

Shen, W., Li, Z., & Liu, Y. (2008). Surface Chemical Functional Groups Modification of Porous Carbon. *Recent Patents on Chemical Engineeringe*, *1*(1), 27–40. doi:10.2174/2211334710801010027

Shen, Z., Wang, W., Jia, J., Feng, X., Hu, W., & Peng, A. (2002). Catalytically assisted electrochemical oxidation of dye Acid Red B. *Water Environment Research*, *74*(2), 117–121. doi:10.2175/106143002X139811 PMID:12043967

Shimin, T., Zhaoyin, W., & Hongxia, S. (2011). *Study on the Self-purification of Juma River. Procedia Environmental Sciences*, *11*, 1328–1333.

Shim, W. J., Hong, S. H., & Eo, S. (2017). Identification methods in microplastic analysis: A review. *Analytical Methods*, *9*(9), 1384–1391. doi:10.1039/C6AY02558G

Shirmohammadi, M., Chupani, L., & Salamat, N. (2017b). Responses of immune organs after single-dose exposure to phenanthrene in yellowfin seabream (*Acanthopagrus latus*): CYP1A induction and oxidative stress. *Chemosphere*, *186*, 686–694. doi:10.1016/j.chemosphere.2017.08.051 PMID:28826128

Shirmohammadi, M., Salamat, N., Ronagh, M. T., Movahedinia, A., & Hamidian, G. (2017a). Assessment of immune status of yellowfin seabream (*Acanthopagrus latus*) during short term exposure to phenanthrene. *Comparative Biochemistry and Physiology Part C: Toxicology & Pharmacology*, *195*, 78–90. doi:10.1016/j.cbpc.2017.02.012 PMID:28257924

Shreeshivadasan, C., & Paul, J. S. (2011). Application of Anaerobic Biotechnology for Pharmaceutical Wastewater Treatment. *Special Issue on Environmental Management For Sustainable Development.*, *2*(1), 13–21.

Siddique, A., Saied, S., Alam Zaigham, N., Mumtaz, M., Ali Mahar, G., & Mohiuddin, S. (2012). Temporal variability of disinfection by-products concentration in urban public water system. *Global NEST Journal*, *14*(4), 393–398.

Siddique, S., Kubwabo, C., & Harris, S. A. (2016). A review of the role of emerging environmental contaminants in the development of breast cancer in women. *Emerging Contaminants*, *2*(4), 204–219. doi:10.1016/j.emcon.2016.12.003

Silva, B., Bastos, S., Justino, I. L., da Costa, P., Duarte, C., & Teresa, A. P. (2018). Microplastics in the environment: Challenges in analytical chemistry -A review. *Analytica Chimica Acta*, *1017*, 1–19. doi:10.1016/j.aca.2018.02.043 PMID:29534790

Simpson, S. L., Batley, G. E., Chariton, A. A., Stauber, J. L., King, C. K., Chapman, J. C., ... Maher, W. A. (2005). *Handbook for Sediment Quality Assessment*. Bangor, NSW: CSIRO.

Sim, W. J., Lee, J. W., Lee, E. S., Shin, S. K., Hwang, S. R., & Oh, J. E. (2011). Occurrence and distribution of pharmaceuticals in wastewater from households, livestock farms, hospitals and pharmaceutical manufactures. *Chemosphere*, *82*(2), 179–186. doi:10.1016/j.chemosphere.2010.10.026 PMID:21040946

Singh, A. P., Kodan, N., Mehta, B. R., Held, A., Mayrhofer, L., & Moseler, M. (2016). Band Edge Engineering in $BiVO_4$/TiO_2 Heterostructure: Enhanced Photoelectrochemical Performance through Improved Charge Transfer. *ACS Catalysis*, *6*(8), 5311–5318. doi:10.1021/acscatal.6b00956

Sizirici, B., & Yildiz. (2018). Adsorption capacity of iron oxide-coated gravel for landfill leachate: Simultaneousstudy. *I. International Journal of Environmental Science and Technology*. doi:10.100713762-018-1644-8

Slooff, W. (1983). Benthic macroinvertebrates and water quality assessment:Some toxicological considerations. *Aquatic Toxicology (Amsterdam, Netherlands)*, *4*(1), 73–82. doi:10.1016/0166-445X(83)90062-0

Smeti, E., Samios, S., Nikolaou, A., Golfinopoulos, S., & Lekkas, T. D. (2011). Statistical analysis of trihalomethanes in treated-water tanks: Seasonality, local variability and correlations. *Global NEST Journal*, *13*(1), 41–49.

Smith, A. (2006). Nanotech- the way forward for clean water. *Filtration & Separation*, *43*(8), 32–33. doi:10.1016/S0015-1882(06)70976-4

Sogbanmu, T., Osibona, A., Oguntunde, O., & Otitoloju, A. (2018). Biomarkers of toxicity in *Clarias gariepinus* exposed to sublethal concentrations of polycyclic aromatic hydrocarbons. *African Journal of Aquatic Science*, *43*(3), 1–12. doi:10.2989/16085914.2018.1491825

Song, K.-M., Jeong, E., Jeon, W., Jo, H., & Ban, C. (2012). A coordination polymer nanobelt (CPNB)-based aptasensor for sulfadimethoxine. *Biosensors & Bioelectronics*, *33*(1), 113–119. doi:10.1016/j.bios.2011.12.034 PMID:22244734

Song, Y. K., Hong, S. H., Jang, M., Han, G. M., Rani, M., Lee, J., & Shim, W. J. (2015). A comparison of microscopic and spectroscopic identification methods for analysis of microplastics in environmental samples. *Marine Pollution Bulletin*, *93*(1-2), 202–209. doi:10.1016/j.marpolbul.2015.01.015 PMID:25682567

Song, Y. K., Hong, S. H., Jang, M., Kang, J. H., Kwon, O. Y., Han, G. M., & Shim, W. J. (2014). Large accumulation of micro-sized synthetic polymer particles in the sea surface microlayer. *Environmental Science & Technology*, *48*(16), 9014–9021. doi:10.1021/es501757s PMID:25059595

Sorensen, J. P. R., Lapworth, D. J., Nkhuwa, D. C. W., Stuart, M. E., Gooddy, D. C., Bell, R. A., ... Pedley, S. (2015). Emerging contaminants in urban groundwater sources in Africa. *Water Research*, *72*, 51–63. doi:10.1016/j.watres.2014.08.002 PMID:25172215

Sourav, M., & Alok, S. (2014). Treatment of Pharmaceutical Waste with Special Emphasis to Treatment Processes - A Review. *International Journal of Environmental Research and Development*, *4*(2), 171–176.

Souza-Silva, E. A., Jiang, R., Rodriguez-Lafuente, A., Gionfriddo, E., & Pawliszyn, J. (2015). A critical review of the state of the art of solid-phase microextraction of complex matrices I. Environmental analysis. *Trends in Analytical Chemistry*, *71*, 224–235. doi:10.1016/j.trac.2015.04.016

Spring, A. J., Bagley, D. M., Andrews, R. C., Lemanik, S., & Yang, P. (2007). Removal of endocrine disrupting compounds using a membrane bioreactor and disinfection. *Journal of Environmental Engineering and Science*, *6*(2), 131–137. doi:10.113906-049

Ssebugere, P., Kiremire, B. T., Henkelmann, B., Bernhöft, S., Wasswa, J., Kasozi, G. N., & Schramm, K. W. (2013, June). PCDD/Fs and dioxin-like PCBs in surface sediments from Lake Victoria, East Africa. *The Science of the Total Environment*, *454–455*(1), 528–533. doi:10.1016/j.scitotenv.2013.03.046 PMID:23567173

Stackelberg, P. E., Gibs, J., Furlong, E. T., Meyer, M. T., Zaugg, S. D., & Lippincott, R. L. (2007). Efficiency of conventional drinking-water-treatment processes in removal of pharmaceuticals and other organic compounds. *The Science of the Total Environment*, *377*(2-3), 255–272. doi:10.1016/j.scitotenv.2007.01.095 PMID:17363035

Stalder, T., Alrhmoun, M., Louvet, J.-N., Casellas, M., Maftah, C., Carrion, C., ... Dagot, C. (2013). Dynamic Assessment of the Floc Morphology, Bacterial Diversity, and Integron Content of an Activated Sludge Reactor Processing Hospital Effluent. *Environmental Science & Technology*, *47*(14), 7909–7917. doi:10.1021/es4008646 PMID:23789899

Stamatelatou, K., Pakou, C., & Lyberatos, G. (2011). *Comprehensive biotechnology* (2nd ed.; Vol. 6). Elsevier. doi:10.1016/B978-0-08-088504-9.00496-7

Stefanakis, A., & Becker, J. (2016). A Review of Emerging Contaminants in Water. In *Impact if Water Pollution on Human Health and Environmental Sustainability* (pp. 55–80). IGI Global. doi:10.4018/978-1-4666-9559-7.ch003

Stefanakis, A., & Human, J. B. (2015). *A Review of Emerging Contaminants in Water: Classification, Sources, and Potential Risks. In Impact of Water Pollution on Human Health and Environmental Sustainability*. IGI Global.

Stefano, M., & Elisabetta, G. (2008). Water quality modelling for small river basins. *Environmental Modelling & Software*, *23*(4), 451–463. doi:10.1016/j.envsoft.2007.06.008

Stuart, M., Lapworth, D., Crane, E., & Hart, A. (2012). *Review of risk from potential emerging contaminants in UK groundwater*. Academic Press.

Stuart, M., Lapworth, D., Crane, E., & Hart, A. (2012). Review of risk from potential emerging contaminants in UK groundwater. *The Science of the Total Environment*, *416*, 1–21. doi:10.1016/j.scitotenv.2011.11.072 PMID:22209399

Stylianos, K. S., Katarzyna, S., Ioannis, A. K., & Anastasios, I. Z. (2015). Novel Water Treatment Processes Based on Hybrid Membrane-Ozonation Systems: A Novel Ceramic Membrane Contactor for Bubbleless Ozonation of Emerging Micropollutants. *Journal of Chemistry*, 1–12.

Styszko, K., Nosek, K., Motak, M., & Bester, K. (2015). Preliminary selection of clay minerals for the removal of pharmaceuticals, bisphenol A and triclosan in acidic and neutral aqueous solutions. *Comptes Rendus. Chimie*, *18*(10), 1134–1142. doi:10.1016/j.crci.2015.05.015

Subedi, B., & Kannan, K. (2014). Fate of artificial sweeteners in wastewater treatment plants in New York State, U.S.A. *Environmental Science & Technology*, *48*(23), 13668–13674. doi:10.1021/es504769c PMID:25365516

Sudardsha, N., Hoover, G., & Knorr, D. (1992). Antibacterial action of chitosan. *Food Biotechnology*, *6*(3), 257–272. doi:10.1080/08905439209549838

Sudheesh, K., Ajay, K., Omotayo, A., & Bhekie, B. (2013). Chitosan-base nanomaterials: A state-of-the-art review. *International Journal of Biological Macromolecules*, *59*, 46–58. doi:10.1016/j.ijbiomac.2013.04.043 PMID:23608103

Sui, Q., Huang, J., Deng, S., Yu, G., & Fan, Q. (2010). Occurrence and removal of pharmaceuticals, caffeine and DEET in wastewater treatment plants of Beijing, China. *Water Research*, *44*(2), 417–426. doi:10.1016/j.watres.2009.07.010 PMID:19674764

Sujana, M. G., & Anand, S. (2013). Ferric hydroxide: Preparation, characterization and fluoride removal studies from water. *Desalination and Water Treatment*, *52*(34-36), 6453–6463. doi:10.1080/19443994.2013.822633

Sun, B., Qiao, F., Chen, L., Zhao, Z., Yin, H., & Ai, S. (2014). Effective signal-on photoelectrochemical immunoassay of subgroup J avian leukosis virus based on Bi_2S_3 nanorods as photosensitizer and in situ generated ascorbic acid for electron donating. *Biosensors & Bioelectronics*, *54*, 237–243. doi:10.1016/j.bios.2013.11.021 PMID:24287410

Sun, J. Y., Ge, J. C., Liu, W. M., Fan, Z. Y., Zhang, H. Y., & Wang, P. F. (2011). Highly sensitive and selective colorimetric visualization of streptomycin in raw milk using Au nanoparticles supramolecular assembly. *Chemical Communications*, *47*(35), 9888–9890. doi:10.1039/c1cc12910d PMID:21811731

Sun, J., Ni, H., & Zeng, H. (2012). Ecological risk assessment of parent and halogenated polycyclic aromatic hydrocarbons in surface sediments from an urban river in south China. *Environmental Toxicology and Chemistry*, *31*(8), 1867–1873. doi:10.1002/etc.1890 PMID:22619089

Sun, L., Zuo, Z., Chen, M., Chen, Y., & Wang, C. (2015a). Reproductive and transgenerational toxicities of phenanthrene on female marine medaka (*Oryzias melastigma*). *Aquatic Toxicology (Amsterdam, Netherlands)*, *162*, 109–116. doi:10.1016/j.aquatox.2015.03.013 PMID:25805703

Sun, Y. Y., Xu, H. X., Li, J. H., Shi, X. Q., Wu, J. C., Ji, R., & Guo, H. Y. (2015b). Phytoremediation of soils contaminated with phenanthrene and cadmium by growing willow (Salix×Aureo-PendulaCL "j1011"). *International Journal of Phytoremediation*, *18*(2), 150–156. doi:10.1080/15226514.2015.1073668 PMID:26247604

Sun, Y., Fang, Q., Dong, J., Cheng, X., & Xu, J. (2011). Removal of fluoride from drinking water by natural stilbite zeolite modified with Fe(III). *Desalination*, *277*(1-3), 121–127. doi:10.1016/j.desal.2011.04.013

Supalkova, V., Beklova, M., Baloun, J., Singer, C., Sures, B., Adam, V., ... Kizek, R. (2008). Affecting of aquatic vascular plant Lemna minor by cisplatin revealed by voltammetry. *Bioelectrochemistry (Amsterdam, Netherlands)*, *72*(1), 59–65. doi:10.1016/j.bioelechem.2007.11.012 PMID:18201947

Sutton, R., Mason, S. A., Stanek, S. K., Willis-Norton, E., Wren, I. F., & Box, C. (2016). Microplastic contamination in the san Francisco Bay, California, USA. *Marine Pollution Bulletin*, *109*(1), 230–235. doi:10.1016/j.marpolbul.2016.05.077 PMID:27289280

Taghdisi, S. M., Danesh, N. M., Nameghi, M. A., Ramezani, M., & Abnous, K. (2016). A label-free fluorescent aptasensor for selective and sensitive detection of streptomycin in milk and blood serum. *Food Chemistry*, *203*, 145–149. doi:10.1016/j.foodchem.2016.02.017 PMID:26948599

Tahrani, L., Sou, L., Mehri, I., Najjari, A., Hassan, A., Van Loco, J., & Ben, H. (2015). Isolation and characterization of antibiotic-resistant bacteria from pharmaceutical industrial wastewaters. *Microbial Pathogenesis Journal*, *89*, 54–61. doi:10.1016/j.micpath.2015.09.001 PMID:26343496

Taiwo, A. M., Olujimi, O. O., Bamgbose, O., & Arowolo, T. A. (2012). Surface Water Quality Monitoring in Nigeria: Situational Analysis and Future Management Strategy. Water Quality Monitoring and Assessment.

Talsness, C. E., Andrade, A. J. M., Kuriyama, S. N., Taylor, J. A., & vom Saal, F. S. (2009). Components of plastic: Experimental studies in animals and relevance for human health. *Philos. Trans. R. Soc. B*, *364*(1526), 2079–2096. doi:10.1098/rstb.2008.0281 PMID:19528057

Tang, L., Hamid, Y., Zehra, A., Sahito, Z. A., He, Z., Hussain, B., ... Yang, X. (2019). Characterization of fava bean (Vicia faba L.) genotypes for phytoremediation of cadmium and lead co-contaminated soils coupled with agro-production. *Ecotoxicology and Environmental Safety*, *171*, 190–198. doi:10.1016/j.ecoenv.2018.12.083 PMID:30605848

Tangri, N. (2003). *Waste Incineration: A Dying Technology: Essential Action for GAIA*. Retrieved from www.no-burn.org

Tang, T., Guan, X., Wang, J., Gao, N., McPhail, M., & Chusuei, C. (2009). Fluoride adsorption onto granular ferric hydroxide: Effects of ionic strength, pH, surface loading, and major co-existing anions. *Journal of Hazardous Materials*, *171*(1-3), 774–779. doi:10.1016/j.jhazmat.2009.06.079 PMID:19616377

Tankiewicz, M., Fenik, J., & Biziuk, M. (2011). Solventless and solvent-minimized sample preparation techniques for determining currently used pesticides in water samples: A review. *Talanta*, *86*, 8–22. doi:10.1016/j.talanta.2011.08.056 PMID:22063507

Tanoue, R., Sato, Y., Motoyama, M., Nakagawa, S., Shinohara, R., & Nomiyama, K. (2012). Plant uptake of pharmaceutical chemicals detected in recycled organic manure and reclaimed wastewater. *Journal of Agricultural and Food Chemistry*, *60*(41), 10203–10211. doi:10.1021/jf303142t PMID:23003104

Tarassov, V. J., Perlis, H. J., & Davidson, B. (1969, June). Optimization of a class of river aeration problems by the use of multivariable distributed parameter control theory. *Water Resources Research*, *5*(3), 563–573. doi:10.1029/WR005i003p00563

Tellan, A. C., & Owalude, S. O. (2007). Some Langmuir and Freundlich parameters of adsorption studies of chlorpheniramine maleate. *Res. J. Appl. Sci.*, *2*, 875–878.

Teuten, E. L., Rowland, S. J., Galloway, T. S., & Thompson, R. C. (2007). Potential for plastics to transport hydrophobic contaminants. *Environmental Science & Technology*, *41*(22), 7759–7764. doi:10.1021/es071737s PMID:18075085

The World Bank. (2016). *Health Nutrition and Population Statistics | DataBank*. Retrieved March 22, 2018, from http://databank.worldbank.org/data/reports.aspx?source=health-nutrition-and-population-statistics#

The World Bank. (2017). A Wake Up Call: Nigeria Water Supply, Sanitation, and Hygiene Poverty Diagnostic. *WASH Poverty Diagnostic* (Vol. 431). Author.

Theron, J., & Cloete, T. E. (2002). Emerging waterborne infections: Contributing factors, agents and detection tools. *Critical Reviews in Microbiology*, *28*(1), 1–26. doi:10.1080/1040-840291046669 PMID:12003038

Thiviyanathan, V., & Gorenstein, D. G. (2012). Aptamers and the next generation of diagnostic reagents. *Proteomics. Clinical Applications*, *6*(11-12), 563–573. doi:10.1002/prca.201200042 PMID:23090891

Thomaidis, N. S., Asimakopoulos, A. G., & Bletsou, A. A. (2012). Emerging contaminants: A tutorial mini-review. *Global NEST Journal*, *14*(1), 72–79.

Thomaidis, N. S., Asimakopoulos, A. G., & Bletsou, A. A. (2012). Emerging contaminants: A tutorial mini-review. *Global NEST Journal*, *14*, 72–79.

Thomas, K. V., Dye, C., Schlabach, M., & Langford, K. H. (2007). Source to sink tracking of selected human pharmaceuticals from two Oslo city hospitals and wastewater treatment works. *Journal of Environmental Monitoring*, *9*(12), 1410–1418. doi:10.1039/b709745j PMID:18049781

Thome, J. P., Jeuniaux, C., & Weltrowski, M. (1997). Applications of chitosan for the elimination of organochlorine xenobiotics from wastewater. In M. F. A. Goosen (Ed.), *Applications of Chitin and Chitosan* (pp. 309–331). Lancaster, PA: Technomic.

Thompson, J. M., Gundogˇdu, A., Stratton, H. M., & Katouli, M. (2013). Antibiotic resistant Staphylococcus aureus in hospital wastewaters and sewage treatment plants with special reference to methicillin-resistant Staphylococcus aureus (MRSA). *Journal of Applied Microbiology*, *114*(1), 44–54. doi:10.1111/jam.12037 PMID:23057695

Thompson, R. C., Moore, C. J., vom Saal, F. S., & Swan, S. H. (2009). Plastics, the environment and human health: Current consensus and future trends. *Philosophical Transactions of the Royal Society of London. Series B, Biological Sciences*, *364*(1526), 2153–2166. doi:10.1098/rstb.2009.0053 PMID:19528062

Ting, W. T. E., Yuan, S. Y., Wu, S. D., & Chang, B. V. (2011). Biodegradation of phenanthrene and pyrene by *Ganoderma lucidum*. *International Biodeterioration & Biodegradation*, *65*(1), 238–242. doi:10.1016/j.ibiod.2010.11.007

Tirado-Torres, D., Gayosso-Canales, M., Marmolejo-Santillan, Y., Romo-Gomez, C., & Acevedo-Sandoval, O. (2016). Removal of polycyclic aromatic hydrocarbons by *Pleurotus ostreatus* sp. ATCC38540 in liquid medium. *Academia Journal of Scientific Research*, *4*(10), 376–379.

Tomar, V., & Kumar, D. (2013). A critical study on efficiency of different materials for fluoride removal from aqueous media. *Chemistry Central Journal*, *7*(1), 51. doi:10.1186/1752-153X-7-51 PMID:23497619

Tongo, I., Ezemonye, L., & Akpeh, K. (2017). Distribution, characterization, and human health risk assessment of polycyclic aromatic hydrocarbons (PAHs) in Ovia River, Southern Nigeria. *Environmental Monitoring and Assessment*, *189*(6), 247. doi:10.100710661-017-5931-5 PMID:28466449

Torreiro-Melo, A. G. A. G., Silva, J. S., Bianchini, A., Zanardi-Lamardo, E., & de Carvalho, P. S. M. (2015). Bioconcentration of phenanthrene and metabolites in bile and behavioral alterations in the tropical estuarine guppy *Poecilia vivipara. Chemosphere*, *132*, 17–23. doi:10.1016/j.chemosphere.2014.12.079 PMID:25770832

Torrellas, S. Á., García Lovera, R., Escalona, N., Sepúlveda, C., Sotelo, J. L., & García, J. (2015). Chemical-activated carbons from peach stones for the adsorption of emerging contaminants in aqueous solutions. *Chemical Engineering Journal*, *279*, 788–798. doi:10.1016/j.cej.2015.05.104

Trang, C., Thi, T., Dieu, L. V., Thanh, T. D., & Sinh, L. X. (2015). Assessment of Self-Purification Process of Thi Nai lagoon (Binh Dinh Province, Viet Nam). *Environment and Natural Resources Research*, *5*(3), 19–27.

Tran, N. H., Hu, J., Li, J., & Ong, S. L. (2014). Suitability of artificial sweeteners as indicators of raw wastewater contamination in surface water and groundwater. *Water Research*, *48*, 443–456. doi:10.1016/j.watres.2013.09.053 PMID:24156949

Tripathi, A., & Ranjan, M. R. (2015). *Heavy Metal Removal from Wastewater Using Low Cost Adsorbents*. J Bioremed Biodeg. doi:10.4172/2155-6199.1000315

Trovó, A. G., Melo, S. A. S., & Nogueira, R. F. P. (2008). Photodegradation of the pharmaceuticals amoxicillin, bezafibrate and paracetamol by the photo-Fenton process - application to sewage treatment plant effluent. *Journal of Photochemistry and Photobiology A Chemistry*, *198*(2-3), 215–220. doi:10.1016/j.jphotochem.2008.03.011

Tsibart, A., Gennadiev, A., Koshovskii, T., & Watts, A. (2014). Polycyclic aromatic hydrocarbons in post-fire soils of drained peatlands in western Meshchera (Moscow region, Russia). *Solid Earth*, *5*(2), 1305–1317. doi:10.5194e-5-1305-2014

Tsibranska, I., & Hristova, E. (2011). Comparison of different kinetic models for adsorption of heavy metals onto activated carbon from apricot stones. *Izvestiia po Himiia*, *43*(3), 370–377.

Tumeo, E., Gbaguidi-Haore, H., Party, I., Bertand, X., Thouverez, M., & Talon, D. (2008). Are antibiotic-resistant Pseudomonas aeruginosa isolated from hospitalised patients recovered in the hospital effluents? *International Journal of Hygiene and Environmental Health*, *211*(1-2), 200–204. doi:10.1016/j.ijheh.2007.02.010 PMID:17459769

Tu, W., Zhou, Y., & Zou, Z. (2013). Versatile Graphene-Promoting Photocatalytic Performance of Semiconductors: Basic Principles, Synthesis, Solar Energy Conversion, and Environmental Applications. *Advanced Functional Materials*, *23*(40), 4996–5008. doi:10.1002/adfm.201203547

Twesigye, C. K., Simon, M. O., Getenga, Z. M., Mwakalila, S. S., & Nakiranda, J. K. (2011). The Impact of Land Use Activities on Vegetation Cover and Water Quality in the Lake Victoria Watershed. *The Open Environmental Engineering Journal*, *4*(1), 66–77. doi:10.2174/1874829501104010066

U.S. EPA. (2008). *Nanotechnology for Site Remediation Fact Sheet*. EPA 542-08-009. Available: http://www.epa.gov/tio/download/remed/542-f-08-009.pdf

Uhlı'k, O., Leewis, M. C., Kurzawova´, V., Lovecka´, P. Sˇtursa, P., & Demnerova´, K. (2010). Approaches to microbial diversity analysis in contaminated environments: Trends in bioremediation and phytoremediation. Research Signpost, 55–71.

Ullah, I., Nadeem, R., Iqbal, M., & Manzoor, Q. (2013). Biosorption of chromium onto native and immobilized sugarcane bagasse waste biomass. *Ecological Engineering*, *60*, 99–107. doi:10.1016/j.ecoleng.2013.07.028

UN WWAP. (2003). *Water for people, water for life (2003): 3rd World Water Forum in Kyoto, Japan*. Retrieved from www.norman-network.net

UNDP. (2006). *Niger Delta Human Development Report*. Abuja. Retrieved from http://hdr.undp.org/en/content/human-development-report

UNDP. (2016). *Human Development Report 2016 - Human Development for Everyone*. New York. Retrieved from http://hdr.undp.org/sites/default/files/2016_human_development_report.pdf

UNESCO. (2003). *Water for People, Water for Life. The United Nations World Water Development Report*. UNESCO. doi:10.1017/CBO9781107415324.004

Unger, K. K., Ditz, R., Machtejevas, E., & Skudas, R. (2010). Liquid Chromatography-Its Development and Key Role in Life Science Applications. *Liquid Chromatography*, *49*(13), 2300–2312. PMID:20232421

United Nations Sustainable Development Summit. (2015). *United Nations Sustainable Development Goals*. Retrieved from https://www.un.org/sustainabledevelopment/sustainable-development-goals/

United States Environmental Protection Agency. (2007). *Pharmaceuticals in the Environment: Overview of Concerns and Stewardship Activities*. USDA Forest Service 2007 Sustainable Operations Summit.

United States Environmental Protection Agency. (2016). *Pollution Prevention*. Retrieved from https://www.epa.gov/p2/learn-about-pollution-prevention

Urase, T., & Kikuta, T. (2005). Separate estimation of adsorption and degradation of pharmaceutical substances and estrogens in the activated sludge process. *Water Research*, *39*(7), 1289–1300. doi:10.1016/j.watres.2005.01.015 PMID:15862328

Urbansky, E. T. (2000). Perchlorates in the Environment. In H. Rosenkranz (Ed.), Environmental Science Research (Vol. 440, p. 299). New York: Kluwer Academic/Plenum Publishers. doi:10.1007/978-1-4615-4303-9

US EPA. (1998). *Contaminant Candidate List 1 - CCL 1*. Retrieved November 6, 2015, from http://www2.epa.gov/ccl/contaminant-candidate-list-1-ccl-1

US EPA. (2005). *Contaminant Candidate List 2 - CCL 2*. Retrieved from http://www2.epa.gov/ccl/contaminant-candidate-list-2-ccl-2

US EPA. (2009). *Contaminant Candidate List 3 - CCL 3*. Retrieved from http://www2.epa.gov/ccl/contaminant-candidate-list-3-ccl-3

US EPA. (2014). *Technical Fact Sheet – N-nitroso-dimethylamine*. Retrieved from https://www.epa.gov/sites/production/files/2014-03/documents/ffrrofactsheet_contaminant_ndma_january2014_final.pdf

US EPA. (2015a). *Draft Contaminant Candidate List 4-CCL 4*. Retrieved from http://www2.epa.gov/ccl/draft-contaminant-candidate-list-4-ccl-4

US EPA. (2015b). *Fact Sheet: Drinking Water Contaminant Candidate List 4 (CCL 4)– Draft*. Retrieved from http://www2.epa.gov/ccl

US EPA. (2018). *Contaminant Candidate List 5 (CCL 5)*. Retrieved May 17, 2019, from https://www.epa.gov/ccl/contaminant-candidate-list-5-ccl-5

USEPA (United States Environmental Protection Agency). (1987). Health Effects Assessment for Dimethylphenols. EPA/600.8-88/031. Environmental Criteria Assessment Office.

USEPA (United States Environmental Protection Agency). (1988). *Interim sediment criteria values for nonpolar hydrophobic organic contaminants*. Washington, DC: Criteria and Standards Division.

Van Daele, Y., & Thome, J. P. (1986). Purification of PCB contaminated water by chitosan: A biological test of efficiency using the common barbel, Barbus barbus. *Bulletin of Environmental Contamination and Toxicology*, *37*(1), 858–865. doi:10.1007/BF01607850 PMID:3098331

Van Houten, B., Santa-Gonzalez, G. A., & Camargo, M. (2018). DNA repair after oxidative stress: Current challenges. *Current Opinion in Toxicology*, *7*, 9–16. doi:10.1016/j.cotox.2017.10.009 PMID:29159324

Van Loosdrecht, M. C. M., Nielsen, P. H., Lopez-Vazquez, C. M., & Brdjanovic, D. (2016). *Experimental Methods in Wastewater Treatment* (1st ed.). London: International Water Publishing Alliance House. doi:10.2166/9781780404752

Varanasi, U., Stein, J. E., & Nishimoto, M. (1989). Biotransformation and disposition of PAH in fish. In U. Varanasi (Ed.), *Metabolism of Polycyclic Aromatic Hydrocarbons in the Aquatic Environment* (pp. 93–151). Boca Raton, FL: CRC Press.

Vasavi, A., Usha, R., & Swamy, P. M. (2010). Phytoremediation- An overview review. *Journal of Industrial Pollution Control*, *26*.

Velzeboer, I., Kwadijk, C., & Kolemans, A. A. (2014). Strong sorption of PCBs to nanoplastics, microplastics, carbon nanotubes, and fullerenes. *Environmental Science & Technology*, *48*(9), 4869–4876. doi:10.1021/es405721v PMID:24689832

Vergauwen, L., Schmidt, S. N., Stinckens, E., Maho, W., Blust, R., Mayer, P., ... Knapen, D. (2015). A high throughput passive dosing format for the Fish Embryo Acute Toxicity test. *Chemosphere*, *139*, 9–17. doi:10.1016/j.chemosphere.2015.05.041 PMID:26026258

Vidal-Dorsch, D. E., Colli-Dula, R. C., Bay, S. M., Greenstein, D. J., Wiborg, L., Petschauer, D., & Denslow, N. D. (2013). Gene Expression of Fathead Minnows (Pimephales Promelas) Exposed to Two Types of Treated Municipal Wastewater Effluents. *Environmental Science & Technology*, *47*(19), 11268–11277. doi:10.1021/es401942n PMID:23919544

Vidal, L., Riekkola, M.-L., & Canals, A. (2012). Ionic liquid-modified materials for solid-phase extraction and separation: A review. *Analytica Chimica Acta*, *715*, 19–41. doi:10.1016/j.aca.2011.11.050 PMID:22244164

Vié, J.-C., Hilton-Taylor, C., & Stuart, S. N. (2009). *Wildlife in a changing world–an analysis of the 2008 IUCN Red List of Threatened Species*. Gland.

Vieno, N. M., Hrkki, H., Tuhkanen, T., & Kronberg, L. (2007). Occurrence of pharmaceuticals in river water and their elimination in a pilot-scale drinking water treatment plant. *Environmental Science & Technology*, *41*(14), 5077–5084. doi:10.1021/es062720x PMID:17711226

Vijayaraghavan, K., & Yun, Y.-S. (2008). Bacterial biosorbents and biosorption. *Biotechnology Advances*, *26*(3), 266–291. doi:10.1016/j.biotechadv.2008.02.002 PMID:18353595

Viswanathan, Anitha, Amuthan, Rajesh, Veilumuthu, & Narayanan. (2015). Studies on the Role of Bacteria in self- purification of the River Tamirabarani. *Europ. J. Appl. Sci.*, *7*(1), 1-8.

Viswanathan, P., & Christina, E. C. (2015). Adsorptive Fluoride Removal From Aqueous Solution By Using Saponified Orange Peel Residue Immobilized Sorbent. *International Journal of Pharma and Bio Sciences*, *6*(1), 577–588.

Vivek Vardhan, C., & Srimurali, M. (2016). Removal of fluoride from water using a novel sorbent lanthanum-impregnated bauxite. *SpringerPlus*, *5*(1), 1426. doi:10.118640064-016-3112-6 PMID:27625980 vom Saal, F. S., Nagel, S. C., Coe, B. L., Angle, B. M., & Taylor, J. A. (2012). The estrogenic endocrine disrupting chemical bisphenol-A (BPA) and obesity. *Journal of Molecular and Cellular Endocrinology*, *354*(1-2), 74–88. doi:10.1016/j.mce.2012.01.001 PMID:22249005

Von Moos, N., Burkhardt-Holm, P., & Köhler, A. (2012). Uptake and effects of microplastics on cells and tissue of the blue mussel Mytilus edulis L. after an experimental exposure. *Environmental Science & Technology*, *46*(20), 11327–11335. doi:10.1021/es302332w PMID:22963286

Vymazal, J. (2010). Constructed wetlands for wastewater treatment. *Water (Basel)*, *2*(3), 530–549. doi:10.3390/w2030530 PMID:20795704

Waghmare, S.,Lataye, D.H., Arfin, T., & Rayalu, S.(2015). Defluoridation by Nano-Materials, Building Materials and Other Miscellaneous Materials:A Systematic Review. *International Journal of Innovative Research in Science, Engineering and Technology,* (4), 1198-1210.

Waghmare, S. S., & Arfin, T. (2015). Fluoride Removal from Water by various techniques: Review. *International Journal of Innovative Science. Engineering & Technology*, *9*(2), 560–571.

Wambu, E. W., & Muthakia, G. K. (2011). High fluoride water in the Gilgil area of Nakuru County, Kenya. *Journal of the International Society for Fluoride Research*, (1), 37-41.

Wang, J., Mao, D., Mu, Q., & Luo, Y. (2015). Fate and proliferation of typical antibiotic resistance genes in five full-scale pharmaceutical wastewater treatment plants. *The Science of the Total Environment*, *526*, 366–373. doi:10.1016/j.scitotenv.2015.05.046 PMID:25991498

Wang, L. N., Han, D. X., Ni, S., Ma, W. G., Wang, W., & Niu, L. (2015). Photoelectrochemical device based on Mo-doped $BiVO_4$ enables smart analysis of the global antioxidant capacity in food. *Chemical Science (Cambridge)*, *6*(11), 6632–6638. doi:10.1039/C5SC02277K PMID:29435212

Wang, L., Hua, X., Zhang, L., Song, N., Dong, D., & Guo, Z. (2019). Influence of organic carbon fractions of freshwater biofilms on the sorption for phenanthrene and ofloxacin: The important role of aliphatic carbons. *The Science of the Total Environment*, *685*, 818–826. doi:10.1016/j.scitotenv.2019.06.203 PMID:31238285

Wang, M., Yin, H., Shen, N., Xu, Z., Sun, B., & Ai, S. (2014). Signal-on photoelectrochemical biosensor for microRNA detection based on Bi_2S_3 nanorods and enzymatic amplification. *Biosensors & Bioelectronics*, *53*, 232–237. doi:10.1016/j.bios.2013.09.069 PMID:24141112

Wang, P., Wan, Y., Deng, S., Yang, S., Su, Y., Fan, C., ... Zuo, X. (2016). Aptamer-initiated on-particle template-independent enzymatic polymerization (aptamer-OTEP) for electrochemical analysis of tumor biomarkers. *Biosensors & Bioelectronics*, *86*, 536–541. doi:10.1016/j.bios.2016.07.025 PMID:27448543

Wang, Q. Q., Zhou, Z. X., Zhai, Y. L., Zhang, L. L., Hong, W., Zhang, Z. Q., & Dong, S. J. (2015). Label-free aptamer biosensor for thrombin detection based on functionalized graphene nanocomposites. *Talanta*, *141*, 247–252. doi:10.1016/j.talanta.2015.04.012 PMID:25966410

Wang, S., Li, J., Zhou, X., Zheng, C., Ning, J., Zhong, Y., & Hu, Y. (2014). Facile preparation of 2D sandwich-like CdS nanoparticles/nitrogen-doped reduced graphene oxide hybrid nanosheets with enhanced photoelectrochemical properties. *Journal of Materials Chemistry. A, Materials for Energy and Sustainability*, *2*(46), 19815–19821. doi:10.1039/C4TA04624B

Wang, X., Xi, B., Huo, S., Sun, W., Pan, H., Zhang, J., ... Liu, H. (2013). Characterization, treatment and releases of PBDEs and PAHs in a typical municipal sewage treatment plant situated beside an Urban River, East China. *Journal of Environmental Sciences (China)*, *25*(7), 1281–1290. doi:10.1016/S1001-0742(12)60201-0 PMID:24218838

Wang, X., Yan, T., Li, Y., Liu, Y., Du, B., Ma, H., & Wei, Q. (2015). A competitive photoelectrochemical immunosensor based on a CdS-induced signal amplification strategy for the ultrasensitive detection of dexamethasone. *Scientific Reports*, *5*. doi:10.1038rep17945 PMID:26648409

Wasay, S. A., Tokunaga, S., & Park, S.-W. (1996). Removal of hazardous anions from aqueous solutions by la(lll)- and Y(lll)-lmpregnated alumina. *Separation Science and Technology*, *31*(10), 1501–1514. doi:10.1080/01496399608001409

Wassenberg, D. M., & Di Giulio, R. T. (2004). Synergistic Embryotoxicity of Polycyclic Aromatic Hydrocarbon Aryl Hydrocarbon Receptor Agonists with Cytochrome P4501A Inhibitors in *Fundulus heteroclitus*. *Environmental Health Perspectives*, *112*(17), 1658–1664. doi:10.1289/ehp.7168 PMID:15579409

Water Technology Centre. (2014). Engineered Wetland Technology based. Eco-friendly Wastewater Treatment and Reuse. An initiative of Indian Agricultural Research Institute. WTC-IARI.

Watkinson, A. J., Micalizzi, G. B., Graham, G. M., Bates, J. B., & Costanzo, S. D. (2007). Antibiotic-Resistant Escherichia coli in Wastewaters, Surface Waters, and Oysters from an Urban Riverine System. *Applied and Environmental Microbiology*, *73*(17), 5667–5670. doi:10.1128/AEM.00763-07 PMID:17616617

Weidemann, E., Niinipuu, M., Fick, J., & Jansson, S. (2018). Using carbonized low-cost materials for removal of chemicals of environmental concern from water. *Environmental Science and Pollution Research International*, *25*(16), 15793–15801. doi:10.100711356-018-1781-0 PMID:29582326

Weigel, S., Kallenborn, R., & Hühnerfuss, H. (2004). Simultaneous solid-phase extraction of acidic, neutral and basic pharmaceuticals from aqueous samples at ambient (neutral) pH and their determination by gas chromatography-mass spectrometry. *Journal of Chromatography. A*, *1023*(2), 183–195. doi:10.1016/j.chroma.2003.10.036 PMID:14753684

Weng, B., Zhang, X., Zhang, N., Tang, Z. R., & Xu, Y. J. (2015). Two-dimensional MoS_2 nanosheet-coated Bi_2S_3 discoids: Synthesis, formation mechanism, and photocatalytic application. *Langmuir*, *31*(14), 4314–4322. doi:10.1021/la504549y PMID:25625414

Weng, X., & Neethirajan, S. (2016). A microfluidic biosensor using graphene oxide and aptamer-functionalized quantum dots for peanut allergen detection. *Biosensors & Bioelectronics*, *85*, 649–656. doi:10.1016/j.bios.2016.05.072 PMID:27240012

Wen, S., Zhao, J., Sheng, G., Fu, J., & Peng, P. (2003). Photocatalytic reactions of pyrene at TiO2/water interfaces. *Chemosphere*, *50*(1), 111–119. doi:10.1016/S0045-6535(02)00420-4 PMID:12656236

West, C. E., & Rowland, S. J. (2012). Aqueous photo-transformation of diazepam and related human metabolites under simulated sunlight. *Environmental Science & Technology*, *46*(9), 4749–4756. doi:10.1021/es203529z PMID:22489744

WHO. (1983). International Agency for Research on Cancer (IARC) Monographs on the Evaluation of Carcinogenic Risks to Humans: Polynuclear Aromatic Compounds, Part 1, Chemical, Environmental and Experimental Data. IARC.

WHO. (2006). *Guidelines for Drinking-water Quality: First Addendum to Volume 1: Recommendations*. Retrieved November 15, 2016, from http://comparamus.us

Wild, S. R., & Jones, K. C. (1995). Polynuclear aromatic hydrocarbons in the United Kingdom environment: A preliminary source inventory and budget. *Environmental Pollution*, *88*(1), 91–108. doi:10.1016/0269-7491(95)91052-M PMID:15091573

Willis, A. M., & Oris, J. T. (2014). Acute photo-induced toxicity and toxicokinetics of single compounds and mixtures of polycyclic aromatic hydrocarbons in zebrafish. *Environmental Toxicology and Chemistry*, *33*(9), 2028–2037. doi:10.1002/etc.2648 PMID:24889326

Woo, O. T., Chung, W. K., Wong, K. H., Chow, A. T., & Wong, P. K. (2009). Photocatalytic oxidation of polycyclic aromatic hydrocarbons: Intermediates identification and toxicity testing. *Journal of Hazardous Materials*, *168*(2-3), 1192–1199. doi:10.1016/j.jhazmat.2009.02.170 PMID:19361920

World Economic Forum. (2016). *The New Plastics Economy - Rethinking the Future of Plastics, 2016*. Available from: http://www.ellenmacarthurfoundation.org/publications

Wright, S. L., Thompson, R. C., & Galloway, T. S. (2013). The physical impacts of microplastics on marine organisms: A review. *Environ. Pollut.*, *178*, 483–492. doi:10.1016/j.envpol.2013.02.031 PMID:23545014

WSP. (2012). *Economic Impacts of Poor Sanitation in Africa*. Retrieved from https://www.wsp.org/sites/wsp.org/files/publications/WSP-ESI-Nigeria-brochure.pdf

Wu, J. W., Chen, H. C., & Ding, W. H. (2013). Ultrasound-assisted dispersive liquid-liquid microextraction plus simultaneous silylation for rapid determination of salicylate and benzophenone-type ultraviolet filters in aqueous samples. *Journal of Chromatography. A*, *1302*, 20–27. doi:10.1016/j.chroma.2013.06.017 PMID:23831000

Wu, J. Y., Yan, Z. G., Liu, Z. T., Liu, J., Liang, F., Wang, X. N., & Wang, W. L. (2015). Development of water quality criteria for phenanthrene and comparison of the sensitivity between native and non-native species. *Environmental Pollution*, *196*, 141–146. doi:10.1016/j.envpol.2014.09.024 PMID:25463707

Wu, J., Zhang, L., & Yang, Z. (2010). A Review on the Analysis of Emerging Contaminants in Aquatic Environment. *Critical Reviews in Analytical Chemistry*, *40*(4), 234–245. doi:10.1080/10408347.2010.515467

Wu, M., Xu, Y., Ding, W., Li, Y., & Xu, H. (2016). Mycoremediation of manganese and phenanthrene by *Pleurotus eryngii mycelium* enhanced by Tween 80 and saponin. *Applied Microbiology and Biotechnology*, *100*(16), 7249–7261. doi:10.100700253-016-7551-3 PMID:27102128

Xu, L., Li, H., Yan, P., Xia, J., Qiu, J., Xu, Q., ... Yuan, S. (2016). Graphitic carbon nitride/BiOCl composites for sensitive photoelectrochemical detection of ciprofloxacin. *Journal of Colloid and Interface Science*, *483*, 241–248. doi:10.1016/j.jcis.2016.08.015 PMID:27552431

Xu, N., Xu, Y. F., Xu, S., Li, J., & Tao, H. C. (2012). Removal of estrogens in municipal wastewater treatment plants: A Chinese perspective. *Environmental Pollution*, *165*(0), 215–224. doi:10.1016/j.envpol.2011.12.025 PMID:22226508

Yadav, A. K., Naresh, K., Sreerishnan, T. R., Santosh, S., & Narsi, R. B. (2010). Removal of Cr and Ni from acqueous solution in constructed wetland: Mass balance, adsorption-desorption and FTIR study. *Chemical Engineering Journal*, *160*(1), 122–128. doi:10.1016/j.cej.2010.03.019

Yamina, B., Tahar, B., Lila, M., Hocine, H., & Laure, F. M. (2014). Study on Cadmium Resistant-BacteriaIsolated from Hospital Wastewaters. *Advances in Bioscience and Biotechnology*, *5*(08), 718–726. doi:10.4236/abb.2014.58085

Yang, C., Wang, Q., Xiang, Y., Yuan, R., & Chai, Y. (2014). Target-induced strand release and thionine-decorated gold nanoparticle amplification labels for sensitive electrochemical aptamer-based sensing of small molecules. *Sensors and Actuators. B, Chemical*, *197*, 149–154. doi:10.1016/j.snb.2014.02.036

Yangming, L., Zhemin, S., Xuejun, C., Jinpin, J., & Wenhua, W. (2006). Preparation and application of nano-TiO2 catalyst in dye electrochemical treatment. *Water S.A.*, *32*(2), 205–210.

Yang, S., Hai, F. I., Nghiem, L. D., Nguyen, L. N., Roddick, F., & Price, W. E. (2013). Removal of Bisphenol A and Diclofenac by a Novel Fungal Membrane Bioreactor Operated under Non-Sterile Conditions. *International Biodeterioration & Biodegradation*, *85*, 483–490. doi:10.1016/j.ibiod.2013.03.012

Yang, X., Chen, Z., Wu, Q., & Xu, M. (2018). Enhanced phenanthrene degradation in river sediments using a combination of biochar and nitrate. *The Science of the Total Environment*, *619-620*, 600–605. doi:10.1016/j.scitotenv.2017.11.130 PMID:29156278

Yan-long, W., Jun-hui, C., Li-yuan, G. A. O., Shuai, W., Xiao-ling, Z., Cheng-jun, S. U. N., & Xiao-ru, W. (2016). Determination of Eight Typical Lipophilic Algae Toxins in Particles Suspended in Seawater by Ultra Performance Liquid Chromatography-Tandem Mass Spectrometry. *Chinese Journal of Analytical Chemistry*, *44*(3), 335–341. doi:10.1016/S1872-2040(16)60911-8

Yan, Y., Liu, Q., Du, X., Qian, J., Mao, H., & Wang, K. (2015). Visible light photoelectrochemical sensor for ultrasensitive determination of dopamine based on synergistic effect of graphene quantum dots and TiO2 nanoparticles. *Analytica Chimica Acta*, *853*, 258–264. doi:10.1016/j.aca.2014.10.021 PMID:25467467

Yao, W., Iwai, H., & Ye, J. (2008). Effects of molybdenum substitution on the photocatalytic behavior of $BiVO_4$. *Dalton Transactions (Cambridge, England)*, (11): 1426–1430. doi:10.1039/b713338c PMID:18322621

Yin, C., Zhu, S. M., Chen, Z. X., Zhang, W., Gu, J. J., & Zhang, D. (2013). One step fabrication of C-doped $BiVO_4$ with hierarchical structures for a high-performance photocatalyst under visible light irradiation. *Journal of Materials Chemistry. A, Materials for Energy and Sustainability*, *1*(29), 8367–8378. doi:10.1039/c3ta11833a

Yi, S., Gao, B., Sun, Y., Wu, J., Shi, X., Wu, B., & Hu, X. (2016). Removal of levofloxacin from aqueous solution using rice-husk and wood-chip biochars. *Chemosphere*, *150*, 694–701. doi:10.1016/j.chemosphere.2015.12.112 PMID:26796588

Yonik, M. Y., & Imas, K. (2017). Investigation on the biodegradation capacity of urban rivers in Jakarta, Indonesia. *International Journal of GEOMATE*, *12*(34), 45–50.

Yonik, M. Y., Mia, N., Neneng, S., & Annisa, N. (2018). Influencing parameter of self purification process in the urban area of Cikapundung River, Indonesia. *International Journal of GEOMATE*, *14*(43), 50–54.

Yoo, C. K., Villez, K., Van Hulle, S. W. H., & Vanrolleghem, P. A. (2007). Enhanced process monitoring for wastewater treatment systems. *Environmetrics*, *19*(6), 602–617. doi:10.1002/env.900

Yoon, Y., Westerhoff, P., Snyder, S. A., Wert, E. C., & Yoon, J. (2007). Removal of endocrine disrupting compounds and pharmaceuticals by nanofiltration and ultrafiltration membranes. *Desalination*, *202*(1-3), 16–23. doi:10.1016/j.desal.2005.12.033

Yotinov, I., Lincheva, S., & Topalova, Y. (2014). Vertical fl uctuations of key parameters of the self-purification in the "Middle Iskar" cascade's reservoir Lakatnik. *Bulgarian Journal of Agricultural Science*, (Supplement 1), 113–117.

Yu, B., Zhang, R., Liu, P., Zhang, Y., Zhang, Y., & Bai, Y. (2015). Determination of nine hydroxylated polybrominated diphenyl ethers in water by precolumn derivatization-gas chromatography mass spectrometry. *Journal of Chromatography. A*, *1419*, 19–25. doi:10.1016/j.chroma.2015.09.081 PMID:26454789

Yu, G., Wang, X., Cao, J., Wu, S., Yan, W., & Liu, G. (2016). Plasmonic Au nanoparticles embedding enhances the activity and stability of CdS for photocatalytic hydrogen evolution. *Chemical Communications*, *52*(11), 2394–2397. doi:10.1039/C5CC10066F PMID:26732587

Yu, P., Liu, Y., Zhang, X., Zhou, J., Xiong, E., Li, X., & Chen, J. (2016). A novel electrochemical aptasensor for bisphenol A assay based on triple-signaling strategy. *Biosensors & Bioelectronics*, *79*, 22–28. doi:10.1016/j.bios.2015.12.007 PMID:26686919

Yu, S. L. (1970). Aeration performance in natural streams. *Journal of the Sanitary Engineering Division*, *96*, 1099–1113.

Zaman, D., Tiwari, M. K., & Mishra, S. (2020). Low-Cost Adsorptive Removal Techniques for Pharmaceuticals and Personal Care Products. In T. Gupta, S. Singh, P. Rajput, & A. Agarwal (Eds.), *Measurement, Analysis and Remediation of Environmental Pollutants. Energy, Environment and Sustainability* (pp. 397–421). Singapore: Springer. doi:10.1007/978-981-15-0540-9_19

Zang, Y., Lei, J., Hao, Q., & Ju, H. (2014). "Signal-On" Photoelectrochemical Sensing Strategy Based on Target-Dependent Aptamer Conformational Conversion for Selective Detection of Lead(II) Ion. *ACS Applied Materials & Interfaces*, *6*(18), 15991–15997. doi:10.1021/am503804g PMID:25170538

Zarfl, C., & Matthies, M. (2010). Are marine plastic particles transport vectors for organic pollutants to the Arctic? *Marine Pollution Bulletin*, *60*(10), 1810–1814. doi:10.1016/j.marpolbul.2010.05.026 PMID:20579675

Zejiao, L., & Ebiare, E. (2010). Water quality monitoring in Nigeria; Case Study of Nigeria's industrial cities. *Journal of American Science, 6*(4). Retrieved from http://www.americanscience.org/journalseditor@americanscience.org

Zelada-Guillen, G. A., Luis Sebastian-Avila, J., Blondeau, P., Riu, J., & Xavier Rius, F. (2012). Label-free detection of Staphylococcus aureus in skin using real-time potentiometric biosensors based on carbon nanotubes and aptamers. *Biosensors & Bioelectronics*, *31*(1), 226–232. doi:10.1016/j.bios.2011.10.021 PMID:22154169

Zelinkova, Z., & Wenzl, T. (2015). The Occurrence of 16 EPA PAHs in Food – A Review. *Polycyclic Aromatic Compounds*, *35*(2-4), 248–284. doi:10.1080/10406638.2014.918550 PMID:26681897

Zeng, X., Tu, W., Li, J., Bao, J., & Dai, Z. (2014). Photoelectrochemical Biosensor Using Enzyme-Catalyzed in Situ Propagation of CdS Quantum Dots on Graphene Oxide. *ACS Applied Materials & Interfaces*, *6*(18), 16197–16203. doi:10.1021/am5043164 PMID:25154012

Zenobio, J. E., Sanchez, B. C., Leet, J. K., Archuleta, L. C., & Sepúlveda, M. S. (2015). Presence and effects of pharmaceutical and personal care products on the Baca National Wildlife Refuge, Colorado. *Chemosphere*, *120*, 750–755. doi:10.1016/j.chemosphere.2014.10.050 PMID:25465958

Zhang, Y., Huang, L., Zuo, Z., Chen, Y., Wang, C. (2013). Phenanthrene exposure causes cardiac arrhythmia in embryonic zebrafish via perturbing calcium handling. *Aquatic Toxicology, 142– 143*, 26– 32.

Zhang, C., Liang, Y., & Wang, F. (2007). Microbial degradation of sulfonamides in soils. *Nongye Huanjing Kexue Xuebao*, *26*, 1658–1662.

Zhang, C., Lu, J., & Wu, J. (2019). Adsorptive removal of polycyclic aromatic hydrocarbons by detritus of green tide algae deposited in coastal sediment. *The Science of the Total Environment*, *670*, 320–327. doi:10.1016/j.scitotenv.2019.03.296 PMID:30904645

Zhang, D., Bond, T., Krasner, S. W., Chu, W., Pan, Y., Xu, B., & Yin, D. (2019). Trace determination and occurrence of eight chlorophenylacetonitriles: An emerging class of aromatic nitrogenous disinfection byproducts in drinking water. *Chemosphere*, *220*, 858–865. doi:10.1016/j.chemosphere.2018.12.127

Zhang, G., Ji, S., & Xi, B. (2006). Feasibility study of treatment of amoxicillin wastewater with a combination of extraction, Fenton oxidation and reverse osmosis. *Desalination*, *196*(1-3), 32–42. doi:10.1016/j.desal.2005.11.018

Zhang, H., Bayen, S., & Kelly, B. C. (2015). Multi-residue analysis of legacy POPs and emerging organic contaminants in Singapore's coastal waters using gas chromatography – triple quadrupole tandem mass spectrometry. *The Science of the Total Environment*, *523*(1), 219–232. doi:10.1016/j.scitotenv.2015.04.012 PMID:25863512

Zhang, H., & Lee, H. K. (2012). Simultaneous determination of ultraviolet filters in aqueous samples by plunger-in-needle solid-phase microextraction with graphene-based sol-gel coating as sorbent coupled with gas chromatography-mass spectrometry. *Analytica Chimica Acta*, *742*, 67–73. doi:10.1016/j.aca.2012.03.016 PMID:22884209

Zhang, J., Chen, N., Tang, Z., Yu, Y., Hu, Q., & Feng, C. (2015). A study of the mechanism of fluoride adsorption from aqueous solutions onto Fe-impregnated chitosan. *Physical Chemistry Chemical Physics*, *17*(18), 12041–12050. doi:10.1039/C5CP00817D PMID:25872764

Zhang, J., Tu, L., Zhao, S., Liu, G., Wang, Y., & Yue, Z. (2015). Fluorescent gold nanoclusters based photoelectrochemical sensors for detection of H2O2 and glucose. *Biosensors & Bioelectronics*, *67*, 296–302. doi:10.1016/j.bios.2014.08.037 PMID:25190086

Zhang, K., Xiong, X., Hu, H., Wu, C., Bi, Y., Wu, Y., ... Liu, J. (2017). Occurrence and characteristics of microplastic pollution in Xiangxi Bay of three gorges reservoir, China. *Environmental Science & Technology*, *51*(7), 3794–3801. doi:10.1021/acs.est.7b00369 PMID:28298079

Zhang, W., Sun, Y., Wu, C., Xing, J., & Li, J. (2009). Polymer-functionalized single-walled carbon nanotubes as a novel sol-gel solidphase micro-extraction coated fiber for determination of poly-brominated diphenyl ethers in water samples with gas chromatography-electron capture detection. *Analytical Chemistry*, *81*(8), 2912–2920. doi:10.1021/ac802123s PMID:19364140

Zhang, X., Xia, X., Dong, J., Bao, Y., & Li, H. (2014). Enhancement of toxic effects of phenanthrene to *Daphnia magna* due to the presence of suspended sediment. *Chemosphere*, *104*, 162–169. doi:10.1016/j.chemosphere.2013.10.086 PMID:24275150

Zhang, X., & Yadavalli, V. K. (2011). Surface immobilization of DNA aptamers for biosensing and protein interaction analysis. *Biosensors & Bioelectronics*, *26*(7), 3142–3147. doi:10.1016/j.bios.2010.12.012 PMID:21227676

Zhang, X., Zhang, Y., Zhao, H., He, Y., Li, X., & Yuan, Z. (2013). Highly sensitive and selective colorimetric sensing of antibiotics in milk. *Analytica Chimica Acta*, *778*, 63–69. doi:10.1016/j.aca.2013.03.059 PMID:23639400

Zhang, Z., Zhang, D., & Zhang, X. (2015). Simultaneous determination of pharmaceutical and personal care products in wastewater by capillary electrophoresis with head-column field-amplified sample stacking. *Analytical Methods*, *6*(19), 7978–7983. doi:10.1039/C4AY01562B

Zhao, M., Fan, G.-C., Chen, J.-J., Shi, J.-J., & Zhu, J.-J. (2015). Highly Sensitive and Selective Photoelectrochemical Biosensor for Hg^{2+} Detection Based on Dual Signal Amplification by Exciton Energy Transfer Coupled with Sensitization Effect. *Analytical Chemistry*, *87*(24), 12340–12347. doi:10.1021/acs.analchem.5b03721 PMID:26599580

Zhao, R.-S., Liu, Y. L., Chen, X.-F., Yuan, J.-P., Bai, A.-Y., & Zhou, J.-B. (2013). Preconcentration and determination of polybrominated diphenyl ethers in environmental water samples by solid-phase microextraction with Fe_3O_4-coated bamboo charcoal fibers prior to gas chromatography-mass spectrometry. *Analytica Chimica Acta*, *769*, 65–71. doi:10.1016/j.aca.2013.01.027 PMID:23498122

Zhao, W., Liu, Y., Wei, Z. B., Yang, S. G., He, H., & Sun, C. (2016). Fabrication of a novel p-n heterojunction photocatalyst n-$BiVO_4$@p-MoS_2 with core-shell structure and its excellent visible-light photocatalytic reduction and oxidation activities. *Applied Catalysis B: Environmental*, *185*, 242–252. doi:10.1016/j.apcatb.2015.12.023

Zhao, X., Wang, J., Wu, F., Wang, T., Cai, Y., Shi, Y., & Jiang, G. (2010). Removal of fluoride from aqueous media by Fe3O4@Al(OH)3 magnetic nanoparticles. *Journal of Hazardous Materials*, *173*(1-3), 102–109. doi:10.1016/j.jhazmat.2009.08.054 PMID:19747775

Zhao, Y., Wang, X., Lin, X., Zhao, S., & Lin, J. (2016). Comparative developmental toxicity of eight typical organic pollutants to red sea bream (*Pagrosomus major*) embryos and larvae. *Environmental Science and Pollution Research International*, *24*(10), 9067–9078. doi:10.100711356-016-6282-4 PMID:27130341

Zheng, T., Wang, J., Wang, Q., Meng, H., & Wang, L. (2015). Research trends in electrochemical technology for water and waste water treatment. *Applied Water Science*, 1–18.

Zhen, T., Wang, J., Wang, Q., Meng, H., & Wang, L. (2015). Research trends in electrochemical technology for water and wastewater treatment. *Applied Water Science*. doi:10.100713201-015-0280-4

Zhimin, Q., John, J. M., Melanie, R. M., Rao, S., & Craig, D. A. (2006). Treatment of Antibiotics and Antibiotic Resistant Bacteria in Swine Wastewater with Free Chlorine. *Journal of Agricultural and Food Chemistry*, 8144–8154. PMID:17032022

Zhou, J., Chang, Q., Wang, Y., Wang, J., & Meng, G. (2010). Separation of stable oil–water emulsion by the hydrophilic nano-sized ZrO2 modified Al2O3 microfiltration membrane. *Separation and Purification Technology*, *75*(3), 243–248. doi:10.1016/j.seppur.2010.08.008

Zhou, L., Wang, J., Li, D., & Li, Y. (2014). An electrochemical aptasensor based on gold nanoparticles dotted graphene modified glassy carbon electrode for label-free detection of bisphenol A in milk samples. *Food Chemistry*, *162*, 34–40. doi:10.1016/j.foodchem.2014.04.058 PMID:24874354

Zhou, N. D., Wang, J. Y., Zhang, J., Li, C., Tian, Y. P., & Wang, J. (2013). Selection and identification of streptomycin-specific single-stranded DNA aptamers and the application in the detection of streptomycin in honey. *Talanta*, *108*, 109–116. doi:10.1016/j.talanta.2013.01.064 PMID:23601877

Zhou, S., Kong, Y., Shen, Q., Ren, X., Zhang, J. R., & Zhu, J. J. (2014). Chronic myeloid leukemia drug evaluation using a multisignal amplified photoelectrochemical sensing platform. *Analytical Chemistry*, *86*(23), 11680–11689. doi:10.1021/ac502969x PMID:25372503

Zhou, Y., Gao, B., Zimmerman, A. R., Fang, J., Sun, Y., & Cao, X. (2013). Sorption of heavy metals on chitosan-modified biochars and its biological effects. *Chemical Engineering Journal*, *231*, 512–518. doi:10.1016/j.cej.2013.07.036

Zhou, Y., Yu, C., & Shan, Y. (2004). Adsorption of fluoride from aqueous solution on La3+ impregnated crosslinked gelatin. *Separation and Purification Technology*, *36*(2), 89–94. doi:10.1016/S1383-5866(03)00167-9

Zhu, B.-S., Jia, Y., Jin, Z., Sun, B., Luo, T., Kong, L.-T., & Liu, J.-H. (2015). A facile precipitation synthesis of mesoporous 2-line ferrihydrite with good fluoride removal properties. *RSC Advances*, *5*(103), 84389–84397. doi:10.1039/C5RA15619J

Zoeller, R. T., & Vandenberg, L. N. (2015). Assessing dose–response relationships for endocrine disrupting chemicals (EDCs): A focus on non-monotonicity. *Environmental Health*, *14*(42), 1–5. PMID:25971795

About the Contributors

Victor Shikuku is currently a lecturer in the Department of Physical Sciences at Kaimosi Friends University College (Kenya). Expert areas include Physical, Materials and Environmental Chemistry. Research interests involve Drinking and Wastewater treatment processes, development of materials for water purification, and sustainable development. Holds a Ph.D. in Chemistry from Maseno University (Kenya) supported by the Excellence Centre for Development Cooperation (EXCEED-SWINDON) and Post-doctoral research experience at the University of Trier (Germany) funded by the Alexander Humboldt Foundation.

* * *

Sinmi Abosede teaches Introduction to Quantitative Reasoning and Introduction to Mathematical Methods to first-year undergraduates at PAU. Her current research activities are centered on the water and wastewater sector in Nigeria, specifically on various issues such as water scarcity, water regulation, water resources management and the relationship between water and other sectors such as energy and agriculture. Prior to joining PAU, she worked as a Chemical Engineer in the UK water industry for companies such as Thames Water and Atkins Limited and is a technical specialist in water regulation, audit, engineering design and engineering project management. Currently, she provides consultancy services to various clients providing various engineering solutions for homes, offices, estates or entire districts. Dr Abosede holds a BEng degree in Chemical Engineering from the University of Leeds and a PhD in Chemical Engineering from Imperial College London.

Adekunbi Enoch Adedayo is currently a lecturer of Chemistry in Adeyemi College of Education Ondo. After his primary education at Providence School Fagba, Ifako Lagos State. He went to Christ Apostolic Grammar School. Iperu remo Ogun State for his junior secondary school education and thereafter to Adeyemi College of Education Demonstration Secondary School Ondo, Ondo State for his secondary education. He has his first degree certificate in Chemistry from University of Ilorin and his master degree in Organic Chemistry from University of Ibadan. He is currently on his Ph.D program also in organic chemistry in the Federal University of Technology Akure.

Obijole Olumuyiwa Adewale is a seasoned lecturer and researcher of over two decades. He is a graduate of the University of Ibadan where he obtained Bachelors of Science degree in Chemistry in 1988 and masters degree in Inorganic Chemistry in 1995. He has published over 30 articles in reputable international journals and has attended many international conferences globally. He is a staff of Adeyemi College of Education, Nigeria and presently a research assistant at the University of Venda, South Africa where his research interest covers environmental and water remediation and nanoscience He is a member of different professional bodies including Institute of Chartered Chemist of Nigeria, Chemical societies of Nigeria and South Africa and Water Institute of South Africa.

Temitope A. Aladesanmi has PhD in Technology Management with emphasis on IT Security from Obafemi Awolowo University (OAU), Ile-Ife, Nigeria and an advanced Certificate in Leadership and E-governance from Syracuse University, NY. He is a European Council certified IT security expert. He is a research partner in the Cyber Security Research Laboratory of the university.

Justinah Solayide Amoko was born in 6th May,1978. After her secondary education in 1995, she attended Obafemi Awolowo University, Ile-Ife, Osun state, Nigeria where she graduated with a creditable second class (Hons) Lower division in B.Sc (Ed) Chemistry in 2000. Having completed her compulsory National Youth Service Corps in 2002, she proceeded to University of Ibadan for her master's programme in Chemistry (M.Sc Physical Chemistry) in 2004. She is currently undergoing Ph.D programme in Physical Chemistry at Adekunle Ajasin University, Akungba Akoko. Also, she is currently a Senior lecturer in the Department of Chemistry, Adeyemi College of Education, Ondo, Nigeria.

Krishna Bhuyan is a senior research scholar working in the Laboratory of Environment and Human Toxicology, Department of Life Science and Bioinformatics, Assam University, Silchar, India. She is currently working on the field of environmental toxicology dealing with the potential toxicity of polycyclic aromatic hydrocarbons in anuran amphibians.

Celestin Defo received the MSc in water engineering from the University of Dschang in Cameroon in 2006 and a Ph. D in water science and technology in 2016 at the Indian Agricultural Research Institute, New Delhi, India. In 2010, he was recruited as an Assistant Lecturer at the School of Wood, Water and Natural Resources, Faculty of Agronomy and Agricultural Science, Ebolowa Branch of the University of Dschang. He became a Senior Lecturer in May 2017. Dr. Defo is the author of one book and numerous articles published in peer review journals. He has contributed for more than 10 national and international conferences. In 2013, he was selected as Mwalimu Nyerere African Union Fellow. His research interests cover several aspects across water engineering, mainly community water supply, water pollution, wastewater recycling and computer application.

Anirudha Giri is the principal investigator of the Environmental and Human Toxicology Laboratory, Department of Life Science and Bioinformatics, Assam University, Silchar, India. Prof. Giri got his masters in 1989 and Ph.D. in 1996. He has been a visiting post-doctoral scholar at the University of California, Berkley, CA and visiting faculty in University of Florida, Gainesville; FL. Prof. Giri is particularly interested in how the emerging pollutants of concern could affect the physiology, development, reproduction, genetic and life-history traits of amphibians. The primary focus of his research group is to understand and predict how contaminants could shape the biodiversity in future and devise methods for their conservation.

Oloyede Hammed was born in August, 1980. After his secondary education in 1999, he attended Obafemi Awolowo University, Ile-Ife, Osun state, Nigeria where he graduated with a creditable second class (Hons) upper division in B.Sc (Ed) Chemistry in 2005. Having completed his compulsory National Youth Service Corps in 2007, he proceeded to University of Ibadan for his master's programme in Chemistry (M.Sc Inorganic Chemistry) between 2007 and 2009. He is currently undergoing his Ph.D programme in Inorganic Chemistry at the same University. He is a seasoned lecturer in the field of Inorganic Chemistry with adequate international exposure. He has lectured at the Faculty of Natural Science, Wesley University of Science and Technology (now Wesley University) Ondo, Ondo State, Nigeria between 2011 and 2012. He is currently a lecturer I in the Department of Chemistry, Adeyemi College of Education, Ondo, Nigeria.

Ravinder Kaur, Principal Scientist & Former Project Director, Water Technology Centre and Former Director, Indian Agricultural Research Institute is a doctorate (Gold medalist & Jawahar Lal Nehru awardee) in Agricultural Physics with specialization in Soil Physics & Soil – Water Conservation, from Indian Agricultural Research Institute. She has 27 years of rich experience in applied and extension research on several global/ domestic priority areas, with specific reference to abiotic systems Dr. Kaur has been honored with several prestigious national/ international awards (viz. ICAR Young Scientist Award, ICAR Punjab Rao Deshmukh Women Agricultural Scientist Award, ICAR National Fellow position, NESA Fellowship, Fulbright and IHELP fellowships from Fulbright Commission and CGHERA, Visiting Scientist position under India-Taipei Program of Co-operation in Science & Technology).

Magali Kemmerich has experience in analytical methods development by experimental designs, sample preparation and gas/liquid chromatographic analysis of organic contaminants residues in food and environmental samples. Also, worked with laboratory quality control (ISO 17025:2005 and 19011:2012) and measurement of uncertainty. In recent years has been teaching Chemistry at College in Brazil. Currently works with herbicides dynamics in environmental samples, with analysis by high resolution UHPLC-Q-Orbitrap.

Bowa Kwach is a Lecturer of Inorganic Chemistry at Maseno University. His research interests span from the occurrence and fate of organic and inorganic pollutants in the environment, wastewater treatment among other environmental topics.

Emily Ng'eno is a Ph.D. student at Masinde Muliro University of Science and Technology. Her research topics cover the occurrence and sources of emerging contaminants in the environment, the performance of wastewater treatment facilities in the removal of emerging contaminants from water and the development of low-cost adsorbents for remediation of the pollutants from wastewater.

Maurice Odago is an associate professor of chemistry at the State University of New York College at Oneonta, New York. His areas of interests include; organic synthesis, supramolecular chemistry, molecular sensing, and spectroscopic chemical analysis.

Isaiah Oke holds a PhD degree in Civil Engineering. His main research Interest is in water Engineering. He has published a number of articles in reputable journals and Conference Proceedings. He is an academic in the Department of Civil Engineering, Obafemi Awolowo University, Ile-Ife, Nigeria.

Otieno Kevin Okoth is a lecturer at Tom Mboya University College, Kenya. He received his PhD degree from Huazhong University of Science and Technology, P.R. China in 2017. His research interests include electrochemical sensors, environmental chemistry and chemically modified electrodes.

Babajide Jelili Olalekan is a graduate with a PhD degree in Medicinal Chemistry from the University of the Western Cape, Cape Town, South Africa. Having over twenty years of teaching/research experience in more than three tertiary institutions. Teaches science education and chemistry courses. Having keen interest in science education, synthesis of nano fluid, organic synthesis and medicinal plant research with Herbal option using different chemical and biological investigative techniques for identification of numerous phytochemicals. A member of different professional bodies both in Nigeria and abroad. He has attended many conferences globally with many publications.

Fehintola Olumuyiwa is currently a lecturer of Chemistry with 17 years of teaching/research experience at Adeyemi College of Education, Ondo. After his secondary education in 1988 at Baptist High School, Saki, Oyo State, Nigeria, he attended Obafemi Awolowo University, Adeyemi College of Education, Ondo Campus, Ondo, Nigeria where he graduated with a second class (Hons) upper division in B.Sc (Ed) Chemistry in 1998 and his master degree in Physical Chemistry from University of Ibadan in 2000. He is currently on his Ph.D programme in Physical chemistry at the Federal University of Technology, Akure. He is a member of different professional bodies in Nigeria. He has attended many conferences and has many publications to his credit.

Lukman Salihu is a lecturer in Civil Engineering Department, University of Hafr Al-Batin, Hafr Al-Batin, Saudi Arabia.

Ruth Nduta Wanjau is an associate professor at the School of Pure and Applied Science, Kenyatta University. She obtained her PHD from Wuhan University. Her research interests include: Analysis foods, medicine and other consumer products, for heavy metals, essential elements, vitamins, environmental analysis of air-indoor and outdoor, lake-, river-, and well water and development of analytical methods for determination of trace elements.

Index

U

V

W

www.ingramcontent.com/pod-product-compliance
Lightning Source LLC
LaVergne TN
LVHW081251011026
849162LV00006BA/1448